T0342560

Origin of Power Converters

Origin of Power Converters

Decoding, Synthesizing, and Modeling

Tsai-Fu Wu
National Tsing Hua University, Taiwan, ROC

Yu-Kai Chen
National Formosa University, Taiwan, ROC

This edition first published 2020
© 2020 John Wiley & Sons, Inc.

All rights reserved. No part of this publication may be reproduced, stored in a retrieval system, or transmitted, in any form or by any means, electronic, mechanical, photocopying, recording or otherwise, except as permitted by law. Advice on how to obtain permission to reuse material from this title is available at http://www.wiley.com/go/permissions.

The right of Tsai-Fu Wu and Yu-Kai Chen to be identified as the authors of this work has been asserted in accordance with law.

Registered Office
John Wiley & Sons, Inc., 111 River Street, Hoboken, NJ 07030, USA

Editorial Office
111 River Street, Hoboken, NJ 07030, USA

For details of our global editorial offices, customer services, and more information about Wiley products visit us at www.wiley.com.

Wiley also publishes its books in a variety of electronic formats and by print-on-demand. Some content that appears in standard print versions of this book may not be available in other formats.

Limit of Liability/Disclaimer of Warranty
While the publisher and authors have used their best efforts in preparing this work, they make no representations or warranties with respect to the accuracy or completeness of the contents of this work and specifically disclaim all warranties, including without limitation any implied warranties of merchantability or fitness for a particular purpose. No warranty may be created or extended by sales representatives, written sales materials or promotional statements for this work. The fact that an organization, website, or product is referred to in this work as a citation and/or potential source of further information does not mean that the publisher and authors endorse the information or services the organization, website, or product may provide or recommendations it may make. This work is sold with the understanding that the publisher is not engaged in rendering professional services. The advice and strategies contained herein may not be suitable for your situation. You should consult with a specialist where appropriate. Further, readers should be aware that websites listed in this work may have changed or disappeared between when this work was written and when it is read. Neither the publisher nor authors shall be liable for any loss of profit or any other commercial damages, including but not limited to special, incidental, consequential, or other damages.

Library of Congress Cataloging-in-Publication data applied for
ISBN: 9781119632986

Cover design by Wiley
Cover image: © zf L/Getty Images

Set in 9.5/12.5pt STIXTwoText by SPi Global, Pondicherry, India

To Our Families
Ching-Ying, Charles and Jerry
Yun-Wen and Allen

Contents

Preface *xv*
Acknowledgments *xvii*
About the Authors *xviii*

Part I Decoding and Synthesizing *1*

1 Introduction *3*
1.1 Power Processing Systems *4*
1.2 Non-PWM Converters Versus PWM Converters *7*
1.2.1 Non-PWM Converters *7*
1.2.2 PWM Power Converters *9*
1.3 Well-Known PWM Converters *10*
1.4 Approaches to Converter Development *17*
1.5 Evolution *25*
1.6 About the Text *26*
1.6.1 Part I: Decoding and Synthesizing *26*
1.6.2 Part II: Modeling and Applications *28*
Further Reading *28*

2 Discovery of Original Converter *31*
2.1 Creation of Original Converter *31*
2.1.1 Source–Load Approach *32*
2.1.2 Proton–Neutron–Meson Analogy *32*
2.1.3 Resonance Approach *33*
2.2 Fundamental PWM Converters *34*
2.2.1 Voltage Transfer Ratios *35*
2.2.2 CCM Operation *36*
2.2.3 DCM Operation *38*

2.2.4 Inverse Operation *39*
2.3 Duality *40*
Further Reading *41*

3 Fundamentals *43*
3.1 DC Voltage and Current Offsetting *43*
3.1.1 DC Voltage Offsetting *44*
3.1.2 DC Current Offsetting *47*
3.2 Capacitor and Inductor Splitting *49*
3.3 DC-Voltage Blocking and Pulsating-Voltage Filtering *51*
3.4 Magnetic Coupling *55*
3.5 DC Transformer *58*
3.6 Switch Grafting *62*
3.7 Diode Grafting *67*
3.8 Layer Scheme *72*
Further Reading *74*

4 Decoding Process *77*
4.1 Transfer Ratios (Codes) *77*
4.2 Transfer Code Configurations *82*
4.2.1 Cascade Configuration *82*
4.2.2 Feedback Configuration *82*
4.2.3 Feedforward Configuration *83*
4.2.4 Parallel Configuration *85*
4.3 Decoding Approaches *86*
4.3.1 Factorization *86*
4.3.2 Long Division *88*
4.3.3 Cross Multiplication *89*
4.4 Decoding of Transfer Codes with Multivariables *91*
4.5 Decoding with Component-Interconnected Expression *93*
Further Reading *94*

5 Synthesizing Process with Graft Scheme *95*
5.1 Cell Approaches *95*
5.1.1 P-Cell and N-Cell *96*
5.1.2 Tee Canonical Cell and Pi Canonical Cell *97*
5.1.3 Switched-Capacitor Cell and Switched-Inductor Cell *98*
5.1.4 Inductor–Capacitor Component Cells *100*
5.2 Converter Grafting Scheme *101*
5.2.1 Synchronous Switch Operation *101*
5.2.2 Grafting Active Switches *103*

5.2.3 Grafting Passive Switches *108*
5.3 Illustration of Grafting Converters *110*
5.3.1 Grafting the Well-Known PWM Converters *110*
5.3.1.1 Graft Boost on Buck *111*
5.3.1.2 Graft Buck on Boost *112*
5.3.1.3 Graft Buck on Buck–Boost *114*
5.3.1.4 Graft Boost on Boost–Buck *116*
5.3.1.5 Buck in Parallel with Buck–Boost *119*
5.3.1.6 Grafting Buck on Buck to Achieve High Step-Down Voltage Conversion *119*
5.3.1.7 Grafting Boost on Boost to Achieve High Step-up Voltage Conversion *120*
5.3.1.8 Grafting Boost (CCM) on Buck (DCM) *121*
5.3.1.9 Cascode Complementary Zeta with Buck *123*
5.3.2 Grafting Various Types of Converters *124*
5.3.2.1 Grafting Half-Bridge Resonant Inverter on Dither Boost Converter *124*
5.3.2.2 Grafting Half-Bridge Resonant Inverter on Bidirectional Flyback Converter *124*
5.3.2.3 Grafting Class-E Converter on Boost Converter *125*
5.3.3 Integrating Converters with Active and Passive Grafted Switches *127*
5.3.3.1 Grafting Buck on Boost with Grafted Diode *128*
5.3.3.2 Grafting Half-Bridge Inverter on Interleaved Boost Converters in DCM *128*
5.3.3.3 Grafting *N*-Converters with TGS *130*
5.3.3.4 Grafting *N*-Converters with ΠGS *130*
Further Reading *132*

6 Synthesizing Process with Layer Scheme *133*
6.1 Converter Layering Scheme *133*
6.2 Illustration of Layering Converters *135*
6.2.1 Buck Family *135*
6.2.2 Boost Family *138*
6.2.3 Other Converter Examples *142*
6.3 Discussion *146*
6.3.1 Deduction from Ćuk to Buck–Boost *146*
6.3.2 Deduction from Sepic to Buck–Boost *148*
6.3.3 Deduction from Zeta to Buck–Boost *149*
6.3.4 Deduction from Sepic to Zeta *150*
Further Reading *151*

7 Converter Derivation with the Fundamentals *153*
7.1 Derivation of Buck Converter *153*
7.1.1 Synthesizing with Buck–Boost Converter *154*
7.1.2 Synthesizing with Ćuk Converter *154*
7.2 Derivation of z-Source Converters *154*
7.2.1 Voltage-Fed z-Source Converters *155*
7.2.1.1 Synthesizing with Sepic Converter *157*
7.2.1.2 Synthesizing with Zeta Converter *160*
7.2.2 Current-Fed z-Source Converters *161*
7.2.2.1 Synthesizing with SEPIC Converter *162*
7.2.2.2 Synthesizing with Zeta Converter *162*
7.2.3 Quasi-z-Source Converter *162*
7.2.3.1 Synthesizing with Sepic Converter *164*
7.2.3.2 Synthesizing with Zeta Converter *165*
7.3 Derivation of Converters with Switched Inductor or Switched Capacitor *166*
7.3.1 Switched-Inductor Converters *167*
7.3.1.1 High Step-Down Converter with Transfer Code $D/(2-D)$ *167*
7.3.1.2 High Step-Down Converter with Transfer Code $D/(2(1-D))$ *173*
7.3.2 Switched-Capacitor Converters *178*
7.3.2.1 High Step-Up Converter with Transfer Code $(1+D)/(1-D)$ *178*
7.3.2.2 High Step-Up Converter with Transfer Code $2D/(1-D)$ *181*
7.3.2.3 High Step-Up Converter with Transfer Code $D/(1-2D)$ *184*
7.4 Syntheses of Desired Transfer Codes *185*
7.4.1 Synthesis of Transfer Code: $D^2/(D^2-3D+2)$ *186*
7.4.1.1 Synthesizing with Buck–Boost Converter *187*
7.4.1.2 Synthesizing with Zeta Converter *188*
7.4.1.3 Synthesizing with Ćuk Converter *189*
7.4.2 Synthesizing Converters with the Fundamentals *191*
7.4.2.1 DC Voltage and DC Current Offsetting *191*
7.4.2.2 Inductor and Capacitor Splitting *192*
7.4.2.3 DC Voltage Blocking and Filtering *192*
7.4.2.4 Magnetic Coupling *193*
7.4.2.5 DC Transformer *194*
7.4.2.6 Switch and Diode Grafting *195*
7.4.2.7 Layer Technique *195*
Further Reading *198*

8 Synthesis of Multistage and Multilevel Converters *199*
8.1 Review of the Original Converter and Its Variations of Transfer Code *199*

8.2 Syntheses of Single-Phase Converters *201*
8.3 Syntheses of Three-Phase Converters *203*
8.4 Syntheses of Multilevel Converters *207*
8.5 *L–C* Networks *210*
Further Reading *212*

9 Synthesis of Soft-Switching PWM Converters *215*
9.1 Soft-Switching Cells *215*
9.1.1 Passive Lossless Soft-Switching Cells *216*
9.1.1.1 Near-Zero-Current Switching Mechanism *216*
9.1.1.2 Near-Zero-Voltage Switching Mechanism *218*
9.1.2 Active Lossless Soft-Switching Cells *220*
9.1.2.1 Zero-Voltage Switching Mechanism *222*
9.1.2.2 Zero-Current Switching Mechanism *226*
9.2 Synthesis of Soft-Switching PWM Converters with Graft Scheme *230*
9.2.1 Generation of Passive Soft-Switching PWM Converters *230*
9.2.2 Generation of Active Soft-Switching PWM Converters *234*
9.3 Synthesis of Soft-Switching PWM Converters with Layer Scheme *240*
9.3.1 Generation of Passive Soft-Switching PWM Converters *240*
9.3.2 Generation of Active Soft-Switching PWM Converters *245*
9.4 Discussion *247*
Further Reading *251*

10 Determination of Switch-Voltage Stresses *255*
10.1 Switch-Voltage Stress of the Original Converter *255*
10.2 Switch-Voltage Stresses of the Fundamental Converters *257*
10.2.1 The Six Well-Known PWM Converters *257*
10.2.1.1 Boost Converter *257*
10.2.1.2 Buck–Boost Converter *258*
10.2.1.3 Ćuk, Sepic, and Zeta Converters *259*
10.2.2 z-Source Converters *260*
10.2.2.1 Voltage-Fed z-Source Converter *260*
10.2.2.2 Current-Fed z-Source Converter *261*
10.2.2.3 Quasi-z-Source Converter *262*
10.3 Switch-Voltage Stresses of Non-Fundamental Converters *263*
10.3.1 High Step-Down Switched-Inductor Converter *263*
10.3.2 High Step-Down/Step-Up Switched-Inductor Converter *264*
10.3.3 Compound Step-Down/Step-Up Switched-Capacitor Converter *265*

10.3.4 High Step-Down Converter with Transfer Ratio of D^2 267
10.3.5 High Step-Up Converter with Transfer Ratio of $1/(1 - D)^2$ 268
Further Reading 270

11 Discussion and Conclusion 271
11.1 Will Identical Transfer Code Yield the Same Converter Topology? 271
11.2 Topological Duality Versus Circuital Duality 274
11.3 Graft and Layer Schemes for Synthesizing New Fundamental Converters 277
11.3.1 Synthesis of Buck–Boost Converter 278
11.3.2 Synthesis of Boost–Buck (Ćuk) Converter 279
11.3.3 Synthesis of Buck–Boost–Buck (Zeta) Converter 280
11.3.4 Synthesis of Boost–Buck–Boost (Sepic) Converter 282
11.3.5 Synthesis of Buck-Family Converters with Layer Scheme 284
11.3.6 Synthesis of Boost-Family Converters with Layer Scheme 286
11.4 Analogy of Power Converters to DNA 289
11.4.1 Replication 291
11.4.2 Mutation 291
11.5 Conclusions 295
Further Reading 296

Part II Modeling and Application 299

12 Modeling of PWM DC/DC Converters 301
12.1 Generic Modeling of the Original Converter 302
12.2 Series-Shunt and Shunt-Series Pairs 303
12.3 Two-Port Network 308
12.4 Small-Signal Modeling of the Converters Based on Layer Scheme 315
12.5 Quasi-Resonant Converters 323
Further Reading 326

13 Modeling of PWM DC/DC Converters Using the Graft Scheme 329
13.1 Cascade Family 330
13.2 Small-Signal Models of Buck-Boost and Ćuk Converters Operated in CCM 332
13.2.1 Buck-Boost Converter 336
13.2.2 Boost-Buck Converter 338
13.3 Small-Signal Models of Zeta and Sepic Operated in CCM 340

13.3.1 Zeta Converter *344*
13.3.2 Sepic Converter *346*
Further Reading *349*

14 Modeling of Isolated Single-Stage Converters with High Power Factor and Fast Regulation *351*
14.1 Generation of Single-Stage Converters with High Power Factor and Fast Regulation *352*
14.2 Small-Signal Models of General Converter Forms Operated in CCM/DCM *355*
14.3 An Illustration Example *361*
Further Reading *365*

15 Analysis and Design of an Isolated Single-Stage Converter Achieving Power Factor Correction and Fast Regulation *367*
15.1 Derivation of the Single-Stage Converter *368*
15.1.1 Selection of Individual Semi-Stages *369*
15.1.2 Derivation of the Discussed Isolated Single-Stage Converter *369*
15.2 Analysis of the Isolated Single-Stage Converter Operated in DCM + DCM *369*
15.2.1 Buck-Boost Power Factor Corrector *370*
15.2.2 Flyback Regulator *372*
15.3 Design of a Peak Current Mode Controller for the ISSC *373*
15.4 Practical Consideration and Design Procedure *377*
15.4.1 Component Stress *377*
15.4.2 Snubber Circuit *378*
15.4.3 Design Procedure *379*
15.5 Hardware Measurements *380*
15.6 Design of an H^{∞} Robust Controller for the ISSC *382*
15.6.1 H^{∞} Control *382*
15.6.2 An Illustration Example of Robust Control and Hardware Measurements *386*
Further Reading *392*

Index *395*

Preface

This book is divided into two parts. Part I presents evolution and development of power converters from the original converter. Hundreds of power converter topologies have been developed over past one century by many researchers. However, there is no single systematic approach to developing the converters. Inspired by Charles Darwin who published the book entitled *The Origin of Species* and based on the principle of resonance, we identify the original converter, on which we develop the mechanisms of evolution, decoding, and synthesizing processes, to derive PWM power converters systematically. With the decoding process, the input-to-output transfer codes (ratios) are decoded into code configurations in terms of the transfer codes derived from the original converter. With the synthesizing process, we have developed the graft and layer schemes, which are used in growing plants, along with circuit fundamentals to synthesize the code configurations into converters. With these two processes, illustrations of the existing and newly developed hard-switching and soft-switching PWM converters, including the well-known z-source converters, Vienna converters, modular multilevel converters, switched-inductor/switched-capacitor converters, *etc.*, are presented in detail. Additionally, determination of converters' switch-voltage stresses based on their transfer codes is addressed. Moreover, based on the principle of resonance, the well-known six PWM converters are reconfigured, and analogy of PWM converters to DNA is presented, from which mutation and replication of PWM converters are discussed.

Part II presents modeling and applications of power converters based on the original converter and the developed graft and layer schemes. The six PWM converters can be modeled into families represented in two-port networks. Therefore, relationships among the converters can be identified and the modeling processes can be simplified. In addition, single-stage converters to fulfill multiple functions are derived and modeled, on which two application examples are presented and verified with experimental results.

Since Charles Darwin in 1859 initiated an evolution principle, through around one hundred years and many researchers' study, Gregor J. Mendel developed the laws of inheritance in 1866, Boveri-Sutton developed chromosome theory in 1902, and James D. Watson discovered the double-helix structure of DNA in 1953, affecting significantly the followed genetic engineering innovations. Like Charles Darwin, we initiate an evolution of power converters, and we do expect other researchers can follow this stepstone to go further. This does not conclude the work, but just gets started.

Tsai-Fu Wu
National Tsing Hua University, Taiwan, ROC

Yu Kai Chen
National Formosa University, Taiwan, ROC

Acknowledgments

This book collects most of our work in converter development and modeling over past 25 years. We are grateful to our former PhD and Master students who have contributed to this book, especially Dr Te-Hung Yu and Dr Frank Liang. We are also thankful to the Ministry of Science and Technology, Taiwan, for constantly funding our research work. Our special appreciation goes to Cecilia Wang and Ya-Fen Cheng who edit the book prudently to meet the requirements from Wiley Publisher.

Tsai-Fu Wu
Yu-Kai Chen

About the Authors

Tsai-Fu Wu is a professor in the Department of Electrical Engineering at National Tsing Hua University, Taiwan, ROC where he is the director of Elegant Power Electronics Applied Research Lab (EPEARL). Since 1993, he has worked on more than 100 power electronics research projects sponsored by the Ministry of Science and Technology, ROC, and industry. He has published more than 300 referred journal and conference papers. Under his supervision, more than 30 PhD and 200 master students have graduated. His current research interests include development and modeling of power converters, design and development of Direct Digital Control with D–Σ processes for single-phase and three-phase converters with grid connection, rectification, APF, power balancing and UPS functions, and design of resonant converters for ultrasonic cutter, ozone generator, remote-plasma-source, and electrical surgery unit applications.

Yu-Kai Chen is a professor in the Department of Aeronautical Engineering, National Formosa University, Taiwan, ROC where he is the director of Innovative Design and Energy Application Lab. (IDEAL). In 2015, he received the outstanding industry collaboration award from National Formosa University. His research interests include modeling and control of power converters, design of solar panel-supplied inverters for grid connection, and DSP- and microprocessor-based application systems with fuzzy and robust controls.

Part I

Decoding and Synthesizing

1

Introduction

Electrical energy has been widely applied, and its growth rate has been increasing dramatically over the past two decades. In particular, renewable energy coming to play has driven electricity utilization and processing needs to reach another growth peak. Additionally, machine electrification and factory automation have also increased the demand of electricity. With increasing use of sophisticated equipment and instruments, high power quality becomes just essential. To supply sufficient, of high quality, and stable electrical power in desired voltage or current forms, power processing systems are indispensable. Meanwhile, they also play an important role in supporting continuous growth of human beings' civilization, environmental conservation, and energy harvesting. In designing a power processing system, the first step needs to select a power converter topology since the converter topology mainly governs the fundamental properties, such as step-up, step-down, bipolar operation, component stresses, etc. Converters come out with very diversified configurations. How to derive or develop them systematically without trial and error is an interesting topic. Thus, many researchers have devoted in developing power converter topologies for various types of applications.

In this chapter, configuration of a power processing system is first addressed. Fundamental two types of power converter classifications, general pulse-width modulated (PWM) converters and non-PWM ones, are presented. Then, the well-known PWM converters are introduced for later comparison and illustration. In literature, there are many approaches to developing power converters, and their fundamental principles will be described briefly. In addition, an evolution concept is presented for illustrating later converter derivation. A section introducing the overall organization of this book will be included in the end of this chapter.

Origin of Power Converters: Decoding, Synthesizing, and Modeling, First Edition.
Tsai-Fu Wu and Yu-Kai Chen.
© 2020 John Wiley & Sons, Inc. Published 2020 by John Wiley & Sons, Inc.

1.1 Power Processing Systems

Configuration of a power processing system can be illustrated in Figure 1.1, which mainly includes input filter, power converter, feedback/feedforward circuits, controller, gate driver, and protection circuit. The input source can be obtained from utility outlet/grid or renewable energy generators, such as photovoltaic panel, wind turbine, geothermal heat pump, etc. Its voltages and currents can be any form, and their amplitudes might vary with time or fluctuate frequently. At the output side, the load may require various voltage and current forms, too. Thus, a proper power converter topology along with a promising controller is usually required to realize a power processing system.

Conventionally, the conceptual block diagram of a power processing system shown in Figure 1.1 can be realized by the circuit shown in Figure 1.2, which, as an example, is a linear regulator. In the circuit, semiconductor switch Q_N is operated in linear region to act as a variable resistor, which can absorb the voltage difference between input voltage V_i and output voltage V_o and in turn regulate V_o under load variation. The primary merits of a linear regulator include low output voltage ripple and low noise interference. However, it has many drawbacks, such

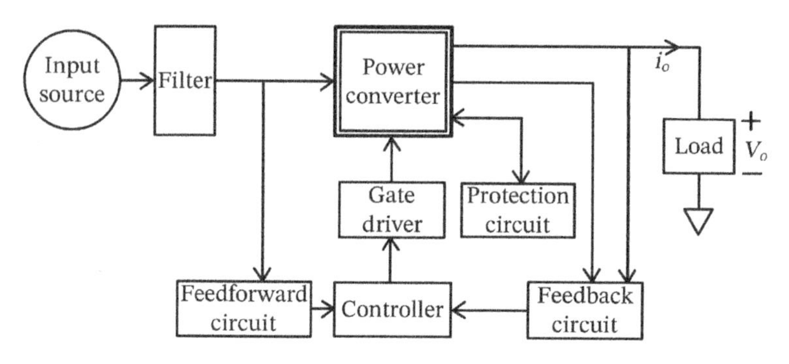

Figure 1.1 Configuration of a power processing system.

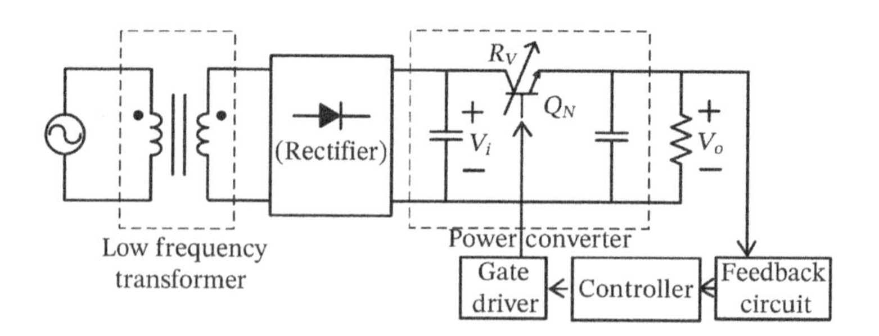

Figure 1.2 Block diagram of a linear regulator.

as the transformer with low operating frequency results in bulky size and heavy weight, semiconductor Q_N operating in linear region results in high power loss and low efficiency, and the low efficiency requires a large, heavy heat sink and even needs forced ventilation. These drawbacks have limited its wide applications to compact electronic products, renewable power generators, and energy harvesters, where efficiency and size are the essential concerns.

To improve efficiency and release the aforementioned limitations, switching power regulators were developed. A typical configuration of the switching regulators is shown in Figure 1.3, where in the power converter, switch M_1 is operated in saturation region (if using BJT as a switch) or in ohmic region (if using MOSFET as a switch), reducing conduction loss dramatically. At the input side, the corner frequency of the filter is close to switching frequency, and its size and weight can be also reduced significantly. If isolation is required, high frequency transformer will be introduced to the converter, and its size and weight are relatively small as compared with a low-frequency one (operating at 50/60 Hz). In general, a switching regulator has the merits of high power density, small volume, low weight, improved efficiency, and cost and component reduction. There still exist several limitations, such as resulting in high switching noise, increasing analysis and design complexity, and requiring sophisticated control. Although switching regulators have the limitations, thanks to recent advances in high efficiency and high frequency component development, nanoscale integrated circuit (IC) fabrication technique, and analysis tool, they have been widely applied to electronic products, energy harvesting, and power quality improvement. For further discussion, we will focus on switching regulators only.

For a switching regulator or a more general term, switching power converter, the input source can be either AC or DC form, and the output load can be also supplied by either AC or DC form. Thus, there are four types of combinational forms in classifying power converter topologies, which are AC to DC, AC to AC, DC to AC, and DC to DC. In Figure 1.3, the *rectifier* converts AC to DC, and the *power converter* converts DC to DC. Typically, a power processing system may

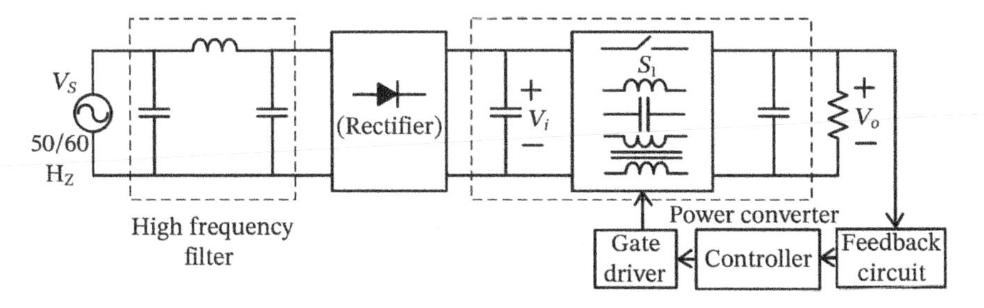

Figure 1.3 Block diagram of a switching regulator.

need multiple power converters to collaborate each other, but they might be integrated into a single power stage for certain applications, such as a notebook adapter consisting of a rectifier (AC to DC), a power factor corrector (DC to DC), and an isolated regulator (DC to DC), which can be integrated into a bridgeless isolated regulator. A power converter requires at least a control gear or switch to control power flow between source and load, and it might need some buffers or filters to smooth and hold up voltage and current, which can be illustrated in Figure 1.4. The switch can be realized with BJT, MOSFET, IGBT, GTO, etc. along with freewheeling diodes. It is worth noting that recent advances in wide-band-gap switching device development, such as SiC and GaN, have merited to switching power converters because their switching losses have been reduced significantly. The buffer or filter is realized with capacitor alone or capacitor–inductor pair. If it requires galvanic isolation, a transformer is introduced into the converter. Additionally, the transformer provides another degree of freedom in tuning input-to-output voltage ratio and can implement multiple outputs readily. To fulfill multiple functions or increase power capacity, converters can be connected in series or parallel, which will complicate analysis, design, and control.

As shown in Figure 1.4, connecting switch(es) and capacitors/inductors to form a power converter sounds simple. However, how to configure a power converter to achieve step-up, step-down, and step-up/step-down DC output, AC output, PWM control, variable frequency control, etc. is not an easy task. Even with the same step-up/step-down transfer ratio, there exist different converter topologies, and they might have different dynamic performances and different component stresses. Among the four types of power converter topologies is the DC to DC, simplified to DC/DC, converter type relatively popular. In the following, we will first present how to figure out the derivation of DC/DC converter topologies, on which the rest of converter types will be discussed. Exploring systematic approaches to developing power converter topologies is the unique feature of this book.

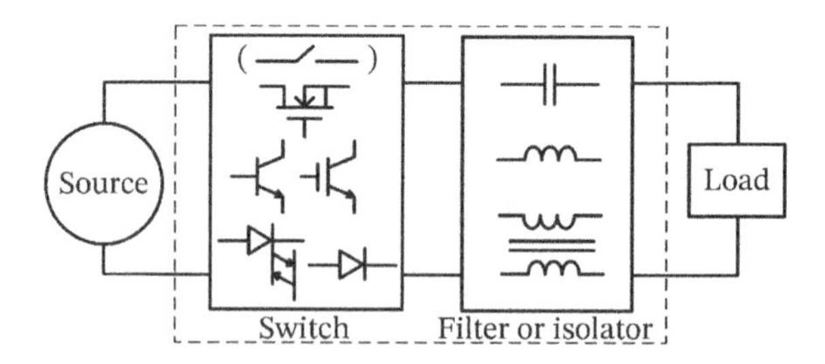

Figure 1.4 Possible components in a power converter.

1.2 Non-PWM Converters Versus PWM Converters

In power converters, when switch turns on with infinite current through or infinite voltage across components, this is because there is no current-limiting or voltage-blocking components in the conduction path, resulting in severe electromagnetic interference (EMI) problems. This type of power converter cannot be controlled with PWM and is called a non-PWM converter. On the contrary, there exist current-limiting and voltage-blocking components in the conduction path of a power converter, and it can be controlled with PWM, which is called a PWM converter. This claim will be presented and illustrated with some power converter examples, as follows.

1.2.1 Non-PWM Converters

The major concern of a power converter is its input–output conversion efficiency. In practice, there is no resistor allowed in a converter configuration. A qualified converter includes only ideal switch(es) and capacitor(s)/inductor(s). However, even with these components only, there might still exist loss during power transfer, such as the converters shown in Figure 1.5a and b. Figure 1.5a shows power transfer between two capacitors, and it is controlled by switch S_1. Assuming capacitor C_1 is associated with an initial voltage of V_o and C_2 is with zero voltage, and capacitance $C_1 = C_2$, it can be shown that there is an electrical energy loss, $(1/4)C_1V_o^2$, which is half the initially stored energy in C_1. Moreover, when switch S_1 is turned on, an inrush current flows from C_1 to C_2 and through S_1 in almost no time, which may damage the components and cause EMI problems. For this type of circuit configuration, the only current limiter is the equivalent inductance and resistance of the components and the circuit path. It can be said that there is no control on the capacitor currents and voltages, and the voltages of both capacitors C_1 and C_2 will be always balanced at $(1/2)V_o$. This type of power converter configuration is classified as a non-PWM converter.

Similarly, the conceptual inductor–inductor–switch configuration shown in Figure 1.5b has the same limitations. If, initially, inductor L_1 carries a current of I_o

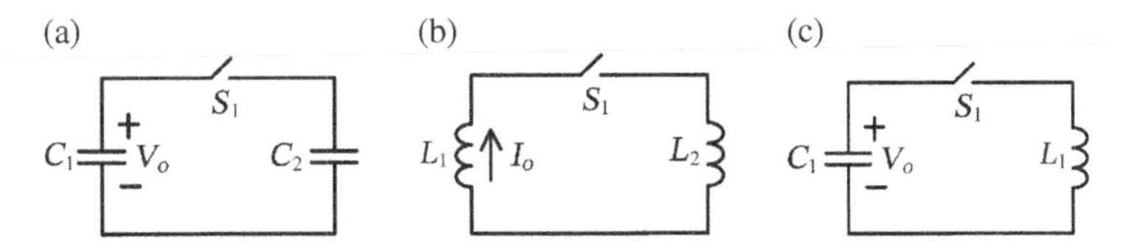

Figure 1.5 (a) Capacitor–capacitor–switch, (b) inductor–inductor–switch, and (c) capacitor–inductor–switch networks.

but there is no current in L_2, after turning on switch S_1, there will be an extremely high impulse voltage across the inductors and the switch, causing EMI problems and damage to the components. Again, there is half electrical energy loss $(1/4)L_1I_o^2$ if $L_1 = L_2$, and there is no control on the inductor voltages and currents at all. It is also a non-PWM converter.

In summary, non-PWM converters come out high inrush current or high impulse voltage, resulting in high EMI, as well as high component stress, and they could yield low conversion efficiency even with ideal components. In particular, under large initial voltage difference, the maximum electrical energy loss can be as high as 50%.

Other examples adopting the configuration shown in Figure 1.5a are shown in Figure 1.6. Figure 1.6a shows a two-lift converter. When switches S_1 and S_2 are turned on, capacitor C_1 will charge C_2 directly. On the other hand, when switch S_3 and S_4 are turned on, capacitors C_1 and C_2 are connected in series to charge capacitor C_3 and lift the output voltage V_o to be twice the input voltage V_i. It can be seen

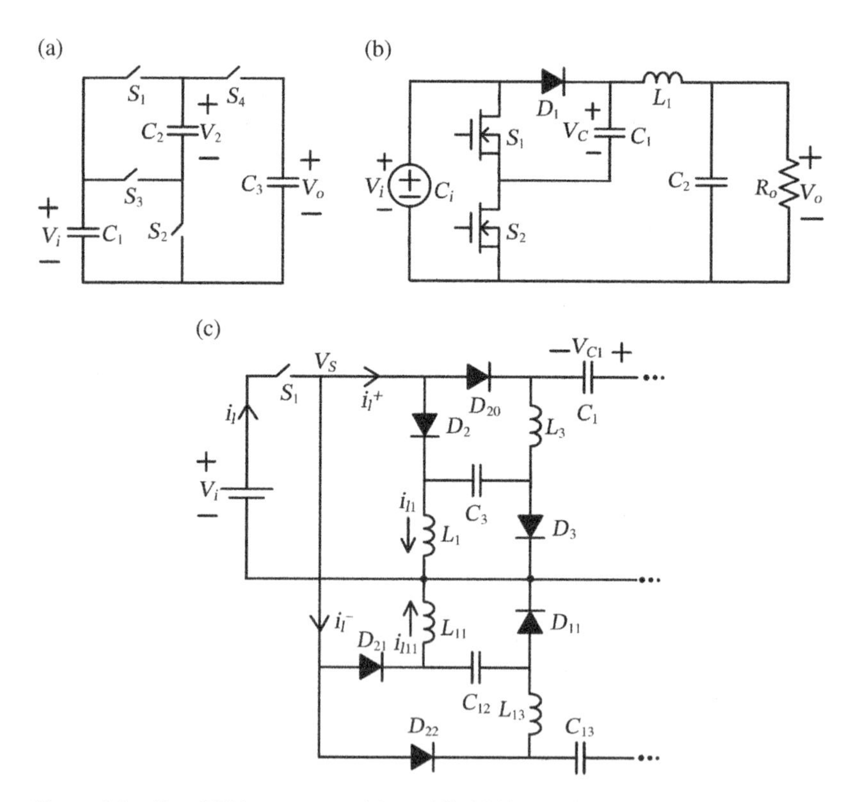

Figure 1.6 Non-PWM converters: (a) two lift, (b) KY, and (c) re-lift circuit.

that during capacitor charging, there is no current limiter, resulting in high inrush current. Figure 1.6b shows the KY converter. When switch S_2 is turned on, input voltage V_i will charge capacitor C_1 through diode D_1 but again without current limiter. When switch S_2 is turned off and S_1 is turned on, input voltage V_i together with capacitor voltage V_C will magnetize inductor L_1 through the output path. This path of power flow is with the current limiter of inductor L_1. Figure 1.6c shows a re-lift converter. When switch S_1 is turned on, there are two capacitor charging paths without current limiter, V_i-S_1-D_2-C_3-D_3-V_i and V_i-S_1-D_{21}-C_{12}-D_{11}-V_i. When switch S_1 is turned off, the energy stored in capacitors C_3 and C_{12} will be released to the output through the inductors and capacitors, which are the current limiters.

With a non-PWM converter, the processed power level is usually pretty low because of high inrush current or high pulse voltage. It can be used for supplying integrated circuits, which require low power consumption, of which the low current rating switches have high conduction resistance and act as current limiters. For high power processing, we need PWM power converters.

1.2.2 PWM Power Converters

Power transfer between a capacitor and an inductor can be modulated by a switch, as shown in Figure 1.5c, and their total electrical energy is always conserved to their initially stored energy. In the network, capacitor C_1 limits the slew rate of voltage variation, inductor L_1 limits that of current variation, and switch S_1 controls the time interval of power transfer, i.e., pulse-width modulation. Thus, component stresses can be properly controlled, and high conversion efficiency can be insured. Additionally, EMI level can be also reduced significantly. Power converter configurations based on this type of network are called PWM power converters. Note that it requires an additional freewheeling path when switch S_1 is turned off, which will be discussed in later section. For simplicity while without confusion in power electronics area, the short-form PWM converters or converters will be used to represent the PWM power converters. They have been widely applied to various types of power conversion for their controllable power transfer, theoretically no loss, and finite component stresses.

The minimum-order network of a PWM converter is a second-order LC network, and it must at least include a switch to control power flow. The order of network can be increased to third, fourth, and even higher. For a valid PWM converter, the network must be always in resonant manner at either switch turn-on or turn-off.

Over the past century, PWM converters have been well developed and have diversified configurations, such as buck, boost, buck-boost, Ćuk, sepic, Zeta, flyback, forward, push-pull, half-bridge, full-bridge, Z-source, neutral-point clamped (NPC), modular multilevel, quasi-resonant, and LLC resonant converters. They can be classified into non-isolated and isolated configurations. Typically, the isolated versions can be derived from the non-isolated ones by inserting a DC transformer or an AC transformer to a proper location of the converter. Thus, we will first introduce non-isolated converters, which can lay out a firm foundation for later discussions on isolated converters.

1.3 Well-Known PWM Converters

Almost all people entering power electronics field know about buck, boost, and buck-boost converters, as shown in Figure 1.7. To my best knowledge, it is unknown that who invented the buck converter and when it was invented. Since electricity started to be used frequently between the late nineteenth century and the early twentieth century, the invention of the buck converter was designated as year 1900. The boost converter was invented during World War II, which was used to boost voltage for transmitting radio signals across Atlantic Ocean. The buck-boost converter was invented around 1950.

Analyzing their operational principles will realize that the buck, boost, and buck-boost converters can achieve step-down, step-up, and step-down/step-up input-to-output voltage conversions, respectively. They all have a second-order *LC* network and a pair of active–passive switches but have different circuit configurations.

If we explore further, there are another three famous converters, and each of which has a fourth-order *LC* network and a pair of active–passive switches, as shown in Figure 1.8, in which they have different circuit configurations, but they all can fulfill the same step-down/step-up voltage conversion. Ćuk converter was

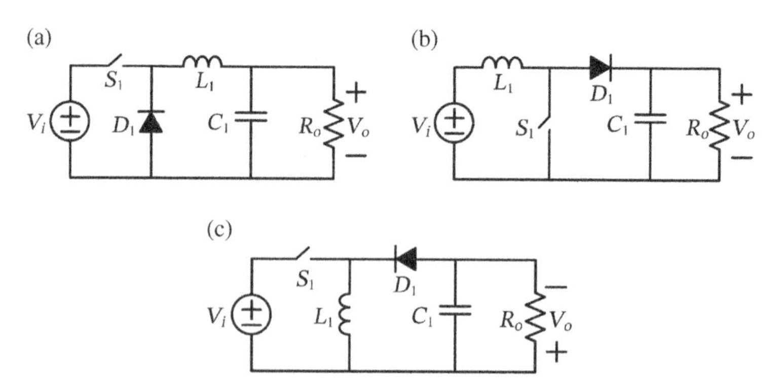

Figure 1.7 Power converters with a second-order *LC* network and a pair of active–passive switches: (a) buck converter, (b) boost converter, and (c) buck-boost converter.

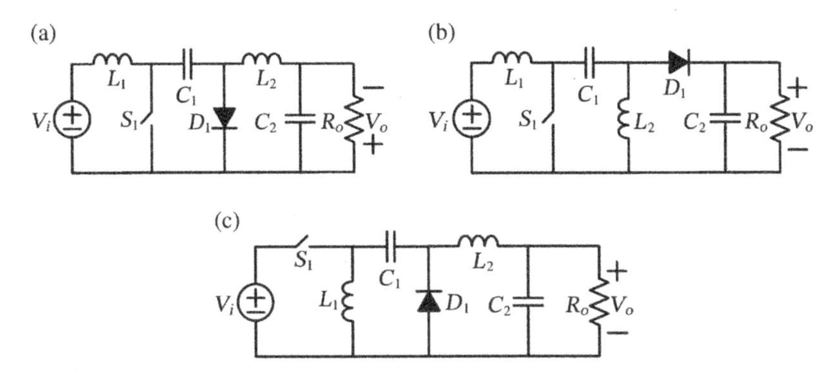

Figure 1.8 Power converters with a fourth-order LC network and a pair of active–passive switches: (a) Ćuk converter, (b) sepic converter, and (c) Zeta converter.

invented by Prof. S. Ćuk in 1975. Sepic is an acronym of single-ended primary inductor converter, which was invented in 1977. Zeta (dual sepic) converter was introduced in 1989.

Couples of questions come to our minds. Converter configurations are so diversified: thus, how to connect the components to become a converter, how to know ahead that the converter can achieve a step-down or step-up voltage conversion, why researchers spent around one century to develop these six PWM converters shown in Figures 1.7 and 1.8, does there exist an origin of power converters from which the rest of PWM converters can be evolved and derived systematically, and so on?

Based on the three PWM converters shown in Figure 1.7, three types of converters with a fourth-order *LC* network can be derived, as shown in Figure 1.9. Again,

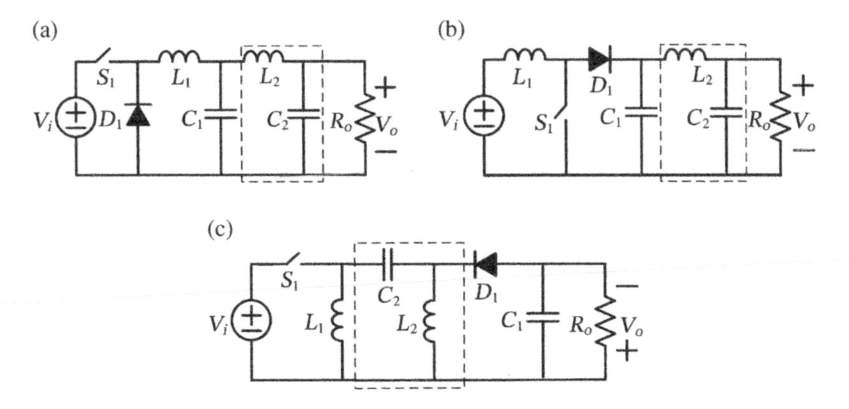

Figure 1.9 Converters with a fourth-order network: (a) buck derived, (b) boost derived, and (c) buck-boost derived.

some questions come to our minds: What is the difference between the converters shown in Figures 1.7 and 1.9, can we generate new converters by keeping on introducing extra *LC* networks into the old converters, what is the role of L_2C_2 network in Figure 1.9, how to verify a valid converter, etc.?

With switched inductors or capacitors, some of the PWM converters shown in Figures 1.7 and 1.8 can be modified to the ones shown in Figure 1.10, which are called switched-inductor/switched-capacitor hybrid converters. They can achieve higher step-down or step-up voltage conversion than their original counterparts. In each of the converters, there are one active switch and two passive diodes with either a third-order or a fifth-order *LC* network. It looks like that a diode-inductor or diode-capacitor cell is inserted into a certain PWM converter to form a new one. It is curious to ask why the concept cannot be applied to all of the six PWM converters shown in Figures 1.7 and 1.8, and how do the inventors know ahead they can achieve higher step-down or step-up voltage conversion? Moreover, can this concept be extended to all of other PWM converters, and what is the converter derivation mechanism behind?

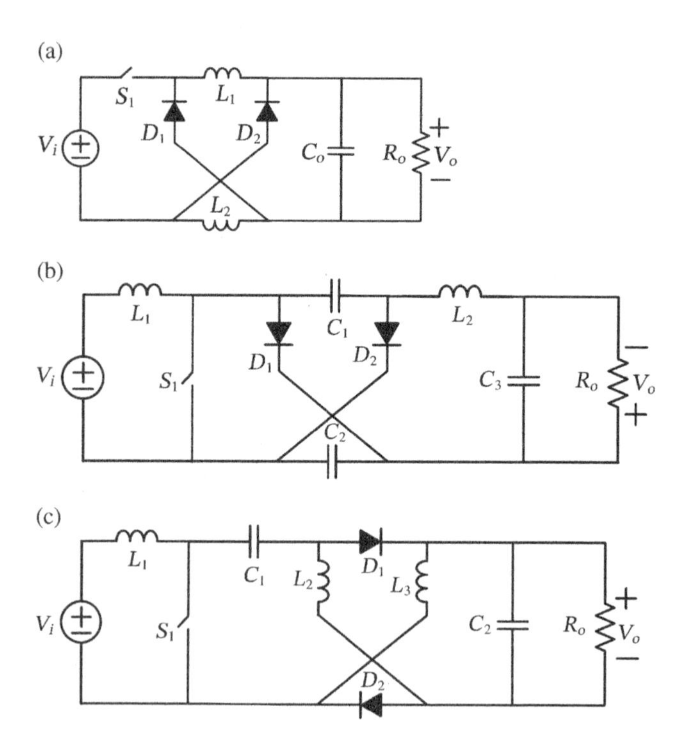

Figure 1.10 Converters with a switched inductor/capacitor: (a) buck derived, (b) Ćuk derived, and (c) sepic derived.

In literature, there are several types of Z-source converters, which have been widely applied to DC/DC and DC/AC power conversion. They are voltage-fed, current-fed, and quasi-Z-source converters, as shown in Figure 1.11, and each of which includes only one active–passive switch pair but has higher order *LC* network. The circuit configurations look quite different from the ones shown in Figures 1.7–1.10 and somehow look weird. For instance, a rectifier diode D_1 is connected in series with a DC voltage source V_i, as shown in Figure 1.11a, and the inductor-diode pair shown in Figure 1.10a is replaced with an *LC* network pair. Moreover, the output voltage of a Z-source converter becomes negative under certain range of duty ratios, which will be discussed in Chapter 7. If the converter derivation is just based on trial and error, there are thousands of circuit combinations, and thus, it is almost impossible to derive a valid converter without a systematic mechanism.

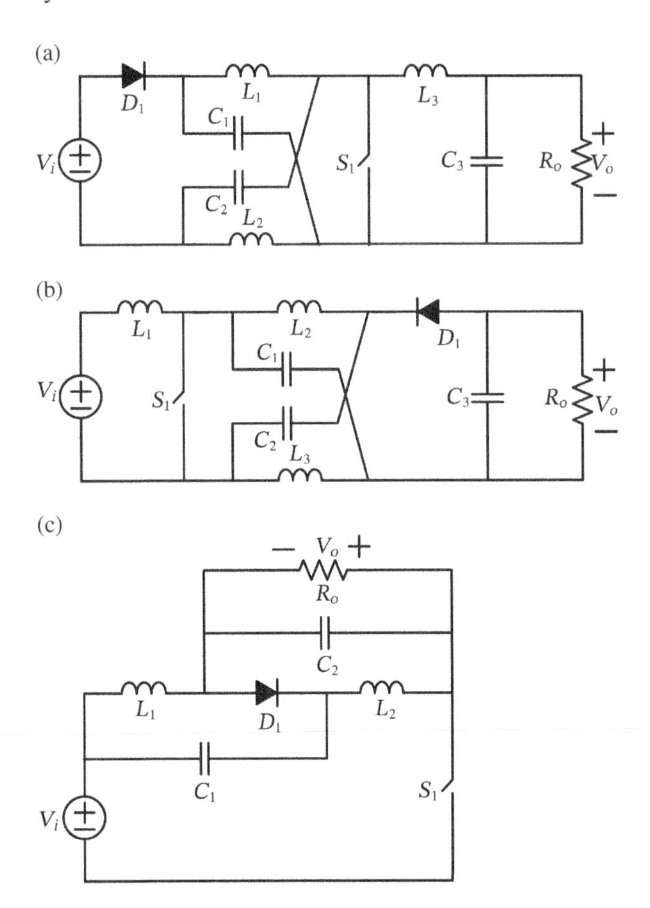

Figure 1.11 (a) Voltage-fed, (b) current-fed, and (c) quasi-Z-source converters.

Quasi-resonant converters were developed in the earlier 1980s by introducing LC resonant cells to PWM converters. Figure 1.12 shows quasi-resonant buck, quasi-resonant boost, and quasi-resonant Zeta converters, which can achieve zero-voltage switching at switch turn-on transition. By following the same mechanism, the rest of PWM converters shown in Figures 1.7 and 1.8 can be transformed to their counterparts, quasi-resonant converters. In Figure 1.12a and b, there are two LC pairs, L_RC_S and L_1C_1, in each converter, but their natural resonant frequencies are in different orders. They play different roles in the converter operation. Without the component values and without specifying the operational principle, it is hard to tell the difference between L_RC_S and L_1C_1 from the circuit configuration, although they are derived from the conventional PWM converters with L_1C_1 network only. It increases one more degree of difficulty in developing power converters.

For the quasi-resonant converters, the power transfer from input to output is still based on LC network and active–passive switch pair, and it can be pulse-width modulated. However, the current flow in L_R can be bidirectional and has higher resonant frequency, while the one in inductor L_1 is unidirectional only. How to construct this type of quasi-resonant converters is worth further discussing. In literature, there are similar converters, such as zero-current switching

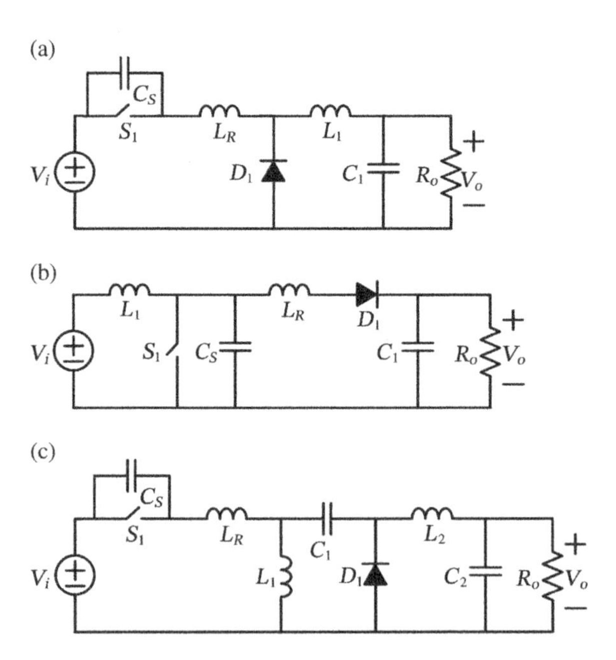

Figure 1.12 Quasi-resonant converters: (a) buck type, (b) boost type, and (c) Zeta type.

quasi-resonant converters and multi-resonant converters. Can they be developed with a systematical approach?

PWM converters can have more pairs of active–passive switches, such as the half-bridge and full-bridge converters shown in Figure 1.13. Figure 1.13a shows a half-bridge configuration, which has two pairs of switches. The two switches take turn conducting, and each one takes care of one-half switching cycle. In each half cycle, the switch is pulse-width modulated to control power flow from the input to the output. If the natural frequency of the L_1C_1 network is designed to be far below the switching frequency, the converter is just like a conventional PWM converter. On the other hand, if the frequency is close to the switching frequency, the current and voltage waveforms are sinusoidal-like, and it is called a resonant converter. In fact, it is still belonged to a PWM converter but just with variable frequency operation. In general, it is also classified as a PWM converter, because its power transfer is still limited by an LC network. Figure 1.13b shows a full-bridge converter, in which there are four switches and they form two pairs, S_1&S_4 and S_2&S_3. When these two pairs of switches take turn conducting or are in bipolar operation, the converter is the same as the half-bridge one. Again, it can act as a conventional PWM or a resonant converter depending on the order of the LC network natural frequency. This is also classified as a PWM converter.

All of the converters discussed above are non-isolated. By introducing transformers into the non-isolated versions of PWM converters, they can be transformed to their isolated counterparts. Figure 1.14 shows four isolated converters, flyback, forward, push-pull, and quasi-resonant flyback. With a transformer, several secondary windings can be wound on the same core to form multiple outputs, such as the ones shown in Figure 1.14a and b. The one shown in Figure 1.14c is derived from a buck converter with a DC transformer, and Figure 1.14d shows a

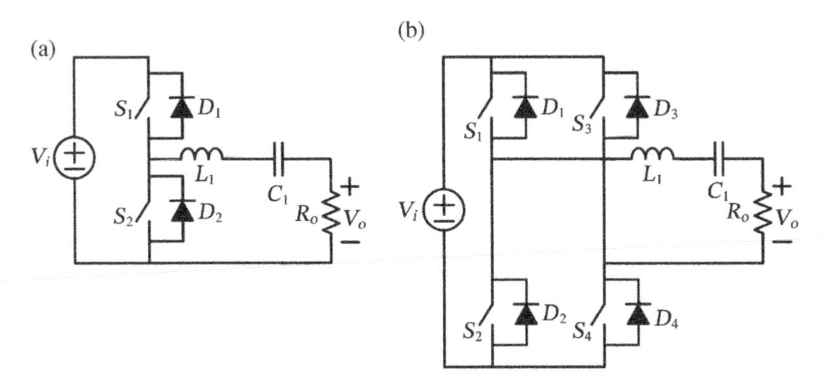

Figure 1.13 Converters with multiple pairs of active–passive switches: (a) half-bridge and (b) full-bridge configurations.

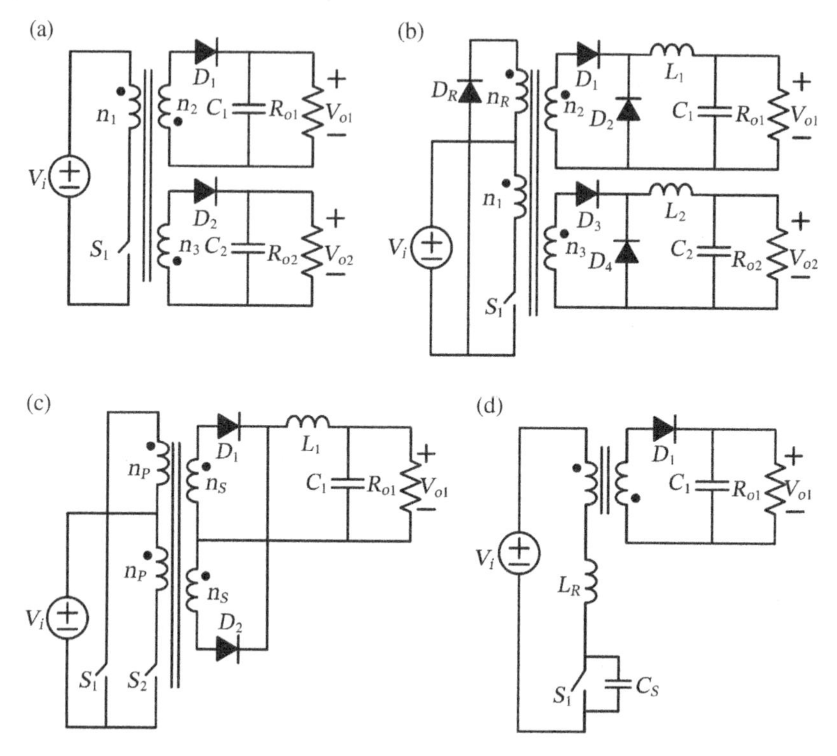

Figure 1.14 Isolated PWM converters: (a) flyback, (b) forward, (c) push-pull, and (d) quasi-resonant flyback.

flyback with an L_RC_S resonant network to form a quasi-resonant converter. Thus, it can be observed that combining the fundamental PWM converters with other components can yield new converters.

Converters with isolation transformers have many unique features, which include realizing multiple outputs, achieving galvanic isolation, providing one more degree of freedom in stepping down or stepping up voltage ratio, and protecting the components on the secondary side from damage by the high input voltage on the primary side. In literature, there are a big bunch of converters with isolation.

Each PWM converter has at least an inductor. With a coupled inductor, the converter can be modified to a new version. Figure 1.15 shows four PWM converters with coupled inductors, and they are derived from buck, boost, Ćuk, and buck-flyback converters. In the converters shown in Figure 1.15a and b, they just simply introduce a secondary winding into the converter itself and place at a proper path where the magnetization and demagnetization of the inductor satisfies the volt-second balance principle. Figure 1.15c shows the Ćuk converter in the form with a coupled inductor. Originally, the Ćuk converter has two separate inductors. Analyzing the operation of the converter will realize that the two inductors can be

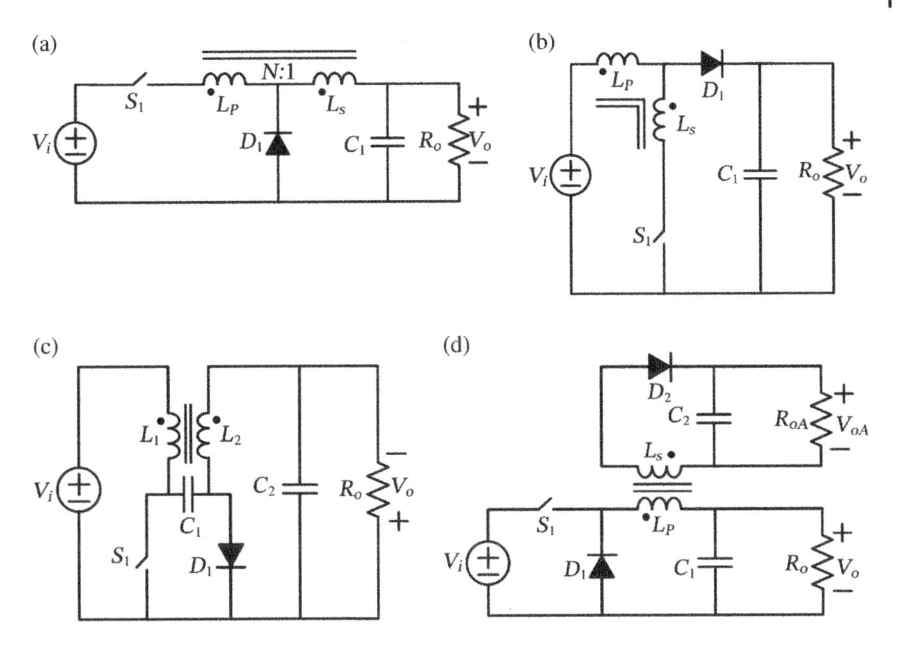

Figure 1.15 PWM converters with coupled inductors: (a) buck type, (b) boost type, (c) Ćuk type, and (d) buck-flyback type.

coupled with each other. Other examples are sepic and Zeta converters, of which there are two inductors in each converter and they can be coupled and wound on the same core. A converter with coupled inductors will reduce one degree of dynamic order. Can all of converters with two or more inductors be constructed with coupled inductors? How to place a secondary winding in a proper path in the converter is another issue, which needs to discuss further.

Figure 1.15d shows a buck and a flyback combined converter with coupled inductors to form an isolated output V_{oA}, which is the flyback-type output, while output V_o is just the regular buck output without isolation. Comparing the converters shown in Figure 1.15a and d reveals that the coupled winding can be connected back to the converter itself or to a separate network, which can be isolated from the primary side. It is quite diversified when introducing coupled winding(s) to the converters. What is the mechanism behind in developing such kind of converters?

1.4 Approaches to Converter Development

In last section, many well-known PWM converters were introduced, but many questions were also brought up. A general question is that *how to develop the converters systematically*. In this section, several typical approaches are described briefly for later discussion.

From the converters shown in Figures 1.7 and 1.8a, one can observe that the active and passive switches have a common node. Thus, a switching cell concept was introduced to explain the configurations of the converters. There are two types of switching cells, P cell and N cell, as shown in Figure 1.16, and each has two terminals for connecting to source or output and one current terminal for connecting to inductor to form a PWM converter. For examples, buck converter can be derived from P cell, while boost converter can be derived from N cell. This is a kind of intuitive approach with induction but without manipulation on the converters. In fact, with a little bit of manipulation by relocating capacitor C_1 in the sepic and Zeta converters shown in Figure 1.8b and c from the forward path to the return path, a P cell and an N cell can be identified, and the converters can be derived accordingly. Typically, this approach is used to explain the existing converter configurations, but it is hard to develop new converters. It is based on a cell level.

Similarly, based on observation and induction, the converters shown in Figures 1.7 and 1.8 can be explained with two canonical switching cells, namely, Tee canonical cell and Pi canonical cell, as shown in Figure 1.17. By exhaustively enumerating all of possible combinations of Z_{in}, Z_{out}, and Z_x, and including *LC* network, source, and load, the converters shown in Figures 1.7 and 1.8 can be derived. Additionally, the converters with extra *LC* filters and inverse versions of the converters can be derived. Figure 1.18 shows the buck and boost converters

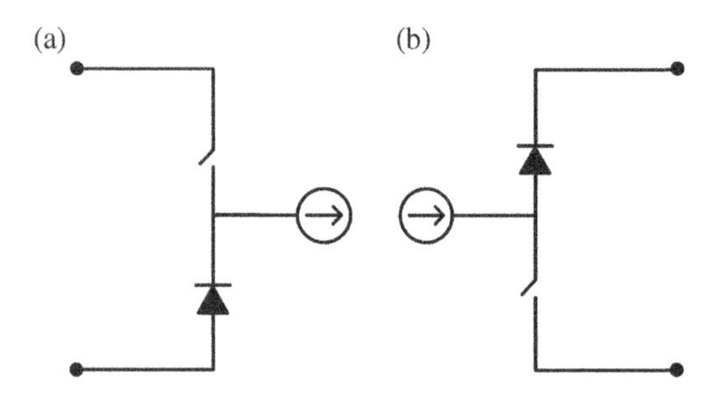

Figure 1.16 (a) P cell and (b) N cell.

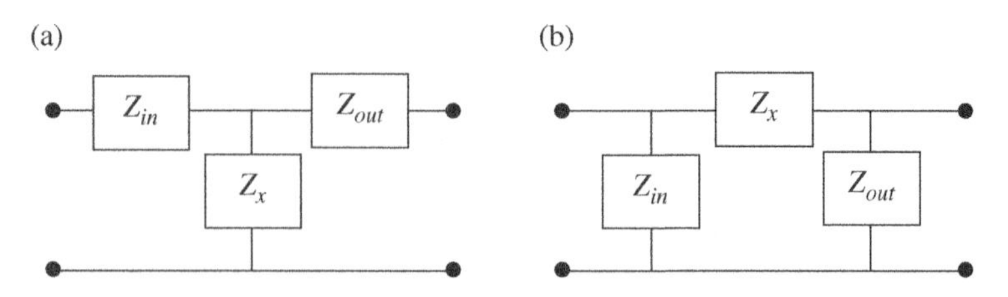

Figure 1.17 (a) Tee canonical cell and (b) Pi canonical cell.

and their inverse versions, in which the input-to-output transfer ratios shown in the bottom of the converters are corresponding to continuous conduction mode, and D is the duty ratio of the active switch. Moreover, the converters with the same higher step-up voltage transfer ratio but with different circuit configurations, as shown in Figure 1.19, can be developed. Although this approach can

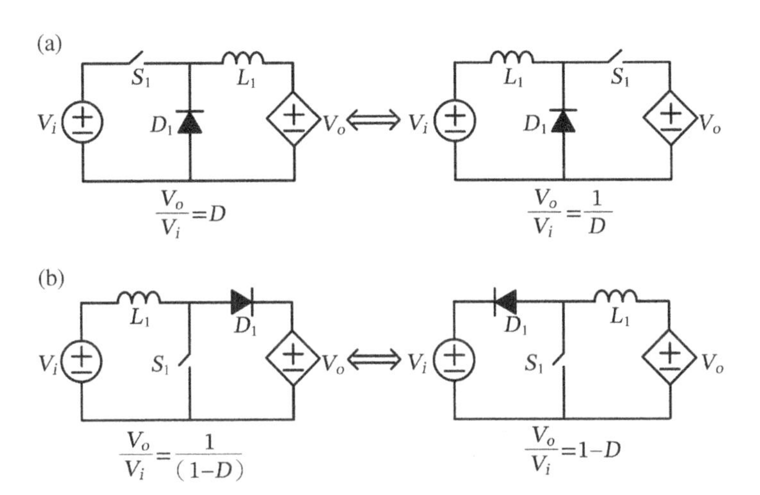

Figure 1.18 (a) Buck and inverse buck and (b) boost and inverse boost.

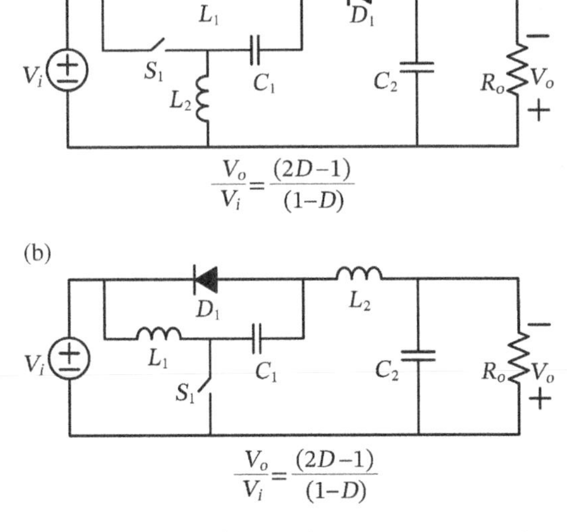

Figure 1.19 With the same input-to-output transfer ratio of $(2D - 1)/(1 - D)$ but with different configurations (a) and (b).

derive new PWM converters and is straightforward, it is still tedious and lacks of mechanism to explain the converters with identical transfer ratio but with different configurations. Again, it is based on a cell level.

Based on a cell level, another approach to developing new converters with higher step-up and step-down voltage ratios by introducing switched-capacitor or switched-inductor cells to the PWM converters shown in Figures 1.7 and 1.8 was proposed. Typical switched-capacitor and switched-inductor cells are shown in Figure 1.20, and their derived converters have been shown in Figure 1.10. Applications of this approach are quite limited, and the chance of deriving new converters is highly depending on experience. Otherwise, it might need many trial and errors. When inserting a cell into a PWM converter, one has to use volt-second balance principle to verify if the converter is valid. This approach still needs a lot of ground work to derive a valid converter.

Another synthesis approach based on a converter cell concept, 1L and 2L1C, was proposed. The synthesis procedure is supported with graph theory and matrix representation and based on a prescribed set of properties or constraints as criteria to extract a converter from all of the possible combinations, reducing the number of trial and errors. A structure of PWM DC–DC converters included in the synthesis procedure is depicted in Figure 1.21, and possible positions of inserting an inductor into a second-order PWM converter are shown in Figure 1.22. Typical converter properties include number of capacitors and inductors, number of active–passive switches, DC voltage conversion ratio, continuous input and/or output current, possible coupling of inductors, etc. This approach seems general and has a broad vision in synthesizing converters, but it needs a lot of efforts or even trial and errors in selecting a valid converter when considering many properties simultaneously. For the main purpose of deriving a converter, maybe, we have

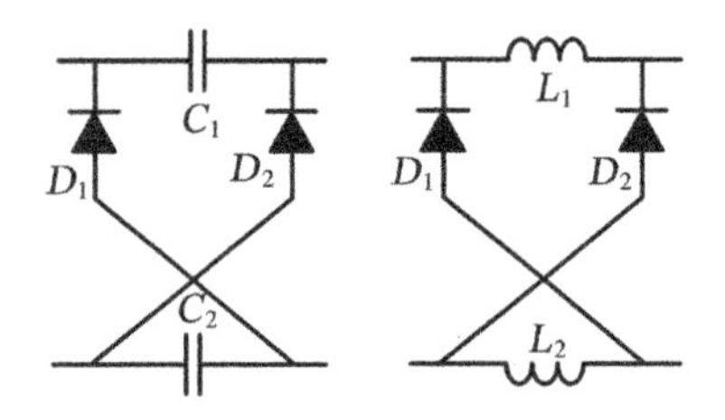

Figure 1.20 (a) Switched-capacitor and (b) switched-inductor cells.

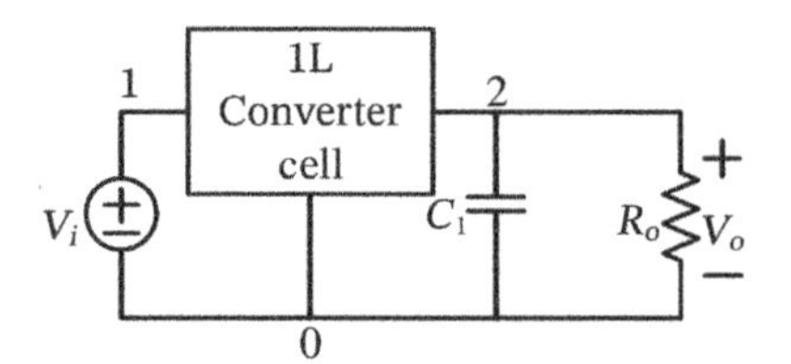

Figure 1.21 Structure of PWM converters used in the derivation procedure.

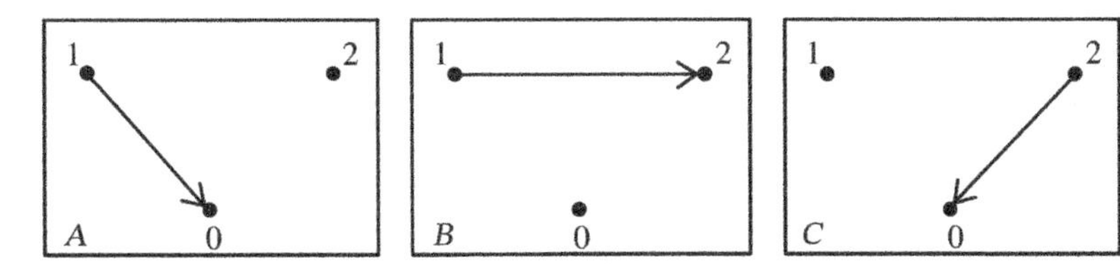

Figure 1.22 Possible positions of the inductor in a second-order PWM converter based on 1L converter cell.

to consider the static performance, such as input-to-output voltage transfer ratio and continuous inductor current, first, and if users would like to know more about the dynamics of the converter, they can analyze them further.

In the above discussed approaches, the converters are derived or synthesized based on cell or component levels. They select a proper converter configuration and add certain cell or component to the converter to form a new converter topology. Essentially, they exhaustively enumerate all of possible combinations and extract converters based on certain constraints or properties. Valid converters are verified with the volt-second balance principle. Applications of these approaches to developing new converters are quite limited because the chance of obtaining a valid converter is depending highly on experience. Is it possible to start from valid converters and with certain manipulation to develop new converters? To answer this question, several viable approaches are briefly discussed.

The three well-known valid PWM converters, buck, boost, and buck-boost, are shown in Figure 1.7. With a synchronous switch technique, the buck-boost converter can be derived from buck and boost converters in cascade connection. The derivation procedure is illustrated in Figure 1.23, in which the buck and boost converters in cascade connection is shown in Figure 1.23a. Without considering ripple current, it can be proved that capacitor C_1 can be eliminated, and inductors

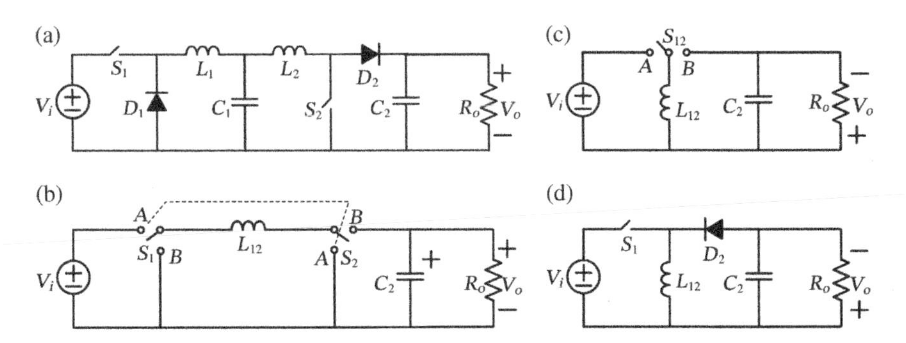

Figure 1.23 Evolution of the buck-boost converter from the buck and boost converters with a synchronous switch technique.

L_1 and L_2 are just connected in series to become L_{12}, as shown in Figure 1.23b. If switches S_1 and S_2 are synchronized and have identical duty ratio, the active–passive switch pairs, S_1&D_1 and S_2&D_2, can be replaced with two single-pole double-throw (SPDT) switches, as also shown in Figure 1.23b, in which node "A" corresponds to an active switch and node "B" is to a passive switch. Thus, the circuit shown in Figure 1.23b can be simplified to that shown in Figure 1.23c, and the two switch pairs can be combined to S_{12}. Replacing the switch pair with an active switch and a passive one yields the buck-boost converter shown in Figure 1.23d. Note that at the output of Figure 1.23b, the positive polarity is located at the upper node, while that in Figure 1.23c and d, the positive polarity is in the lower node. How to determine the polarity is not straightforward. And it usually needs several words to explain the polarity transition. Similarly, the Ćuk converter that can be proved to be a cascade connection of boost and buck converter can be also derived with the same procedure. Again, the change of output polarity needs extra explanation, and it is not so obvious and convincible.

The derivation procedure based on the synchronous switch technique is so far only applied to two switch pairs, because its combination of switch pairs, location of inductor/capacitor, and determination of output voltage polarity are not straightforward. This approach is essentially based on a preliminary observation of converter operation and configuration, but it lacks of principle or mechanism in decoupling and decoding PWM converters. Thus, it cannot be extended to derive other PWM converters, such as the sepic and Zeta converters shown in Figure 1.8b and c.

Based on the synchronous switch concept, the graft switch technique (GST) was proposed. Instead of starting from converter manipulation, the GST starts to deal with how to graft two switches operated in unison or synchronously and with at least a common node, from which four types of grafted switches are developed, as shown in Figure 1.24. They are T-type, inverse T-type, Π-type, and inverse Π-type grafted switches, which can be used to integrate the active switches in the converters. An illustration example in deriving the buck-boost converter is shown in Figure 1.25. Again, the buck and boost converters in cascade connection shown in Figure 1.23a is still adopted. After simplifying the $L_1C_1L_2$ filter, we can obtain a circuit shown in Figure 1.25a. By exchanging the connection of source V_i and switch S_1, we can create a common D–S node for switches S_1 and S_2, as shown in Figure 1.25b. Then, we replace S_1 and S_2 with a Π-type grafted switch S_{12}, as shown in Figure 1.25c. Since the currents through switches S_1 and S_2 are identical when they are operated in unison, the two circulating-current diodes D_{F1} and D_{F2} can be removed from the Π-type grafted switch, and the circuit becomes the one shown in Figure 1.25d. Note that detailed explanation for the diode degeneration will be presented in later chapter. From the circuit shown in Figure 1.25d, we can recognize that diodes D_1 and D_2 are just in series connection, and they can be

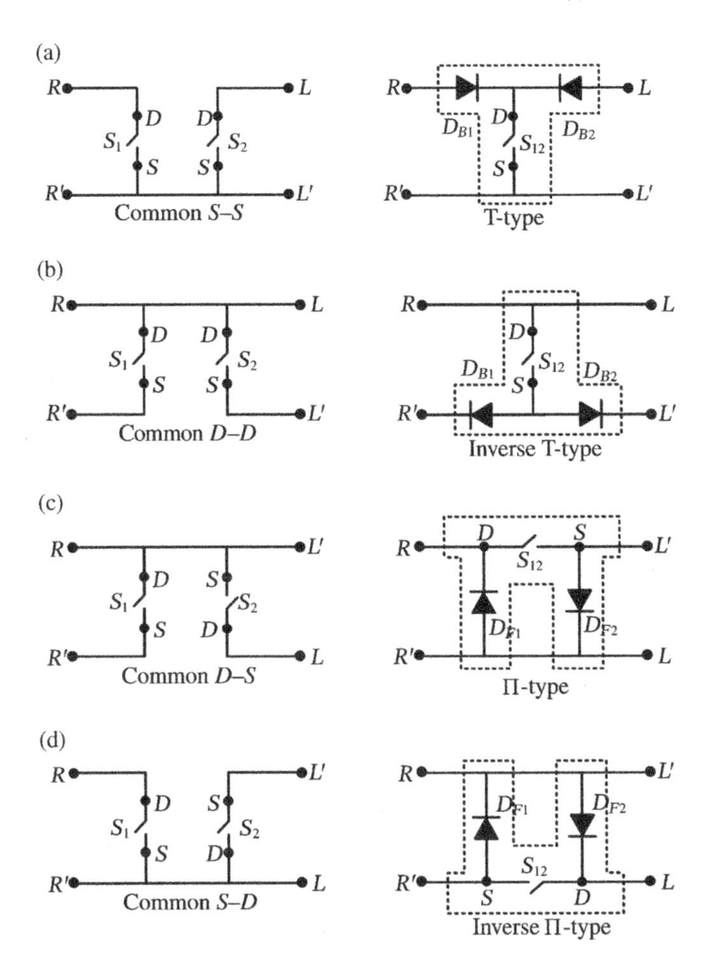

Figure 1.24 Four types of grafted switches: (a) T-type, (b) inverse T-type, (c) Π-type, and (d) inverse Π-type.

replaced with a single one D_{12}, as shown in Figure 1.25e. By redrawing the circuit, we can have the one shown in Figure 1.25f. Since switches S_{12} and diode D_{12} can be moved from the return path to the forward path without changing its operational principle, we can have a well-known form of the buck-boost converter shown in Figure 1.25g. Note that the output voltage polarity is naturally different from that of the input through the switch integration, without the need of extra words to explain that.

Note that for converter applications, the buck and the boost converters in cascade and with a simplified filter shown in Figure 1.25a are also feasible for their lower voltage stresses imposed on switches S_1&S_2 and diodes D_1&D_2, and V_i and

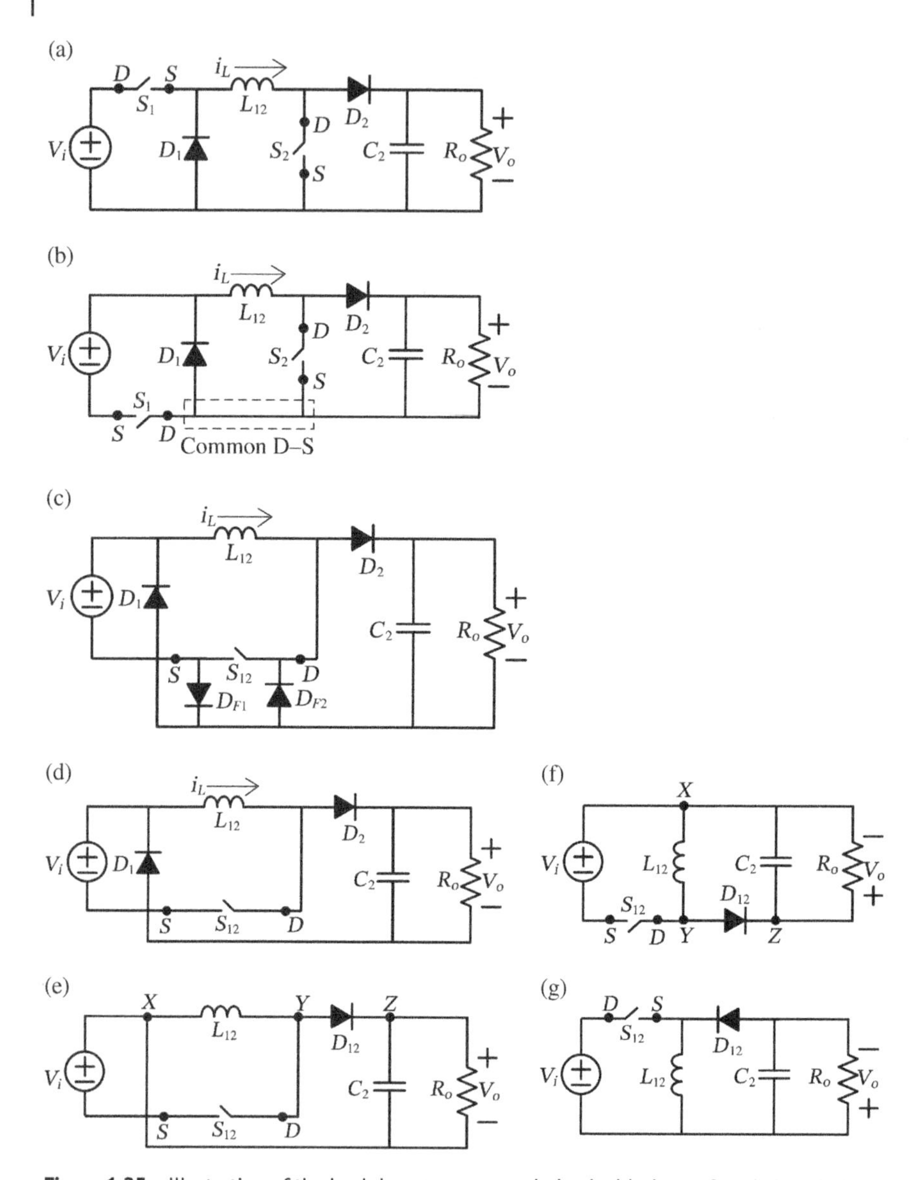

Figure 1.25 Illustration of the buck-boost converter derived with the graft switch technique.

V_o have a common voltage polarity, even though the cascaded converter requires two pairs of active–passive switches and more switch drivers.

The GST starts from dealing with two active switches, but in fact, it can be extended to any number of switches operated synchronously and with at least a common node. Moreover, with a transfer-ratio decoding process, the GST can be

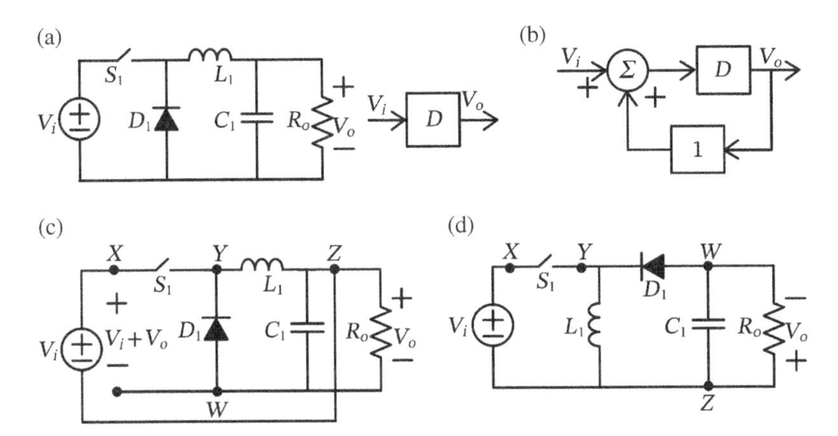

Figure 1.26 Derivation of the buck-boost converter with the converter layer technique.

applied to derive other PWM converters, including the converters shown in Figures 1.8–1.13.

Based on a transfer-ratio decoding process, the converter layer technique (CLT) was proposed. With the CLT, the buck-boost converter can be derived readily, as shown in Figure 1.26. Figure 1.26a shows a buck converter with its input-to-output voltage transfer ratio D. With a positive unity output feedback to the input, the input-to-output voltage transfer ratio can be determined as $V_o/V_i = D/(1-D)$, as shown in Figure 1.26b, which can be realized with the buck converter and a unity output feedback shown in Figure 1.26c. Redrawing the circuit in a well-known form can be recognized as a buck-boost converter, as shown in Figure 1.26d.

The GST can be equivalent to a converter feedforward approach, while the CLT is a feedback scheme. With these two techniques together, many new PWM converters can be derived. When further associated with the transfer-ratio decoding process, more converters can be synthesized, and readers can understand the converter evolution mechanism or principle comprehensively.

1.5 Evolution

In "On the Origin of Species" by Charles Darwin, evolution is the primary principle of generating offsprings from the original species through natural selection. Evolution is the process of change in all forms of life over generations. Biological populations evolve through genetic changes that correspond to changes in the organisms' observable traits. Genetic changes include mutations, which are caused by damage or replication errors in organisms' DNA. As the genetic variation of a population drifts randomly over generations, natural selection gradually leads traits to become more or less common based on the relative reproductive success of organisms with those traits. The above statements are digested from "Introduction to Evolution" in Wikipedia. DNA carries codes for heredity through the mechanisms of replication, mutation, and natural selection in diverging species from the original one. Offsprings decode parental DNA for synthesizing all tissues and organs. In general, we are all a family and have the same ancestor.

Analogously, this book is entitled "Origin of Power Converters," and we are searching for possible similar mechanisms for evolving power converters from the original converter. We will develop the decoding and synthesizing mechanisms for evolving power converters artificially. Additionally, we will make use of fundamental circuit theories to extend the converters with soft-switching features and isolation.

1.6 About the Text

The objective of this book is to present approaches to decoding, synthesizing, and modeling PWM converters systematically and to provide readers a comprehensive understanding of converter evolution from the original converter. This book is divided into two parts.

1.6.1 Part I: Decoding and Synthesizing

Part I includes 11 chapters. They present an introduction, discovery of the original converter, some fundamentals related to power converter synthesis and evolution, illustration of converter synthesis approaches, synthesis of multistage/multilevel converters, extension of hard-switching converters to soft-switching ones, and determination of switch-voltage stresses in the converters. Converters evolved from the original converter are the primary concept developed in this book.

Chapter 2 presents three approaches to creating the origin of power converters, including source–load, proton–neutron–meson, and resonant approaches. In addition, it reviews the properties and typical operation of three conventional PWM converters. Moreover, the conventional topological duality and current source are re-examined to set up a foundation for later discussions on the development of new PWM converters.

During converter synthesis and evolution, several fundamental circuit theories and principles are used frequently, and they are briefly reviewed and presented in Chapter 3. The fundamentals include DC voltage/current offsetting, capacitor/inductor splitting, DC voltage blocking and filtering, magnetic coupling, DC transformer, switch/diode grafting, and layer scheme.

Chapter 4 first reviews several typical transfer codes, such as step-down, step-up, step-up and step-down, and ±step-up and step-down, from which a proper transfer code derived from the typical codes can be designed for the desired applications. Given a transfer ratio or code, we have to first decode it in terms of the codes of the original converter and its derivatives. Thus, the code configurations and decoding approaches are also presented in Chapter 4.

In Chapter 5, the conventional cell and synchronous switch approaches to developing power converters are first reviewed, from which their limitations are addressed. Then, we describe the principles of the proposed graft switch/diode techniques, and along with the code configurations, syntheses of power converters with the graft switch/diode techniques are illustrated.

Chapter 6 adapts the layer scheme, which is a kind of technique used in growing new plants, to synthesize power converters. We identify the buck and boost families and their DNAs. Moreover, syntheses of new PWM converters have been presented in detail. Furthermore, different converter configurations with identical voltage transfer code are proved to be identical by just changing their capacitor DC offset voltages.

In Chapter 7, we present the syntheses of the well-known PWM converters, such as voltage-fed Z-source, current-fed Z-source, quasi-Z-source, switched-capacitor, and switched-inductor converters. Additionally, the syntheses of PWM converters with the graft switch technique, the converter layer technique, and the fundamentals are also presented.

There are many multistage and multilevel power converters, such as single-phase converters, three-phase converters, flywheeling capacitor converters (FCC), Vienna rectifier, neutral-point clamped converters (NPC), modular multilevel converters (MMC), etc. Systematical syntheses of the converters based on the discussed approaches is presented in Chapter 8. In general, resonant converters, which include quasi-resonant, resonant, and multi-resonant converters, are belonged to PWM converters with two half discontinuous operation modes. Their syntheses are also addressed in Chapter 8.

Based on the syntheses of hard-switching PWM converters, the discussed approaches can be extended to the syntheses of soft-switching ones, including near-zero-voltage switching, near-zero-current switching, zero-voltage switching, and zero-current switching PWM converters, which are presented in Chapter 9.

Chapter 10 presents the determination of switch-/diode-voltage stresses based on the voltage transfer code. This is an additional feature of the code configuration approaches.

Finally, discussions about topological and circuital dualities, identification of new PWM converters based on circuital duality, and analogy of PWM converters to DNA are presented in Chapter 11. A brief conclusion that hopefully no more trial and error is needed in developing power converters is addressed. Moreover, our mindsets in doing research over past 25 years are summarized in a free-style poem and presented in the end of Part I.

1.6.2 Part II: Modeling and Applications

Part II includes Chapters 12–15. They first review conventional two-port network theory and state-space averaged (SSA) modeling approach, from which systematical modeling approaches based on the graft switch technique, the converter layer scheme, and some fundamental circuit theories are presented. Basically, they model the PWM converters out of the original converter in two-port networks. Additionally, two application examples are presented to illustrate the discussed modeling approaches.

Chapter 12 first reviews the two-port network theory and discusses the SSA modeling of PWM converters with the converter layer technique. It includes the modeling of the original converter, two typical feedback configurations, and possible extension to quasi-resonant converters.

Chapter 13 presents modeling of PWM converters with the graft switch technique, which includes modeling of the buck family, buck, buck-boost and Zeta converters, and boost family, boost, boost-buck, and sepic converters; out of the basic converter units; and representations of small-signal transfer characteristics in terms of two-port network circuits.

Chapters 14 and 15 present two application examples based on the proposed modeling approaches. One is presenting design and modeling of an isolated single-stage converter for power factor correction and fast regulation; the other is to develop another single-stage converter for achieving high power factor and fast regulation with either peak current control or robust control. Simulated and experimental results have been presented to verify the feasibility of the proposed approaches.

Further Reading

Anderson, J. and Peng, F.Z. (2008a). Four quasi-Z source inverters. *Proceedings of the IEEE Power Electronics Specialists Conference*, pp. 2743–2749.

Anderson, J. and Peng, F.Z. (2008b). A class of quasi-Z source inverters. *Proceedings of the IEEE Industrial Application Meeting*, pp. 1–7.

Axelrod, B., Borkovich, Y., and Ioinovici, A. (2008). Switched-capacitor/switched-inductor structures for getting transformerless hybrid dc-dc PWM converters. *IEEE Trans. Circuits Syst. I* 55 (2): 687–696.

Berkovich, Y., Shenkman, A., Ioinovici, A., and Axelrod, B. (2006). Algebraic representation of DC-DC converters and symbolic method of their analysis. *Proceedings of the IEEE Convention. Electrical and Electronics Engineers*, pp. 47–51.

Berkovich, Y., Axelrod, B., Tapuchi, S., and Ioinovici, A. (2007). A family of four-quadrant, PWM DC-DC converters. *Proceedings of the IEEE Power Electronics Specialists Conference*, pp. 1878–1883.

Bhat, A.K.S. and Tan, F.D. (1991). A unified approach to characterization of PWM and quasi-PWM switching converters: topological constraints, classification, and synthesis. *IEEE Trans. Power Electron.* 6 (4): 719–726.

Bryant, B. and Kazimierczuk, M.K. (2002). Derivation of the buck-boost PWM DC-DC converter circuit topology. *Proc. IEEE Int. Symp. Circuits Syst.* 5: 841–844.

Bryant, B. and Kazimierczuk, M.K. (2003). Derivation of the Ćuk PWM DC-DC converter circuit topology. *Proc. IEEE Int. Symp. Circuits Syst.* 3: 292–295.

Cao, D. and Peng, F.Z. (2009). A family of Z source and quasi-Z source DC-DC converters. *Proceedings of the IEEE Applied Power Electronics Conference*, pp. 1097–1101.

Ćuk, S. (1979). General topological properties of switching structures. *Proceedings of the IEEE Power Electronics Specialists Conference*, pp. 109–130.

Erickson, R.W. (1983). Synthesis of switched-mode converters. *Proceedings of the IEEE Power Electronics Specialists Conference*, pp. 9–22.

Freeland, S.D. (1992). Techniques for the practical applications of duality to power circuits. *IEEE Trans. Power Electron.* 7 (2): 374–384.

Hopkins, D.C. and Root, D.W. Jr. (1994). Synthesis of a new class of converters that utilize energy recirculation. *Proceedings of the IEEE Power Electronics Specialists Conference*, pp. 1167–1172.

Khan, F.H., Tolbert, L.M., and Peng, F.Z. (2006). Deriving new topologies of DC-DC converters featuring basic switching cells. *Proceedings of the IEEE Workshops on Computers in Power Electronics*, pp. 328–332.

Lee, F.C. (1989). *High-Frequency Resonant, Quasi-Resonant and Multi-Resonant Converters.* Virginia Power Electronics Center.

Liu, K.-H. and Lee, F.C. (1988). Topological constraints of basic PWM converters. *Proceedings of the IEEE Power Electronics Specialists Conference*, pp. 164–172.

Makowski, M.S. (1993). On topological assumptions on PWM converters: a re-examination. *Proceedings of the IEEE Power Electronics Specialists Conference*, pp. 141–147.

Maksimovic, D. and Ćuk, S. (1989). General properties and synthesis of PWM DC-to-DC converters. *Proceedings of the IEEE Power Electronics Specialists Conference*, pp. 515–525.

Ogata, M. and Nishi, T. (2003). Topological criteria for switched mode DC-DC converters. *Proc. IEEE Int. Symp. Circuits Syst.* 3: 184–187.

Peng, F.Z. (2003). Z source inverter. *IEEE Trans. Ind. Appl.* 39 (2): 504–510.

Peng, F.Z., Joseph, A., Wang, J. et al. (2005a). Z source inverter for motor drives. *IEEE Trans. Power Electron.* 20 (4): 857–863.

Peng, F.Z., Tolbert, L.M., and Khan, F.H. (2005b). Power electronic circuit topology – the basic switching cells. *Proceedings of the IEEE Power Electronics Education Workshop*, pp. 52–57.

Qian, W., Peng, F.Z., and Cha, H. (2011). Trans-Z source inverters. *IEEE Trans. Power Electron.* 26 (12): 3453–3463.

Severns, R.P. and Bloom, G.E. (1985). *Modern DC-to-DC Switch Mode Power Converter Circuits.* New York: Van Nonstrand Reinhold Co.

Tolbert, L. M., Peng, F.Z., Khan, F.H., and Li, S. (2009). Switching cells and their implications for power electronic circuits. *Proceedings of the IEEE International Power Electronics and Motion Control Conference*, pp. 773–779.

Tymerski, R. and Vorperian, V. (1986). Generation, classification and analysis of switched-mode DC-to-DC converters by the use of converter cells. *Proceedings of the International Telecommunications Energy Conference*, pp. 181–195.

Williams, B.W. (2008). Basic DC-to-DC converters. *IEEE Trans. Power Electron.* 23 (1): 387–401.

Williams, B.W. (2014). Generation and analysis of canonical switching cell DC-to-DC converters. *IEEE Trans. Ind. Electron.* 61: 329–346.

Wu, T.-F. and Chen, Y.-K. (1996). A systematic and unified approach to modeling PWM DC/DC converters using the layer scheme. *Proceedings of the IEEE Power Electronics Specialists Conference*, pp. 575–580.

Wu, T.-F. and Yu, T.-H. (1998). Unified approach to developing single-stage power converters. *IEEE Trans. Aerosp. Electron. Syst.* 34 (1): 221–223.

Wu, T.-F., Liang, S.-A., and Chen, Y.-K. (2003). A structural approach to synthesizing soft switching PWM converters. *IEEE Trans. Power Electron.* 18 (1): 38–43.

2

Discovery of Original Converter

A general question was brought up in last chapter on how to develop or derive PWM converters systematically. There are several approaches to developing converters based on switching cells, canonical converter cells, and switched capacitor/inductor cells, but they left a lot of questions behind without answers. In this book, our attempt is adopting from Charles Darwin's believe of evolution on which he published the book entitled *On the Origin of Species*. Similarly, we are intended to identify the origin of power converters, from which all of PWM power converters are evolved. The evolution mechanism and principle will be explored in later chapters. In this chapter, three approaches to creating the original converter, the buck converter, are discussed. Based on the original converter, we present three fundamental PWM converters that are used frequently in decoding and synthesizing converters. Their operational modes are discussed correspondingly to verify the evolved converters.

2.1 Creation of Original Converter

Charles Darwin's dilemma is that if species are evolved from their ancestor species and if we keep tracking back to the origin, who is the original one and how to create or generate it? The same questions bother us: if PWM converters are evolved, which one is the origin and what is the mechanism to create it? In literature, there are many converters with various transfer codes, such as $D/(1-2D)$, $(1-D)/(1-2D)$, $D/(2-D)$, $D/2(1-D)$, $D^2/(1-D)$, etc., in which all of the combinational codes include the duty ratio D and it is just the transfer code of the buck converter in continuous conduction mode (CCM). Thus, we can intuitively reveal that buck converter is the original converter. In this book, we propose three approaches to creating the original converter.

Origin of Power Converters: Decoding, Synthesizing, and Modeling, First Edition.
Tsai-Fu Wu and Yu-Kai Chen.
© 2020 John Wiley & Sons, Inc. Published 2020 by John Wiley & Sons, Inc.

2.1.1 Source–Load Approach

Based on the natural law, Faraday's law, there exists only voltage source, and the output load is usually supplied with voltage form. Thus, the source and load can be configured by the circuit shown in Figure 2.1a. To control the power flow from the source to the load, a control gear or a switch is adopted to link the source and load, as shown in Figure 2.1b. Since source voltage V_i is not necessarily always equal to load voltage V_o, there might exist an inrush current through the switch when switch S_1 is turned on. An inductor is therefore inserted in series with the switch, as shown in Figure 2.1c, to limit current slew rate. However, when switch S_1 is turned off, there is no path for inductor current to flow continuously, resulting in high voltage spike imposing on the other components. Thus, a freewheeling diode D_1 is introduced to provide a path for the inductor current, as shown in Figure 2.1d, and it is the buck converter. How to prove that the buck converter is the origin of converters is left for later discussion.

2.1.2 Proton–Neutron–Meson Analogy

There was a legend about Dr. Hideki Yukawa who was the 1949 Nobel Prize winner in Physics area. Dr. Yukawa used to get stuck at the interaction between protons inside nuclei. Protons carrying positive charges are supposed to repel each other, but how come they stick together tightly? One day, when he walked from his home to laboratory, he found out that the two dogs used to bark each other, but today they stick to each other without barking or fighting. He was curious and walked closely to take a look, and he revealed that they were biting on the same pig's bone. Finally, he realized that the neutrons inside a nucleus play the role of pig's bone to tight protons (dogs) closely. However, if they always stick to

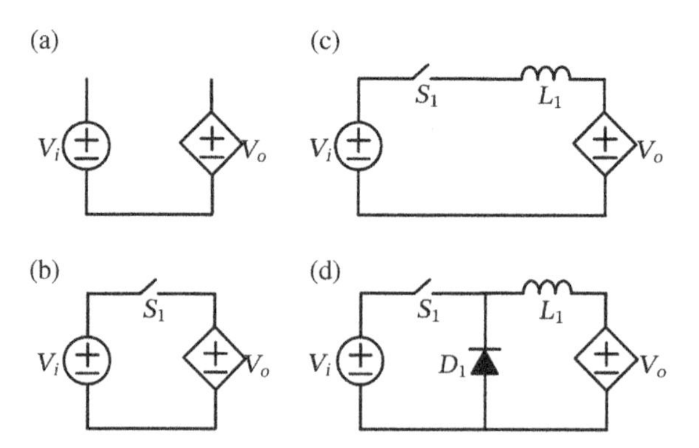

Figure 2.1 Derivation of the original converter, buck converter, with a source–load approach.

each other, which is equivalent to a static balance, how can they interact? This became his research topic, and he figured out that there exist mesons appearing in pair (π^+, π^-), which govern the interactions between protons and neutrons inside nuclei. When meson π^+ is in action, meson π^- will be deactivated, and vice versa. In other words, when π^- is in action, π^+ will be deactivated.

With this understanding, derivation of the buck converter can be analogous as follows. The input voltage source and output voltage sink can be treated as two protons, as shown in Figure 2.2a, since they have the same polarity. An inductor, which looks like a rawhide bone, plays the role of a neutron to tight two voltages (similar to protons or dogs), as shown in Figure 2.2b. With this configuration, the output voltage will finally equal to the input voltage, which is a static balance, and there is no further interaction. Thus, it requires to introduce an active–passive switch pair (S_1, D_1), like meson pair (π^+, π^-), into the circuit to control power flow from the input to the output, as shown in Figure 2.2c. When switch S_1 is turned on, diode D_1 is in reverse bias, and on the other way, when S_1 is turned off, D_1 will conduct to freewheel the inductor current. The configuration and action are similar to those of the proton–neutron–meson model. Thus, the buck converter can be derived accordingly.

It can be noted that with a source–load approach, the buck converter derivation is constructed one component by one component. While, with the proton–neutron–meson approach, the active–passive switch pair is introduced to the converter at a time, like a meson pair.

2.1.3 Resonance Approach

Power transfer between energy storage elements, capacitors and inductors, has three types of configurations, as shown in Figure 2.3. The two types shown in Figure 2.3a and b will result in electrical energy loss up to half the initially stored energy when turning on switch S_1 and under the condition of capacitance $C_1 = C_2$ or inductance $L_1 = L_2$. To conserve the total electrical energy during power transfer, the only valid configuration is shown in Figure 2.3c where the power transfer

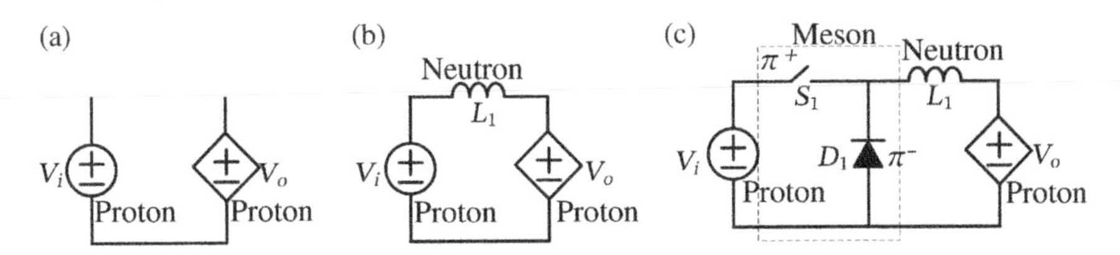

Figure 2.2 Analogy of the buck converter derivation to proton–neutron–meson model of a nucleus.

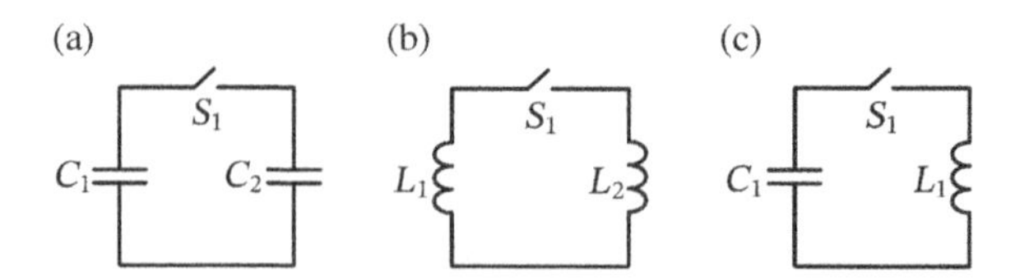

Figure 2.3 Three types of configurations of power transfer between capacitors and inductors.

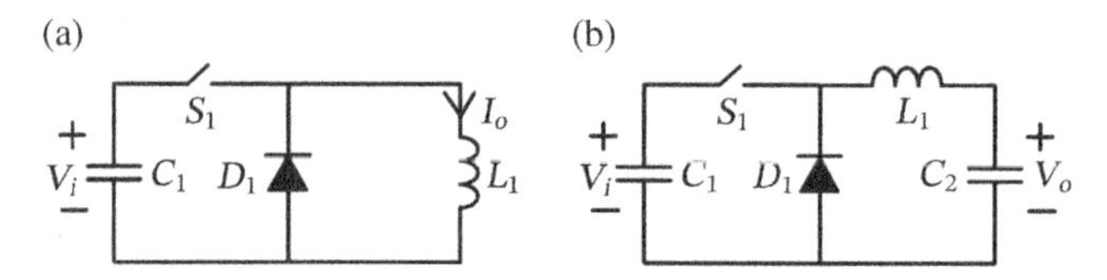

Figure 2.4 Practical examples applying the configuration shown in Figure 2.3c: (a) with current output and (b) with voltage output.

is from capacitor to inductor or vice versa and it is conducted in resonant manner. Practical examples applying this configuration are shown in Figure 2.4, in which Figure 2.4a shows a current output and Figure 2.4b shows a voltage one, and the diode D_1 is introduced to the converter to circulate the energy stored in inductor L_1 when switch S_1 turns off. With the freewheeling diode D_1, the converter shown in Figure 2.4b can be controlled with PWM to tune its input-to-output voltage transfer gain. This converter, namely, buck converter, consists of the minimum number of components for power transfer with resonance.

In the above discussions, the buck converter has been derived with different approaches. Its power transfer is straightforward from input to output when turning on active switch S_1, and when turning off the switch, the energy stored in inductor L_1 is continuously releasing to the output. The power flow can be controlled with PWM, and its output is always limited within the input voltage in the steady state. From dynamic point of view, the buck converter is a kind of minimum-phase system, and it is easy to achieve high stability margin. With all of these positive natural properties together, the buck converter has the potential to be the original converter for evolving the rest of PWM converters. This viewpoint will be proved through decoding and synthesizing processes in later chapters.

2.2 Fundamental PWM Converters

Before embarking on the proof of the original converter, we review the operational principle of the buck converter and derive its input-to-output voltage transfer ratios in CCM and discontinuous conduction mode (DCM). Additionally, we show the derivation of buck-boost and boost converters from the buck converter to get some feeling about the potential of the buck converter acting as the original converter.

2.2.1 Voltage Transfer Ratios

From power transfer point of view, the resonance approach can describe the derivation of the buck converter with more physical insight. For resonance, it requires at least a second-order *LC* network. In addition to the buck converter, there are other two well-known PWM converters, boost and buck-boost, each of which is also with a second-order *LC* network. As discussed previously, the buck converter is considered as the candidate of the original converter. Thus, let us see if it is possible to evolve buck-boost and boost converters from the buck converter. For illustrating the evolution, operation mode and transfer ratio of the buck converter need to be discussed first.

Given a buck converter shown in Figure 2.5a and assuming all of the components are ideal, when switch S_1 turns on, the voltage across inductor L_1 is $V_i - V_o$, while when switch S_1 turns off, diode D_1 will conduct, and the voltage across L_1 is $-V_o$, as illustrated in Figure 2.5b. Thus, based on the volt-second balance principle, we can have the following equation:

$$(V_i - V_o)DT_S + (-V_o)(1-D)T_S = 0, \tag{2.1}$$

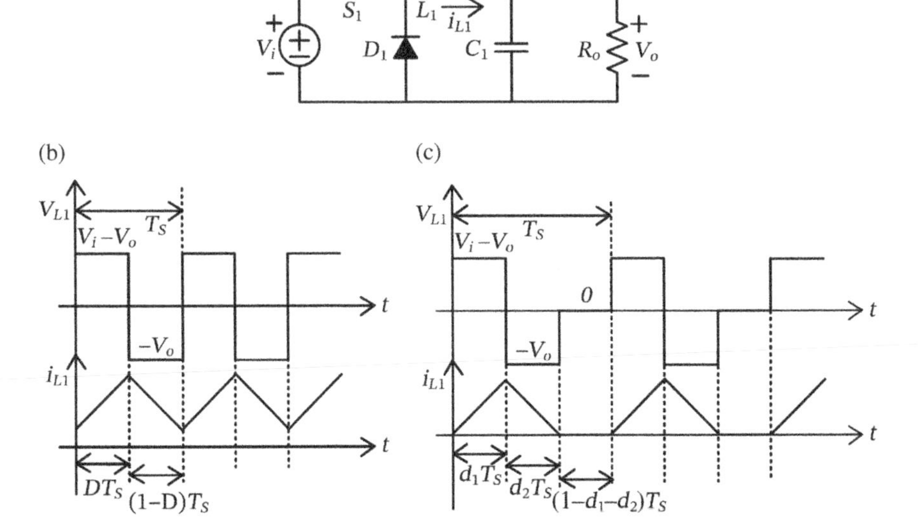

Figure 2.5 (a) The buck converter, (b) inductor voltage V_{L1} and current i_{L1}, and (c) those in DCM operation.

where D is the duty ratio of switch S_1 and T_S is the switching period. Since inductor current i_{L1} never drops to zero, this operation mode is called CCM. From (2.1), we can derive the input-to-output voltage transfer ratio below:

$$\frac{V_o}{V_i} = D. \tag{2.2}$$

If inductor current i_{L1} drops to zero before turns on switch S_1 again, as shown in Figure 2.5c, the operation mode is called DCM. Again, based on the volt-second balance principle, we can have the following equation:

$$\left(V_i - V_o\right) d_1 T_S + \left(-V_o\right) d_2 T_S + 0\left(1 - d_1 - d_2\right) T_S = 0, \tag{2.3}$$

where d_1 is the duty ratio of switch S_1, d_2 is the duty ratio of diode D_1, and $(1 - d_1 - d_2)T_S$ is the dead time. From (2.3), we have the following input-to-output voltage transfer ratio under DCM operation:

$$\frac{V_o}{V_i} = \frac{d_1}{\left(d_1 + d_2\right)}. \tag{2.4}$$

If

$$\left(d_1 + d_2\right) = 1, \tag{2.5}$$

the operation will be CCM and $d_1 = D$. Thus, the CCM can be considered a special case of DCM. No matter what mode of operation, the configuration of buck converter keeps unchanged. For simplicity and considering most of converters operated in CCM, we will first discuss the evolution of converters based on the voltage transfer ratio under CCM operation.

2.2.2 CCM Operation

A buck converter and its voltage transfer ratio under CCM operation are shown in Figure 2.6a. When taking the output from capacitor C_1, we can have the following voltage transfer ratio:

$$\frac{V_o}{V_i} = D. \tag{2.6}$$

If taking the output from capacitor C_2, we will have a transfer ratio of

$$\frac{V_o'}{V_i} = 1 - D. \tag{2.7}$$

The voltage transfer ratio of a buck-boost converter is

$$\frac{V_o}{V_i} = \frac{D}{(1-D)}, \tag{2.8}$$

and it can be decoded into the form shown in Figure 2.6b, in which the forward-path gain is D and the feedback-path gain is unity. The gain D can be synthesized with a buck converter for its single-ended characteristic and meeting the requirement of a forward path in a control system with forward and feedback paths. The unity-gain feedback is synthesized by feeding back the output voltage V_o to the input, and there is no power flow from this feedback path to the output. Thus, the overall converter configuration depicted in Figure 2.6c can synthesize the control block diagram shown in Figure 2.6b satisfactorily and correctly. Redrawing the

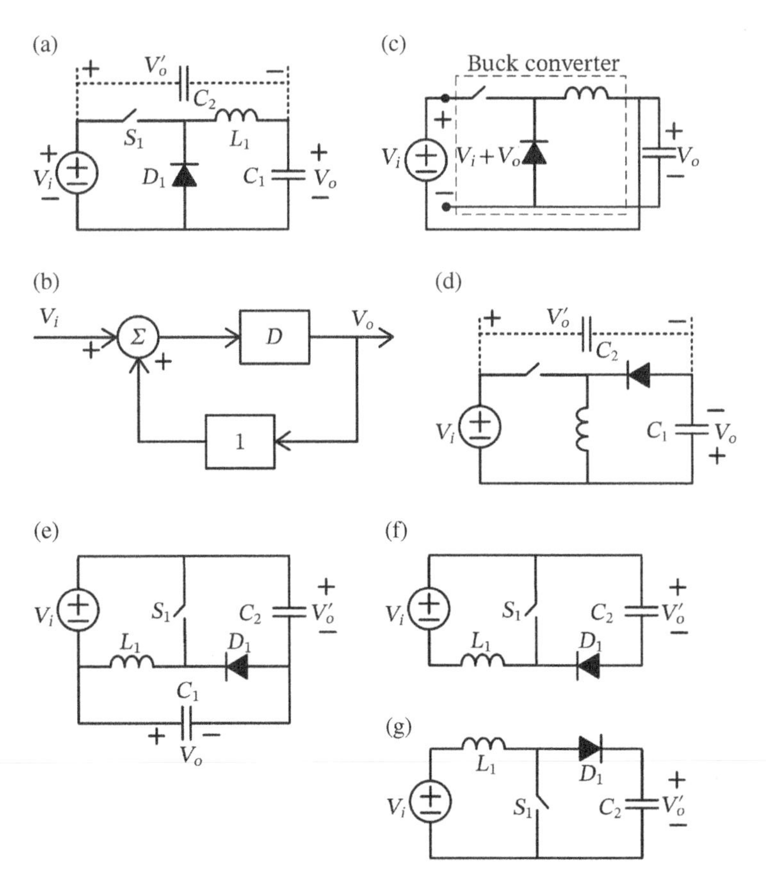

Figure 2.6 Decoding, synthesizing, and evolution of buck-boost and boost converters from the buck converter.

circuit shown in Figure 2.6c yields the buck-boost converter, as shown in Figure 2.6d, which is in the form that people are familiar with.

Similarly, we can take the output from capacitor C_2 of the buck-boost converter depicted in Figure 2.6d, and we have the following voltage transfer ratio or code:

$$\frac{V_o'}{V_i} = \frac{V_i + V_o}{V_i} = 1 + \frac{D}{(1-D)} = \frac{1}{(1-D)}, \tag{2.9}$$

which is the input–output voltage transfer ratio or code of the boost converter in CCM operation. Redrawing the circuit of Figure 2.6d yields the one shown in Figure 2.6e, in which voltages V_i, V_o, and V_o' form a loop and one of the voltages is a dependent variable. Since we do not take output from capacitor C_1, it can be removed from the circuit, as shown in Figure 2.6f. When moving inductor L_1 and diode D_1 from the return path to the forward, we can obtain the boost converter, as shown in Figure 2.6g. With the decoding and synthesizing processes, the buck-boost and boost converters can be evolved from the buck converter. Thus, the buck converter can be considered preliminarily the origin of power converters.

2.2.3 DCM Operation

The above decoding process is based on CCM operation of the converters. In order to confirm the decoding process is also working for DCM operation, the input–output voltage transfer ratios of the converters in DCM operation are discussed as follows.

For the buck converter shown in Figure 2.6a and operated in DCM, the input–output voltage transfer ratio was derived and expressed in (2.4). Similarly, the transfer ratio of the buck-boost converter in DCM operation can be derived as

$$\frac{V_o}{V_i} = \frac{d_1}{d_2}, \tag{2.10}$$

by substituting the transfer ratio D shown in (2.8) with the ratio, $d_1/(d_1 + d_2)$, of the buck converter in DCM operation. The transfer ratio of the boost converter, therefore, can be derived as

$$\frac{V_o}{V_i} = 1 + \frac{d_1}{d_2} = \frac{(d_1 + d_2)}{d_2}. \tag{2.11}$$

Again, if

$$d_1 + d_2 = 1, \tag{2.12}$$

and d_2 is replaced with $(1-D)$, the boost converter is, therefore, in CCM operation, and the transfer ratio will become the one shown in (2.9).

Based on the transfer ratio of the buck converter in DCM operation, those of the buck-boost and boost converters can be derived correspondingly. The transfer ratios of the buck-boost and boost converters in DCM operation can be also derived directly based on volt-second balance principle, and they come out the same expressions as those shown in (2.10) and (2.11). This confirms that the decoding and synthesizing processes can be applied for both DCM and CCM operations. In the rest of the chapters, for simplicity, the discussion of decoding and synthesizing processes will be based on CCM operation only. From now on, the transfer ratio will be treated as a transfer code for further decoding processing.

From the above discussion, we observed that the original PWM code is D, and the derived codes include $(1-D)$, $1/(1-D)$, and $D/(1-D)$, which can be adopted as fundamental codes in decoding transfer codes. Buck converter is the origin, and the evolved converters are buck-boost and boost converters up to this moment. In the evolution process, the evolved converters are not always directly evolved from the original converter, but they can be evolved from the evolved converters or their descendant converters, like that the boost converter is evolved from the buck-boost converter instead of the buck converter, while the buck-boost converter is evolved from the buck converter.

2.2.4 Inverse Operation

By exchanging the roles of the active and passive switches in the converters shown in Figure 2.6, we have the inverse converters, as shown in Figure 2.7, and their corresponding transfer ratios can be derived as $1/D$, $(1-D)$, and $(1-D)/D$, which are the reciprocals of D, $1/(1-D)$, and $D/(1-D)$, respectively. These codes provide more choices for decoding transfer codes. Illustrations of decoding the transfer

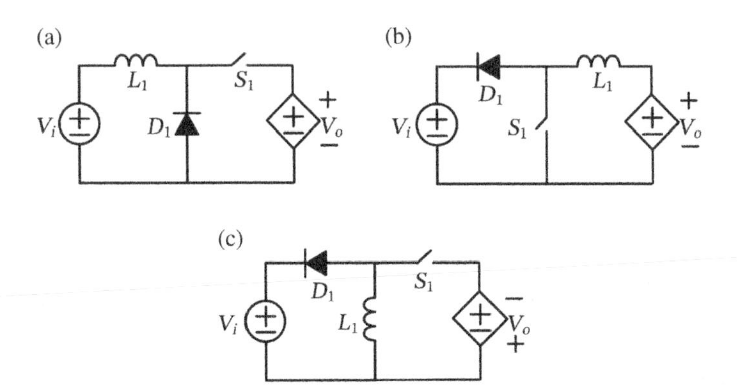

Figure 2.7 (a) Inverse buck, (b) inverse boost, and (c) inverse buck-boost with the transfer ratios of $1/D$, $(1-D)$, and $(1-D)/D$, respectively.

codes of PWM converters in terms of the fundamental codes discussed previously will be presented in later chapters.

Typically, the inverse converters do not operate independently since its output sink will transfer power back to the input source in unidirection. They usually work with other regular converters to control power flow between input and output, which can achieve higher step-up or step-down power conversion. The regular and the inverse buck, buck-boost, and boost converters are considered the fundamental converters since in the decoding process, their transfer codes will be used frequently and they are with second-order filters only.

2.3 Duality

In mathematical optimization theory, duality means that optimization problems may be viewed from either of two perspectives, the primal problem or the dual problem. Similarly, in circuit theory, duality means that viewing from either of the two variables, voltage and current, can yield the same output expression for the two dual circuits. Typical examples of two *RLC* networks are shown in Figure 2.8, in which the series *RLC* network is driven by a voltage source, while the parallel *RLC* network is driven by a current source. The circuits shown in Figure 2.8a and b are dual. It can be observed that when replacing the voltage source with a current source and the components in series with the ones in parallel, we can obtain a dual circuit of the other and we can determine the branch currents in parallel from the voltages across the branches in series. The dual networks shown in Figure 2.8 have one-to-one correspondence.

Conventionally, with the duality theory, the boost converter can be derived from the buck converter. The buck converter is first represented in the form shown in Figure 2.9a, in which the output inductor–capacitor filter is simplified to a current sink. Then, replacing the voltage source V_i with a current one I_i, the switch S_1 in series with a parallel one, the diode D_1 in parallel with a series one, and the output current sink I_o with a voltage sink V_o can yield the boost converter shown in Figure 2.9b. They have one-to-one correspondence. Since the converters are

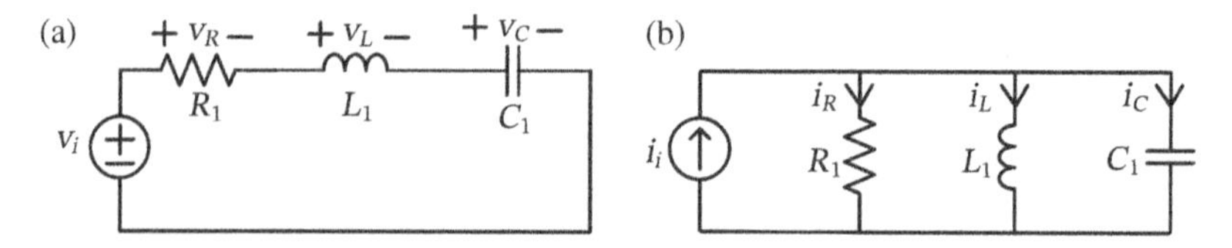

Figure 2.8 Dual RLC networks (a) in series and driven by a voltage source and (b) in parallel and driven by a current source.

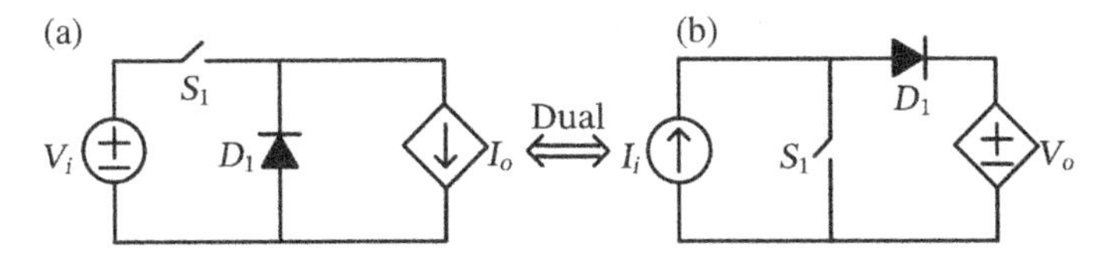

Figure 2.9 Topologically dual converters: (a) buck and (b) boost.

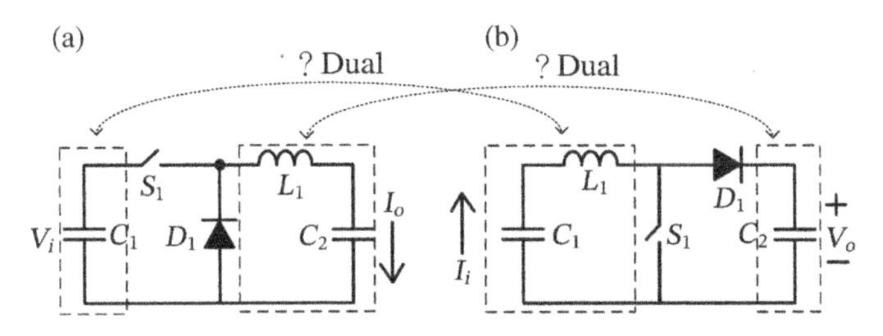

Figure 2.10 Component realization of (a) buck and (b) boost converters.

represented in voltage/current sources and voltage/current sinks, they are called topologically dual converters. When comparing the topological converters with the component converters shown in Figure 2.10, one can reveal that they do not have one-to-one correspondence and, in fact, they are not dual converters. In Figure 2.10a, capacitor C_1 realizes the voltage source V_i, while in Figure 2.10b, capacitor C_1 and inductor L_1 together realize the current source I_i. Similarly, the output current sink I_o is realized with capacitor C_2 and inductor L_1 in Figure 2.10a. In the buck and boost converters, the voltage source or sink is realized with a single capacitor. Why is a current source or sink realized with a capacitor and an inductor rather than a single inductor? Is it because there is no current source or sink, or even the configurations of the buck and boost converters are incorrect? The answers to these questions will be left for later discussions in Chapter 11.

Evolution of power converters from the buck converter, the original converter, is based on a converter level, and it requires having the desired voltage transfer code decoded in terms of the transfer codes of the fundamental converters. How to manipulate the converters from the decoded or factorized transfer codes requires using some fundamental circuit theories and principles. They will be addressed in Chapter 3.

Further Reading

Ćuk, S. (1979). General topological properties of switching structures. *Proceedings of the IEEE Power Electronics Specialists Conference*, pp. 109–130.

Darwin, C. (1993). *On The Origin of Species*. Norwalk, CT: The Easton Press (MBI, Inco).

Erickson, R.W. (1983). Synthesis of switched-mode converters. *Proceedings of the IEEE Power Electronics Specialists Conference*, pp. 9–22.

Freeland, S.D. (1992). Techniques for the practical applications of duality to power circuits. *IEEE Trans. Power Electron.* 7 (2): 374–384.

Hopkins, D.C. and Root, D.W. Jr. (1994). Synthesis of a new class of converters that utilize energy recirculation. *Proceedings of the IEEE Power Electronics Specialists Conference*, pp. 1167–1172.

Lin Wang, Y.-Y., Chang, S.-L., Wu, Y.-E. et al. (1991). Resonance-the missing phenomena in hemodynamics. *Circ. Res.* 69: 246–249.

Liu, K.-H. and Lee, F.C. (1988). Topological constraints on basic PWM converters. *Proceedings of the IEEE Power Electronics Specialists Conference*, pp. 164–172.

Maksimovic, D. and Ćuk, S. (1989). General properties and synthesis of PWM DC-to-DC converters. *Proceedings of the IEEE Power Electronics Specialists Conference*, pp. 515–525.

Schrödinger, E. (1944). *What Is Life?* Cambridge University Press.

Severns, R.P. and Bloom, G.E. (1985). *Modern DC-to-DC Switch Mode Power Converter Circuits*. New York: Van Nostrand Reinhold Co.

Tymerski, R. and Vorperian, V. (1986). Generation, classification and analysis of switched-mode DC-to-DC converters by the use of converter cells. *Proceedings of the International Telecommunications Energy Conference*, pp. 181–195.

Wu, T.-F. and Yu, T.-H. (1998). Unified approach to developing single-stage power converters. *IEEE Trans. Aerosp. Electron. Syst.* 34 (1): 221–223.

3

Fundamentals

For developing various power converters and identifying their similarity or equivalence, some fundamentals are used frequently. They include DC voltage and DC current offsetting, capacitor and inductor splitting, DC voltage blocking and pulsating voltage filtering, magnetic coupling, switch grafting, diode grafting, and layer scheme.

To conserve the total electrical energy, power transfer between energy storage elements, capacitor and inductor, should be configured as shown in Figure 3.1a where the power transfer is from capacitor to inductor or vice versa and it is conducted in resonant manner. A practical example applying this configuration is shown in Figure 3.1b and c, in which Figure 3.1b shows a current output and Figure 3.1c shows a voltage one. With the freewheeling diode D_1, the converter shown in Figure 3.1c can be operated with PWM to tune its input-to-output transfer ratio or code. After decoding and synthesizing the PWM power converters, we will recognize that this converter, buck converter, is the original converter. In the following discussions, the original converter will be used for illustrating the fundamentals.

3.1 DC Voltage and Current Offsetting

In PWM converters, capacitors can buffer DC voltages and inductors can buffer DC currents, while they can still have resonant manner. In literature, converters with different DC offsets are considered different topologies even with the same operational principle and transfer ratio. This section discusses the two fundamental DC voltage and DC current offsetting schemes that will be used for deriving converters from one to the others, and it can be recognized that the converters are

Origin of Power Converters: Decoding, Synthesizing, and Modeling, First Edition.
Tsai-Fu Wu and Yu-Kai Chen.
© 2020 John Wiley & Sons, Inc. Published 2020 by John Wiley & Sons, Inc.

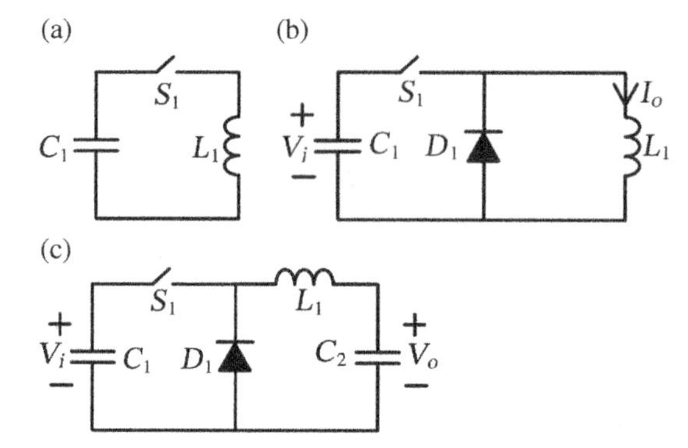

Figure 3.1 (a) Circuit configuration of power transfer between capacitor and inductor in resonant manner and two practical examples applying the configuration with (b) current output and (c) voltage output.

equivalent but just with different DC voltage offsets in buffer capacitors. Since the offsets do not affect power conversion, their transfer ratios will be proved to be identical.

3.1.1 DC Voltage Offsetting

A voltage source V_S is buffered with a capacitor C_S, and when it is in series with another capacitor C_1, the two capacitors can be equivalent to a single one C_1', as illustrated in Figure 3.2. The only difference is that the capacitor has a DC offset voltage V_S. A typical example applying the DC voltage offsetting concept is illustrated in Figure 3.3. In Figure 3.3a, one terminal of capacitor C_1 is connected to voltage source V_i, and they can be separated to be a new branch, as shown in Figure 3.3b. With a DC voltage offset, voltage source V_i can be merged into capacitor C_1 to become C_1', as shown in Figure 3.3c. It has been shown that the converter in Figure 3.3a is equivalent to that in Figure 3.3c in the aspects of transfer gain and operational principle and they should be considered the same converter topology, except that capacitors C_1 and C_1' have different voltage offsets.

Another example adopting the DC voltage offsetting concept is illustrated in Figure 3.4, in which the buck converter shown in Figure 3.4a has input voltage V_i and output voltage V_o sharing a common node. With a $-V_o$ offset, capacitor C_1 can

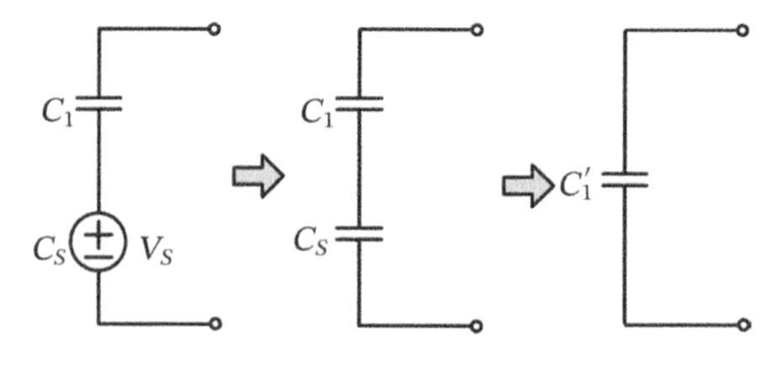

Figure 3.2 A voltage source in series with a capacitor is equivalent to a single capacitor with a DC offset voltage.

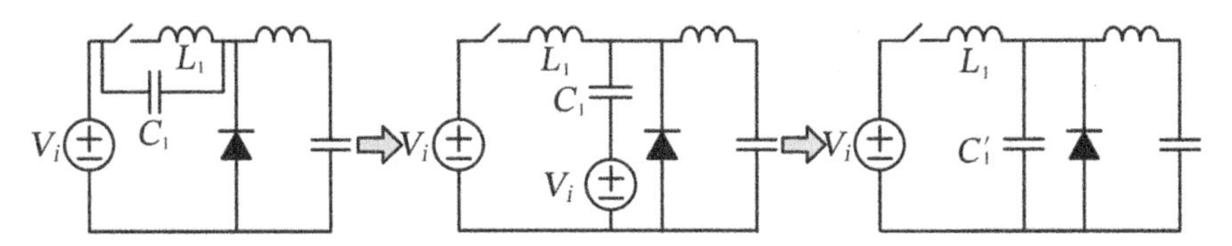

Figure 3.3 Illustration of capacitor C_1 with different DC offset voltages in a quasi-resonant buck converter to form different circuit configurations.

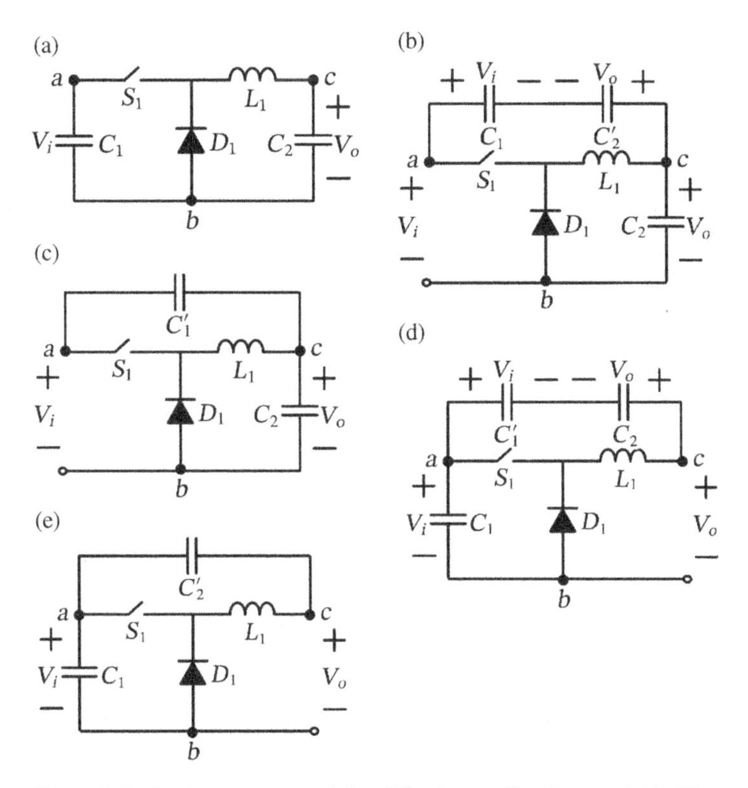

Figure 3.4 Buck converter applying DC voltage offsetting to yield different configurations but identical input-to-output voltage conversion ratio.

be connected across the positive input and output terminals, i.e. port a–c, as shown in Figure 3.4b. By combining capacitors C_1 and C_2', we can obtain the converter configuration shown in Figure 3.4c. Similarly, with a $-V_i$ offset, capacitor C_2 can be connected across port a–c, as shown in Figure 3.4d, and it can be simplified to the one shown in Figure 3.4e. It can be shown that all of the input-to-output voltage transfer ratios of the converters shown in Figure 3.4 are identical, $V_o/V_i = D$ in continuous conduction mode. For example, based on volt-second balance principle, we can write down the following equation for inductor L_1 in the converter shown in Figure 3.4e:

$$(V_i - V_o)DT_s + (-V_o)(1-D)T_s = 0. \tag{3.1}$$

From the above equation, we can obtain the relationship of $V_o/V_i = D$. In fact, based on KVL, the total voltage around loop *a–b–c–a* is zero, and one of the three possible capacitors ($C_{ab} = C_1$, $C_{bc} = C_2$, and $C_{ca} = C_1'$ or C_2') is dependent on the other two. Thus, only are two capacitors good enough to buffer energy for the converters shown in Figure 3.4.

The input current of buck converter is pulsating, and a C-L filter is usually added at the input terminals of the converter to smooth out input current. A C-L filter can be added to each of the converters shown in Figure 3.4a, c, and e to become the ones shown in Figure 3.5a–c, respectively. They are all equivalent converters in operational principle and transfer ratio.

In literature, there are many well-known PWM converters that can adopt this DC voltage offsetting concept to yield their equivalent converter configurations with the same transfer ratio and operational principle, such as the equivalent converters derived from Ćuk converter and shown in Figure 3.6.

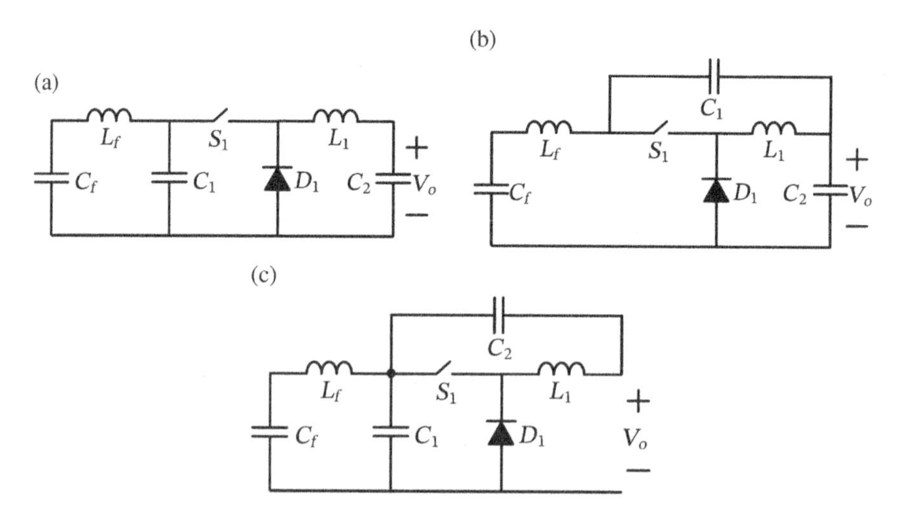

Figure 3.5 Buck and its equivalent configurations with C_f–L_f filters at their input terminals.

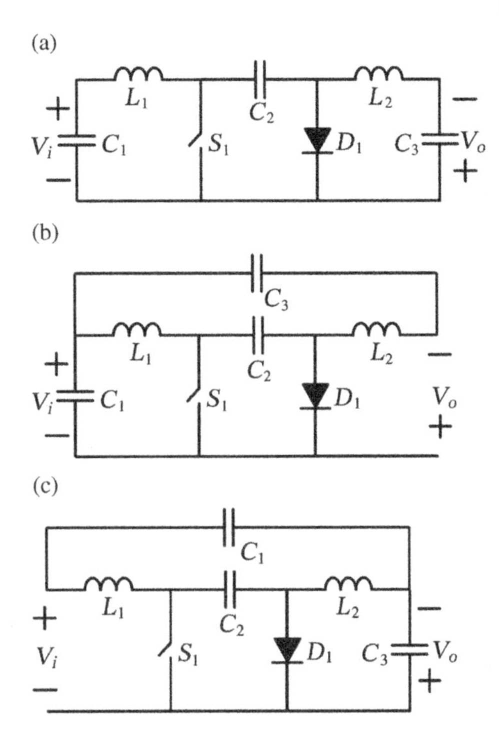

Figure 3.6 Equivalent converters derived from Ćuk converter.

3.1.2 DC Current Offsetting

According to duality, the DC voltage offsetting concept can be converted to a DC current one. Figure 3.7 shows a current source I_s in parallel with an inductor L_1, and they are equivalent to a single inductor L_1' with DC offset current. An example adopting this offsetting is shown in Figure 3.8, which is a quasi-resonant boost converter. The current source I_i in Figure 3.8a can be represented by two identical currents connected in series, as shown in Figure 3.8b. Since the voltage across a current source can be any value, node X and node Y can be connected together without changing its operational principle. Additionally, there is no current flow between the nodes. Now, inductor L_1 is in parallel with current source I_i, and they can be merged into an inductor L_1' with DC current offset, as shown in Figure 3.8c. The converter in Figure 3.8a is equivalent to that in Figure 3.8c, and they are considered equivalent topologies.

It should be noted that realization of a current source is not as straightforward as that of a voltage one. Thus, the application of DC current offsetting is not so popular. For example, a boost converter can have different configurations with different DC voltage offsets, as shown in Figure 3.9. However, if the input of the boost converter is represented in current form, it only has a kind of configuration, as shown in Figure 3.10.

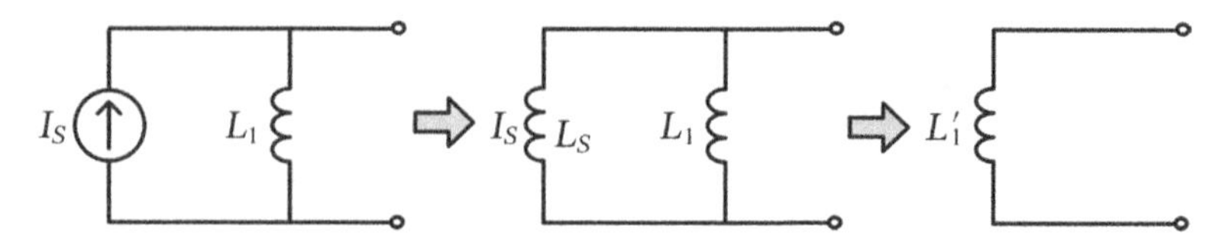

Figure 3.7 A current source in parallel with an inductor is equivalent to a single inductor with Dc offset current.

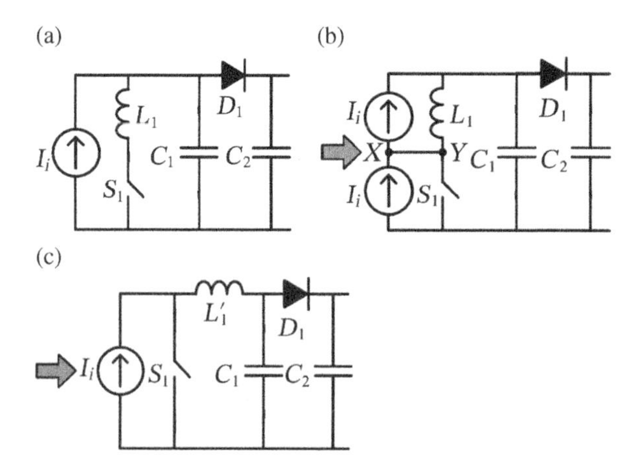

Figure 3.8 Illustration of inductor L_1 with different DC offset currents in a quasi-resonant boost converter.

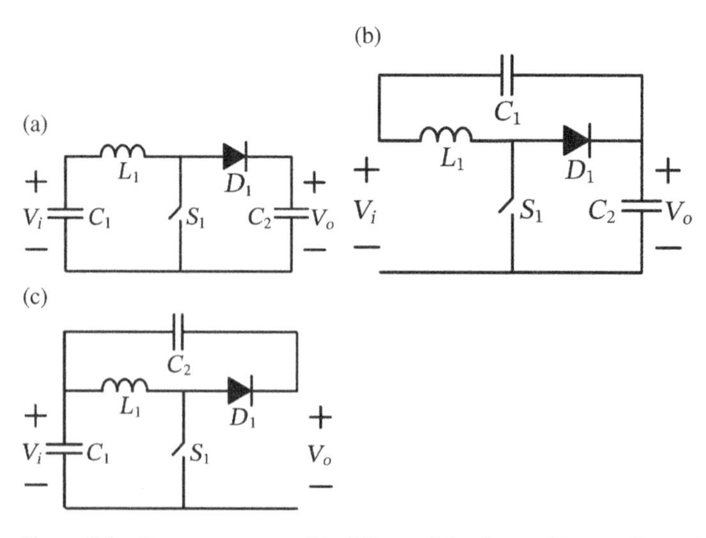

Figure 3.9 Boost converter with different DC voltage offset configurations.

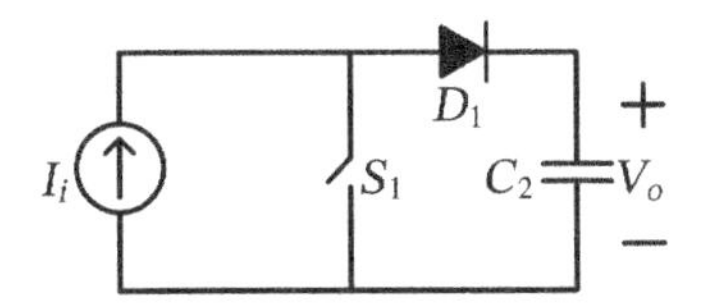

Figure 3.10 The input of the boost converter represented in current form.

3.2 Capacitor and Inductor Splitting

Capacitors and inductors are utilized to buffer energy during power conversion or transfer. For reducing voltage or current stresses under high-power ratings, a single capacitor or a single inductor is split into two or more. In fact, they are either in series or in parallel, and they can be combined back to a single one. In literature, there are several converter topologies, such as voltage-fed Z-source converters, derived based on the inductor or capacitor splitting, but they are just equivalent to those without splitting.

For capacitor splitting, if the voltage difference between nodes can be kept identical and the capacitor voltage complies with ampere-second balance, a capacitor can be split into two capacitors, as shown in Figure 3.11a, in which $V_{XZ} = V_{YZ} = V_C$ and $V_{XY} = 0$ in both networks. Since voltage $V_{XY} = 0$, capacitors C_{11} and C_{12} are equivalent to be in parallel connection, and they can be combined back to a single capacitor, C_1. Similarly, if there is one more branch, W–Z, capacitor C_1 can be split into three, C_{11}, C_{12}, and C_{13}, as shown in Figure 3.11b. To comply with the ampere-second balance, terminals X, Y, and W cannot be in series with switches or diodes if the switches might block the capacitor charging and discharging.

An example to illustrate the capacitor splitting technique, as well as the DC voltage offsetting technique, is shown in Figure 3.11c–f, in which the flyback converter is incorporated with an active clamp snubber, including switch S_s, diode D_s, and capacitor C_s. First of all, we utilize a DC voltage offsetting for capacitor C_s, as shown in Figure 3.11d. Then, capacitor C_s is split into two capacitors, C_{s1} and C_{s2}, as shown in Figure 3.11e. Note that capacitor C_{s1} is in series with the voltage source V_i and capacitor C_{s2} is in series with inductor L_k that allow the capacitors to be charged and discharged freely. Combining capacitor C_{s1} with the input voltage source V_i will yield the final version of the converter topology, as shown in Figure 3.11f. Note again that since capacitor C_s is originally connected to the bidirectional switch S_s and D_s, it can be charged/discharged to satisfy ampere-second balance.

When a capacitor is split into several branches, one has to check if their charging and discharging are satisfied with ampere-second balance; otherwise, the splitting is invalid and the new branch cannot exist. For example, if a capacitor is connected to a diode or a unidirectional switch, as shown in Figure 3.12a, the capacitor cannot be split into the ones shown in Figure 3.12b where the capacitors can be either charged or discharged only, and it is invalid capacitor splitting.

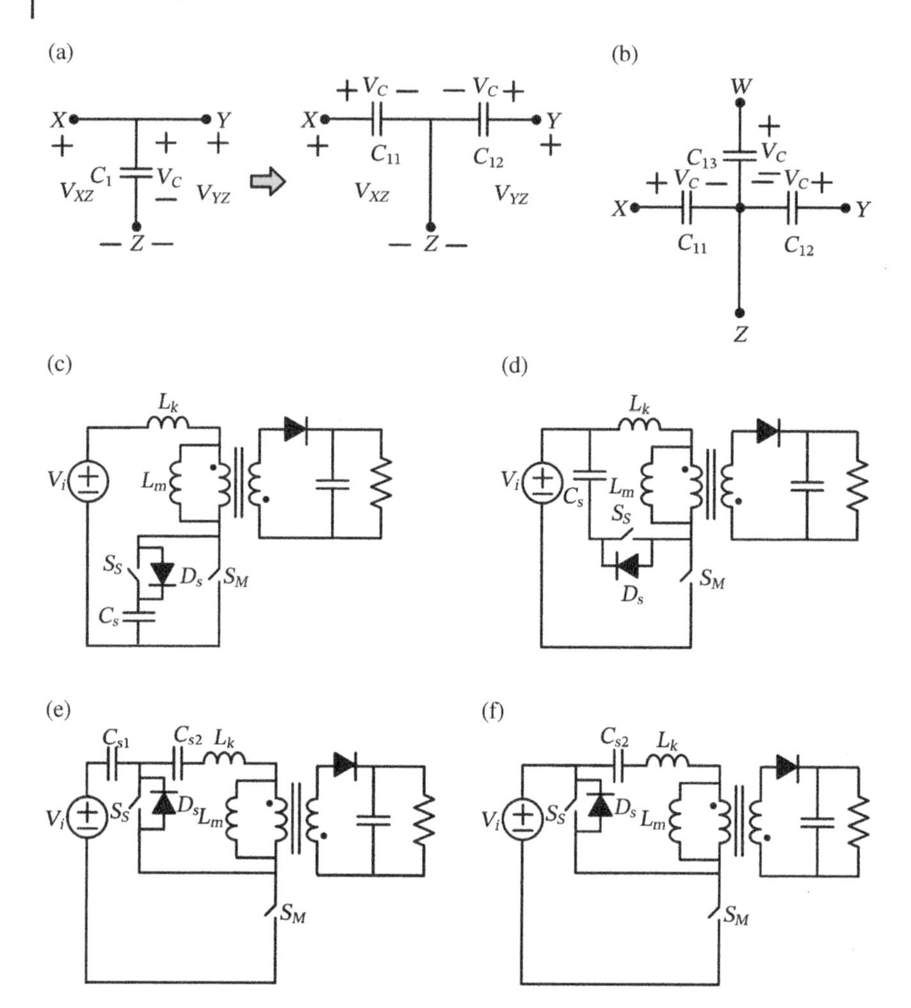

Figure 3.11 A capacitor is split into (a) two and (b) three capacitors with identical node voltages and illustration examples with the DC voltage offsetting (c) and (d) and capacitor splitting (e) and the final version (f).

When an inductor is formed in loops with other linear components denoted as Z_1 and Z_2 and shown in Figure 3.13a, it can be split into two inductors, as shown in Figure 3.13b where the total branch currents $i_{l1} + i_{l2}$ are equal to the original one i_l. In addition, since, for a valid converter topology, inductors must be operated with volt-second balance in the steady state, their average voltages across the inductors over a switching cycle will be zero, $V_{XY} = 0$, and the node voltages in both networks shown in Figure 3.13 are identical in average sense. Similarly, if there are more than two loops formed with the inductor, it can be split into more inductors. It should be noted that one has to check if all of the inductors are satisfied with volt-second balance.

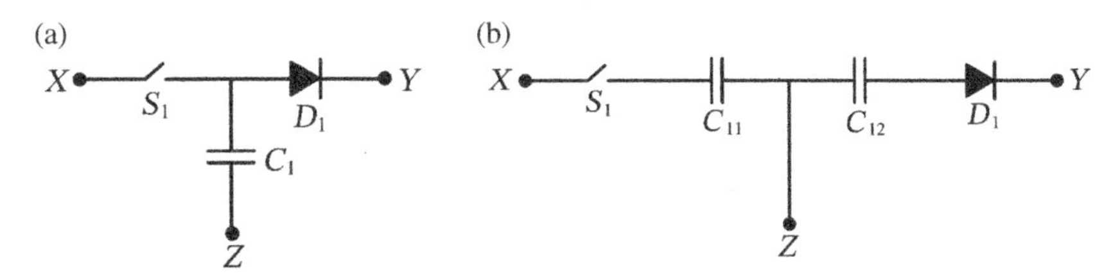

Figure 3.12 Illustration of invalid capacitor splitting. (a) a capacitor is connected to unidirectional switch S_1 and diode D_1 and (b) invalid capacitor splitting circuit.

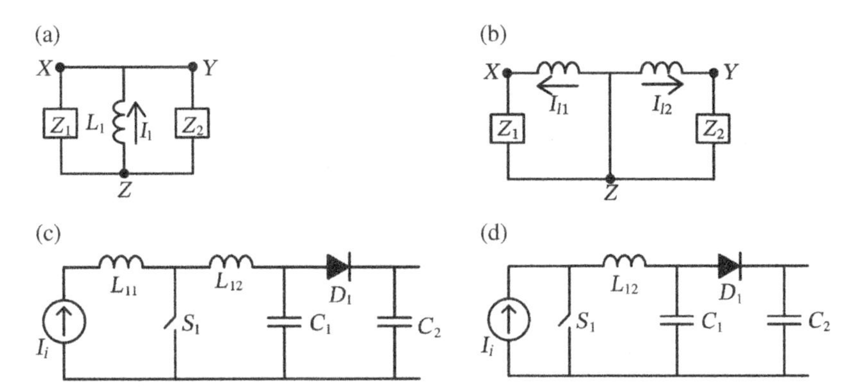

Figure 3.13 An inductor is split into two inductors (a) and (b) with identical total branch current and node voltages and illustration examples (c) and (d) from the converter shown in Figure 3.8a.

Basically, the converter shown in Figure 3.8c can be derived from Figure 3.8a with the inductor splitting technique. Based on the converter shown in Figure 3.8a, we can apply the inductor splitting technique to obtain the one shown in Figure 3.13c, in which inductor L_1 is split into inductors L_{11} and L_{12}. Since inductor L_{11} can be merged with the input current source I_i, the final converter topology becomes the one shown in Figure 3.13d, which is identical to the one shown in Figure 3.8c. Note that a current source is usually realized by an inductor in series to a voltage source; thus, inductor L_{11} can be combined with current source I_i.

3.3 DC-Voltage Blocking and Pulsating-Voltage Filtering

PWM power converters are always configured with switches and LC components. Two typical configurations are shown in Figure 3.14, in which the one shown in Figure 3.14a has a voltage source with a pair of PWM switches and the one in Figure 3.14b has a voltage source, a filter inductor, and a pair of PWM switches.

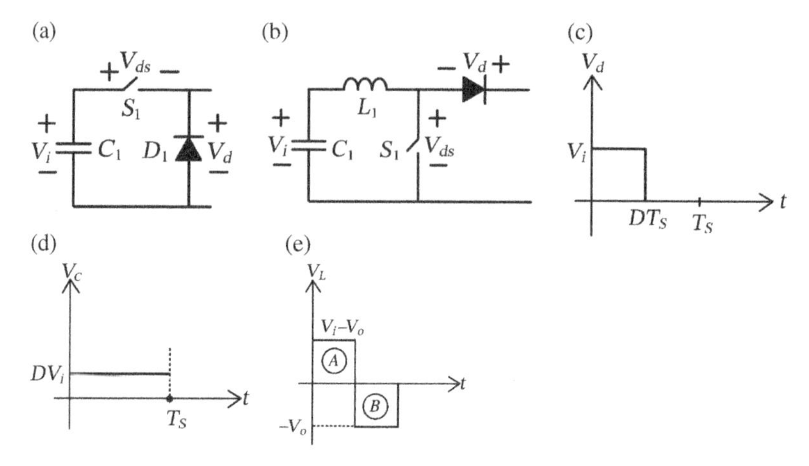

Figure 3.14 Two typical configurations for generating pulsating voltages. (a) Voltage-type and (b) current-type configurations for generating pulsating voltages and (c) pulsating voltage V_d across the diode, (d) the average voltage of V_d blocked by the capacitor in an *LC* filter network, and (e) the voltage across the inductor in an *LC* filter network.

The power transfer in PWM converters is in resonance, and thus, the circuit configurations shown in Figure 3.14 will be connected to other *LC* components.

In a PWM converter, switching action generates pulsating voltage across either active or passive switch, taking buck converter shown in Figure 3.14a as an example in which V_{ds} and V_d are pulsating. The pulsating voltages can be decoupled into DC component (V_C) and AC one (V_L), which is then filtered with an *LC* network to yield a smooth voltage. Figure 3.14c shows pulsating voltage V_d waveform of the buck converter, and its average voltage, the DC component, is $V_C = DV_i$, as shown in Figure 3.14d, which will be blocked by the capacitor in the *LC* network, and the AC voltage V_L, as shown in Figure 3.14e, across the inductor complies with volt-second balance principle; i.e. area *A* equals area *B*.

For generating multiple output voltages, *LC* components are connected to different pulsating voltages. For instance, in Figure 3.15a, which is belonged to the configuration type shown in Figure 3.14a, when switch S_1 is turned on, voltage V_d is equal to V_i and lasts for a time interval of DT_S where *D* is the duty ratio of switch S_1 and T_S is the switching period. When switch S_1 is turned off, voltage V_d is zero. Thus, the average voltage of V_d is DV_i over one switching period, as illustrated in Figure 3.15b. Taking the average value of V_d yields DV_i that is the V_o at the output of the buck converter shown in Figure 3.15a. Alternately, one can take the pulsating voltage V_{ds} filtering through an *LC* network to obtain another output voltage $V_o' = (1-D)V_i$, as shown in Figure 3.15c. It should be noted that power transfer between input and output must be in resonance to insure high efficiency.

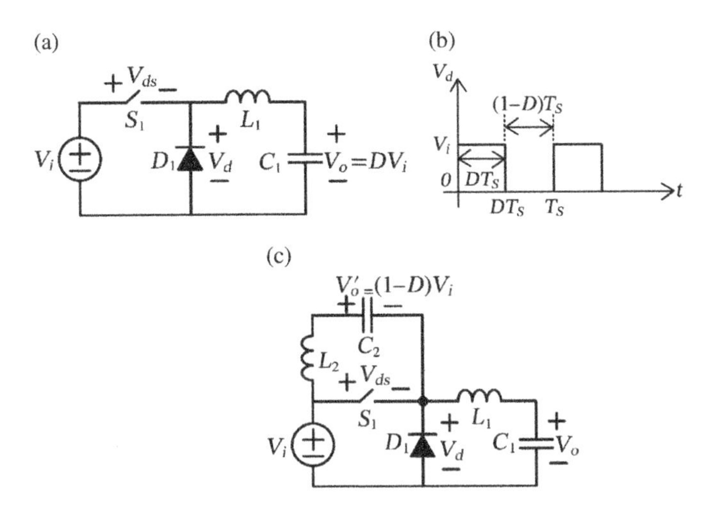

Figure 3.15 (a) Smooth voltage obtained from the pulsating voltage across diode and with an *LC* filter, (b) the voltage waveforms of diode voltage V_d, and (c) another smooth output voltage V_o' obtained from pulsating voltage V_{ds}.

Therefore, filtering of pulsating voltage in Figure 3.15 cannot be achieved with a single capacitor, but an *LC* resonant network.

For a boost converter shown in Figure 3.16a, its output voltage $V_o = V_i/(1 - D)$. The configuration type of a PWM switch pair and LC components is belonged to the one shown in Figure 3.14b, in which the pulsating voltage of V_{ds} is equal to 0 or V_o and its average voltage is $(1 - D)V_o$. Since the input is a voltage source with a filter inductor, the pulsating voltage can be filtered and averaged with a capacitor and a blocking diode. However, if one takes the pulsating voltage, V_d, from the diode and through an *LC* filter network, output voltage V_o' will be

$$V_o' = DV_o = \frac{DV_i}{(1-D)}, \tag{3.2}$$

and its converter configuration is a Ćuk converter with its buffer capacitor C_1 and output filter inductor L_2 located in the return path, as shown in Figure 3.16c. When relocating the capacitor and inductor to the forward path, we can obtain the well-known configuration of Ćuk converter.

The buck-boost converter with an *LC* filter across diode D_1 to smooth the pulsating voltage is shown in Figure 3.16e. With further processing, we can have the Zeta converter with capacitor C_1 in return path, as shown in Figure 3.16f. Moving

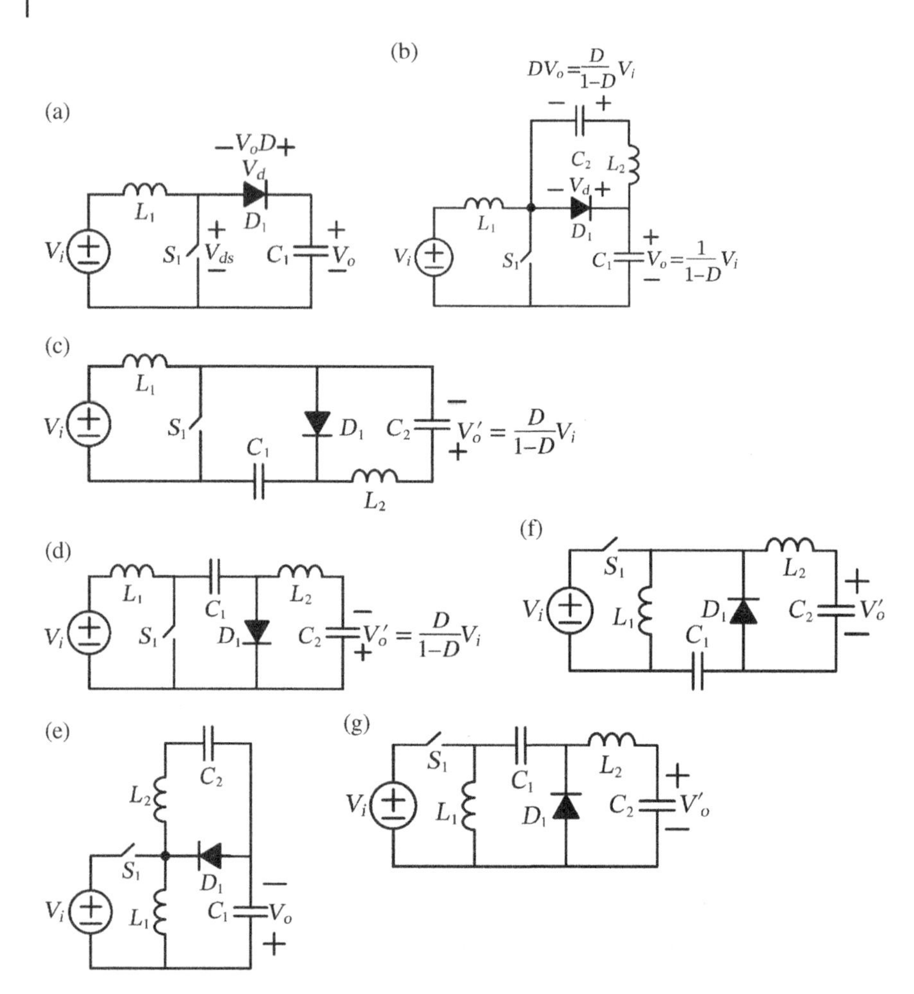

Figure 3.16 (a) Smooth output voltage V_o obtained with a filter capacitor, (b) another smooth output voltage V_o' obtained from V_d with an LC filter, (c) Ćuk converter with C_1 and L_2 in return path, (d) the well-known configuration of Ćuk converter, (e) buck-boost converter with an LC filter across diode, (f) Zeta converter with the buffer capacitor in the return path, and (g) the well-known Zeta converter topology.

capacitor C_1 from return path to the forward path yields the well-known Zeta converter topology, as shown in Figure 3.16g.

In literature, there are many PWM converters, and they may have more than one active and passive switches. When taking any one of the pulsating voltages and filtered with an LC filter or buffered with a capacitor, one can obtain other smooth output voltages. Note that the new output still has to follow a resonant manner to transfer power.

3.4 Magnetic Coupling

Power transfer with resonance has the highest conversion efficiency, and there always exist LC components or filters in PWM converters. In a steady state, the inductors are driven or magnetized/demagnetized with volt-second balance. Thus, when winding another wire(s) on the same core of the inductor, we can obtain an isolation version of the existing converter with magnetic coupling. Figure 3.17a shows a buck converter with its inductor wound with another winding. At the secondary side, its output voltage V_{o2} is a kind of AC waveform with volt-second balance; i.e. area A equals area B, as shown in Figure 3.17b. Since the duty ratio is not always equal to 0.5, the waveform is usually asymmetrical, and it is rectified and filtered to obtain a DC voltage. Two typical rectification types are shown in Figure 3.17c and d, in which the one shown in Figure 3.17c is a forward type and that shown in Figure 3.17d is a flyback type, depending on its dot positions. If the dots are located in the same-side positions, the two coupled windings act like a transformer at switch S_1 turn-on, and it requires an LC filter and a

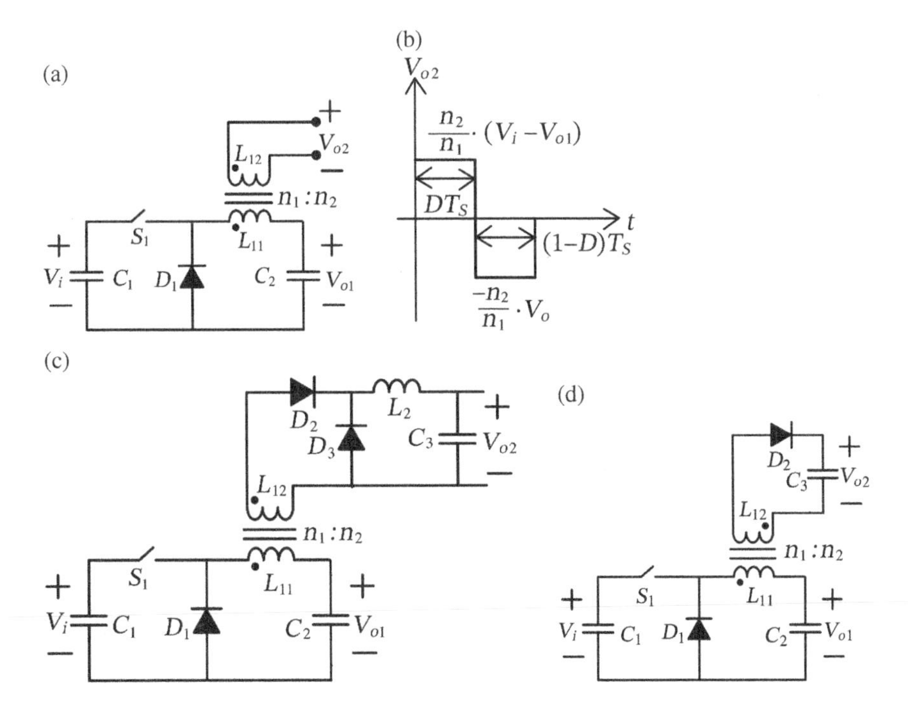

Figure 3.17 (a) Buck converter with a secondary winding coupled from the inductor, (b) the voltage waveforms of output voltage V_{o2} over one switching cycle, (c) the second output voltage with the forward type of rectification, and (d) the second output voltage with the flyback type of rectification.

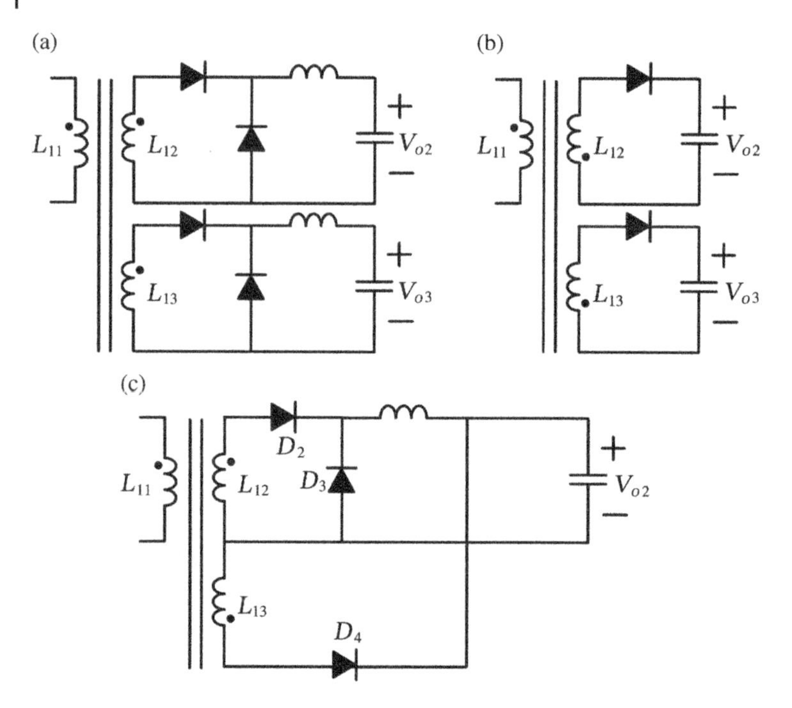

Figure 3.18 The secondary (a) with two forward-type rectifications, (b) with two flyback-type rectifications, and (c) with the combination of one forward-type and one flyback-type rectifications.

freewheeling diode D_3. While if they are located in the diagonal positions, power transfer occurs at switch S_1 turn-off, and a filter capacitor is good enough to buffer output voltage, in which the capacitor will resonate with the magnetizing inductance of the coupled windings. The power transfer in the both secondary outputs, V_{o2}, shown in Figure 3.17c and d still follows resonant manner, and the original output operation can be sustained. The output V_{o2} of a forward-type coupling has the relationship of $n_2(V_i - V_{o1})/n_1$ to V_i and V_{o1}. However, there exists cross-coupling in both V_{o1} and V_{o2} under step load changes, since there is only one degree of freedom, one active switch, for regulating output voltages. For the output V_{o2} of a flyback-type coupling, it has a direct relationship to V_{o1} as $V_{o2} = n_2V_{o1}/n_1$, and its output voltage is limited by V_{o1} through the winding turns ratio.

When there is a magnetic coupling, more windings can be wound on the same core to generate multiple outputs. Figure 3.18 shows several possible combinations of secondary windings. In Figure 3.18a, two forward-type windings are connected to the secondary side; in Figure 3.18b, there are two flyback-type windings; and in Figure 3.18c, they are the combination of one forward-type and one flyback-type windings. If more outputs are required, one can add more

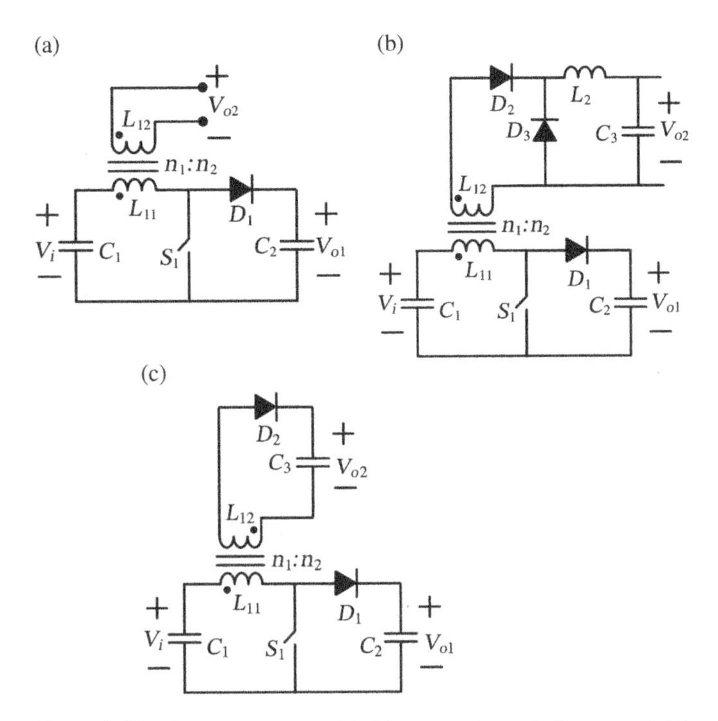

Figure 3.19 Boost converter with (a) a secondary winding coupled from the inductor, (b) a forward-type rectification, and (c) a flyback-type rectification.

than two windings on the same core. That is, one of the benefits is magnetic coupling, which in turn has galvanic isolation. However, the primary penalty is that extra effort might be needed to deal with leakage inductance of the coupled windings.

Similarly, a boost converter can have a magnetic coupling winding, as shown in Figure 3.19a. It can have a forward-type filter at the secondary side, as shown in Figure 3.19b, or it can have a flyback-type one, as shown in Figure 3.19c. Moreover, the boost converter can also have multiple outputs by adopting the coupling configurations shown in Figure 3.18.

With magnetic coupling, the AC output voltage at the secondary side of a PWM converter is limited by the switching action and the input and output voltage relationship. The AC voltages across the inductor comply with volt-second balance, but they are usually asymmetrical. Thus, only are certain types of coupling configurations allowed to isolate the primary and the secondary sides, to step up or step down input voltage, and to create multiple outputs. In the following section, DC transformers are introduced to PWM converters to release the limitations while still fulfill the same functions.

3.5 DC Transformer

Transformers operate with AC voltages only. How can they operate with DC to act as dc transformers? The basic idea is that a DC voltage is first converted to AC; then with a transformer, the AC is transferred to the secondary side to become another AC; and finally, it is rectified to be a DC. The overall circuit operation is equivalent to a DC transformer. Since a DC transformer is to convert a DC voltage to another DC located at the secondary side, there is no need to introduce any extra *LC* filter to the transformer circuit.

In a DC transformer, the DC input voltage can be a DC-source voltage or an intermediate port of DC voltage, like the DC voltage at certain location inside a PWM converter. The output DC voltage can be the terminal voltage or an intermediate voltage inside a converter. Figure 3.20 shows three typical types of DC transformers,

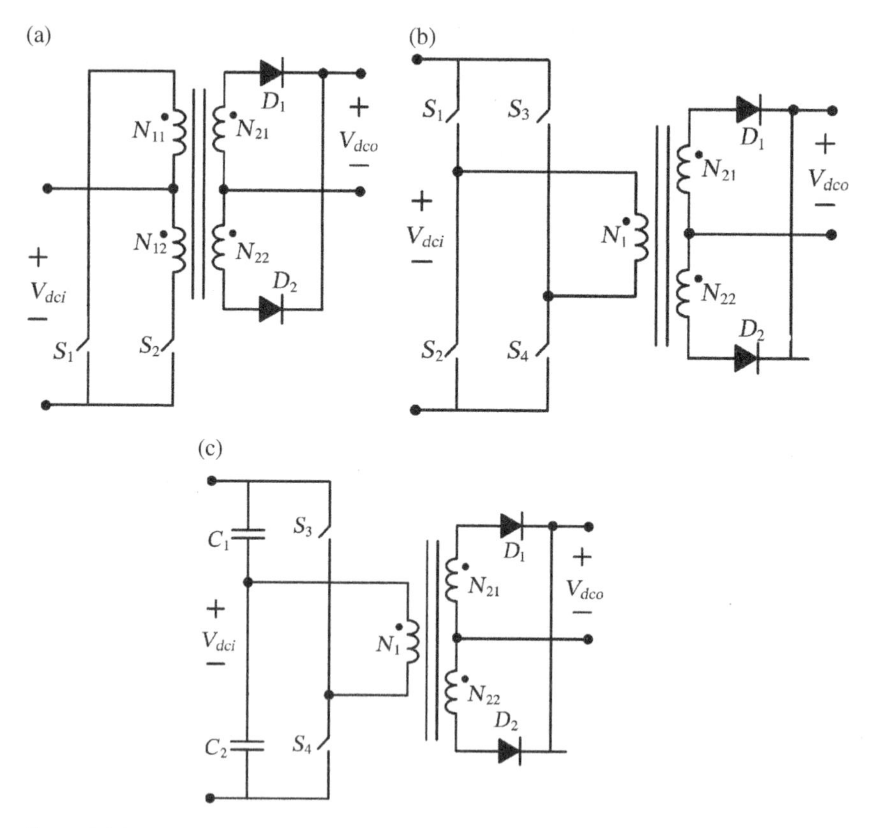

Figure 3.20 DC transformers with (a) push-pull type, (b) full-bridge type, and (c) half-bridge type.

push-pull, full-bridge, and half-bridge, which can be adopted to create an isolation version for a given PWM converter. The secondary side of the dc transformers shown in Figure 3.20b and c can be replaced with the one shown in Figure 3.21a that has a full-bridge diode rectifier. If a DC transformer is combined with filter inductors and capacitors, its output can become a terminal output. Figure 3.21b and c shows two types of current doublers that are usually adopted for an output to increase current rating while reduce current ripple. If the primary side is a current type (with inductor(s)) and its secondary side is the terminal output, the secondary side can be replaced with a voltage doubler, as shown in Figure 3.21d. The terminal output can be also realized with a half-bridge rectifier and an *LC* filter, or a full-bridge one, as shown in Figure 3.21e and f, respectively. These variations provide users more choices in selecting converter topologies.

The DC transformer shown in Figure 3.20a is a push-pull type, in which the two switches, S_1 and S_2, take turn conducting for 50% duty ratio and it is equivalent that 100% of V_{dci} can be transferred to V_{dco}. Thus, it can be inserted into any port location of a PWM converter. Figure 3.22 shows a buck converter partitioned in five portions, from 1–1′ to 5–5′. When inserting a push-pull DC transformer at port 2–2′ location, we obtain the isolated buck converter shown in Figure 3.22b, which is the TRW converter. When inserting it at port 4–4′ location, we can obtain another one shown in Figure 3.22c, in which switch S_1 controls the duty ratio to yield the desired output voltage V_o and switches S_1 and S_2 take turn conducting for 50% duty ratio, and meanwhile, each of diodes D_2 and D_3 also conducts correspondingly for 50%. During the operation of switches S_1 and S_2/S_3, the switch sets (S_1, S_2) and (S_1, S_3) are in series, respectively. If we operate each of switches S_2 and S_3 with the duty ratio $D/2$ of switch S_1, which is ≤50%, switch S_1 can be eliminated. Additionally, diode D_1 provides a freewheeling path for conducting the output inductor current, and in principle, its role can be replaced by diodes D_2 and D_3. Thus, diode D_1 is no longer needed, and the overall simplified version is shown in Figure 3.22d, which becomes the well-known push-pull converter topology. If modifying the DC transformer from two active switches to four and simplifying the switch operation, we can obtain IBM, Cronin, Hunter, and Severns converters. Moreover, if combining a magnetic coupling winding with the push-pull DC transformer, we can obtain the Weinberg converter and its improved version. Derivation of these converters can be found in R. P. Severns' book entitled *Modern DC-To-DC Switchmode Power Converter Circuits.*

Another two types of DC transformers, the full bridge and the half-bridge shown in Figure 3.21b and c, can be inserted in PWM converters to yield their isolation versions. And it is also possible to simplify the switch configurations in the converters further while still keep the same operational principle. Once the converters are associated with transformers, they can have multiple outputs by introducing more secondary windings wound on the same core.

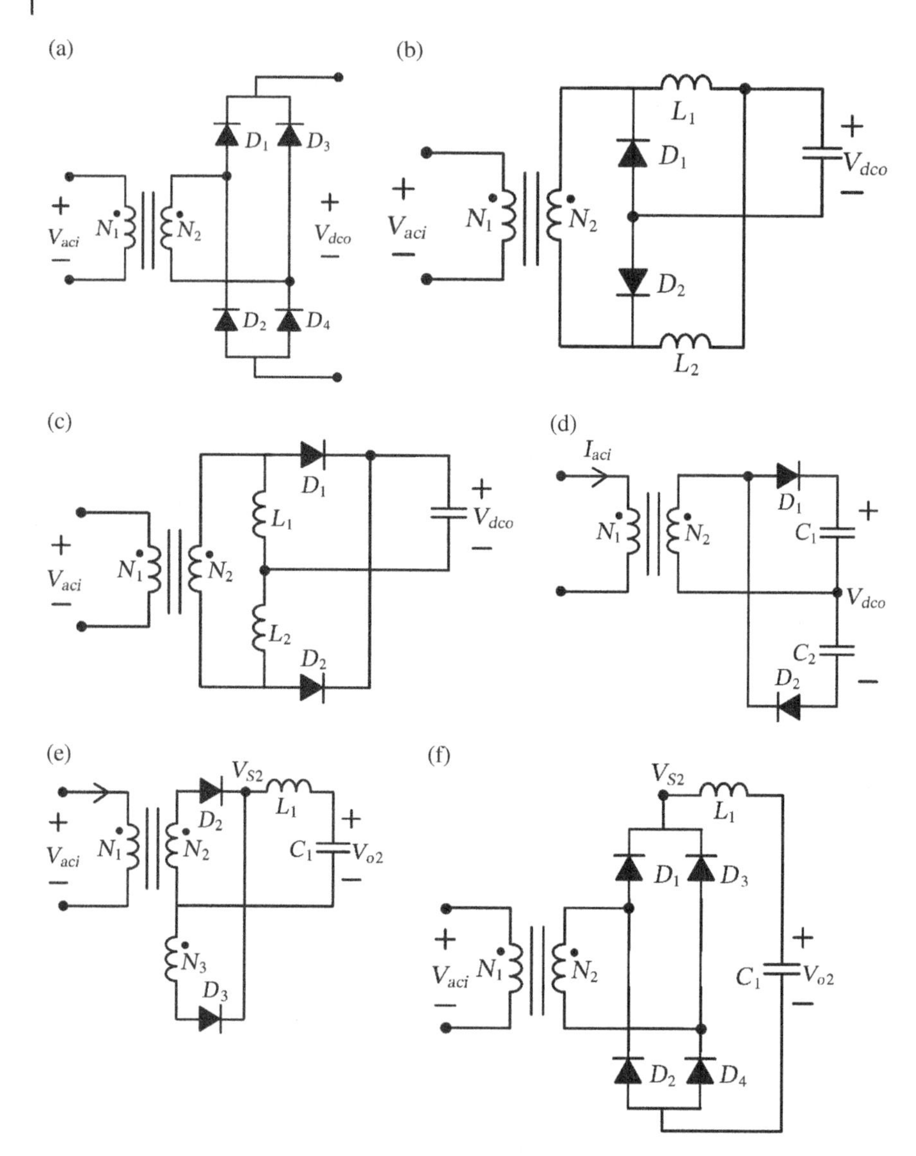

Figure 3.21 (a) Full-bridge diode rectifier, terminal outputs with (b) type-I current doubler, (c) type-II current doubler, (d) voltage doubler, (e) half-bridge rectifier, and (f) full-bridge rectifier.

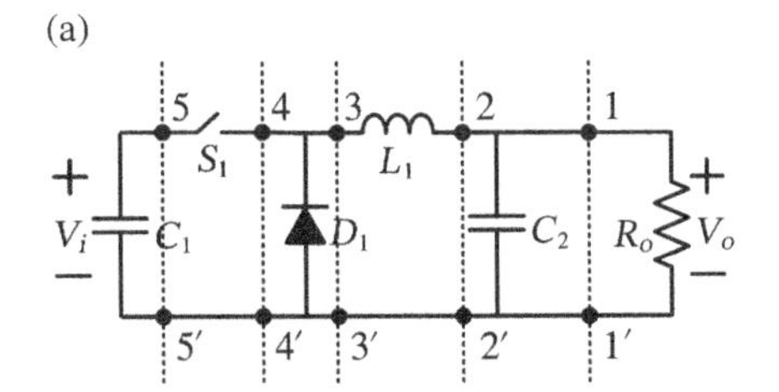

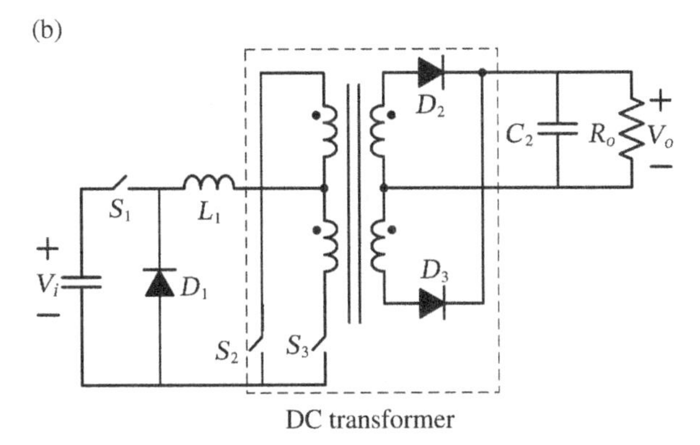

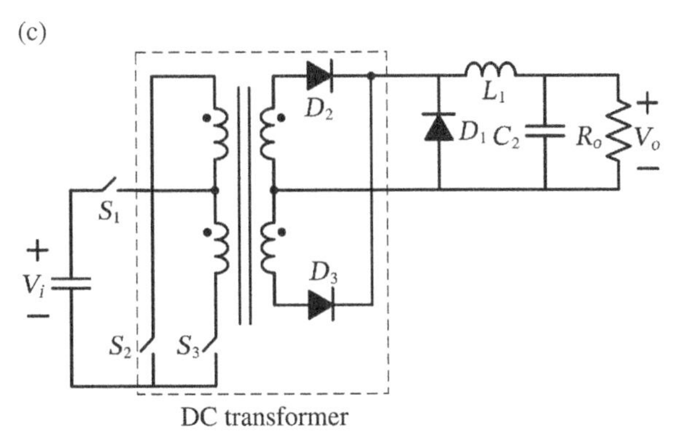

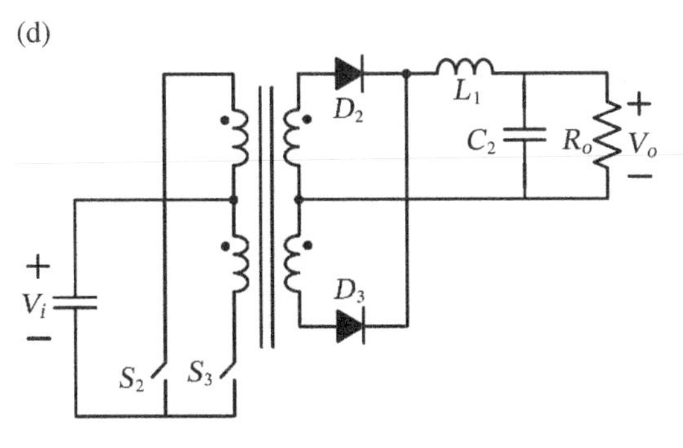

Figure 3.22 (a) Buck converter partitioned in five positions, (b) a DC transformer inserted at position 2–2′, (c) DC transformer inserted at position 4–4′, and (d) push-pull converter.

3.6 Switch Grafting

Grafting an apple tree on a pear tree yields a pear-apple tree. In the process of grafting trees, it usually requires to cut a branch of apple tree or simply chop out the root of an apple tree, leaving only the upper part; chop out the upper or leaf part of the pear tree, leaving only the lower root part; and then graft the apple branch part on the pear root part to form a new tree. When grafting a converter on the other, we frequently have to cut or chop the *LC* filter in the converters, which is similar to cutting a branch or chopping out a root. Thus, we adopt the tree-grafting approach to grafting converters, which mainly includes switch grafting and diode grafting.

In the previous sections, the discussed fundamentals are all related to linear component manipulation. Starting from this section, we will deal with the grafting of nonlinear components, active and passive switches, in PWM converters. Grafting a boost converter on a buck converter, as shown in Figure 3.23a, yields a buck-boost one, as illustrated in Figure 3.23 step by step. In Figure 3.23b, switches S_1 and S_2 share a common node "*C*" and are operated synchronously or in unison, and they can be grafted into a single one, S_{12}, through the process of Figure 3.23c. Since the currents through the two switches are identical, the two diodes D_{F1} and D_{F2} are no longer needed and can be removed from the circuit, as shown in Figure 3.23d. Meanwhile, diodes D_1 and D_2 are in series connection, and they can be integrated into D_{12}, as shown in Figure 3.23e, yielding the buck-boost converter. In the buck converter, there exist an active switch and a passive one, so do the boost converter. However, in the grafted converter, buck-boost, there are also an active and a passive ones only. Thus, one can imagine that there must be existing switch sharing in the two converters. From the above brief description, it reveals that grafting a converter on the other to yield a grafted converter mainly involves the processes of switch grafting and diode grafting.

In the above discussion, an active switch consists of two acting nodes and one control gate node. If the two switches in two separate converters do not operate in unison or have no common node, they are just individually belonged to each individual converter, and it is impossible to graft them into a single one. In other words, the two switches cannot be shared by the two converters. If the two switches have two common acting nodes, they are just in parallel, and of course they can be replaced with a single one but with a higher current rating. For active switch grafting, it requires two switches operating in unison and sharing at least a common acting node. In the following, we use MOSFET switches without body diodes and with unidirectional current flow from drain to source only as an example to illustrate the switch grafting.

Since there are two acting nodes in each MOSFET switch, there are four possible combinations of common nodes in two switches, as shown in Figure 3.24a, c, e, and g corresponding to common nodes *s–s*, *d–d*, *d–s*, and *s–d*, respectively.

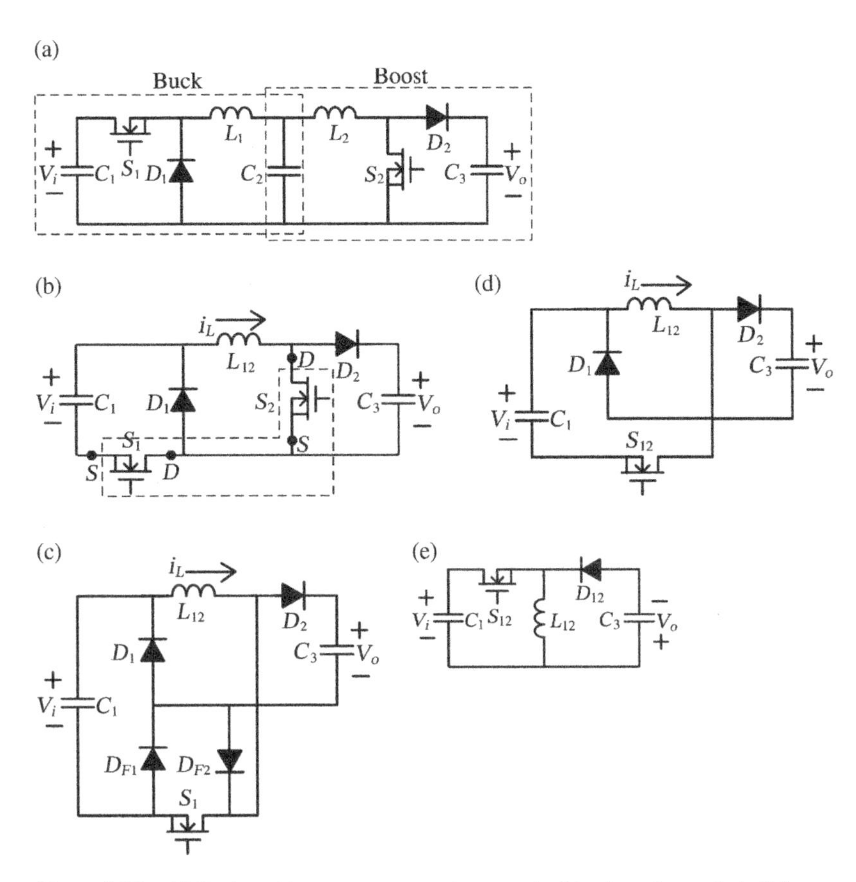

Figure 3.23 (a) Buck and boost converters in cascade, (b) relocating switch S_1 in return path to create a common node $D-S$, (c) switches S_1 and S_2 replaced with a ΠGS, (d) degenerated ΠGS by removing D_{F1} and D_{F2}, and (e) the grafted converter – buck-boost.

When the two switches share with a common node s–s and operate in unison, they can be integrated into T-type grafted switch (TGS), as shown in Figure 3.24b, in which diodes D_{B1} and D_{B2} are introduced to block the voltage difference between V_1 and V_2. During switch on state, voltages $V_1 = V_2 = 0$, node voltages $d_x = d_y$, and the two nodes can be tied together. While during switch off state, voltage V_1 is not necessarily equal to voltage V_2, and thus, it requires diodes to block the voltage difference. If voltage V_1 always equals voltage V_2 over a whole switching cycle, both D_{B1} and D_{B2} can be removed (shorted) from the TGS. If V_1 is always greater than V_2, diode D_{B1} is in forward bias and it can be shorted. However, diode D_{B2} is in reverse bias, and it is needed to block the voltage difference, $(V_1 - V_2)$. On the other hand, if V_2 is always greater than V_1, diode D_{B2} is no longer needed. Similarly, when the two switches share a common node d–d, as shown in Figure 3.24c, they

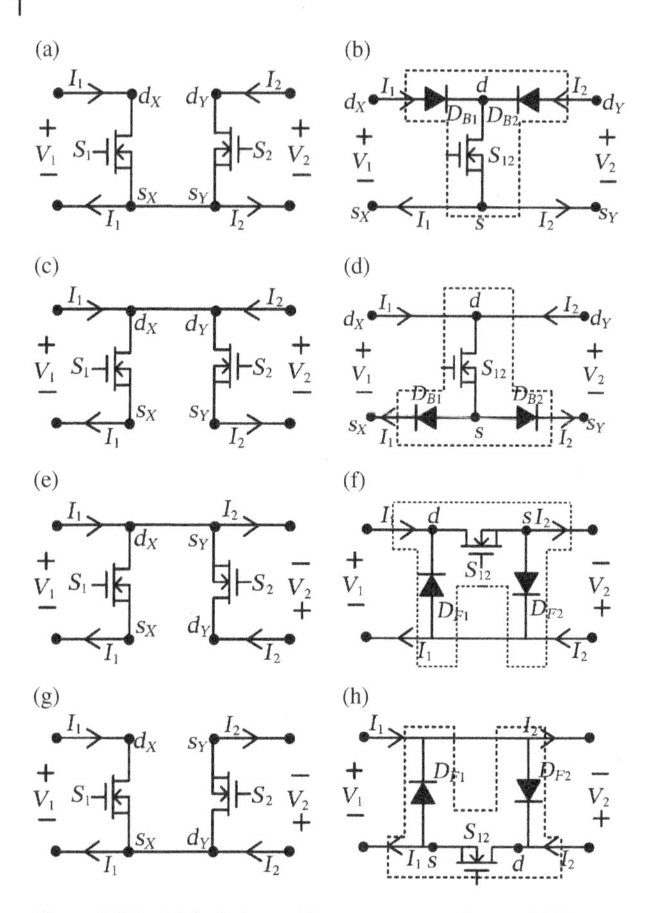

Figure 3.24 (a) Switches with a common node *s*–*s*, (b) T-type grafted switch (TGS), (c) switches with a common node *d*–*d*, (d) inverse T-type grafted switch (ITGS), (e) switches with a common node *d*–*s*, (f) Π-type grafted switch (ΠGS), (g) switches with a common node *s*–*d*, and (h) inverse Π-type grafted switch (IΠGS).

can be integrated into an inverse T-type grafted switch (ITGS), as shown in Figure 3.24d, in which voltage blocking diodes D_{B1} and D_{B2} can be further degenerated, depending on the relationship between the voltages V_1 and V_2 across the switches in the off state. The degenerated TGS and ITGS are collected in Table 3.1. Note that the current through the grafted switches is equal to $I_1 + I_2$ during switch on state.

When grafting switches, we assume the current flow is unidirectional from drain to source. Thus, under common node *s*–*s* or *d*–*d*, the two switches can be connected in parallel and grafted into a single one with two possible blocking diodes. When the two switches have either common node *d*–*s* or *s*–*d*, they can be

Table 3.1 Degeneration of TGS and ITGS based on the relationship of voltages V_1 and V_2.

Relationship between voltages across switches in off state	Degenerated T-type grafted switch (TGS)	Degenerated inverse T-type grafted switch (ITGS)
$V_1 > V_2$	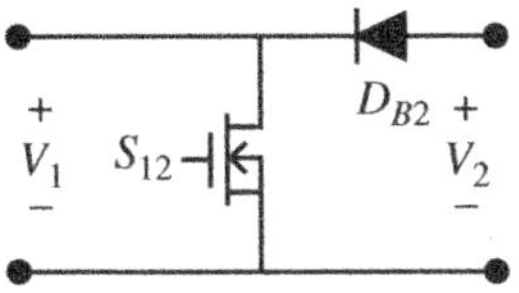 V_1 S_{12} D_{B2} V_2	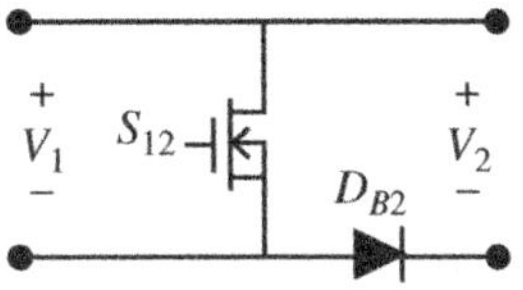 V_1 S_{12} D_{B2} V_2
$V_1 < V_2$	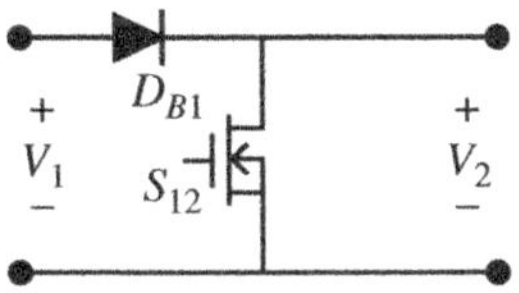 V_1 D_{B1} S_{12} V_2	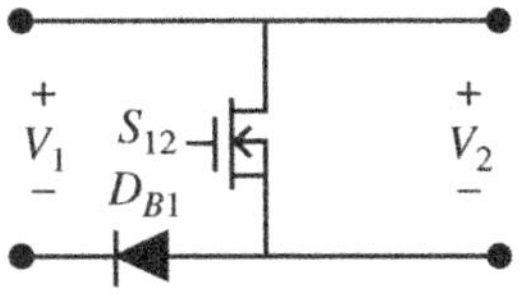 V_1 S_{12} D_{B1} V_2
$V_1 = V_2$	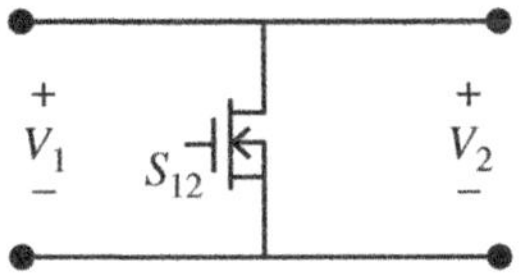 V_1 S_{12} V_2	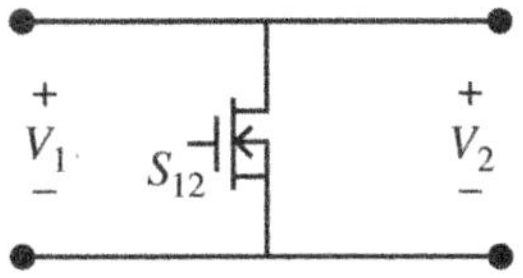 V_1 S_{12} V_2

connected in series and integrated into Π-type grafted switch (ΠGS) or inverse Π-type grafted switch (IΠGS) with two current circulating diodes D_{F1} and D_{F2}. Figure 3.24e shows two switches with a common node *d–s*, and they are integrated into a ΠGS, as shown in Figure 3.24f, from which it can be observed that when the grafted switch is turned off, it has to block the voltage of $V_1 + V_2$. When the grafted switch is turned on, the higher current of I_1 and I_2 will flow through the switch. If current I_1 is always greater than current I_2, diode D_{F2} will circulate the current difference, $I_1 - I_2$, but diode D_{F1} conducts no current, and it can be removed (open) from the ΠGS. On the other hand, if I_2 is greater than I_1, diode D_{F2} can be removed from the switch. If both I_1 and I_2 are always identical during switch on state, diodes D_{F1} and D_{F2} are no longer needed. Similarly, when the two switches are with a common node *s–d* shown in Figure 3.24g, they can be integrated into an IΠGS, as shown in Figure 3.24h. The degenerated ΠGS and IΠGS are collected in Table 3.2.

The TGS and ΠGS are dual with the properties summarized in Table 3.3. Basically, the TGS is formed with two switches in parallel, and it will conduct the total current from the original two switches in the on state. However, the ΠGS is formed with two switches in series, and it has to sustain the total voltages imposed on the two original switches in the off state. Figure 3.23 illustrates an example of

Table 3.2 Degeneration of ΠGS and IΠGS based on the relationship of currents I_1 and I_2.

Relationship between currents through switches in on state	Degenerated Π-type grafted switch (ΠGS)	Degenerated inverse Π-type grafted switch (IΠGS)
$I_1 > I_2$	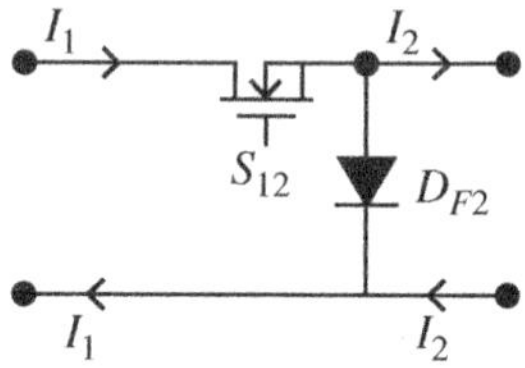 I_1, I_2, S_{12}, D_{F2}, I_1, I_2	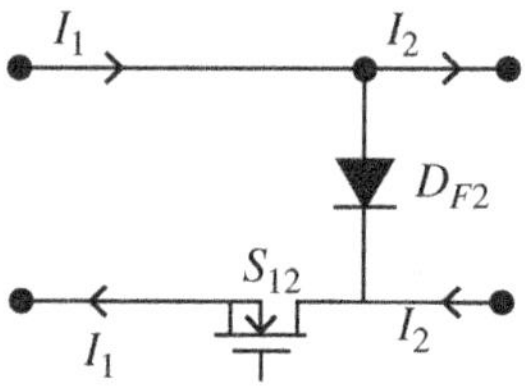 I_1, I_2, D_{F2}, S_{12}, I_1, I_2
$I_1 < I_2$	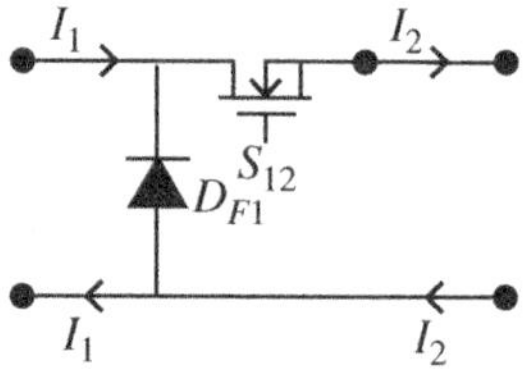 I_1, I_2, S_{12}, D_{F1}, I_1, I_2	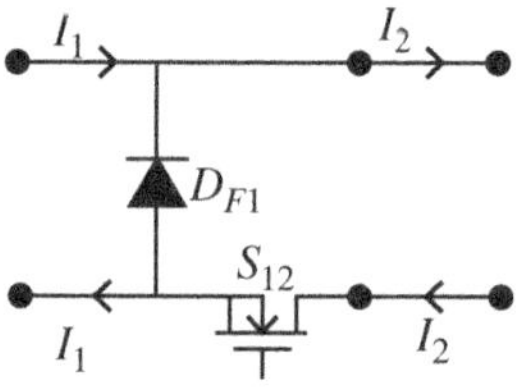 I_1, I_2, D_{F1}, S_{12}, I_1, I_2
$I_1 = I_2$	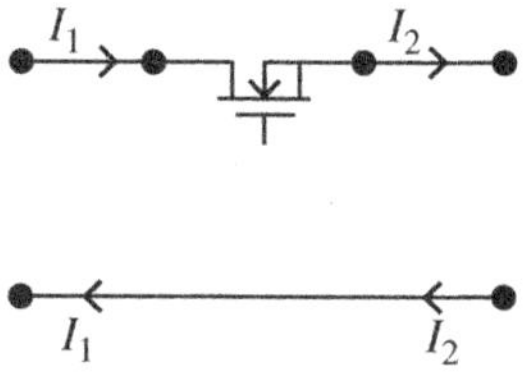 I_1, I_2, I_1, I_2	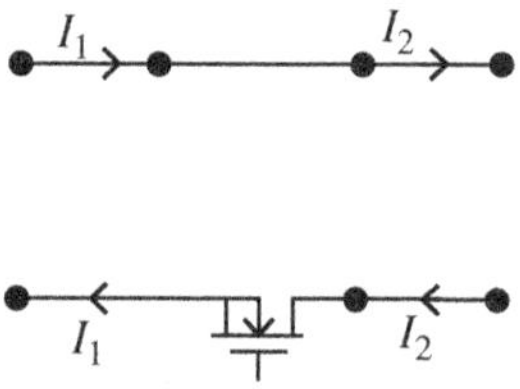 I_1, I_2, I_1, I_2

Table 3.3 Duality between T-type and Π-type grafted switches.

TGS and ITGS	ΠGS and IΠGS
1) D_{B1}, D_{B2}, and S_{12} share a node	1) D_{F1}, D_{F2}, and S_{12} form a loop
2) D_{Bi} blocks the voltage difference between V_1 and V_2 when both switches S_1 and S_2 are in the off states	2) D_{Fi} circulates the current difference between I_1 and I_2 when both switches S_1 and S_2 are in the on states
3) Required to determine the V_1 and V_2 when S_1 and S_2 are in the off states	3) Required to determine the I_1 and I_2 when S_1 and S_2 are in the on states
4) The grafted switch conducts current $I_1 + I_2$ in the on state	4) The grafted switch blocks voltage $V_1 + V_2$ in the off state

adopting ΠGS to generate a buck-boost converter. First, one has to identify the type of common node, but it usually has to relocate one switch, as shown in Figure 3.23b, to create a common node for the two switches in the converters, and to operate the two active switches in unison. Then, replace the two switches with a proper type of grafted switch, as shown in Figure 3.23c, and degenerate the switch based on the rules presented in Tables 3.1 and 3.2. Finally, one can obtain a grafted converter with only one active switch, as shown in Figure 3.23e. Figure 3.25 illustrates an example of grafting buck on boost converter with a TGS to yield boost-buck (Ćuk) converter. Figure 3.25a shows a boost converter and a buck converter connected in cascade. When relocating switch S_2 from the forward path to the return path, as shown in Figure 3.25b, we can identify a common node *s–s* for the two switches, and they can be replaced with a TGS, as shown in Figure 3.25c. Since the voltages, V_1 and V_2, imposed on switches S_1 and S_2, respectively, during off state are identical, blocking diodes D_{B1} and D_{B2} are no longer needed, and the simplified circuit is shown in Figure 3.25d, from which it can be observed that diode D_1 is in parallel with diode D_2 and they can be replaced with a single one, as shown in Figure 3.25e. After redrawing the configuration, the Ćuk or boost-buck converter is shown in Figure 3.25f.

In the above discussion, switch MOSFET is adopted and assumed to be unidirectional for current conduction. In fact, IGBT and BJT can be also utilized as switches for the same grafting processes since they are also unidirectional switches.

When switches can conduct bidirectional current, such as with body diodes, extra diodes D_{C1} and D_{C2} are needed to conduct reverse current, as shown in Figure 3.26. For TGS, when $V_1 > V_2$, diode D_{C1} can be just the body diode of switch S_{12}, but it requires a diode D_{C2} connected across V_2 terminal, as shown in Figure 3.26a. When $V_1 < V_2$, diode D_{C1} is connected across V_1 terminal, as shown in Figure 3.26b. When $V_1 = V_2$, there is no need of extra diode, but we can just use the body diode, D_C, of switch S_{12} to conduct bidirectional current, as shown in Figure 3.26c. For ΠGS, diodes D_{F1} and D_{F2}, as shown in Figure 3.26d, can serve for conducting bidirectional currents no matter what is the relationship between I_1 and I_2. The same discussion can be applied to the inverse TGS and ΠGS, yielding the degenerated grafted switches shown in Figure 3.27.

3.7 Diode Grafting

In the previous section, generation and degeneration of TGS and ΠGS were addressed. They can be adopted to integrate power converters. Similarly, the diodes in the converters operated in unison and sharing at least a common node can be also grafted to simplify the converter configurations. If two diodes are in

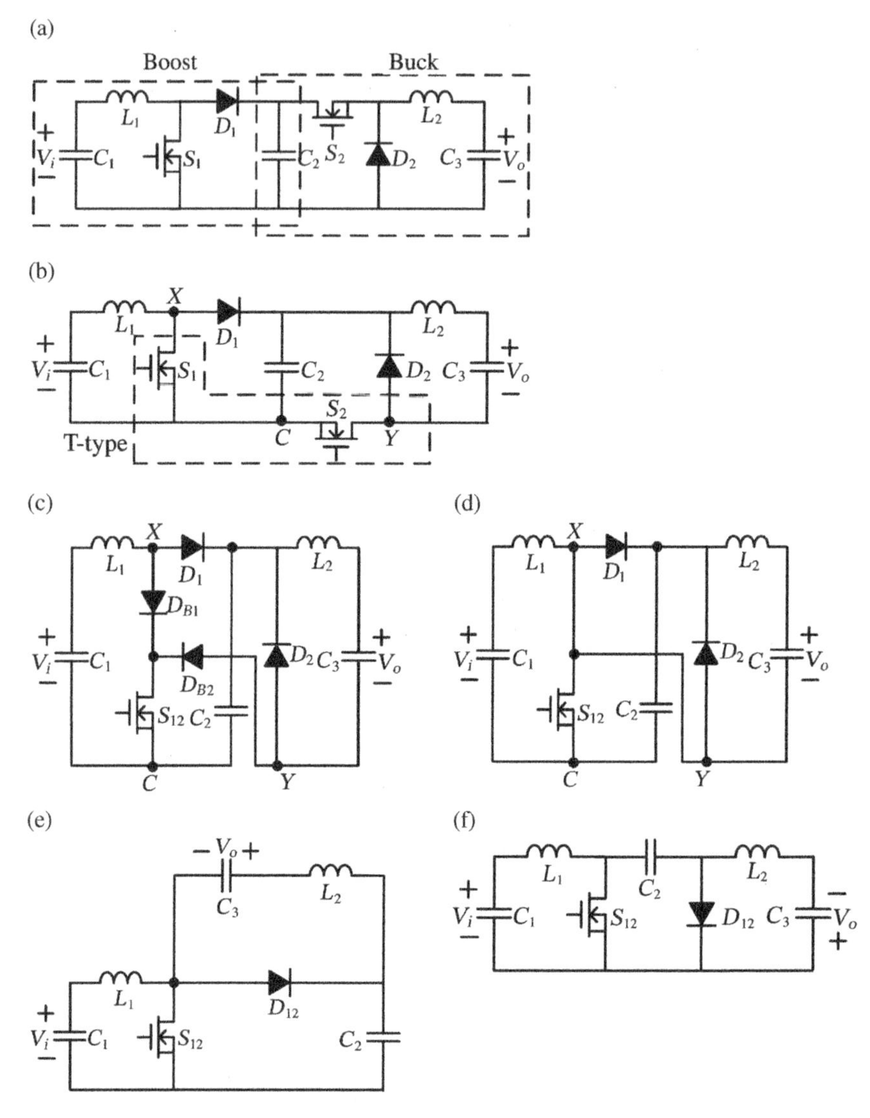

Figure 3.25 (a) Boost and buck converters in cascade connection, (b) relocating switch S_2 to the return path, (c) switches S_1 and S_2 replaced with a TGS, (d) degenerating the TGS into a single switch S_{12}, (e) paralleled diodes D_1 and D_2 replaced with a single one D_{12}, and (f) Ćuk converter.

series or in parallel, they can be simply replaced with a single one. However, if they share only a common node, common *N* or common *P*, as shown in Figures 3.28 and 3.29, they can be grafted depending on their node voltages. Figure 3.28a shows two diodes with a common node *N*, and they can be grafted into a single

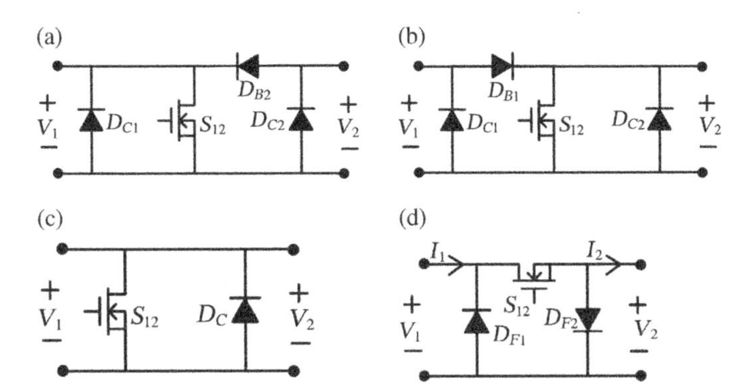

Figure 3.26 Degeneration of TGS and ΠGS bidirectional current switches under the conditions of (a) $V_1 > V_2$, (b) $V_1 < V_2$, (c) $V_1 = V_2$, and (d) any relationship between I_1 and I_2.

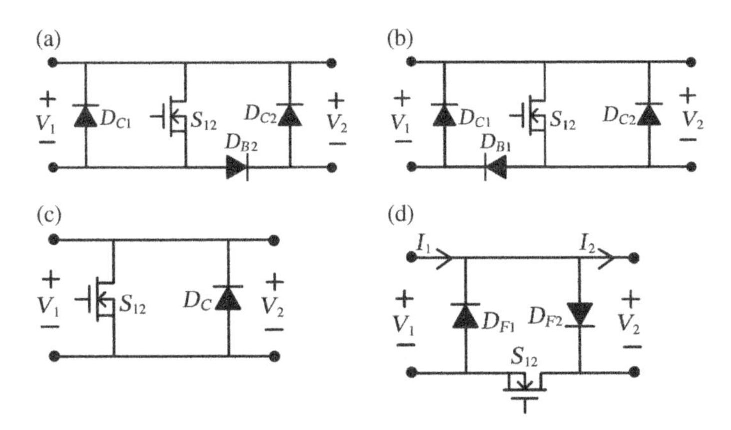

Figure 3.27 Degeneration of ITGS and IΠGS bidirectional current switches under the conditions of (a) $V_1 > V_2$, (b) $V_1 < V_2$, (c) $V_1 = V_2$, and (d) any relationship between I_1 and I_2.

one D_{12}. However, it might require two blocking diodes, D_{B1} and D_{B2}, to block the voltage difference between V_x and V_y, as shown in Figure 3.28b, namely, N-type grafted diode (NGD). If voltages $V_x > V_y$, diode D_{B1} is always in forward bias, and it can be shorted, as shown in Figure 3.28c. On the contrary, if voltages $V_x < V_y$, diode D_{B2} can be shorted, as shown in Figure 3.28d. If voltages $V_x = V_y$, both D_{B1} and D_{B2} can be shorted, and there is only D_{12} left, as shown in Figure 3.28e. Similarly, two diodes share a common node P, as shown in Figure 3.29a, which can be grafted and degenerated into the configurations shown in Figure 3.29b–e. The one shown in Figure 3.29b is named as P-type grafted diode (PGD).

Figure 3.25 illustrates the derivation of boost-buck (Ćuk) converter with TGS. It can be also derived with a grafted diode, as shown in Figure 3.30, in which the boost and buck converters connected in cascade and with switch S_2 in the return

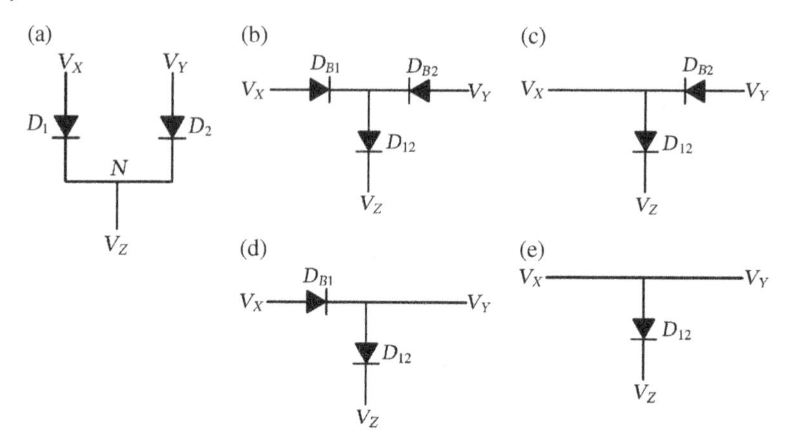

Figure 3.28 (a) Two diodes with common node N, (b) N-type grafted diode (NGD), and degenerated NGD under (c) $V_X > V_Y$, (d) $V_X < V_Y$, and (e) $V_X = V_Y$.

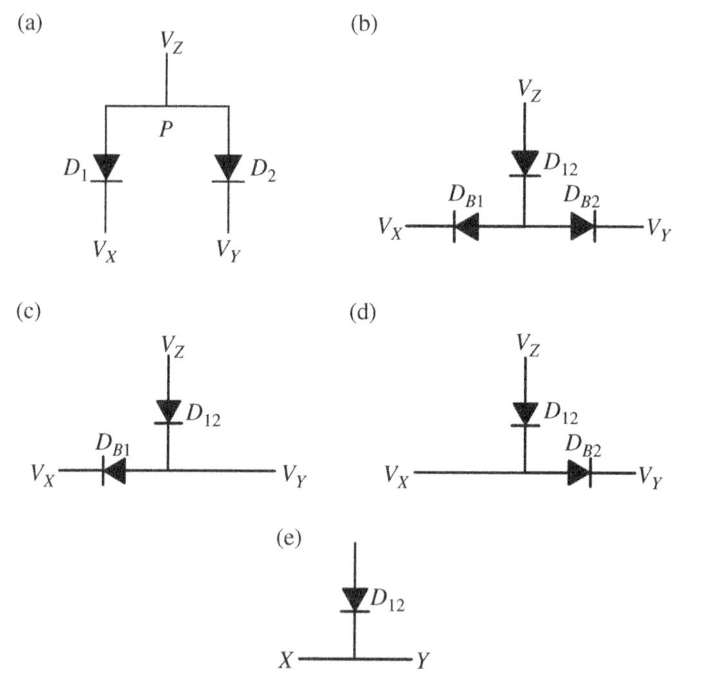

Figure 3.29 (a) Two diodes with common node P, (b) P-type grafted diode (PGD), and degenerated PGD under (c) $V_X > V_Y$, (d) $V_X < V_Y$, and (e) $V_X = V_Y$.

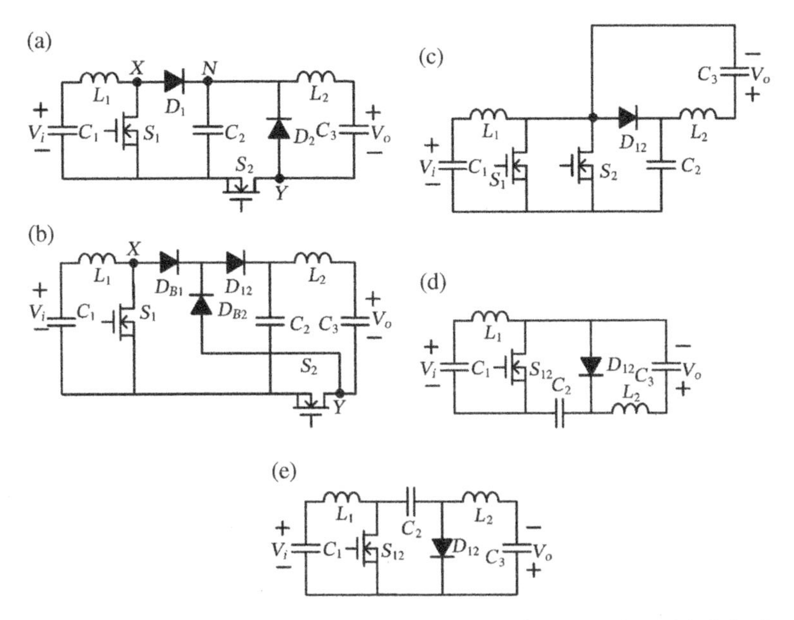

Figure 3.30 (a) Boost and buck converters in cascade connection with S_2 in the return path, (b) diodes D_1 and D_2 replaced with a NGD, (c) degenerated NGD without blocking diodes, (d) Ćuk converter with C_2 and L_2 in the return path, and (e) Ćuk converter in the well-known configuration.

path are shown in Figure 3.30a. It can be seen that diodes D_1 and D_2 are with a common node N, and they can be replaced with an NGD, as shown in Figure 3.30b. Since the reverse bias voltages V_X and V_Y of the original diodes D_1 and D_2, respectively, are all equal to the voltage across capacitor C_2, blocking diodes D_{B1} and D_{B2} are no longer needed, and the circuit can be simplified to the one shown in Figure 3.30c. Thus, the two switches S_1 and S_2 in the converter are in parallel, and they can be replaced with a single one, as shown in Figure 3.30d. After moving capacitor C_2 and inductor L_2 from the return path to the forward path, we can obtain the Ćuk converter in the well-known configuration, as shown in Figure 3.30e.

In the boost-buck converter, the active switch and diode form a PWM switch pair. It can be derived either with grafted switch or grafted diode. In literature, there are many PWM converters that have such property in which there is no need of blocking diodes during degeneration process. Thus, it is usually having switches in parallel or in series, and they can be combined into a single one. However, there are quite a few of other types of PWM converters with a single active switch but with multiple diodes because one or two blocking diodes are required. These types of PWM converters will be discussed in later chapters.

3.8 Layer Scheme

In previous two sections, we adopted the tree-grafting approach. There is another approach to growing new trees, namely, layer scheme. With a layer scheme to grow a new tree, we first bend a branch down to ground, strip part of rind of the branch, and cover with soil or embed it into ground, and finally it will grow to be a new tree. This process is similar to adopting a feedback branch from a converter output to the source ground node, breaking the ground for connecting the feedback branch, and finally forming a new converter. Thus, the layer scheme is adapted to layering converters. Analogy of grafting and layering trees to converters provides converters with life. That is the idea we want to propagate in addition.

In the following, layering buck and boost converters to yield buck-boost and boost-buck converters is presented. Figure 3.31a shows a buck converter, and its input-to-output voltage transfer block diagram is shown in Figure 3.31b where D is the duty ratio of switch S_1 in CCM. With a positive unity feedback, the input-to-output voltage transfer block diagram is shown in Figure 3.31c, yielding the transfer ratio of $V_o/V_i = D/(1-D)$. To realize the transfer block diagram shown in Figure 3.31c, a feedback branch C_f is connected from the output of the buck converter to the ground of its input source, as shown in Figure 3.31d, in which $V_f = V_o$ is added to V_i and the sum of $V_f + V_i$ is applied to the buck con-

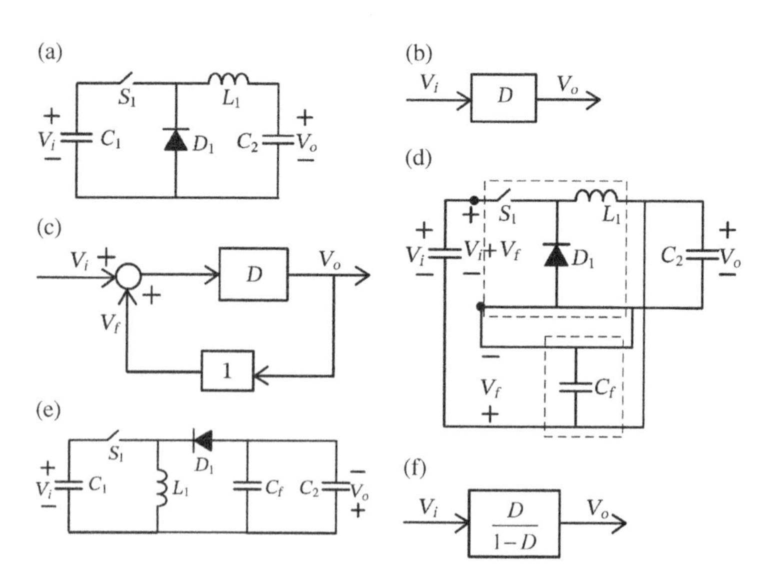

Figure 3.31 (a) Buck converter, (b) input-to-output voltage transfer ratio in CCM, (c) block diagram of transfer ratio D with a positive unity feedback, (d) converter realization of the transfer ratio block diagram shown in (c), (e) buck-boost converter, and (f) input-to-output voltage transfer ratio of the buck-boost converter in CCM.

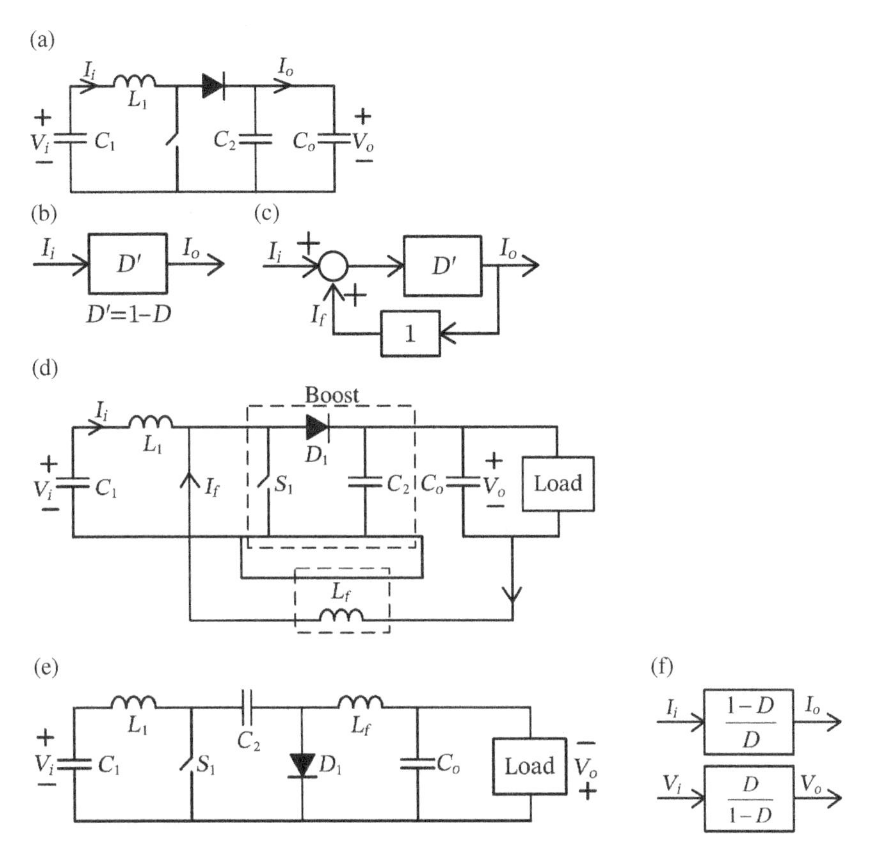

Figure 3.32 (a) Boost converter with split output capacitor C_o, (b) input-to-output current transfer ratio in CCM, (c) block diagram of transfer ratio D' with a positive unity feedback, (d) converter realization of the transfer-ratio block diagram shown in (c), (e) boost-buck (Ćuk) converter, and (f) input-to-output voltage and current transfer ratios of the Ćuk converter in CCM.

verter. Basically, a buck converter consists of a PWM switch pair, an active switch and a passive one, and an inductor. Redrawing the circuit configuration and moving diode D_1 from the return path to the forward path will yield the buck-boost converter in the well-known configuration, as shown in Figure 3.31e with capacitors C_f and C_2 in parallel, and its voltage transfer ratio $D/(1-D)$ in CCM is shown in Figure 3.31f.

Analogously, with an output current feedback, we can layer a boost converter to yield a boost-buck converter, as illustrated in Figure 3.32. Figure 3.32a shows a boost converter, and its input to output current transfer block diagram is shown in Figure 3.32b where $D' = 1-D$ and D is the duty ratio of switch S_1 in CCM. With a positive unity output current feedback I_f, the input to output current transfer block diagram is shown in Figure 3.32c that is realized with the boost converter and a feedback branch L_f connected to input inductor L_1, as shown in Figure 3.32d. Basically, a boost converter consists of a PWM switch pair and a capacitor, and therefore, an output capacitor C_o is separately introduced to the output side. Redrawing the circuits and moving capacitor C_2 and inductor L_2 from the return path to the forward path will yield the well-known boost-buck (Ćuk) converter configuration shown in Figure 3.32e, and its input-to-output current transfer ratio of $I_o/I_i = (1-D)/D$ is shown in Figure 3.32f. Based on power balance principle, the input-to-output voltage transfer ratio of $V_o/V_i = D/(1-D)$ is also shown in the figure.

From Figures 3.31 and 3.32, we can observe that summation of voltages must be with capacitor feedback, while summation of current must use inductor feedback. Moreover, only can capacitor be added up with capacitor and inductor with inductor. Since voltage feedback is added with another voltage, the capacitors are connected in series. However, since current feedback is added with another current, the inductors are joined at the same node.

With a layer scheme to generate new converters, we deal with only linear components, and the number of switches in the converters is not changed. In literature, there are many PWM converters that can be layered to yield new converters. However, one has to make sure that the power flow from input to output has to go through the forward path, while the power flow from the output to the input has to go through the feedback path; and thus, the derivation of input-to-output transfer ratio based on the block diagram with a feedback path will be correct. In other words, the power flow should be unidirectional over a switching cycle. For some converters, such as Z-source voltage-fed converters, with bidirectional power flow, if their power flow in one switching cycle is always unidirectional, they can be still derived with the layer scheme, but their switches are replaced with bidirectional ones.

Further Reading

Ćuk, S. (1979). General topological properties of switching structures. *Proceedings of the IEEE Power Electronics Specialists Conference*, pp. 109–130.

David Irwin, J. (2002). *Basic Engineering Circuit Analysis*, 7e. Wiley.

Erickson, R.W. (1983). Synthesis of switched-mode converters. *Proceedings of the IEEE Power Electronics Specialists Conference*, pp. 9–22.

Kraus, J.D. (1984). *Electromagnetics*, 3e. McGraw-Hill Book Co.

Lee, F.C. (1989). *High-Frequency Resonant, Quasi-Resonant and Multi-Resonant Converters*. Virginia Power Electronics Center.

Liu, K.-H. and Lee, F.C. (1988). Topological constraints of basic PWM converters. *Proceedings of the IEEE Power Electronics Specialists Conference*, pp. 164–172.

Makowski, M.S. (1993). On topological assumptions on PWM converters: a re-examination. *Proceedings of the IEEE Power Electronics Specialists Conference*, pp. 141–147.

Maksimovic, D. and Ćuk, S. (1989). General properties and synthesis of PWM DC-to-DC converters. *Proceedings of the IEEE Power Electronics Specialists Conference*, pp. 515–525.

Severns, R.P. and Bloom, G.E. (1985). *Modern DC-to-DC Switch Mode Power Converter Circuits*. New York: Van Nostrand Reinhold Co.

Tymerski, R. and Vorperian, V. (1986). Generation, classification and analysis of switched-mode DC-to-DC converters by the use of converter cells. *Proceedings of the International Telecommunications Energy Conference*, pp. 181–195.

Wu, T.-F. and Chen, Y.-K. (1996). A systematic and unified approach to modeling PWM DC/DC converters using the layer scheme. *Proceedings of the IEEE Power Electronics Specialists Conference*, pp. 575–580.

Wu, T.-F. and Chen, Y.-K. (1998a). A systematic and unified approach to modeling PWM DC/DC converter based on the graft scheme. *IEEE Trans. Ind. Electron.* 45 (1): 88–98.

Wu, T.-F. and Chen, Y.-K. (1998b). Modeling PWM DC/DC converter out of basic converter units. *IEEE Trans. Power Electron.* 13 (5): 870–881.

Wu, T.-F. and Yu, T.-H. (1998). Unified approach to developing single-stage power converters. *IEEE Trans. Aerosp. Electron. Syst.* 34 (1): 221–223.

Wu, T.-F., Chen, Y.-K., and Liang, S.-A. (1999). A structural approach to synthesizing, analyzing and modeling quasi-resonant converters. *Proceedings of the IEEE Power Electronics Specialists Conference*, pp. 1024–1029.

Wu, T.-F., Liang, S.-A., and Chen, Y.-K. (2003). A structural approach to synthesizing soft switching PWM converters. *IEEE Trans. Power Electron.* 18 (1): 38–43.

4

Decoding Process

Conventionally, people develop converters first and derive their input-to-output voltage transfer ratios based on volt-second balance principle. The transfer ratios might not exactly meet their desired needs of applications. Hence, if we can conduct on the other way, obtaining the desired transfer ratio first and then developing its corresponding converter, it will be more feasible and convenient. This chapter presents the decoding process to decode a given transfer ratio, or called code in this book, into a certain code combination or configuration, and in Chapter 5, we will address how to synthesize the code configuration into its corresponding converter.

In the following, several typical input-to-output voltage transfer codes will be presented to lay out a foundation in selecting codes, from which transfer gain block diagrams can be constructed. Decoding approaches to figuring out code configurations are then presented, and code configuration block diagrams can be obtained for converter synthesis.

4.1 Transfer Ratios (Codes)

Transfer ratios are the typical terms for power conversion that are used to represent either voltage transfer gain or current transfer gain of an input to output. Here, we name the transfer ratios transfer codes because we treat the ratios, which are typically represented in arithmetic expressions, such as D, $1/(1-D)$, $D/(1-D)$, *etc.*, as codes.

Power transfers from input to output in a converter can be conceptually illustrated by Figure 4.1. Its input-to-output voltage transfer code can be denoted by the following equation:

$$\frac{V_o}{V_i} = f\left(x_1, x_2, \ldots, x_n\right), \tag{4.1}$$

Origin of Power Converters: Decoding, Synthesizing, and Modeling, First Edition.
Tsai-Fu Wu and Yu-Kai Chen.
© 2020 John Wiley & Sons, Inc. Published 2020 by John Wiley & Sons, Inc.

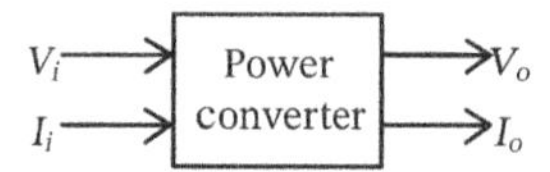

Figure 4.1 Conceptual block diagram of power transfer from input (V_i, I_i) to output (V_o, I_o) with a power converter.

where x_i, $i = 1, 2, ..., n$, are the arguments that could be duty ratio, switching frequency, resonant frequency, quality factor, load resistance, *etc*. Based on power balance ($V_o I_o = V_i I_i$), the input-to-output current transfer ratio can be then derived as

$$\frac{I_o}{I_i} = \frac{1}{f(x_1, x_2, \ldots, x_n)}. \tag{4.2}$$

When the resonant frequency of the *LC* filter in the power converter is much lower than the switching frequency and the voltage and current waveforms of a converter can be approximated as rectangular-like or triangular-like ones with linear slopes, the arguments can be simply reduced to duty ratio only. This type of converter is conventionally called a PWM converter.

For non-isolated PWM converters, typical input-to-output voltage transfer codes can be classified into step-down, step-up, step-up/step-down, and ±step-up/step-down. Even though there are many transfer codes, their expressions in terms of duty ratio cannot be arbitrary due to the limitation of the original converter. For a step-down transfer code, it is typically D, $1 - D$, D^2, or $(1 - D)^2$ where D is the duty ratio of the active switch and the voltage transfer code of the original converter, buck converter, operated in continuous conduction mode (CCM). Designers may choose a certain transfer code to fit their applications. Figure 4.2 shows the plots of gain versus duty ratio D (*Gain-D*) of transfer codes D and D^2 as an example. Typically, the transfer code D is chosen for a step-down application. However, one might need a higher step-down ratio over a specific range of duty ratio, such as $0.2 \leq D \leq 0.7$, and the transfer code of D^2 might be selected, as also illustrated in Figure 4.2. Since duty ratio D is always less than or equal to unity, transfer codes $(1 - D)$ and $(1 - D)^2$ also belong to step-down codes. It is worth noting that in the Gain-D plots shown in Figure 4.2, when duty ratio D increases, the gain increases monotonically.

Figure 4.3 shows the *Gain-D* plots of step-up transfer codes $1/(1-D)$ and $1/(1 - D)^2$, which can be used as a guideline in selecting a desired transfer code for stepping up input voltage applications. Theoretically, when $D = 1$, the two transfer codes yield infinite gains. However, it cannot happen in practice since there always exists equivalent resistance or loss in all components. Again, in the plots, the gain increases monotonically with the increase of D.

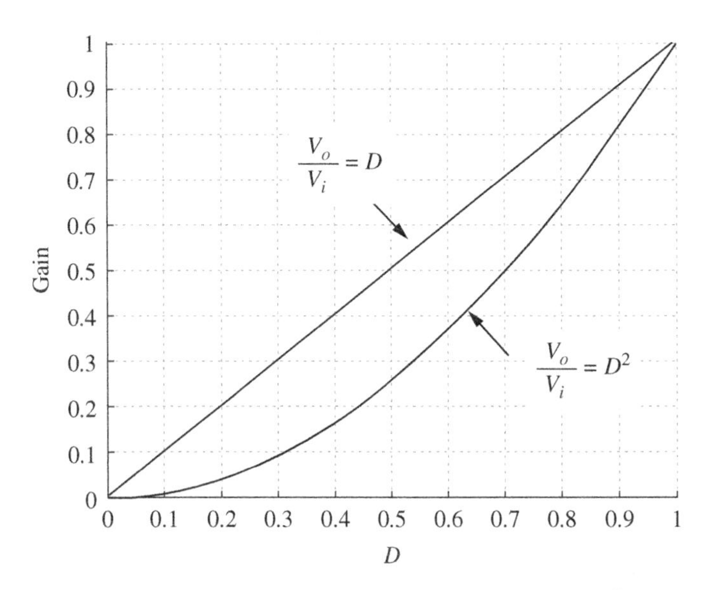

Figure 4.2 *Gain-D* plots of step-down transfer codes D and D^2.

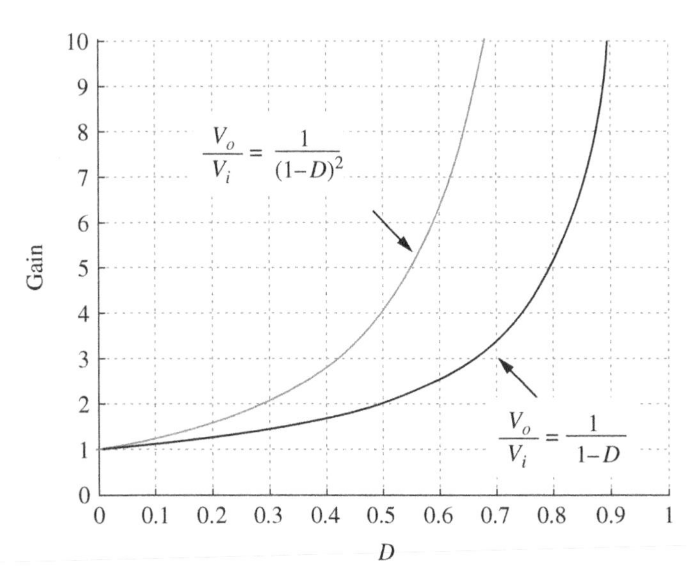

Figure 4.3 *Gain-D* plots of step-up transfer codes $1/(1 - D)$ and $1/(1 - D)^2$.

When input voltages swing over a small range, a converter with pure step-down or step-up transfer code may be good enough to yield a desired output voltage. On the other hand, if input voltages swing over a wide range, such as the terminal voltages of photovoltaic panels, converters with step-up/step-down transfer codes would be needed. They can be the product of step-down and step-up transfer codes, and two examples are shown as follows:

$$\frac{V_o}{V_i} = D \cdot \frac{1}{(1-D)} = \frac{D}{(1-D)}, \tag{4.3}$$

and

$$\frac{V_o}{V_i} = D \cdot D \cdot \frac{1}{(1-D)} = \frac{D^2}{(1-D)}. \tag{4.4}$$

The *Gain-D* plots of the above two transfer codes are shown in Figure 4.4. It can be observed that when duty ratio D is lower than a certain value, its gain is lower than unity, and when it is higher than that value, the gain is higher than unity and increases monotonically. Basically, the two *Gain-D* curves show no significant difference except from $D = 0.4$ to 0.8. When we select a step-up/step-down transfer code, $D/(1 - D)$ is good enough, and it can save one power converter stage.

People might ask the question that why we don't just use the converters with step-up/step-down transfer codes to replace those with pure step-down and step-up ones. Comparing the plots shown in Figures 4.2–4.4 can find some reasons that we still need pure step-down and step-up transfer codes. For the *Gain-D* plot of a step-down transfer code D shown in Figure 4.2, it varies from 0 to 1 to yield the gain from 0 to 1, but that of $D/(1-D)$ shown in Figure 4.4 varies from 0 to 0.5, which has higher slope and means more sensitive to D variation. On the contrary, when D is greater than 0.5, which yields the gain being higher than unity, the slope is lower than that shown in Figure 4.3 with the transfer code of $1/(1-D)$.

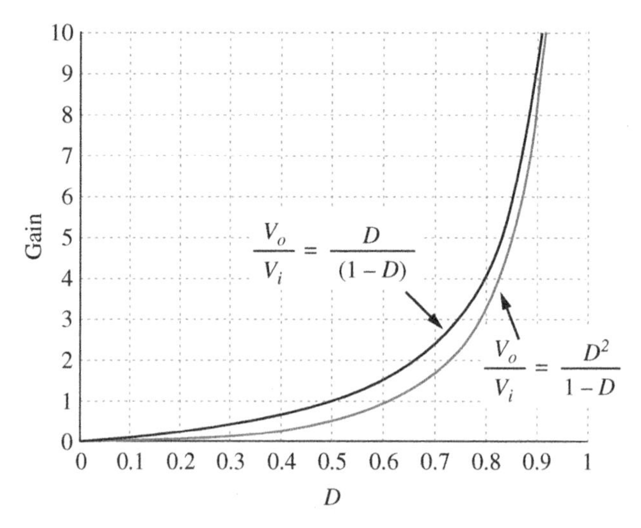

Figure 4.4 *Gain-D* plots of step-up/down transfer ratios $D/(1 - D)$ and $D^2/(1 - D)$.

Even though different transfer codes can yield the same gain, they have different characteristics. Moreover, it will be seen in the next chapter that converters to achieve pure step-down and step-up transfer codes have different component stresses from those achieving step-up/step-down transfer codes. We still need the converters with pure step-down or step-up transfer codes.

The fourth category, ±step-up/step-down transfer codes, can yield positive and negative outputs. The converters with this kind of transfer codes can act like DC–AC inverters. Figure 4.5 shows several *Gain-D* plots of transfer codes $(1-D)/(1-2D)$, $D/(1-2D)$, $(2D-1)/(1-D)$, and $(2D-1)/D(1-D)$ in which the positive and negative gains are not symmetrical. By taking the example of $(1-D)/(1-2D)$, one can observe that when D is less than 0.5, the gain varies from 1 to ∞, while when D is greater than 0.5, the gain varies from $-\infty$ to 0. For inverter applications, one has to choose the range that can yield the same gain. For instance, if the gain is 2, one can select the duty ratio around 0.3 for positive output and choose around 0.6 for negative output. For other transfer codes shown in Figure 4.5, they have similar characteristics, and proper selections are needed.

From the transfer codes denoted in Figure 4.5, one can observe that the coefficients of variable D are not always unity. In other words, they cannot be obtained simply from the product of D and $1/(1-D)$. It is interesting that how designers figure out such kind of asymmetrical transfer codes. As described at the beginning of this chapter, we obtain a transfer code first and then synthesize it into a converter. Thus, we would like to ask that what are the valid transfer codes and how to construct them systematically. It is really hard to have a quick and simple answer. Designers usually select transfer codes based on the existing ones. For a

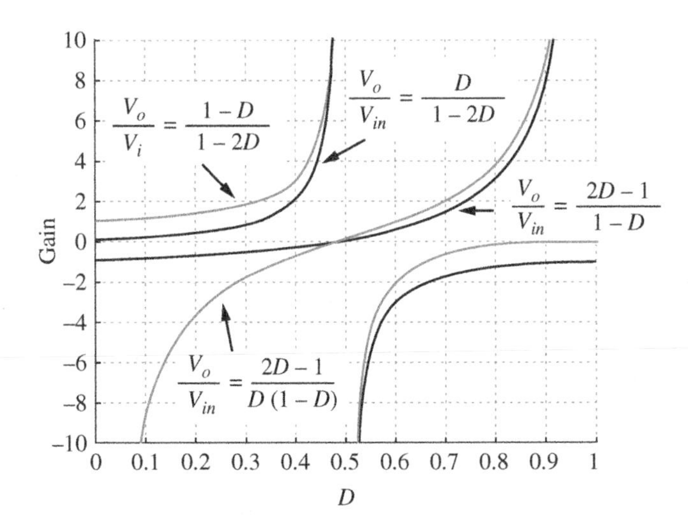

Figure 4.5 *Gain-D* plots of ±step-up/step-down transfer codes $(1-D)/(1-2D)$, $D/(1-2D)$, $(2D-1)/(1-D)$, and $(2D-1)/D(1-D)$.

step-down example, the *Gain-D* plot of transfer code D shown in Figure 4.2 is modified to D^2 to obtain a higher step-down code over a certain duty-ratio range. Without reference transfer codes (typically the existing ones), it is not easy to list one to fit desired applications. In next sections, we are going to decode transfer codes into certain configurations, from which, maybe, we can understand the way to generate valid transfer codes.

4.2 Transfer Code Configurations

Buck converter is the original converter with the transfer code of D, from which we can derive other transfer codes and they can be configured with the following forms: cascade, feedback, feedforward, their combinations, and parallel.

4.2.1 Cascade Configuration

A cascade configuration is that two or more transfer codes are connected in series, as shown in Figure 4.6. The overall transfer code from input to output is equal to the product of all of the individual transfer codes; that is,

$$\frac{V_o}{V_i} = TC_1 \times TC_2 \times \cdots \times TC_n. \tag{4.5}$$

A practical example is illustrated in Figure 4.2 where transfer code $D^2 = D \times D$ consists of two identical transfer codes in cascade. Another example is shown in Figure 4.4 where transfer code $D/(1-D) = D \times 1/(1-D)$ has two different transfer codes in cascade. This configuration is a straightforward and simple yet feasible connection of multiple transfer codes to achieve an overall higher step-up or step-down transfer code. Its power flow is unidirectional, and the overall transfer code is the product of all of the individual codes, as shown in (4.5).

4.2.2 Feedback Configuration

The second type of transfer code configuration is a feedback configuration, as shown in Figure 4.7, where TC_f is the forward transfer code and TC_b is the feedback one. The input-to-output transfer code can be obtained as follows:

$$\frac{V_o}{V_i} = \frac{TC_f}{\left(1 \mp TC_f \times TC_b\right)} \tag{4.6}$$

The above equation is valid if and only if the forward power flows through TC_f exclusively and the return power flows through TC_b only. A practical example with

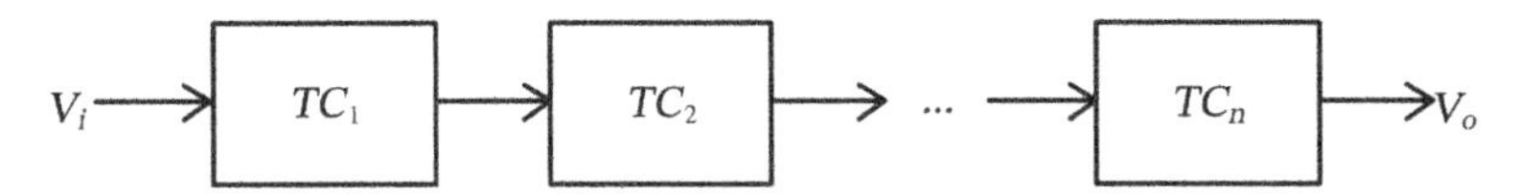

Figure 4.6 Transfer codes in a cascade configuration.

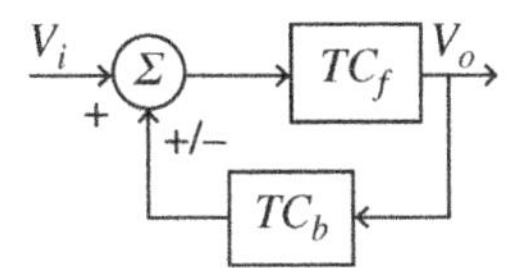

Figure 4.7 Feedback configuration with a forward code TC_f and a feedback code TC_b.

the transfer codes of $TC_f = D$ and $TC_b = 1$, and with a positive feedback (+) is shown in Figure 4.8a, which yields $V_o/V_i = D/(1 - D)$. Another example shown in Figure 4.8b is with $TC_f = D/(1 - D)$, $TC_b = 1$, and a negative feedback (–), which yields $V_o/V_i = D$. In the next chapter, we will use a buck converter with a positive unity feedback to derive a buck–boost converter and then derive the buck converter from the buck–boost converter with a negative unity feedback. These two examples can illustrate the feedback configuration shown in Figure 4.7.

4.2.3 Feedforward Configuration

The third type of transfer code configuration is a feedforward configuration, as shown in Figure 4.9, in which the one in Figure 4.9a has a TC_f forward path and a unity-gain feedforward and the other shown in Figure 4.9b has a forward path TC_f, a feedback TC_b, and a unity-gain feedforward. Figure 4.10a shows an example implementing the configuration shown in Figure 4.9a, which yields $V_o/V_i = 1 - D$. Figure 4.10b shows another example that implements the configuration shown in Figure 4.9b, which yields $V_o/V_i = 1/(1 - D)$.

For forward gain TC_f and feedback gain TC_b, they can be unity or other valid transfer codes. However, without a transformer, the feedforward gain is always

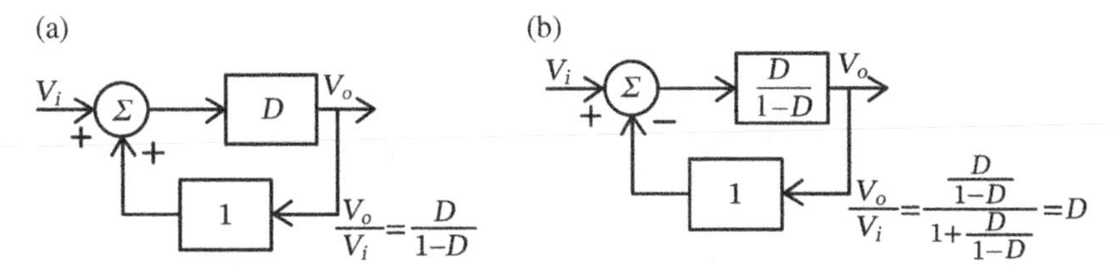

Figure 4.8 Two examples to illustrate the feedback configuration: (a) forward code D with a positive unity-gain feedback code and (b) forward code $D/(1 - D)$ with a negative unity-gain feedback code.

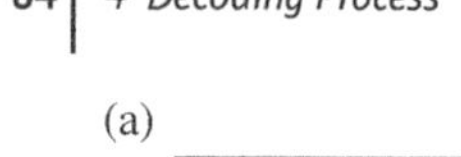

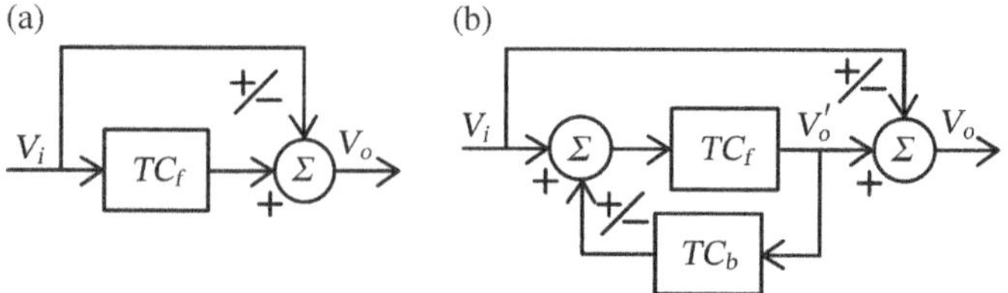

Figure 4.9 Feedforward configuration: (a) a unity-gain feedforward with a forward code TC_f and (b) a unity-gain feedforward with a feedback configuration.

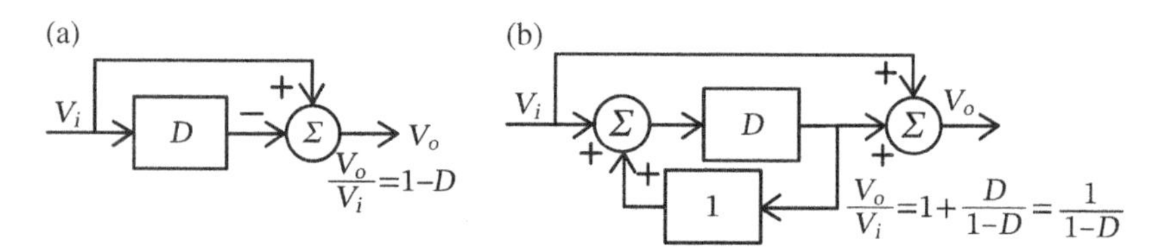

Figure 4.10 Two examples to illustrate the feedforward configuration: (a) a unity-gain feedforward with a forward code D and (b) a unity-gain feedforward with an equivalent transfer code of $D/(1-D)$ derived from a feedback configuration.

unity. It must be combined with forward or/and feedback gains to form a transfer code.

Combinations of all of the aforementioned three types of code configurations can yield other transfer codes. Three examples are shown in Figure 4.11, in which Figure 4.11a shows three forward codes TC_{f1}, TC_{f2}, and TC_{f3}, two feedback codes TC_{b1} and TC_{b2}, and one unity-gain feedforward. Note that forward code TC_{f3} is connected after the unity-gain feedforward. Figure 4.11b shows that in between the two forward codes TC_{f1} and TC_{f2}, it can have a feedback code but without any unity-gain feedforward. The one shown in Figure 4.11b can be combined with a unity-gain feedforward to form the code configuration shown in Figure 4.11c. Of course, there can be more other combinations to derive transfer codes.

From Figure 4.11a, we can derive the overall transfer code as follows:

$$\frac{V_o}{V_i}=\left(\frac{TC_{f1}\times TC_{f2}}{1\mp TC_{f1}\times TC_{f2}\times TC_{b1}\times TC_{b2}}\pm 1\right)\times TC_{f3}. \tag{4.7}$$

From Figure 4.11b, we have

$$\frac{V_o}{V_i}=\frac{TC_{f1}\times TC_{f2}}{1\mp TC_{f1}\times TC_{b1}}. \tag{4.8}$$

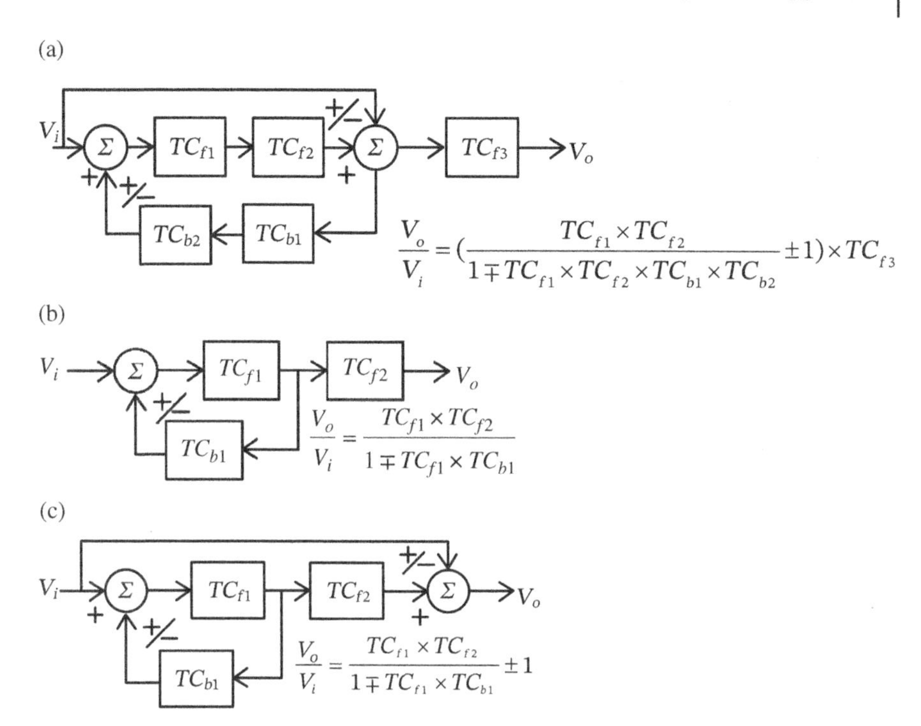

Figure 4.11 Three combinational code configurations: (a) forward, feedback and feedforward configurations, (b) forward and feedback configurations, and (c) those of (b) with a unity-feedforward configuration.

From Figure 4.11c, we have

$$\frac{V_o}{V_i} = \frac{TC_{f1} \times TC_{f2}}{1 \mp TC_{f1} \times TC_{b1}} \pm 1. \tag{4.9}$$

4.2.4 Parallel Configuration

To achieve higher power output, reduce current or voltage ripple, reduce component stress, or increase dynamics, multiple converters are connected in parallel and, usually, in interleaving operation. The converters in the system can have nonidentical power modules or different control algorithms. Their transfer code configuration is conceptually shown in Figure 4.12. Assuming the parallel operation does not affect their transfer codes, the total input-to-output transfer code can be expressed below:

$$\frac{V_o}{V_i} = TC_1 + TC_2 + \cdots + TC_n, \tag{4.10}$$

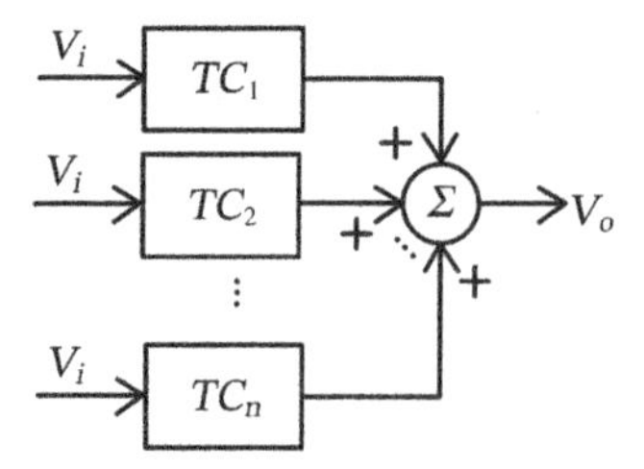

Figure 4.12 Parallel configuration of n transfer codes.

in which TC_1, TC_2, and TC_n are the transfer codes of the converters connected in the system. Basically, it requires galvanic isolation for some of the converters in the parallel system, and then their outputs can be summed up to yield a single output V_o. Parallel configuration can be combined with other configurations to yield desired transfer codes.

Based on the above discussion, we might obtain a desired transfer code from the aforementioned code configurations. However, conventionally, we only have final overall transfer codes, but not in the configurations shown in Figures 4.6–4.12. In other words, given an overall transfer code, we have to decode it into specific transfer codes, which are connected in a certain configuration, and then we could find converters to synthesize their transfer codes in the configuration. In Section 4.3, decoding approaches will be discussed.

With a transformer, the coefficients in transfer codes and feedforward gain can be fractional by tuning the turns ratio of the transformer. However, configurations of the circuits with the transformer must satisfy with the volt-second balance principle.

4.3 Decoding Approaches

In previous section, we have shown several transfer code configurations. Essentially, they are just like encoding approaches to configuring transfer codes. Thus, the decoding approaches discussed in this section are the reverse processes of the encoding approaches. In the following, the decoding approaches of factorization, long division, and cross multiplication are discussed.

4.3.1 Factorization

Factorization of a polynomial with an arbitrary order is not an easy work. In literature, there are classic and modern methods, such as Kronecker, Trager, and Zassenhaus's methods, for factorizing polynomials. In this book, we are not intended to spend much space in discussing the detail while selecting certain examples to illustrate the proposed decoding ideas and algorithms.

With factorization, a transfer code is partitioned into fractions. The numerators and denominators of the fractions are zero-order or first-order polynomials, and then they can be synthesized with fundamental converters readily. Given a transfer code $TC(D)$, its numerator and denominator are expressed in two polynomials, as follows:

$$TC(D)=\frac{\left(a_0+a_1D+a_2D^2+\cdots+a_nD^n\right)}{\left(b_0+b_1D+b_2D^2+\cdots+b_mD^m\right)},\tag{4.11}$$

where a_0, a_1, ..., a_n, and b_0, b_1, ..., b_m are integers. We first examine if it can be factorized into the following expression:

$$TC(D)=\frac{\left(c_1D+d_1\right)\left(c_2D+d_2\right)\ldots\left(c_nD+d_n\right)}{\left(e_1D+f_1\right)\left(e_2D+f_2\right)\ldots\left(e_mD+f_m\right)},\tag{4.12}$$

where c_i, d_i, e_j, and f_j, $i=1, 2, ..., n$ and $j=1, 2, ..., m$ are integers. Then, we can start to divide it into the following fractions (if $n = m$, as an example), which can be encoded into cascade configuration:

$$\begin{aligned}TC(D)&=TC_1\times TC_2\times\cdots\times TC_n\\&=\frac{\left(c_1D+d_1\right)}{\left(e_1D+f_1\right)}\times\frac{\left(c_2D+d_2\right)}{\left(e_2D+f_2\right)}\times\cdots\times\frac{\left(c_nD+d_n\right)}{\left(e_mD+f_m\right)},\end{aligned}\tag{4.13}$$

An example that illustrates the above discussion is shown as follows:

$$TC(D)=\frac{D^2}{2-3D+D^2}.\tag{4.14}$$

By factorization, Eq. (4.14) can be partitioned into the following fractions and two transfer codes:

$$TC(D)=\frac{D}{1-D}\times\frac{D}{2-D}=TC_1\times TC_2.\tag{4.15}$$

These two transfer codes can be connected in cascade configuration shown in Figure 4.6. Another example is shown below

$$TC(D)=\frac{1+D+D^2}{2+D+3D^2}.\tag{4.16}$$

In (4.16), both of the polynomials in the numerator and denominator are prime factors or irreducible quadratic factors, and they cannot be factorized into linear factors with integer coefficients. If a transfer code cannot be factorized, does it mean that the transfer code is invalid? An invalid transfer code means that it cannot be synthesized by PWM converters. Thus, it comes out a question: what is a valid transfer code? The transfer code shown in (4.16) is a second-order fraction. What if it is a first-order one, can it be always a valid transfer code that can be synthesized with PWM converters? For instance, transfer code

$$TC_1(D) = \frac{1-D}{1-2D} \tag{4.17}$$

is a first-order fraction, so is

$$TC_2(D) = \frac{1+D}{1+2D}. \tag{4.18}$$

How to prove a transfer code being valid or invalid is a critical and tough work. Basically, any useful transfer code should be a combination from the codes of the original converter, buck converter, and its derivatives, which typically include the fundamental transfer codes of D, $1/D$, $1-D$, $1/(1-D)$, $D/(1-D)$, $(1-D)/D$, *etc.* Thus, an effective way to prove its validness is to find a combination in terms of the fundamental codes and in a certain code configuration. However, a factorization approach itself is not good enough to derive a proper configuration for synthesizing a converter.

4.3.2 Long Division

In the feedforward configuration shown in Figure 4.9, there is a unity-gain feedforward that is combined with a forward-path transfer code to become an improper fraction. To identify this unity gain and obtain a proper fraction, we adopt a long division approach to divide the transfer code into a unity gain and a residue in a proper fraction format. Thus, the transfer code in (4.11) can be expressed as follows:

$$TC(D) = 1 + TC_r(D), \tag{4.19}$$

where $TC_r(D)$ is a residual transfer code of $TC(D)$ in proper fraction, and it can be factorized further and expressed in terms of the fundamental codes.

For instance, transfer code $TC_1(D)$ in (4.17) can be expressed alternately as follows:

$$\begin{aligned} \frac{V_o}{V_i} &= TC_1(D) = \frac{1-D}{1-2D} \\ &= 1 + \frac{D}{1-2D} \\ &= 1 + TC_r(D). \end{aligned} \tag{4.20}$$

Transfer code $TC_r(D)$ can be further defined as follows:

$$\frac{V_o'}{V_i} = TC_r(D) = \frac{D}{1-2D}, \tag{4.21}$$

where V_o' is a sub-output, which will be combined with the unity gain to yield output voltage V_O. This transfer code is not obviously close to the well-known fundamental codes. Thus, we need to use cross multiplication approach to derive a relation, which can be readily recognized in a feedback configuration.

4.3.3 Cross Multiplication

For the transfer code shown in (4.21), we can conduct a cross multiplication to find a relationship between V_i and V_o' as follows:

$$V_o'(1-2D) = V_i D, \tag{4.22}$$

which is also equal to the following expression:

$$V_o'(1-D) = (V_i + V_o')D. \tag{4.23}$$

That is,

$$V_o' = \frac{(V_i + V_o')D}{1-D}. \tag{4.24}$$

Equation (4.24) can be put into a feedback configuration, as shown in Figure 4.13a, in which $D/(1-D)$ is a fundamental code. Alternately, the expression in (4.22) can be expressed in the following two simultaneous equations:

$$V_o'' = \frac{V_o''D + V_i}{1-D} \tag{4.25}$$

and

$$V_o' = V_o''D. \tag{4.26}$$

The above two equations can be configured into the feedback configuration shown in Figure 4.13b, in which the two forward codes, $1/(1-D)$ and D, and a feedback code, D, are all fundamental codes. Combining the unity gain in (4.20) with (4.24), or (4.25) and (4.26), can yield the transfer code: $(1-D)/(1-2D)$, and its code configurations are shown in Figure 4.14a and b, which are corresponding to Figure 4.13a and b, respectively.

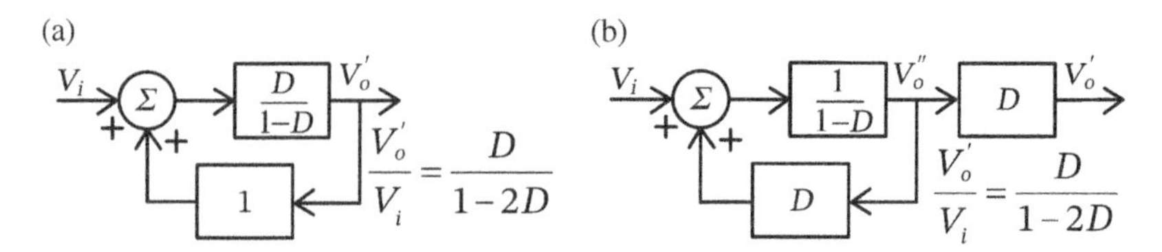

Figure 4.13 Two possible configurations of $D/(1-2D)$: (a) feedback configuration and (b) feedback with forward configuration.

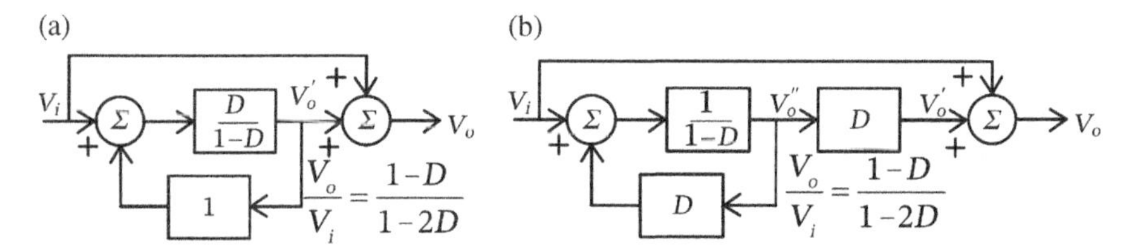

Figure 4.14 The two configurations shown in Figure 4.13 combined with a unity-gain feedforward.

Another example is the transfer code shown in (4.18), which can be expressed as follows by a long division:

$$\begin{aligned} TC_2(D) &= \frac{V_o}{V_i} = \frac{1+D}{1+2D} \\ &= 1 + \frac{-D}{1+2D} \\ &= 1 + TC_{r2}(D), \end{aligned} \tag{4.27}$$

where $TC_{r2}(D) = (-D)/(1+2D)$ is the residual transfer code of $TC_2(D)$. By redefining $TC_{r2}(D)$ as follows:

$$TC_{r2}(D) = \frac{V_o'}{V_i} = \frac{-D}{1+2D} \tag{4.28}$$

we can conduct cross multiplication below

$$-DV_i = V_o'(1+2D) \tag{4.29}$$

or

$$\frac{-D}{1+D} \times (V_i + V_o') = V_o'. \tag{4.30}$$

We can configure (4.30) into the feedback configuration shown in Figure 4.15a, and the one with unity-gain feedforward is shown in Figure 4.15b. Obviously, $(-D)/(1+D)$ is different from the fundamental codes, it is an invalid code, and it cannot be synthesized by the original converter and its derivatives.

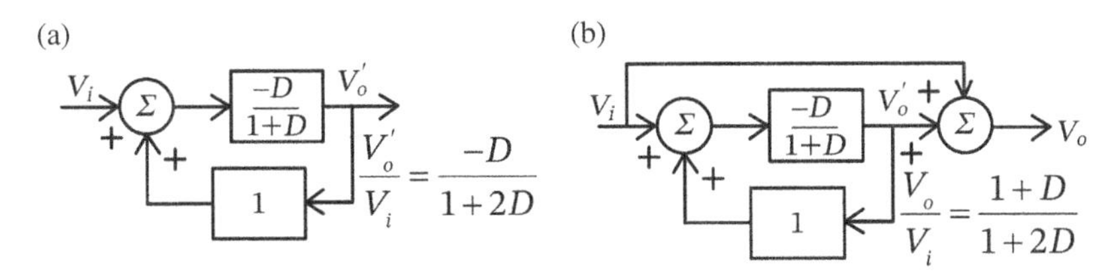

Figure 4.15 (a) feedback configuration of transfer code: $(-D)/(1+2D)$, and (b) the one shown in (a) with a unity-gain feedforward.

On the other hand, a valid code can be expanded to the combination of fundamental codes. For instance, transfer code D of the original converter can be expanded as follows:

$$TC(D) = D = (1-D) \times \frac{D}{1-D}, \tag{4.31}$$

where $(1-D)$ and $D/(1-D)$ are two fundamental codes.

In general, fundamental codes are corresponding to fundamental converters, and they have first-order fractions and simple converter topologies. Thus, when decoding a transfer code, we usually decode it into the combination of fundamental codes, which can be synthesized readily by fundamental converters. In fact, the fundamental codes are the combinations of the transfer code of the original converter and the unity-gain of a source. This is similar to compounds that are obtained from elements. After recognizing some fundamental codes, we can combine them further to form more sophisticated transfer codes to fit the desired applications, and we can find their corresponding converters to synthesize the codes.

4.4 Decoding of Transfer Codes with Multivariables

As claimed before, resonant converters are with power transfer in resonant manner, and thus, they are considered, in general, as a type of PWM converters. However, its transfer code is a function of many variables, such as duty ratio, switching frequency, resonant frequency, quality factor, *etc*. How to decode such a kind of transfer code is a tough yet critical issue.

For a conventional well-known PWM converter, its transfer code is a representation of active and passive switch configuration, while its LC network is just to ensure power transfer with resonant manner. With the LC network, the DC voltage gain of inductor L is unity, and that of capacitor C is its voltage divided by source voltage. Basically, the LC network acts as a filter, and it does not change the transfer code. For a resonant converter that is equivalent to two PWM converters operating in discontinuous conduction mode (DCM), its transfer code, however, is highly related to its LC network and load resistor R, but not its switch configuration when the duty ratio is equal to 50%.

Power transfer through a resonant converter can be conceptually illustrated by Figure 4.16, in which input voltage V_i is in square waveforms generated from a half- or full-bridge switch configuration. When we determine the input-to-output transfer code, the fundamental component of the square waveforms is obtained and applied to the RLC resonant network. We can represent the transfer code of a resonant converter as $f(X)$, where $X = x_1, x_2, ..., x_n$, and its corresponding forward, feedback, and feedforward configurations can be illustrated in Figure 4.17. In Figure 4.17a, $f_1(X)$, $f_2(X)$, and $f_n(X)$ are the transfer codes of cascaded resonant converters; in Figure 4.17b, $f_f(X)$ and $f_r(X)$ are the forward and feedback transfer codes, respectively; and in Figure 4.17c, a unity gain is combined with the output. In practice, since there exists a transformer at the output side to isolate the output from the input, a unity-gain feedforward to the output will break down the isolation, and it is not feasible. Figure 4.18 shows a parallel configuration of multiple transfer codes, which are functions of multivariables. However, the paralleled converters must be independent with each other.

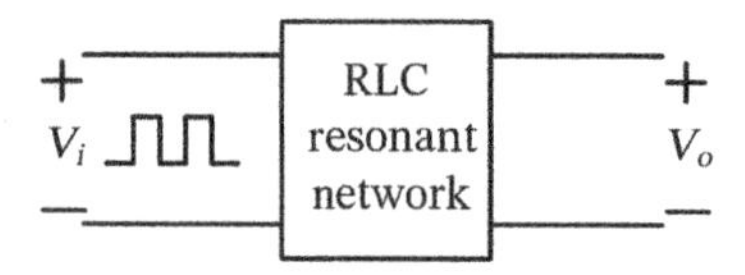

Figure 4.16 Conceptual power transfer through resonant converters, which can be used to determine input-to-output transfer code.

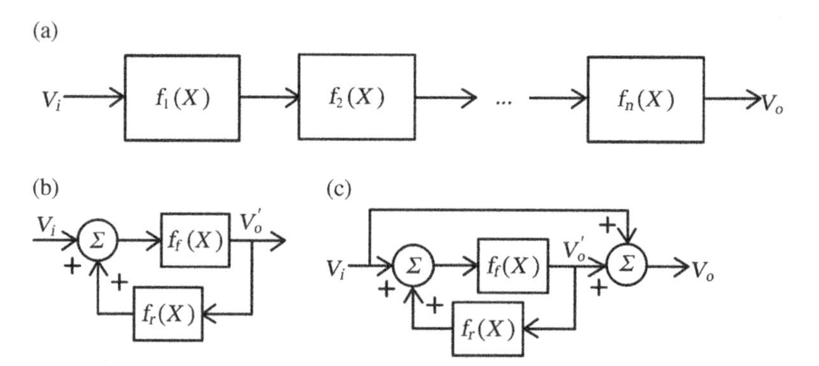

Figure 4.17 Code configurations of transfer codes with multivariables: (a) cascade configuration, (b) feedback configuration, and (c) the one in (b) with a unity-gain feedforward or simply feedforward configuration.

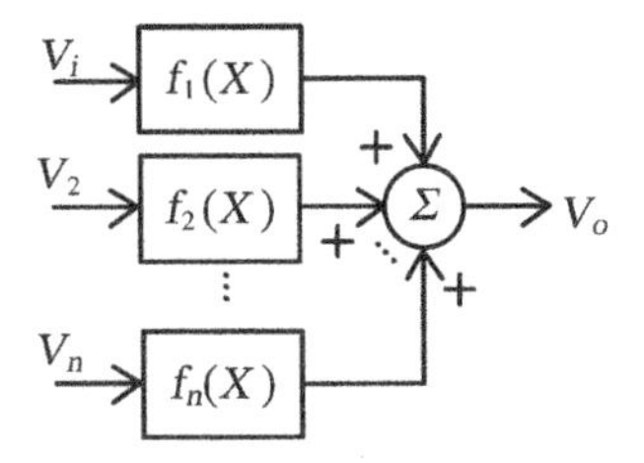

Figure 4.18 Parallel configuration.

Decoding of transfer codes with multivariables, which are all coupled each other, is different from those with a univariable, since it is not possible to separate the variables from the transfer codes one by one. Thus, in code configurations, each transfer code is dealt individually and treated as a single transfer code.

4.5 Decoding with Component-Interconnected Expression

There is one more possible decoding approach, which can be described as follows. First, we represent converters in specific matrix expressions and then introduce a specific operator to manipulate the matrixes, from which we can obtain a new matrix expression for a new converter. This can be conceptually illustrated in Figure 4.19a, in which buck and boost converters are represented in two matrix expressions. When introducing a specific operator "?" to process these two matrixes, we can obtain a specific matrix expression, from which we can conduct an inverse process to yield the buck–boost converter as shown in Figure 4.19b.

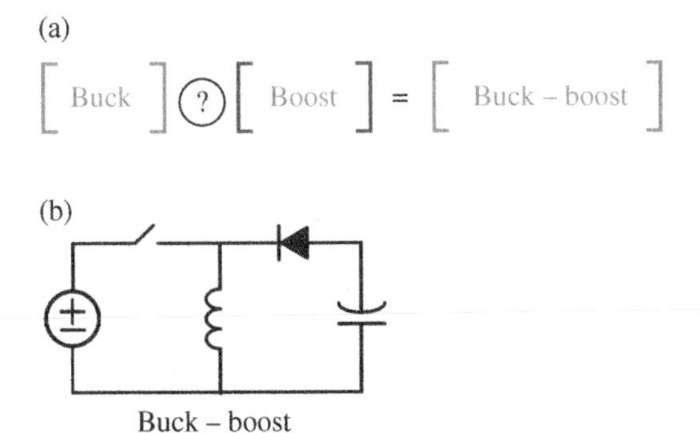

Figure 4.19 Decoding with component-interconnected expression: (a) buck and boost converters in matrix expressions with a specific operator to yield a new expression for the buck–boost converter and (b) the derived buck–boost converter.

Really, so far, we did not find out a proper way to represent the converter in matrix expression because there exist nonlinear components, switch and diode, which cannot be represented in two-port network-like matrix. This approach can be a future study topic.

Further Reading

Anderson, J. and Peng, F.Z. (2008a). A class of quasi-Z source inverters. *Proceedings of the IEEE Industrial Application Meeting*, IEEE, pp. 1–7.

Anderson, J. and Peng, F.Z. (2008b). Four quasi-Z source inverters. *Proceedings of the IEEE Power Electronics Specialists Conference*, IEEE, pp. 2743–2749.

Berkovich, Y., Axelrod, B., Tapuchi, S., and Ioinovici, A. (2007). A family of four-quadrant, PWM DC–DC converters. *Proceedings of the IEEE Power Electronics Specialists Conference*, IEEE, pp. 1878–1883.

Cao, D. and Peng, F.Z. (2009). A family of Z source and quasi-Z source DC–DC converters. *Proceedings of the IEEE Applied Power Electronics Conference*, IEEE, pp. 1097–1101.

Dorf, R.C. and Bishop, R.H. (1998). *Modern Control Systems*, 8e. Addison Wesley.

Kreyszig, E. (1993). *Advanced Engineering Mathematics*, 7e. Wiley.

Kuo, B.C. (1995). *Automatic Control Systems*, 7e. Prentice Hall.

Peng, F.Z. (2003). Z source inverter. *IEEE Trans. Ind. Appl.* 39 (2): 504–510.

Qian, W., Peng, F.Z., and Cha, H. (2011). Trans-Z source inverters. *IEEE Trans. Power Electron.* 26 (12): 3453–3463.

Williams, B.W. (2008). Basic DC-to-DC converters. *IEEE Trans. Power Electron.* 23 (1): 387–401.

Wu, T.-F. and Yu, T.-H. (1998). Unified approach to developing single-stage power converters. *IEEE Trans. Aerosp. Electron. Syst.* 34 (1): 221–223.

Wu, T.-F., Chen, Y.-K., and Liang, S.-A. (1999). A structural approach to synthesizing, analyzing and modeling quasi-resonant converters. *Proceedings of the IEEE Power Electronics Specialists Conference*, IEEE, pp. 1024–1029.

5

Synthesizing Process with Graft Scheme

For power processing applications, designers choose certain transfer codes to achieve desired input-to-output voltage or current conversion ratios. The transfer codes are first decoded into the combinations of fundamental or well-known codes, and they are organized into specific code configurations, as described in Chapter 4. According to the code configurations, the codes are synthesized with their corresponding converters, which may be processed further to simplify their circuit configurations.

This chapter first reviews cell approaches to synthesizing PWM converters, and two schemes, grafting and layering, for synthesizing power converters based on transfer codes are introduced. Examples are presented to illustrate the synthesizing process with these two schemes and the fundamentals presented in Chapter 3.

5.1 Cell Approaches

Since 1980, there have been many researchers trying to identify the similarities of the six well-known PWM converters and high step-down/step-up converters, and then to introduce component cells to explain the derivation of the converters, including P-cell and N-cell, Tee cell and Pi cell, and switched-capacitor cell and switched-inductor cell. Since the cells themselves do not conduct power conversion, they cannot exist independently. In other words, they have to be connected with other sources, outputs, switches, or *LC* filters to form converters. Using the cell approaches to synthesizing or developing PWM power converters is based essentially on observation of component connections in existing converters.

Origin of Power Converters: Decoding, Synthesizing, and Modeling, First Edition.
Tsai-Fu Wu and Yu-Kai Chen.
© 2020 John Wiley & Sons, Inc. Published 2020 by John Wiley & Sons, Inc.

5.1.1 P-Cell and N-Cell

The well-known PWM converters, such as buck, boost, buck–boost, and Ćuk, are shown in Figure 5.1. From these converters, one can observe that the active and passive switches have a common node. Thus, a switching cell concept was introduced to explain the configurations of the converters. There are two types of switching cells, P-cell and N-cell, as shown in Figure 5.2, and each has two terminals for connecting to source or output and one current buffer that is realized with an inductor to form a PWM converter. For example, the buck converter can be derived from P-cell, while the boost converter can be derived from N-cell, as shown in Figure 5.3. In the converters, the current source is realized with an inductor and a capacitor (equivalent to a voltage source or sink). This is a kind of intuitive approach with induction, but without manipulation on the converters.

In fact, with a little bit of manipulation by relocating the buffer capacitor C_1 in the sepic and Zeta converters, as shown in Figure 5.4, from the forward path to the return path, as shown in Figure 5.5, there exists a common node between the active and passive switches. Therefore, an N-cell and a P-cell can be identified,

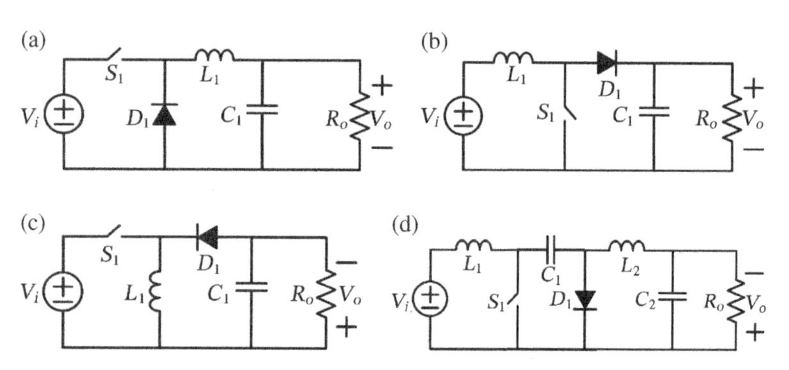

Figure 5.1 (a) Buck converter, (b) boost converter, (c) buck–boost converter, and (d) Ćuk (or boost–buck) converter.

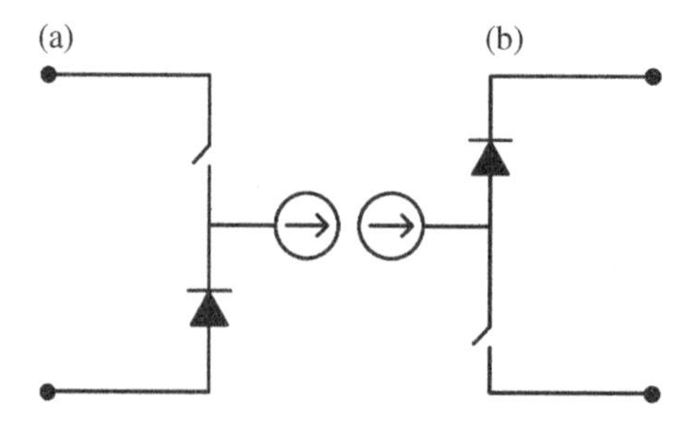

Figure 5.2 (a) P-cell and (b) N-cell.

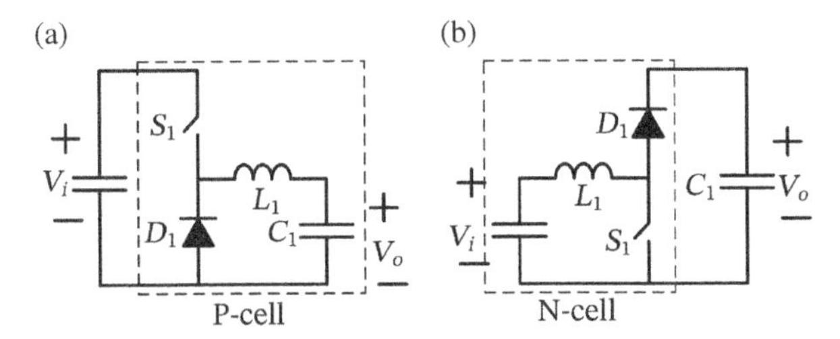

Figure 5.3 (a) Buck converter synthesized with a P-cell and (b) boost converter synthesized with an N-cell.

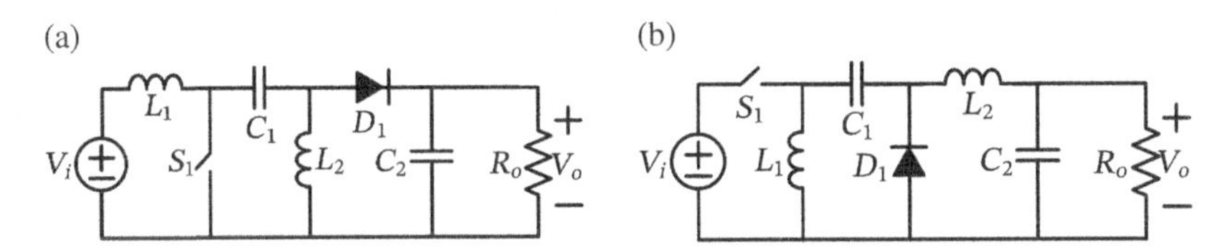

Figure 5.4 (a) Sepic converter and (b) Zeta converter.

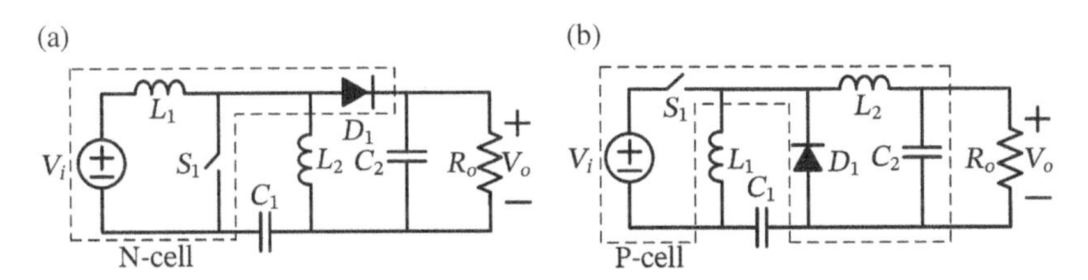

Figure 5.5 (a) Sepic converter synthesized with an N-cell and an *LC* filter and (b) Zeta converter synthesized with a P-cell and an *LC* filter.

respectively, and the converters can be derived accordingly. In the two converters shown in Figure 5.5, an extra *LC* filter network is introduced to each converter, while it does not change its input-to-output voltage transfer code, $D/(1-D)$.

Essentially, this approach is used to explain the component connections of the existing converter configurations. However, it is hard to develop new converters, because it needs to know the component connections of converters ahead.

5.1.2 Tee Canonical Cell and Pi Canonical Cell

Similarly, based on observation and induction, the converters shown in Figures 1.7 and 1.8 can be explained with two canonical switching cells, namely, Tee canonical cell and Pi canonical cell, as shown in Figure 5.6. By exhaustively enumerating all

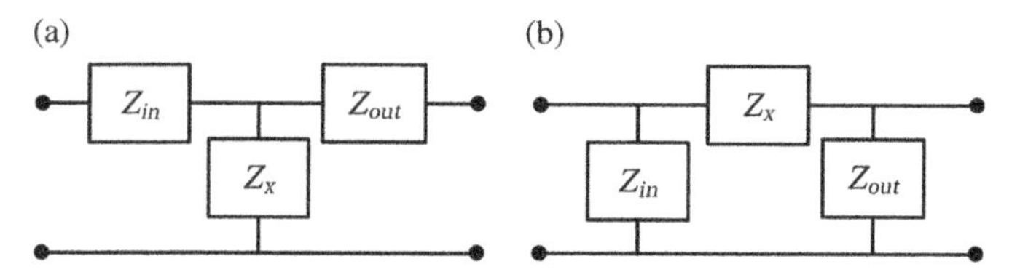

Figure 5.6 (a) Tee canonical cell and (b) *Pi* canonical cell.

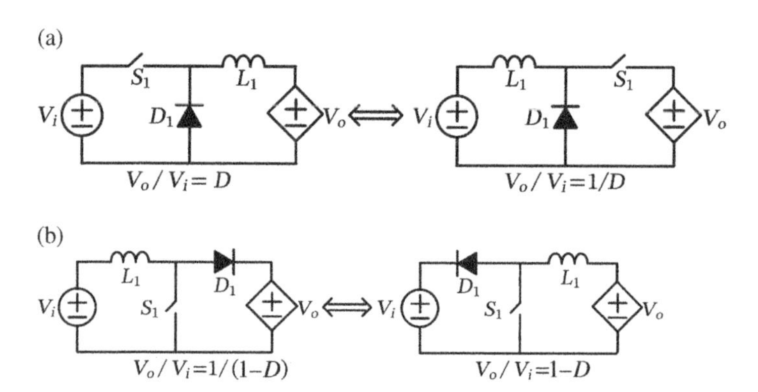

Figure 5.7 (a) Buck and inverse buck and (b) boost and inverse boost.

of possible combinations of Z_{in}, Z_{out}, and Z_x, and including *LC* network, source, and load, the converters shown in Figures 1.7 and 1.8 can be derived. Additionally, the converters with extra *LC* filters and inverse versions of the converters can be derived. Figure 5.7 shows the buck and boost converters and their inverse versions, in which the input-to-output transfer ratios shown in the bottom of the converters are corresponding to continuous conduction mode (CCM), and *D* is the duty ratio of the active switch. Moreover, the converters with the same higher step-up voltage transfer ratio but with different circuit configurations, as shown in Figure 5.8, can be developed. Although this approach can derive new PWM converters and is straightforward, it is still tedious and lacks of mechanism to explain the converters with identical transfer ratio but with different configurations. Again, it is based on cell levels.

5.1.3 Switched-Capacitor Cell and Switched-Inductor Cell

Based on cell levels, another approach to developing new converters with higher step-up and step-down voltage transfer ratios by introducing switched-capacitor or switched-inductor cells to the PWM converters shown in Figures 1.7 and 1.8 was proposed. Two typical switched-capacitor and

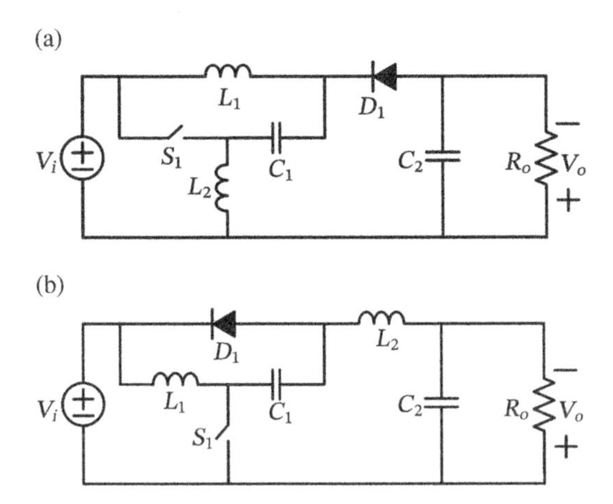

Figure 5.8 With the same input-to-output transfer ratio of $(2D-1)/(1-D)$ but with different configurations: (a) and (b).

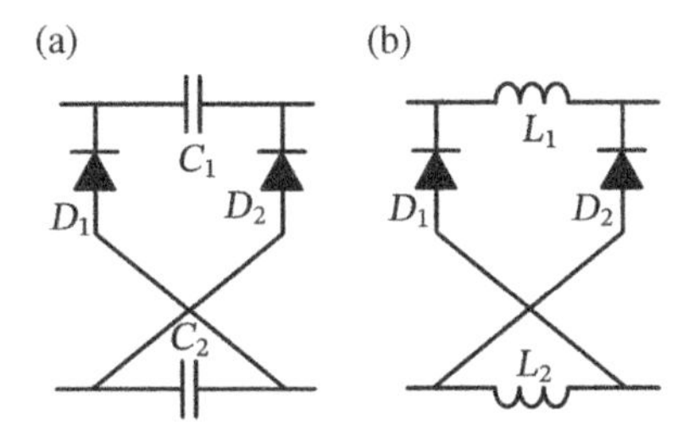

Figure 5.9 (a) Switched-capacitor and (b) switched-inductor cells.

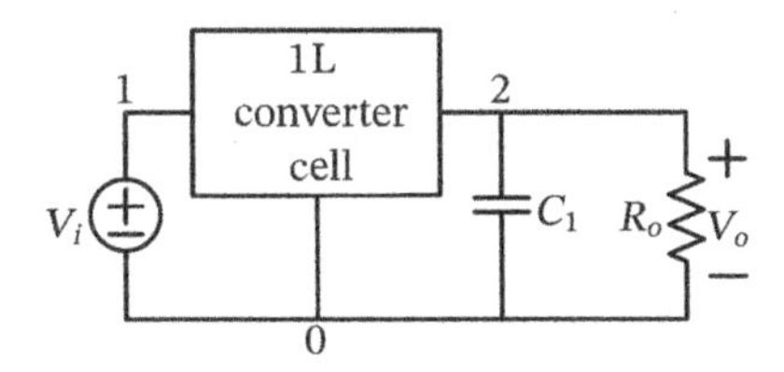

Figure 5.10 Structure of PWM converters in the derivation procedure.

switched-inductor cells are shown in Figure 5.9, and their derived converters have been shown in Figure 1.10. Applications of this approach are quite limited, and the chance of deriving new converters is depending highly on experience. Otherwise, it might need many trial and errors. When inserting a cell into the PWM converter, one has to use volt-second balance principle to verify if the converter is valid. This approach still needs a lot of ground work to derive a valid converter.

5.1.4 Inductor–Capacitor Component Cells

Another synthesis approach based on inductor–capacitor component cell concepts, one inductor (1L) and two inductors and one capacitor (2L1C), was proposed. The synthesis procedure is supported with graph theory and matrix representation, and based on a prescribed set of properties or constraints as criteria to extract a converter from all of the possible combinations, reducing the number of trial and errors. A structure of PWM DC–DC converters included in the synthesis procedure is depicted in Figure 5.10, and possible positions of inserting inductor into a second-order PWM converter is shown in Figure 5.11. Typical converter properties include number of capacitors and inductors, number of active–passive switches, DC voltage conversion ratio, continuous input and/or output current, possible coupling of inductors, *etc*. This approach seems general and has a broad vision in synthesizing converters, but it needs a lot of efforts or even trial and error in selecting a valid converter when considering many properties simultaneously. For the main purpose of deriving a converter, maybe, we have to consider the static performance, such as input-to-output voltage transfer ratio and continuous inductor current, first, and if users would like to know more about the dynamics of the converter, they can analyze them further.

In the above discussed approaches, the converters are derived or synthesized based on cells or component levels. They select a proper converter configuration and add a certain cell or components to the converter to form a new converter topology. Essentially, they exhaustively enumerate all of possible combinations and extract converters based on certain constraints or properties. Valid converters are verified with the volt-second balance principle. Applications of these approaches to developing new converters are quite limited because the chance of obtaining a valid converter is depending highly on designer experiences. Is it possible to start from the transfer codes of the existing converters and with certain manipulation to develop new converters? The answer to this question is presented in the following sections; several approaches are being discussed.

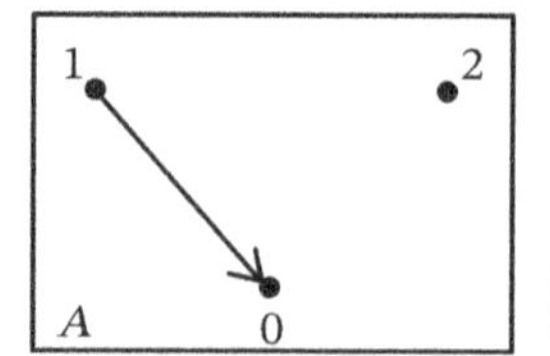

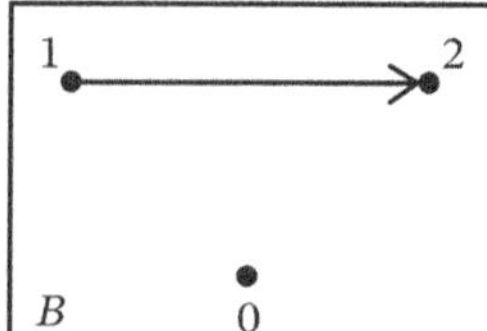

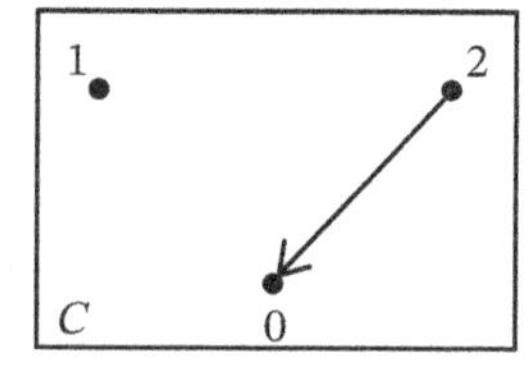

Figure 5.11 Possible positions of the inductor in a second-order PWM converter based on 1L converter cell.

5.2 Converter Grafting Scheme

Given a transfer code, the decoding process is first conducted to figure out a code configuration. Then, a converter is used to synthesize each code in the configuration. When multiple converters are connected in cascade, cascode, or parallel and have switches with common nodes, the switches can be integrated to be a single one basically through switch sharing.

5.2.1 Synchronous Switch Operation

When sharing with a single one, the multiple switches have to be operated synchronously. According to the synchronous switch operation, buck–boost and Ćuk (boost–buck) converters, as shown in Figures 5.12 and 5.13, respectively, have been evolved from buck and boost converters. The procedure was explained by Dr R. Severns and Mr G. Bloom. Figure 5.12a shows buck and boost converters connected in cascade, and Figure 5.12b shows that the two active–passive switch pairs are replaced with two single-pole double-throw (1P2T) switches in which position A represents an active switch in action and position B represents that of a passive one. In the figure, it can be proved that the *LCL* filter can be reduced to *LL* one when the two switches are operated synchronously. The two inductors can be replaced with a single one, L_{12}. When the two switches are operated synchronously, they are both simultaneously connected to either position A or B, as shown in Figure 5.12c. Then, the circuit shown in Figure 5.12c can be replaced with the one shown in Figure 5.12d in which the output voltage polarity is reversed and inductor L_{12} is connected like in parallel with the input source and the two 1P2T switches are replaced with a single one, S_{12}. Finally, the 1P2T switch is replaced with the active–passive switch pair, as shown in Figure 5.12e, to yield a buck–boost converter. Curiously, how can be the circuit shown in Figure 5.12c replaced with the one shown in Figure 5.12d? It is not so straightforward and obvious why we have to reverse the output voltage polarity and to connect the inductor in vertical position by assigning A and B in these positions. Moreover, why is diode D_{12} connected in this direction?

Similarly, evolution of boost–buck (Ćuk) converter is illustrated in Figure 5.13, in which the circuit shown in Figure 5.13b is reconfigured to the one shown in Figure 5.13c. Again, it is hard to understand why we have to reverse the output voltage polarity and connect capacitor C_1 with the switch in this position. How come diode D_{12} is connected in this direction?

How about sepic and Zeta converters? They are evolved from more than two buck and boost converters, and it would be much more complicated to derive the converters with this kind of circuit illustration. Thus, they were not shown in the book entitled *Modern DC-to-DC Switchmode Power Converter Circuits*, which is authored by Dr Severns and Mr Bloom. In recognizing the deficiency, we propose

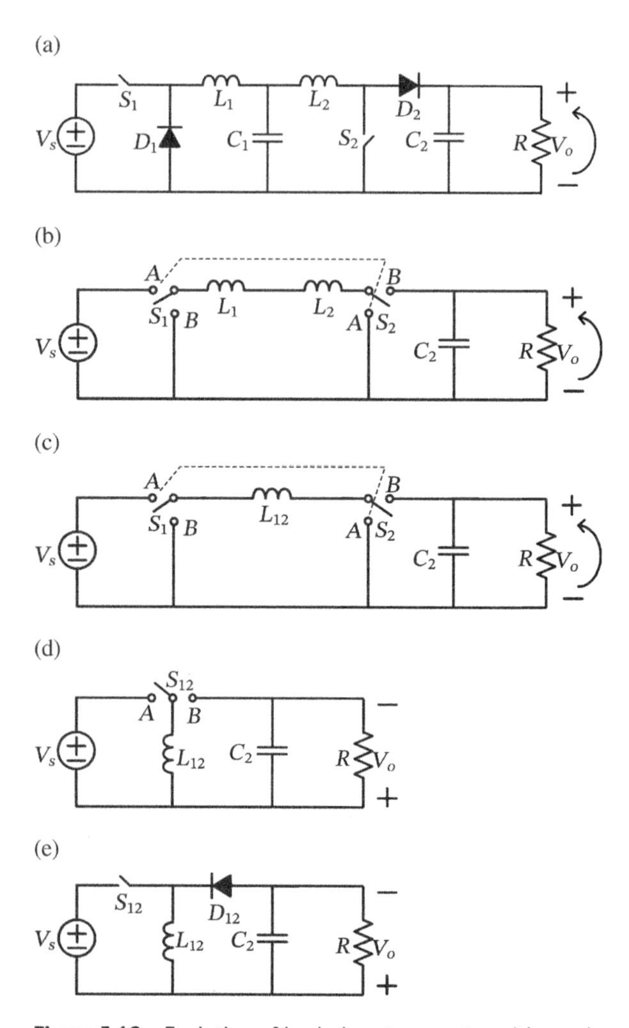

Figure 5.12 Evolution of buck–boost converter with synchronous switches.

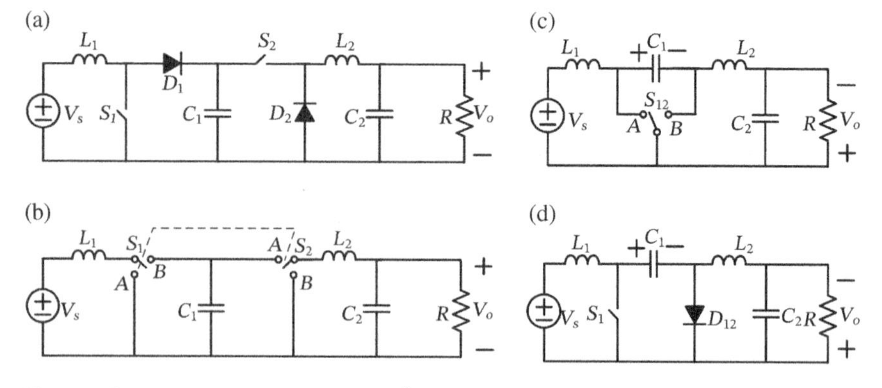

Figure 5.13 Evolution of boost–buck (Ćuk) converter with synchronous switches.

a converter grafting scheme that still follows the synchronous switch operation but starting from converter integration by sharing their switches.

Why do we use the word "Graft"? In observing the circuit evolution shown in Figures 5.12 and 5.13, we recognize that some parts of the original converters have been chopped out, comparing the circuits shown in Figure 5.12a and c, and its process is similar to that grafting an apple tree on a pear tree needs to chop the upper part or leaf part of the pear tree and chop the root part of the apple tree and then to integrate the apple tree on the pear tree. Thus, we use the word "Graft" to graft a converter on the other.

5.2.2 Grafting Active Switches

There is at least an active switch in a converter. When converters are integrated to become a single one with reduced number of active switches, their switches are operated in unison or synchronously and also integrated to share each other. Let us take two converters with two active switches and use MOSFET as an example, as shown in Figure 5.14. The MOSFET discussed here is a kind of unidirectional current flow and unidirectional voltage blocking switch. That is, the switch can only block voltage across node Drain and node Source in the *off* state, and the current can only flow from Drain to Source in the *on* state. When two MOSFETs are connected in parallel or series, they can be simply replaced with a single one with higher current or voltage rating.

However, when the two MOSFET switches are connected by sharing only one common node, they can be integrated to a single switch subject to the type of common node and operation. As shown in Figure 5.15, the common node types

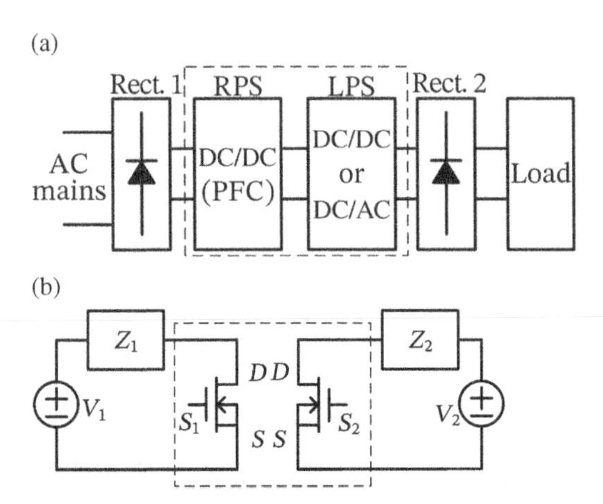

Figure 5.14 (a) Block diagram of two converters connected in series and (b) the two converters represented in Thevenin equivalent circuits and with two active switches.

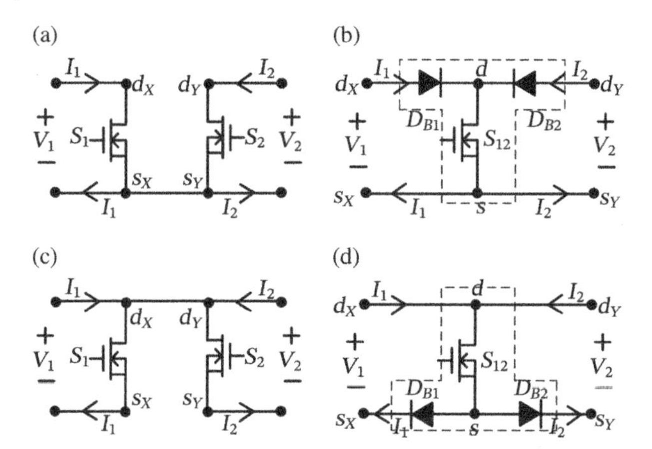

Figure 5.15 (a) Switches with common S–S, (b) TGS, (c) switches with common D–D, and (d) ITGS.

are common source–source (S–S) and common drain–drain (D–D). In the case of common S–S, as shown in Figure 5.15a, if the two switches are operated in unison, they can be replaced with a single active switch and two passive switches (diodes), as shown in Figure 5.15b, and the grafted switch is called T-type grafted switch (TGS). In the TGS, diodes D_{B1} and D_{B2} are introduced into the grafted switch to block the voltage difference between voltages V_1 and V_2 when the two active switches are in the off state. Typically, voltages V_1 and V_2 are the voltage stresses imposed on switches S_1 and S_2, which will be grafted, respectively. If the relationship between voltages V_1 and V_2 can be determined definitely during operation, the TGS can be degenerated to the switches with less number of diodes, as summarized in Table 5.1. Note that the relationship cannot change at certain switching cycles; otherwise, it will cause high current spike or stress without the blocking diode(s). If the relationship changes around during operation, the two diodes are always needed. In certain case, such as $V_1 = V_2$, there is no need of diodes, and the two active switches can be integrated into a single one, saving one switch and its driver.

Similarly, when the two active switches share a common D–D, as shown in Figure 5.15c, and are operated synchronously, they can be replaced with the inverse T-type grafted switch (ITGS), as shown in Figure 5.15d. Again, diodes D_{B1} and D_{B2} are used to block the voltage difference between voltages V_1 and V_2. The ITGS can be degenerated to the switches shown in Table 5.1 if the relationship between voltages V_1 and V_2 can be determined definitely during operation.

In converters, through component relocation while without changing converter operational principle, the switches with common S–S usually can be changed to

Table 5.1 Degeneration of TGS and ITGS based on the relationship between V_1 and V_2.

Relationship between voltages across switches in off state	Degenerated T-type grafted switch (TGS)	Degenerated inverse T-type grafted switch (ITGS)
$V_1 > V_2$	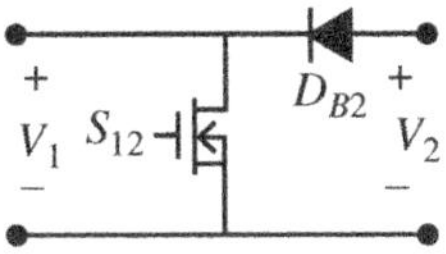 $+$ V_1 $-$ S_{12} D_{B2} $+$ V_2 $-$	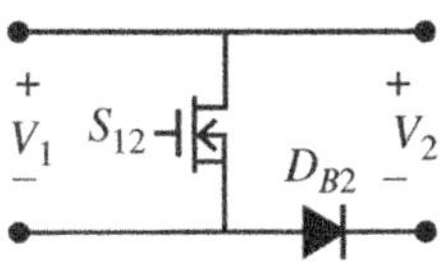 $+$ V_1 $-$ S_{12} D_{B2} $+$ V_2 $-$
$V_1 < V_2$	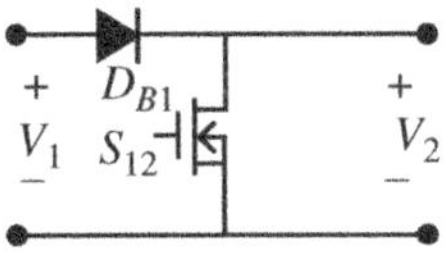 $+$ V_1 $-$ D_{B1} S_{12} $+$ V_2 $-$	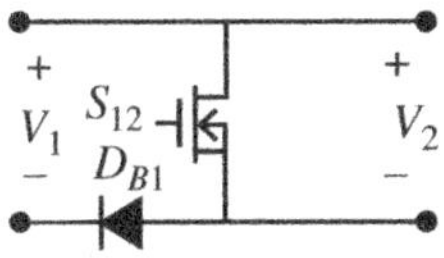 $+$ V_1 $-$ S_{12} D_{B1} $+$ V_2 $-$
$V_1 = V_2$	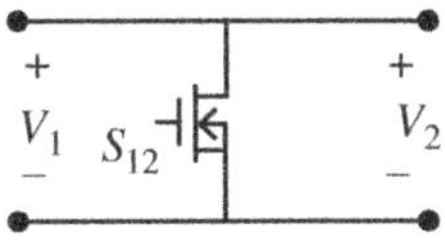 $+$ V_1 $-$ S_{12} $+$ V_2 $-$	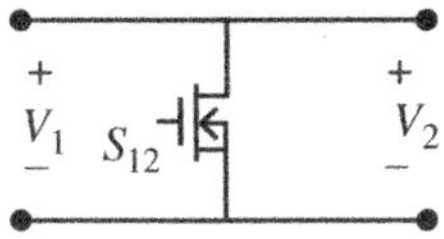 $+$ V_1 $-$ S_{12} $+$ V_2 $-$

common D–D, and TGS can be replaced with ITGS during grafting process. Therefore, both TGS and ITGS are called TGS.

If the two active switches share a common D–S, as shown in Figure 5.16a, and are operated in unison, they can be replaced with a Π-type grafted switch (ΠGS), as shown in Figure 5.16b. Here, I_1 and I_2 are the currents through switches S_1 and S_2, which will be grafted, respectively. Since the current can only flow from node Drain to node Source in the aforementioned MOSFET switch, diodes D_{F1} and D_{F2} are used to circulate the current difference between currents I_1 and I_2. In the circuit shown in Figure 5.16b, if I_1 is greater than I_2, current I_1 flows from node Drain to node Source, and the current difference $(I_1 - I_2)$ will flow through diode D_{F2}. There is no current flowing through diode D_{F1}, and the diode can be removed from the ΠGS. Thus, if the relationship between currents I_1 and I_2 can be determined definitely during operation, the ΠGS can be degenerated to the switches shown in Table 5.2. Otherwise, these two diodes need to be reserved in the ΠGS. Note that the relationship should be kept identical for all time. It cannot be changed for some switching cycles; otherwise, the current gets no path to flow and induce high voltage stress or spike.

In practice, there exists an antiparalleled body diode in a MOSFET with vertical diffusion. The MOSFET will become a bidirectional switch. However, we are

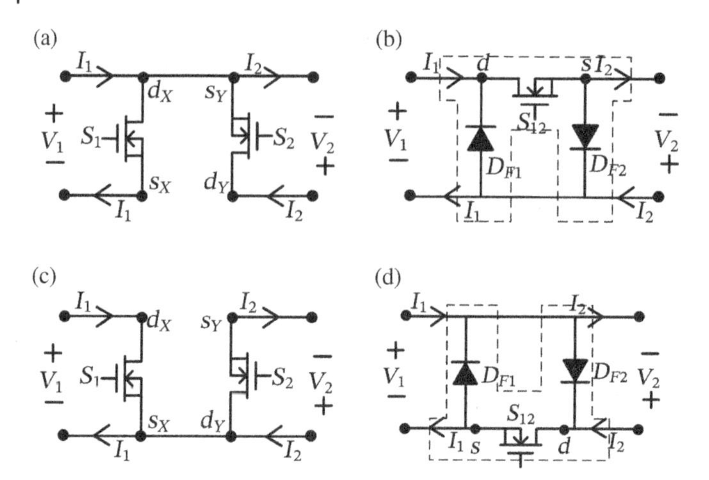

Figure 5.16 (a) Switches with common D–S, (b) ΠGS, (c) switches with common S–D, and (d) IΠGS.

Table 5.2 Degeneration of ΠGS and IΠGS based on the relationship between I_1 and I_2.

Relationship between currents through switches in on state	Degenerated Π-type grafted switch (ΠGS)	Degenerated inverse Π-type grafted switch (IΠGS)
$I_1 > I_2$	I_1, I_2, S_{12}, D_{F2}, I_1, I_2	I_1, I_2, D_{F2}, S_{12}, I_1, I_2
$I_1 < I_2$	I_1, I_2, S_{12}, D_{F1}, I_1, I_2	I_1, I_2, D_{F1}, S_{12}, I_1, I_2
$I_1 = I_2$	I_1, I_2, I_1, I_2	I_1, I_2, I_1, I_2

Table 5.3 Duality between TGS and ΠGS.

TGS and ITGS	ΠGS and IΠGS
1) D_{B1}, D_{B2}, and S_{12} share a node	1) D_{F1}, D_{F2}, and S_{12} form a loop
2) D_{Bi} blocks the voltage difference between V_1 and V_2 when both switches S_1 and S_2 are in the off states	2) D_{Fi} circulates the current difference between I_1 and I_2 when both switches S_1 and S_2 are in the on states
3) Required to determine the V_1 and V_2 when S_1 and S_2 are in the off states	3) Required to determine the I_1 and I_2 when S_1 and S_2 are in the on states
4) The grafted switch conducts current (I_1+I_2) in the *on* state	4) The grafted switch blocks voltage (V_1+V_2) in the *off* state

discussing unidirectional switches in this study first. In the following discussion, the switches will be limited to be unidirectional.

If the two switches share a common *S–D*, as shown in Figure 5.16c, and are operated in unison, they can be replaced with the inverse Π-type grafted switch (IΠGS), as shown in Figure 5.16d. Again, the IΠGS can be degenerated to the switches shown in Table 5.2 if the relationship between currents I_1 and I_2 can be determined definitely during operation.

Again, through component relocation while without changing converter operational principle, the switches with common *D–S* usually can be changed to common *S–D*, and ΠGS can be replaced with IΠGS during grafting process. Therefore, both ΠGS and IΠGS are called Π-type grafted switches.

TGS and ΠGS are dual. The diodes in a TGS block the voltage difference between V_1 and V_2, while they serve for circulating the current difference between I_1 and I_2 in a ΠGS. For a TGS, we have to determine voltages V_1 and V_2 during switch *off* state, while for a ΠGS, we have to determine currents I_1 and I_2 during switch *on* state. Their properties are summarized in Table 5.3.

In quasi-resonant and resonant converters, switches usually conduct bidirectional currents. If a MOSFET including its body diode D_b is used as a bidirectional switch, as shown in Figure 5.17, the TGS and ΠGS will be changed to the configurations shown in Figure 5.18. A bidirectional switch can conduct current either from node Drain to node Source or vice versa during *on* state, but it still can only block the voltage across Drain and Source during *off* state. In TGS, two external diodes, D_{f1} and D_{f2}, are needed to conduct reverse currents I_{r1} and I_{r2}, as

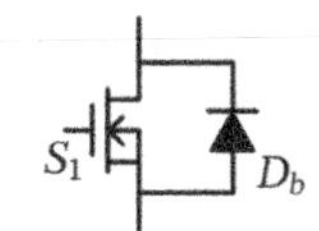

Figure 5.17 A MOSFET with its body diode to function as a bidirectional switch.

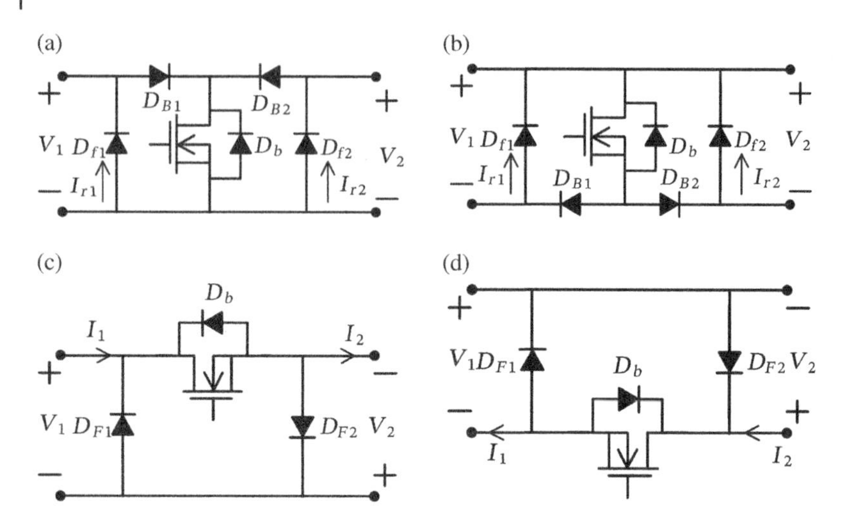

Figure 5.18 Bidirectional grafted switches: (a) TGS, (b) ITGS, (c) ΠGS, and (d) I-ΠGS.

shown in Figure 5.18a and b. However, in ΠGS, there is no need to add any extra diode, as shown in Figure 5.18c and d. If currents I_1 and I_2 are reverse, diodes D_{F1} and D_{F2} still can circulate their current difference. It should be noted that the body diode of the MOSFET usually has longer reverse recovery time than a discrete diode under the same fabrication process.

For TGS, if voltages $V_1 = V_2$, diodes D_{B1} and D_{B2} are no longer needed, and body diode D_b will act the roles of D_{f1} and D_{f2}. Thus, diodes D_{f1} and D_{f2} are no longer needed, either, and the final version becomes a bidirectional MOSFET switch only.

With the derivation processes of TGS and ΠGS, converters can be integrated to be a single one straightforwardly through the graft of their active switches. Illustration examples will be presented in Section 5.3. How about if grafting the diodes (passive switches) of converters, can we also derive a single-stage converter from multiple converters that share a common node and are operated in unison? In the following subsection, grafting diodes will be presented.

5.2.3 Grafting Passive Switches

The prerequisites of grafting switches are that switches (i) share at least a common node and (ii) operate synchronously. Active switches sharing a common node can be controlled to operate in unison, and they can be grafted. However, diodes are passive switches, of which their conduction depends on forward voltage, and they cannot be controlled directly. Diodes can be grafted only when active switches are operated in unison, and then the diodes also conduct synchronously. With this recognition, diodes can be combined to become grafted switches, and under certain conditions, they can be degenerated to reduce the number of diodes in the converters.

As shown in Figure 5.19a, two diodes are sharing a common node N, and they can be integrated to be a grafted diode, as shown in Figure 5.19b, in which D_{B1} and D_{B2} are used to block the voltage difference between voltages V_X and V_Y and diode D_{12} is the integration of diodes D_1 and D_2. If V_X is always greater than V_Y, D_{B1} is in forward bias, and it can conduct all the time; thus, diode D_{B1} is no longer needed and can be removed from the grafted diode, as shown in Figure 5.19c. Similarly, if V_X is less than V_Y, the degenerated grafted diode is shown in Figure 5.19d. If V_X is always equal to V_Y, the two blocking diodes are no longer needed, and the grafted diode becomes the one shown in Figure 5.19e.

If two diodes share a common node P, as shown in Figure 5.20a, they can be integrated to become a grafted diode, as shown in Figure 5.20b. Following the previous discussions can yield the degenerated grafted diodes, as shown in Figure 5.20c–e. It can be observed that the original two diodes share a common node and conduct in unison, and their degenerated grafted diodes may be still consisting of two diodes. What is a big deal? In certain converter configurations, the degenerated grafted diode can effectively use the body diode of a MOSFET switch, saving component count. Illustration of using grafted diodes will be presented in Section 5.3.

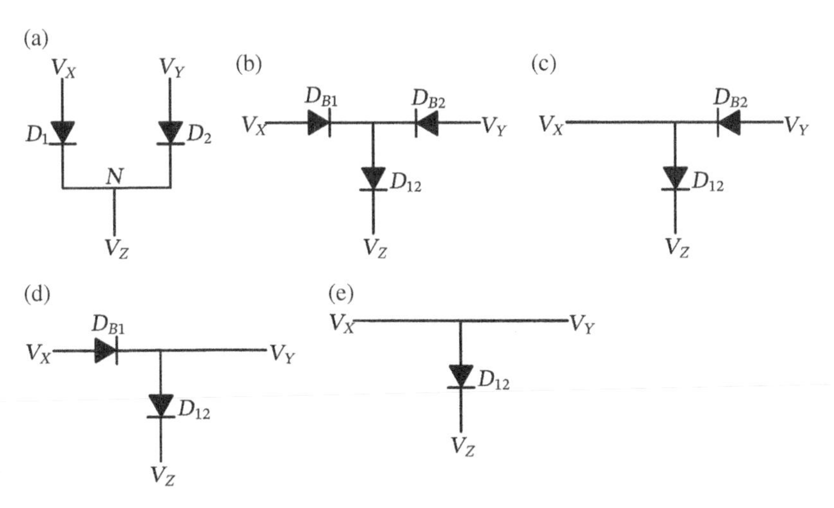

Figure 5.19 (a) Two diodes sharing a common node N, (b) grafted diode, and degenerated grafted diodes under (c) $V_X > V_Y$, (d) $V_X < V_Y$, and (e) $V_X = V_Y$.

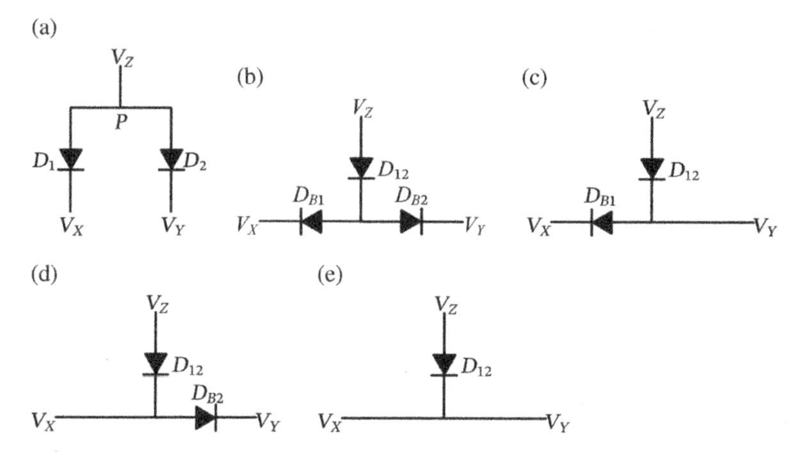

Figure 5.20 (a) Two diodes sharing a common node P, (b) grafted diode, and degenerated grafted diodes under (c) $V_X > V_Y$, (d) $V_X < V_Y$, and (e) $V_X = V_Y$.

5.3 Illustration of Grafting Converters

In the previous section, we discussed two types of switch grafting. They can be applied to graft converters. In the following, several illustration examples are presented, including the well-known PWM converters, buck–boost, boost–buck (Ćuk), sepic and Zeta converters, multiple PWM converters, dither boost + half-bridge, charger/discharger + half-bridge, half-bridge + boost, boost + class E, boost + buck–boost, boost + boost, and buck + buck converters. Essentially, the converters can be operated either in CCM or in discontinuous conduction mode (DCM), and converters operated in unison and sharing a common node can be grafted to become a single-stage converter. Here, only are the converters operated in CCM discussed first and the ones with DCM operation will be addressed later on.

5.3.1 Grafting the Well-Known PWM Converters

The well-known PWM converters include buck, boost, buck–boost, boost–buck (Ćuk), sepic, and Zeta converters. They can be grafted to yield multiple-function converters, such as step-down and step-up functions in one converter. For instance, the transfer code of $D/(1-D)$, which can achieve step-down and step-up conversion, can be decoded into the product of D and $1/(1-D)$, which can be synthesized with the buck converter and the boost converter, respectively. One of the synthesis techniques is grafting. When grafting one on the other, we can yield buck–boost or boost–buck converter, which can fulfill step-down and step-up multiple functions. Grafting buck on buck–boost and boost on boost–buck will yield the other two converters, Zeta and sepic, respectively.

5.3.1.1 Graft Boost on Buck

Figure 5.21a shows the cascade code configuration of D and $1/(1-D)$, which can be synthesized by grafting a boost converter on a buck converter, as shown in Figure 5.21b. As mentioned before, the LCL filter can be simplified to a single inductor L_{12} when the two converters are operated in unison, as shown in Figure 5.21c. When relocating switch S_1 from forward path to return path without changing its operational principle, one can identify that the two switches are with common D–S, and they can be replaced with a ΠGS, as shown in Figure 5.21d. Note that when replacing the two switches with a ΠGS, one can mark the common (C), drain (X)

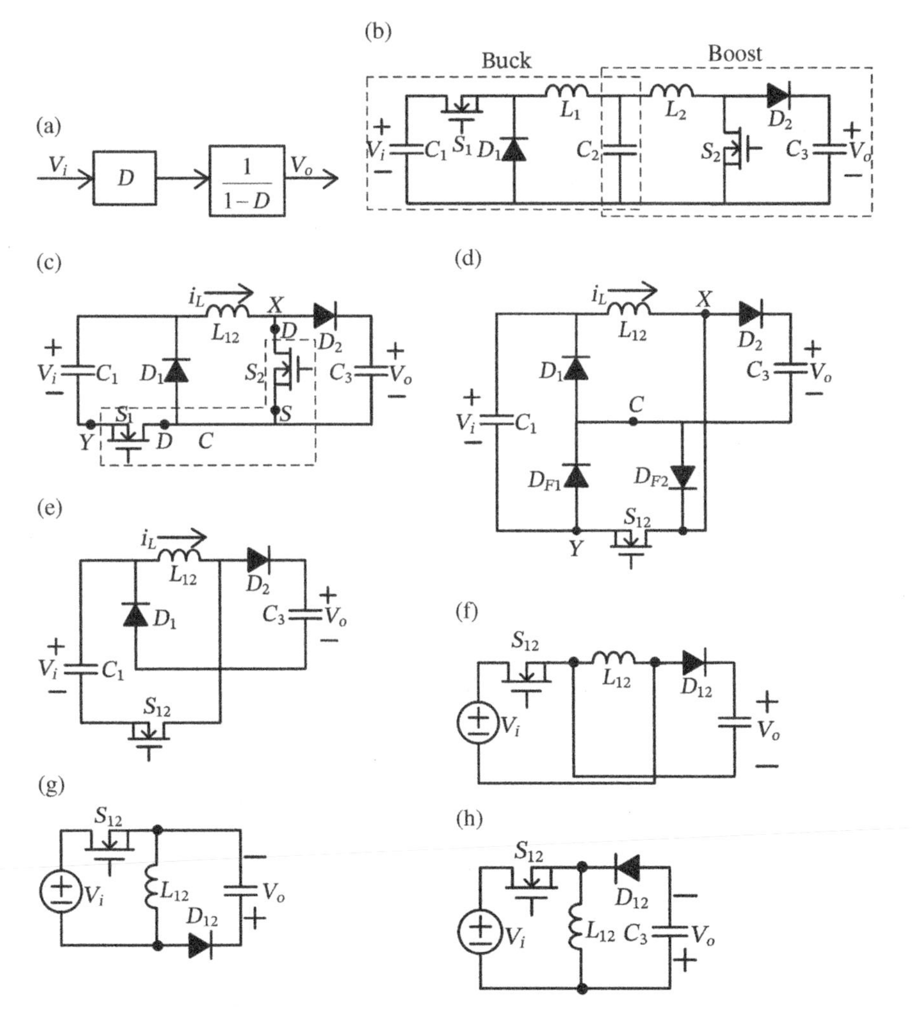

Figure 5.21 Processes of grafting a boost converter on a buck converter to yield the well-known buck–boost converter.

and source (Y) nodes of the two switches first and then replace them node by node with the three-terminal ΠGS. This can facilitate the grafting process.

Next, we have to determine the relationship between currents I_1 and I_2. It can be observed from Figure 5.21c that when the two active switches are turned on synchronously, current i_L will flow through switches S_2 and S_1; thus, $I_1 = I_2$. Under this condition, diodes D_{F1} and D_{F2} are no longer needed, and the circuit can be degenerated to the one shown in Figure 5.21e. In the circuit, diode D_1 becomes in series with diode D_2, and they can be replaced with a single one D_{12} with higher voltage rating, as shown in Figure 5.21f. When redrawing the circuit shown in Figure 5.21f will yield the one shown in Figure 5.21g, in which note that the output voltage polarity is inverted naturally and it is different from the input voltage. And the inductor is connected between the forward and return paths naturally, too. When moving diode D_{12} from return path to forward path without changing operational principle, one can obtain the conventionally well-known buck–boost converter, as shown in Figure 5.21h. Its input-to-output voltage transfer code is $D/(1-D)$, which is equivalent to D times $1/(1-D)$, as shown in Figure 5.21a.

Different from Dr R. Severns' approach, the grafting scheme can derive the buck–boost converter with a straightforward process but without dilemma. Its output voltage polarity is opposite to the input naturally without need of extra explanation. From the converters in cascaded connection till the final configuration of a buck–boost converter, it goes smoothly and naturally.

5.3.1.2 Graft Buck on Boost

Applying the commutative law of multiplication to the code configuration shown in Figure 5.21a yields

$$\frac{D}{1-D} = \left(\frac{1}{1-D}\right)D, \tag{5.1}$$

which can be synthesized with a boost converter and a buck converter to yield the boost–buck (Ćuk) converter. Since it is the dual of a buck–boost converter, to illustrate the synthesis of the boost–buck converter step by step, its input-to-output transfer code is represented in current form, as shown in Figure 5.22a, in which $1-D$ and $1/D$ are the input-to-output current transfer codes of boost and buck converters, respectively. By power conservation, their input-to-output voltage transfer codes will be $1/(1-D)$ and D. Thus, the input-to-output voltage transfer code of a boost–buck converter is supposed to be $D/(1-D)$, which is the same as that of a buck–boost converter. Even though they are dual, their voltage transfer codes are identical.

Based on the cascaded code connection shown in Figure 5.22a, we can draw the converters in series as shown in Figure 5.22b. Relocating switch S_2 to the return

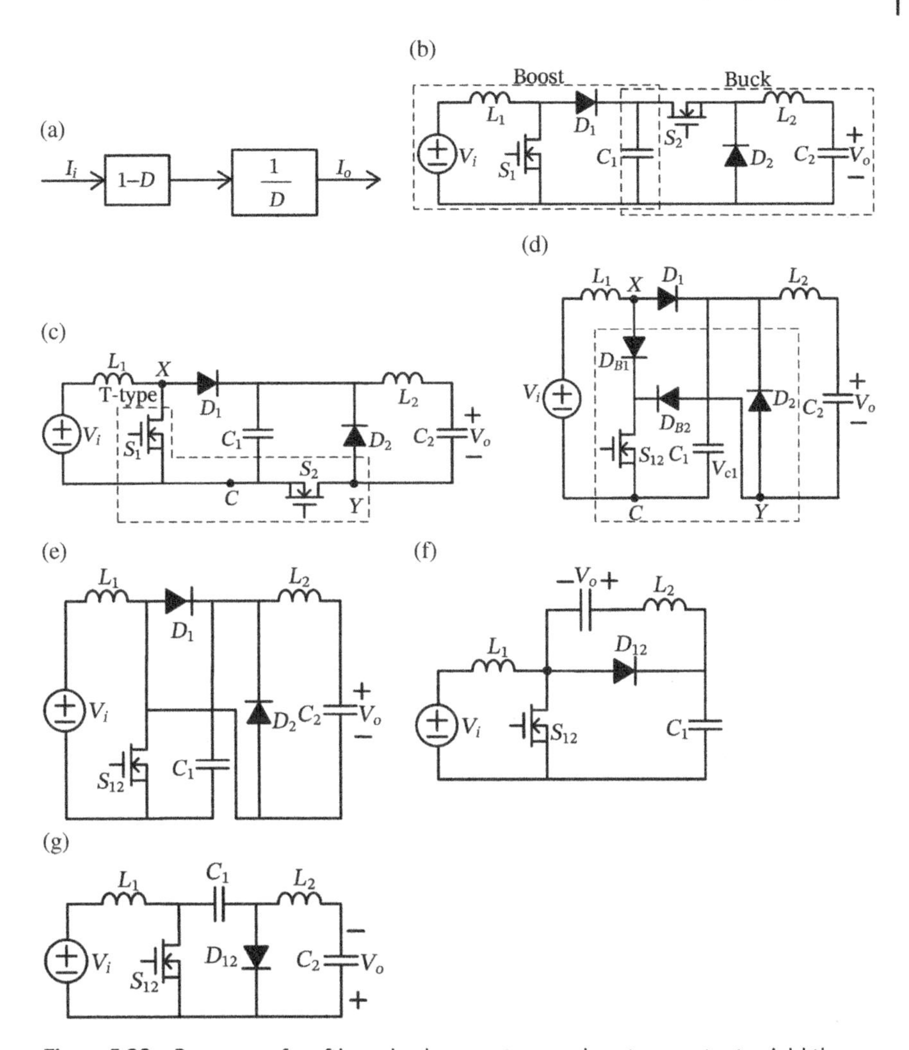

Figure 5.22 Processes of grafting a buck converter on a boost converter to yield the boost–buck (Ćuk) converter.

path can identify a common S–S node, as shown in Figure 5.22c, and then, the two active switches can be replaced with a TGS, as shown in Figure 5.22d.

When both switches S_1 and S_2 are turned off, diodes D_1 and D_2 will conduct simultaneously; the voltage stresses V_1 and V_2 imposed on switches S_1 and S_2, respectively, will be V_{c1}. That is, $V_1 = V_2$. Under this condition, voltage blocking diodes D_{B1} and D_{B2} are no longer needed, and they can be shortened, yielding the circuit shown in Figure 5.22e. Then, diodes D_1 and D_2 become in parallel, and they can be replaced with a single one, D_{12}, with higher current rating, as shown in

Figure 5.22f. Redrawing the circuit and move capacitor C_1 from the return path to the forward path will yield the conventionally well-known boost–buck (Ćuk) converter, as shown in Figure 5.22g.

Again, it can be observed that the output voltage polarity of the boost–buck converter shown in Figure 5.22g is inverted naturally without a need of any extra explanation. And capacitor C_1 is naturally connected in series with the input portion and the output one. Moreover, the original two diodes can be degenerated into a single one. This part has not been explained in Dr R. Severns' book.

There are two more well-known PWM converters, sepic and Zeta, with the same input-to-output voltage transfer code $D/(1-D)$. Their synthesizing process based on grafting scheme will be presented next.

5.3.1.3 Graft Buck on Buck–Boost

Based on the commutative law of multiplication, the voltage transfer codes of D and $1/(1-D)$ can have two straightforward combinations, $D \times 1/(1-D)$ and $1/(1-D)\times D$, to yield $D/(1-D)$. They have been synthesized with buck–boost and boost–buck converters. To synthesize new code configurations to yield Zeta and sepic converters, we have to first decode $D/(1-D)$ into new code configurations. Typically, we follow the decoding process described in Chapter 4. The code $D/(1-D)$ has been a well-known code, and it is not supposed to decode further. However, for deriving new converters, it is decoded as follows:

$$\frac{D}{1-D} = \frac{D\times 1}{1-D}. \tag{5.2}$$

And through long division, $1/(1-D)$ in (5.2) is decoded further as follows:

$$\frac{1}{1-D} = 1 + \frac{D}{1-D}. \tag{5.3}$$

In (5.3), "1" is the quotient and "$D/(1-D)$" is the residue. Thus, the input-to-output voltage transfer code $D/(1-D)$ can be decoded into the code configuration shown in Figure 5.23a, in which $1/(1-D)$ is decoded into $1+D/(1-D)$ and then times D to obtain a code $D/(1-D)$, which can be synthesized with the derived buck–boost converter cascoded with a buck converter. The overall circuit configuration to synthesize code $D/(1-D)$ is shown in Figure 5.23b. As indicated in the figure, the input voltage to the buck converter is $V_i + V_{c1}$ and $V_{c1} = V_i(D/(1-D))$, which is equivalent to $V_i(1+D/(1-D))$.

Next, by following the synthesizing process with a grafted switch, we can derive the Zeta converter. The circuit shown in Figure 5.23b has the two active switches sharing with a common D–D node, and they can be replaced with a TGS, as shown in Figure 5.23c. Investigating the voltages imposed on switches S_1 and S_2 yields that they are all with voltage $V_i + V_{c1}$, and thus, $V_1 = V_2$. Under

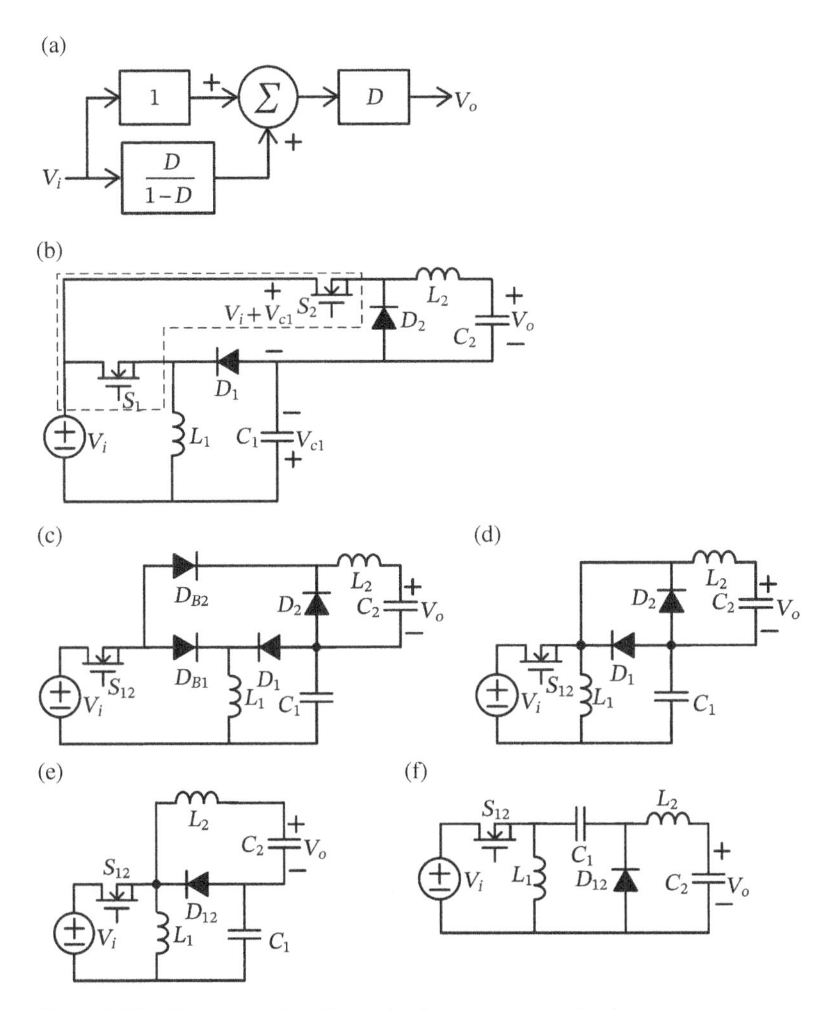

Figure 5.23 Processes of grafting a buck converter on a buck–boost converter to yield the Zeta (buck–boost–buck) converter.

this condition, the blocking diodes D_{B1} and D_{B2} can be shorted or removed from the circuit. The circuit becomes the one shown in Figure 5.23d, from which it can be seen that diodes D_1 and D_2 are connected in parallel and they can be replaced with a single one with higher current rating, as shown in Figure 5.23e. Redrawing the circuit to become the well-known configuration will be the Zeta converter, as shown in Figure 5.23f. From the synthesizing process, it can be recognized that the Zeta converter consists of a unity feedforward, a buck–boost converter and a buck converter, and it can be called a buck–boost–buck converter.

Note that both buck–boost and boost–buck converters have their output voltage polarities being opposite to their inputs. However, with the new code configuration, we have obtained a new converter with the same code as that of the buck–boost or boost–buck converter but with its output voltage polarity being identical to its input. In applications, this feature provides designers a simple way to feedback output voltage signal.

5.3.1.4 Graft Boost on Boost–Buck

According to duality, the voltage transfer code shown in Figure 5.23a can be converted to current transfer code, as shown in Figure 5.24a, in which $(1-D)/D$ and $(1-D)$ are the current transfer codes of boost–buck and boost converters, respectively. A circuit to synthesize this code configuration is shown in Figure 5.24b. Note that the boost–buck (Ćuk) converter is modified to this configuration to ensure that both input current I_i and output current I_{L2} can flow through switches, S_1 and D_1, and this arrangement can really synthesize the code configuration shown in Figure 5.24a. Additionally, since there is no output load in the boost–buck converter, the capacitor in series with inductor L_2 is shorted or removed. Moreover, the inductor in the boost converter is removed since the current through switches S_1 and D_1 also flows through switch S_2.

In the circuit shown in Figure 5.24b, the two active switches share a common S–D node, and they can be replaced with a ΠGS, as shown in Figure 5.24c. It can be recognized that both input current I_i and output current I_{L2} will flow through switches S_1 and S_2 when they are turned on. That is, $I_1 = I_2$, and the two current circulating diodes D_{F1} and D_{F2} can be removed from the circuit. Then, it can be observed that diodes D_1 and D_2 are in series, as shown in Figure 5.24d. Redrawing the circuit can yield the well-known sepic converter, as shown in Figure 5.24e.

Similarly, the converter synthesized from the dual code configuration has its output voltage polarity being identical to its input. This property is the same as that of Zeta converter. Since the sepic converter is derived by grafting a boost converter on a boost–buck converter, it is also called boost–buck–boost converter.

One interesting question is that if figuring out a code configuration, can we synthesize it into a corresponding converter? The answer to this question might not be proved rigorously. However, an example to illustrate this concept is shown as follows. Grafting a Zeta converter on a complementary buck converter is shown in Figure 5.25.

Figure 5.25a shows an input-to-output voltage transfer code expression of the buck converter as follows:

$$\frac{V_o}{V_i} = \frac{(1-D)\times D}{1-D} = D \tag{5.4}$$

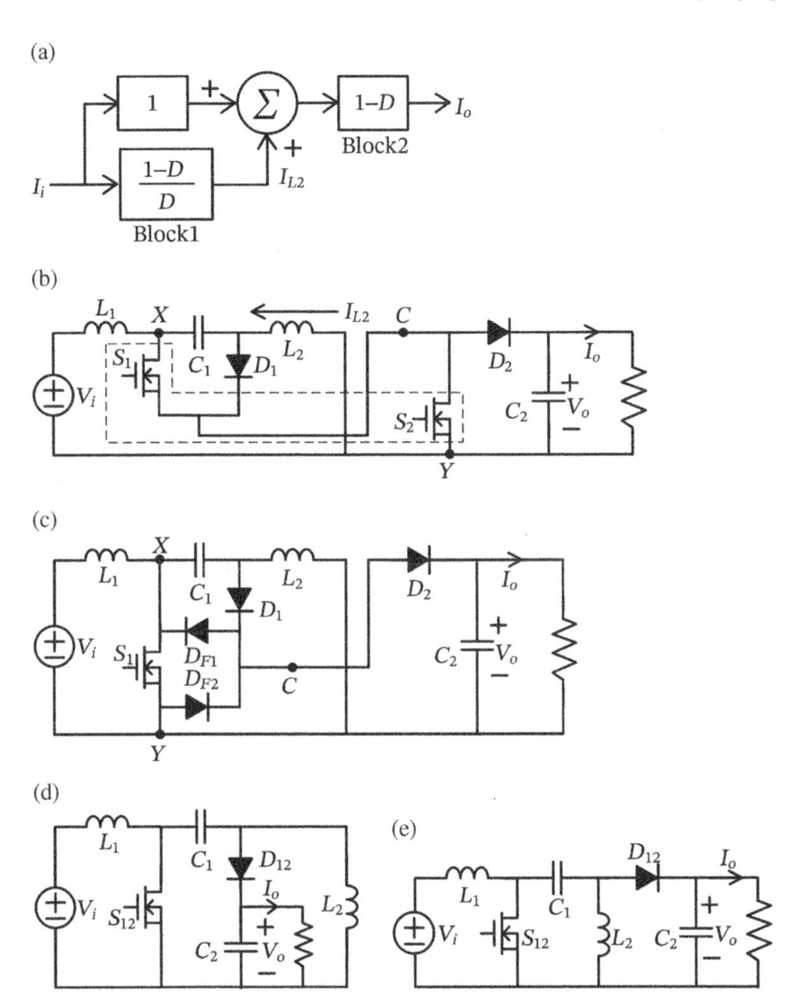

Figure 5.24 Processes of grafting a boost converter on a boost–buck converter to yield the sepic (boost–buck–boost) converter.

in which $(1 - D)$ is the voltage transfer code of the complementary buck converter and $D/(1 - D)$ is that of the Zeta converter. Since the final code is D, we are supposed to yield a buck converter. Following the synthesizing process, first, can obtain a cascade connection of complementary buck converter and Zeta converter, as shown in Figure 5.25b. From the connection of the two active switches S_1 and S_2, we observe that they are sharing with a common D–D node and they can be replaced with a TGS, as shown in Figure 5.25c. Investigating the voltage stresses imposed on switches S_1 and S_2 knows that they are both with voltage V_i; that is, $V_1 = V_2$. Thus, voltage blocking diodes D_{B1} and D_{B2} can be shorted and removed

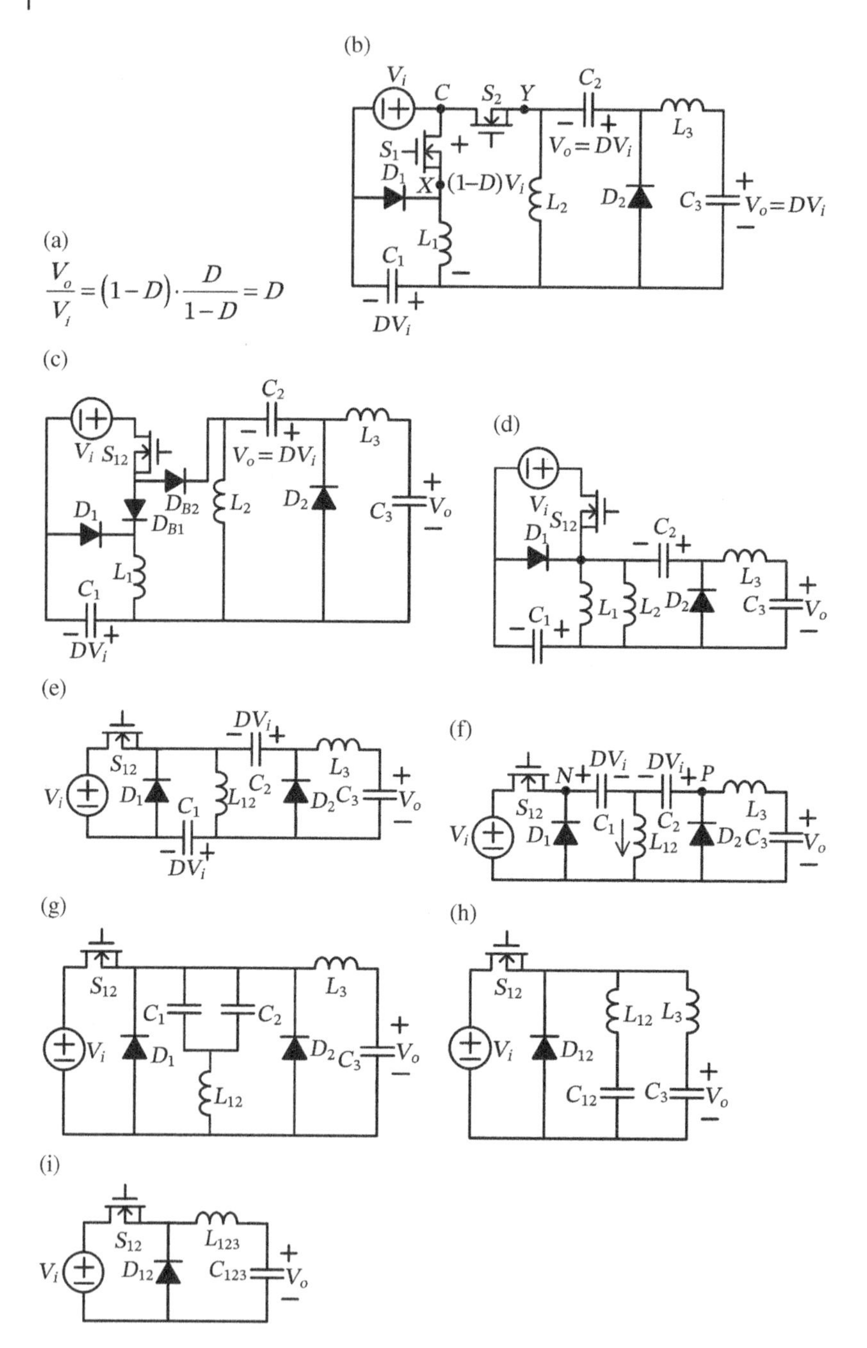

Figure 5.25 Processes of grafting a Zeta converter on a complementary buck converter to yield the buck converter.

from the circuit shown in Figure 5.25c, and it becomes the one shown in Figure 5.25d. Moving input voltage V_i and switch S_1 to the left-hand side without changing its operational principle and replacing the paralleled inductors L_1 and L_2 with a single one, L_{12}, can obtain the circuit shown in Figure 5.25e. Then, moving capacitor C_1 from the return path to the forward path yields the circuit shown in Figure 5.25f. It can be proved that both voltages of capacitors C_1 and C_2 are equal to DV_i and have the assigned polarities shown in the circuit and nodes N and P have the same potential; therefore, they can be tied together, as shown in Figure 5.25g. Then, capacitors C_1 and C_2 are in parallel, so do diodes D_1 and D_2, and they can be replaced with single components C_{12} and D_{12}, respectively. It can be recognized that two sets of LC components are connected in parallel, as shown in Figure 5.25h, and then they can be reduced to a single one from a topological point of view, as shown in Figure 5.25i, which is a buck converter topology.

From the above synthesizing process, we can really derive a buck converter from a special code configuration that yields the transfer code D. This implies that the decoding and synthesizing approaches can work well. How about grafting well-known PWM converters on other PWM converters, such as half-bridge and full-bridge resonant converters? The answers will be illustrated by some synthesizing examples that will be presented later on.

5.3.1.5 Buck in Parallel with Buck–Boost

A PWM converter without transformer can only have one output, and it is usually to parallel multiple converters to achieve multiple outputs. Figure 5.26a shows a buck converter in parallel with a buck–boost converter to yield two outputs. It can be recognized that switch S_1 and switch S_2 have a common D–D node, and they can be replaced with a TGS. Investigating the voltages imposed on S_1 and S_2 understands that switch–voltage stresses $V_1 < V_2$ and blocking diode D_{B1} needs to be kept in the circuit to block the voltage difference between V_1 and V_2 during switch off state. The final circuit becomes the one shown in Figure 5.26b, which has been proposed by Dr R. Severns. However, there was no explanation about how to derive the circuit. Basically, it was derived based on observation and experience.

5.3.1.6 Grafting Buck on Buck to Achieve High Step-Down Voltage Conversion

In voltage conversion applications, if needing a high step-down, we may introduce a high step-down transformer to an isolated converter or use a two-stage step-down converter in a non-isolated converter. Figure 5.27a shows two codes in cascade connection to achieve a high step-down, D^2, voltage conversion. The two buck converters used to realize this conversion is shown in Figure 5.27b. If moving inductor L_1 from the forward path to the return path, one can identify a common S–D node for

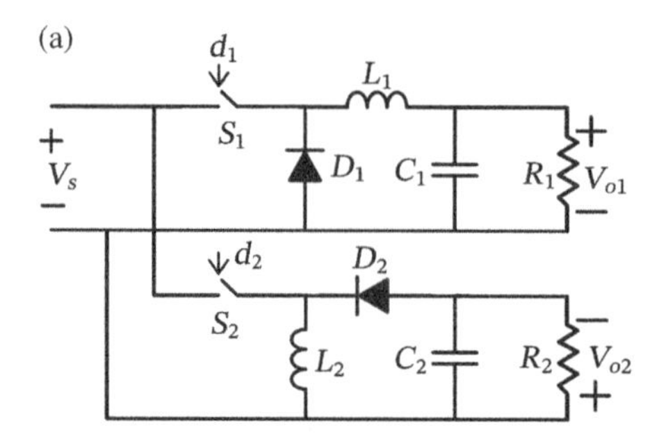

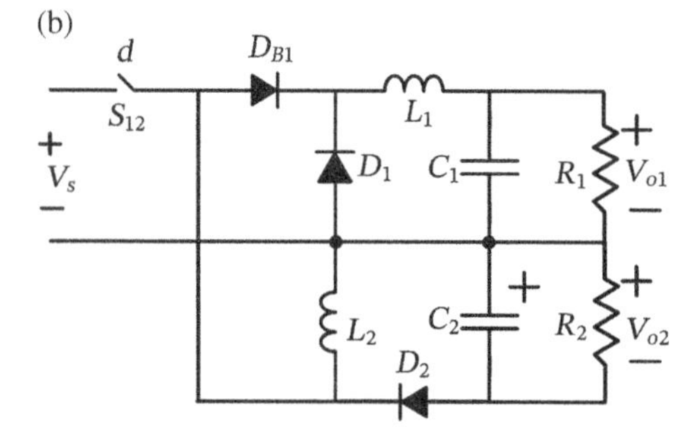

Figure 5.26 Processes of synthesizing a buck converter and a buck–boost converter to yield a compound converter with multiple outputs: (a) With common D–D and (b) $V_2 > V_1 \rightarrow D_{B2}$ is removed.

switches S_1 and S_2, as shown in Figure 5.27c, and they can be replaced with a ΠGS, as shown in Figure 5.27d. Reviewing the currents through switches S_1 and S_2 understands that currents $I_2 > I_1$ because of step-down voltage conversion. Thus, current circulating diode D_{F2} can be removed from the circuit shown in Figure 5.27d, and the final circuit becomes the one shown in Figure 5.27e. The converter has only one active switch but three diodes. It is a PWM converter, but not a regular one.

5.3.1.7 Grafting Boost on Boost to Achieve High Step-up Voltage Conversion

In voltage conversion applications, if needing a high step-up, we may introduce a high step-up transformer to an isolated converter or use a two-stage step-up converter in a non-isolated converter. Figure 5.28a shows two codes in cascade connection to achieve a high step-up, $1/(1-D)^2$, voltage conversion. The two boost converters operated in unison can realize this conversion ratio, and they are shown in Figure 5.28b. It can be recognized that the two active switches in the circuit have a common S–S node, and they can be replaced with a TGS, as shown in Figure 5.28c.

Investigating the voltages imposed on switches S_1 and S_2 yields that $V_1 < V_2$ because of step-up voltage conversion, and blocking diode D_{B2} can be removed

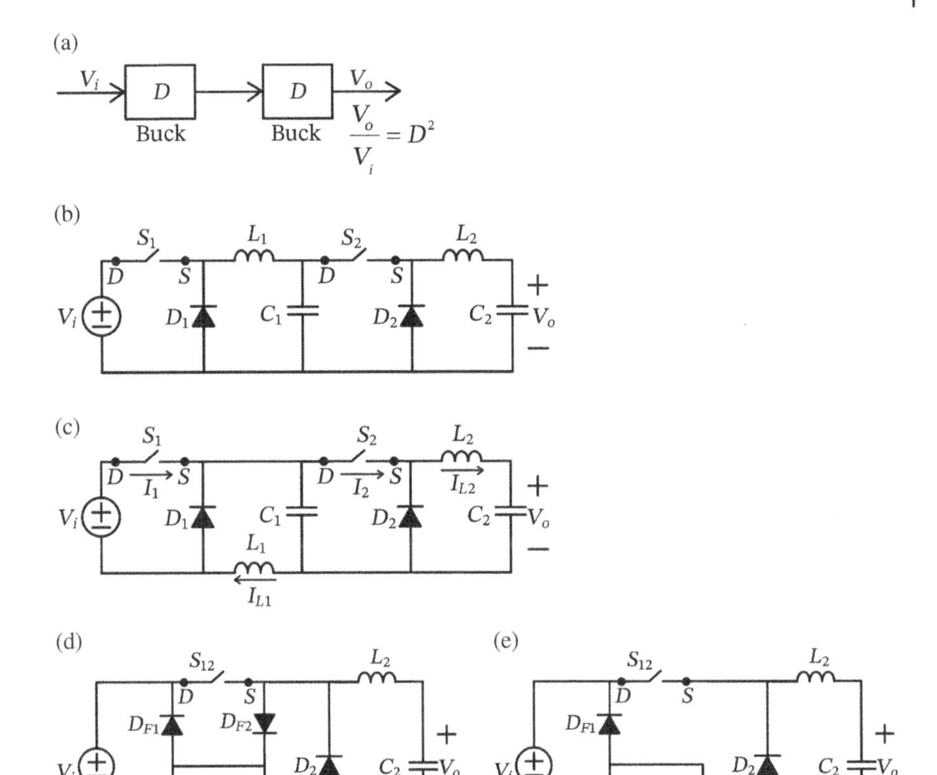

Figure 5.27 Processes of grafting a buck converter on the other buck converter to achieve high step-down voltage conversion. (a) transfer code D X D, (b) Buck+Buck, (c) Common S–D, S_1 and S_2 replaced with a ΠGS, and (e) One switch with three diodes.

from the circuit, as shown in Figure 5.28d. The converter again has only one active switch but three diodes, and it is not a regular PWM converter, with only one active switch and one diode.

5.3.1.8 Grafting Boost (CCM) on Buck (DCM)

To achieve a step-down and step-up voltage converter, we might graft a boost on a buck and operate the boost in CCM but operate the buck in DCM. The voltage transfer code expression is shown in Figure 5.29a, in which d_1 represents the duty ratio of the active switches in the buck and boost converters and d_2 is that of the diode in the buck converter. The two converters in cascade connection used to realize this transfer code expression are shown in Figure 5.29b. When move switch S_1 from the forward path to the return path, as shown in Figure 5.29c, one can identify a common D–S

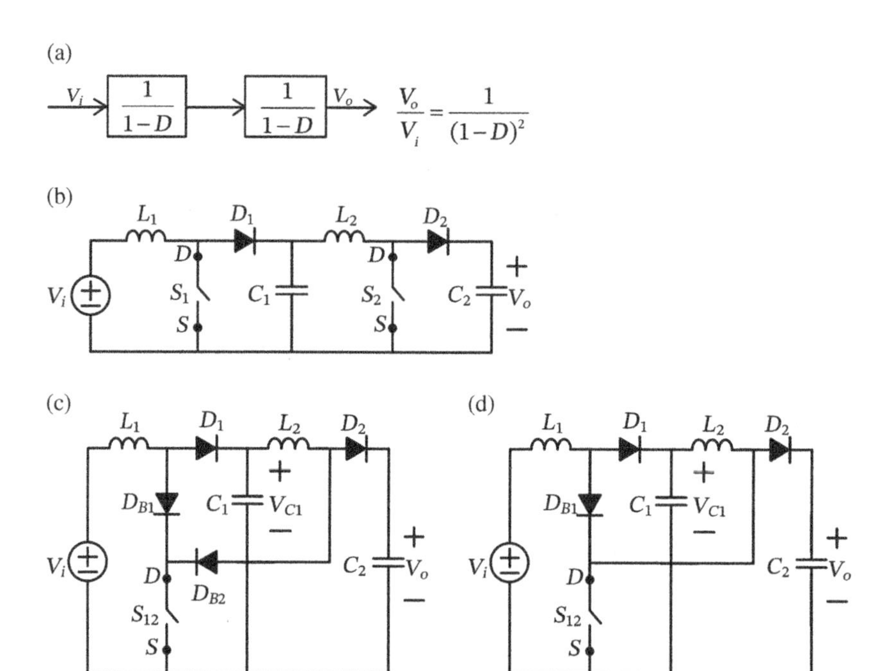

Figure 5.28 Processes of grafting one boost converter on the other boost converter to achieve high step-up voltage conversion. (a) transfer code $1/(1\text{-}D)^2$, (b) Common S–S, (c) $V_2 > V_1$, and (d) one switch with three diodes.

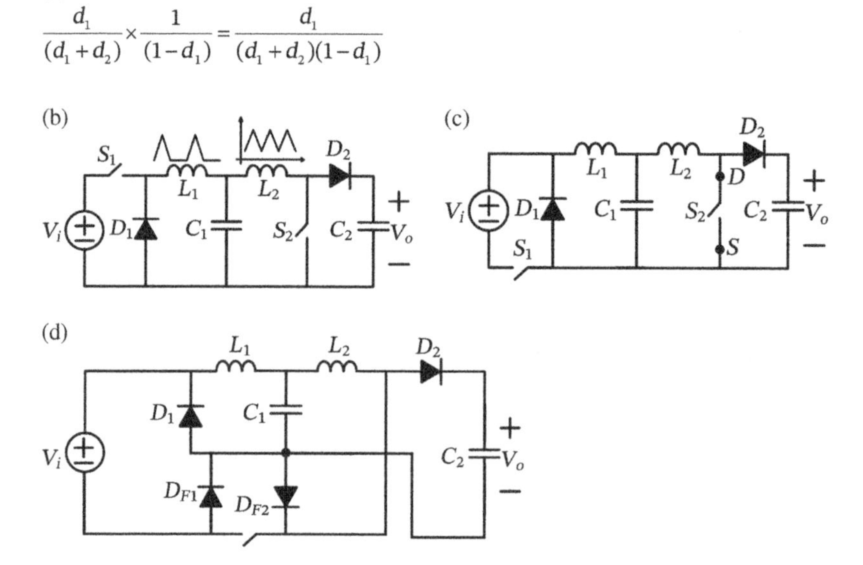

Figure 5.29 Processes of grafting a boost converter in CCM on a buck converter in DCM to achieve step-down and step-up voltage conversion.

node for switches S_1 and S_2, and they can be replaced with a ΠGS, as shown in Figure 5.29d. Since the magnitudes of the two currents flow through switches S_1 and S_2 cannot have a constant relationship over a switching time period, the two current circulating diodes D_{F1} and D_{F2} have to be kept in the converter all the time, and the final circuit becomes the one shown in Figure 5.29d. It has one active switch but four diodes. From cost point of view, it might be cost effective when saving one active switch and its driver but adding two more diodes. Anyway, it is considered a new converter topology.

5.3.1.9 Cascode Complementary Zeta with Buck

When we take an output from the positive polarities of the input and output voltages of a Zeta converter in CCM, the voltage conversion ratio is the complementary of $D/(1-D)$; that is, $1-(D/(1-D))$. When it is multiplied with that of a buck converter, the overall transfer code expression is shown in Figure 5.30a. The converters to realize this code expression are shown in Figure 5.30b. It can be recognized that the two switches S_1 and S_2 share a common D–D node, and they can be replaced with a TGS, as shown in Figure 5.30c. Investigating the voltages imposed

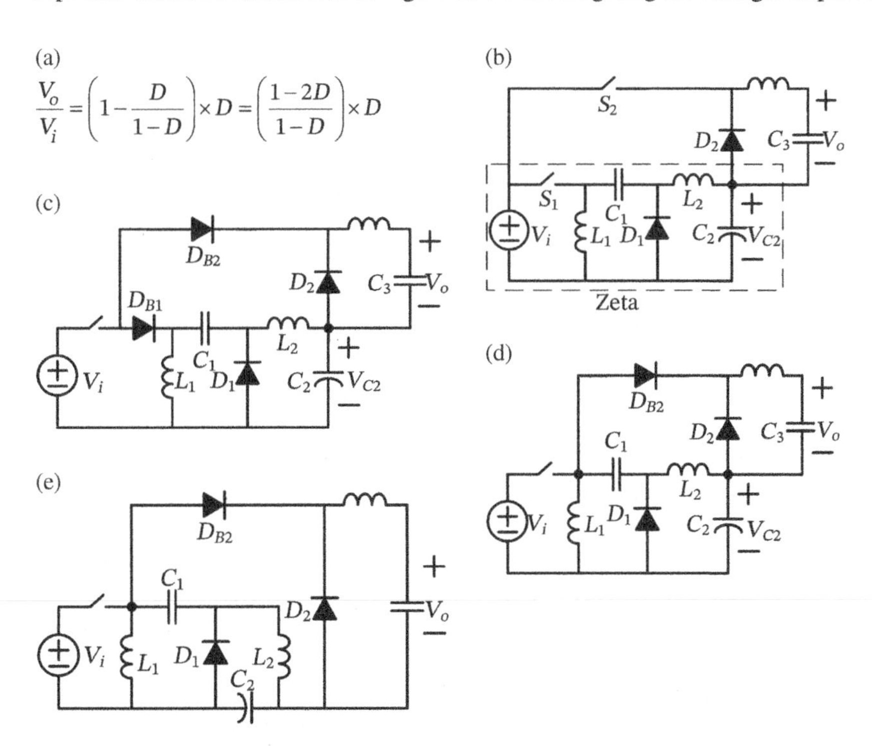

Figure 5.30 Integration processes of cascode complementary Zeta converter with a buck converter and with the grafted switch technique.

on switches S_1 and S_2 reveals that $V_1 = V_i + V_{c2}$ and $V_2 = V_i - V_{c2}$, respectively, and $V_1 > V_2$. Under this condition, blocking diode D_{B1} can be removed from the circuit, as shown in Figure 5.30d. Redrawing the circuit to have the output at the right-hand side yields the one shown in Figure 5.30e.

5.3.2 Grafting Various Types of Converters

Grafting resonant converters on well-known PWM converters can yield single-stage converters to fulfill multiple functions, such as power factor correction (PFC) with electronic ballasting and PFC with voltage regulation.

5.3.2.1 Grafting Half-Bridge Resonant Inverter on Dither Boost Converter

Different from a conventional boost converter, a dither boost converter is using two active switches to replace two diodes in the full-bridge rectifier, as shown in Figure 5.31a, in which inductor L_1 is located in AC side. When grafting a series-parallel resonant inverter on a dither boost converter, we can obtain the circuit shown in Figure 5.31b in which switches S_1 and S_3 can be operated synchronously, so can switches S_2 and S_4. It can be observed that switches S_1 and S_3 share a common D–D node, while switches S_2 and S_4 share a common S–S node, and they can be replaced with TGS, respectively, as shown in Figure 5.31c. Investigating their voltage stresses V_1–V_4 comes out all with V_{dc}; thus, all of the blocking diodes D_{B1}–D_{B4} can be shorted and removed from the circuit, and the final version becomes the one shown in Figure 5.31d. This is the famous single-stage converter proposed by Dr I. Takahashi to fulfill PFC and electronic ballasting. In the two-stage converters, there are four active switches, but in the single-stage one, there are only two active switches left. However, they have to sustain higher current ratings.

It should be noted that the two body diodes D_{b13} and D_{b24} of switches S_{13} and S_{24}, respectively, shown in Figure 5.31d serve to circulate the resonant current in the half-bridge inverter. Thus, switches S_1 and S_2 must be bidirectional.

5.3.2.2 Grafting Half-Bridge Resonant Inverter on Bidirectional Flyback Converter

Another example to fulfill battery charging, discharging, and electronic ballasting for emergency lighting applications is shown in Figure 5.32. Figure 5.32a shows a flyback converter for battery charging. If replacing the diode on the secondary side with an active switch, the flyback converter can be operated bidirectionally to fulfill charging and discharging for batteries, as shown in Figure 5.32b. Reversing the output side can yield the circuit shown in Figure 5.32c, and connecting nodes X and X' to become a non-isolated one, as shown in Figure 5.32d. Grafting a half-bridge series-parallel resonant inverter, as shown in Figure 5.32e, on the bidirectional flyback converter and introducing a transformer with the turns ratio $n_3/n_4 = n_1/n_2$ can yield the one shown in Figure 5.32f. Now, switches S_1 and S_3 and switches S_2 and S_4 can be operated in unison, respectively, and they can be

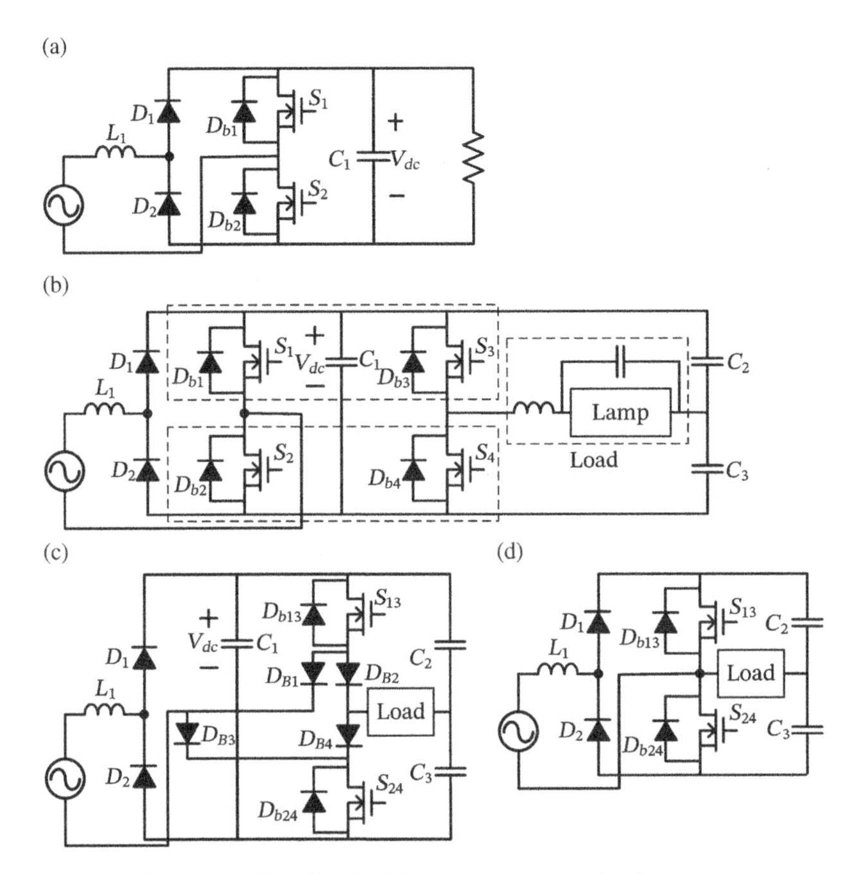

Figure 5.31 Process of grafting half-bridge inverter on dither boost converter.

replaced with grafted switches, as shown in Figure 5.32g. Since both switches S_1 and S_3 are imposed with voltage $V_{dc} + (n_1/n_2)V_B$, voltages $V_1 = V_3$ and their blocking diodes can be shorted. And since both switches S_2 and S_4 are imposed with $V_B + (n_2/n_1)V_{dc}$, voltages V_2 and V_4 are identical and their blocking diodes can be shorted, too, as shown in Figure 5.32h.

Now, the circuit shown in Figure 5.32h has two transformers, T_1 and T_2. Since they have the same turns ratio, $n_1/n_2 = n_3/n_4$, the two transformers can be integrated into a single one with higher current rating, as shown in Figure 5.32i where the two circuits are identical but with different drawings. Besides, the converter can become an isolated one when breaking the connection between windings n_1 and n_2.

5.3.2.3 Grafting Class-E Converter on Boost Converter

When a boost converter functions as a PFC and is integrated with a class-E converter for electronic ballast applications, they are connected in cascade, as shown in Figure 5.33a. In the circuit, the two active switches have a common S–S

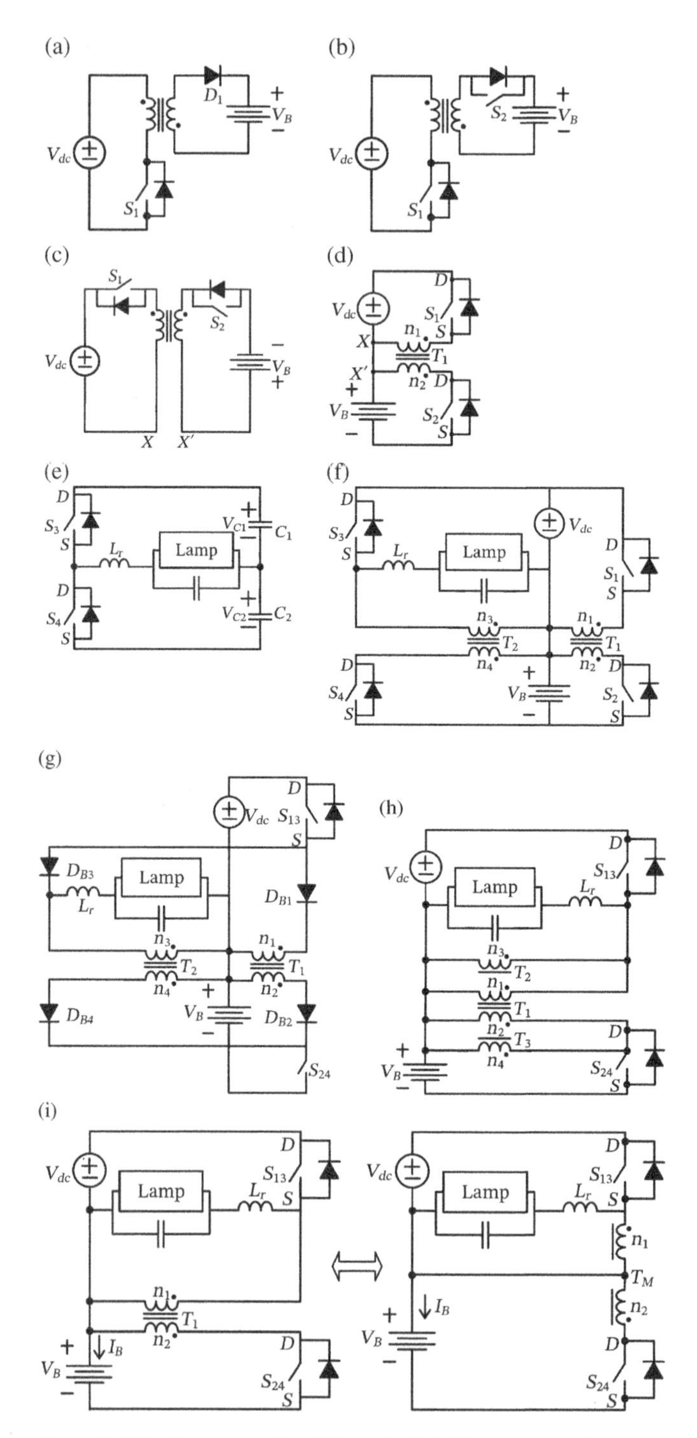

Figure 5.32 Processes of grafting a half-bridge resonant inverter on a bidirectional flyback converter to yield a 3-in-1 emergency lighting converter.

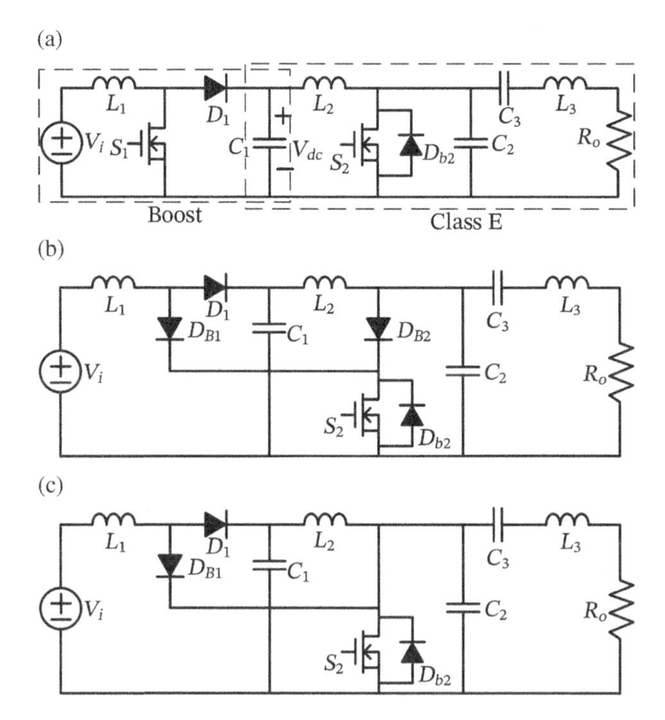

Figure 5.33 Processes of grafting a class-E converter on a boost converter to yield the PFC+electronic ballast.

node, and they can be operated in unison and replaced with a TGS, as shown in Figure 5.33b. It has been proved that the voltage imposed on switch S_1 is V_{dc}, while the one on switch S_2 is around πV_{dc}; thus, $V_1 < V_2$ and D_{B2} is no longer needed and removed from the circuit. Finally, the circuit becomes the one shown in Figure 5.33c. It can be seen that there are two diodes, D_1 and D_{B1}, and the circuit only can save one active switch, but it requires an extra diode. It is different from the previous cases where $V_1 = V_2$ and the two blocking diodes can be all removed.

In literature, many single-stage converters for fulfilling multiple functions have been proposed. Essentially, they are also derived with the grafted switch technique, but authors did not mention this point in their papers. In the following, examples with grafted diodes are illustrated.

5.3.3 Integrating Converters with Active and Passive Grafted Switches

Integrating converters with active and passive grafted switches can generate single-stage converter or further reduce component counts. This section presents two illustration examples: (i) graft a buck on a boost and (ii) graft a half-bridge inverter on two interleaved boosts.

5.3.3.1 Grafting Buck on Boost with Grafted Diode

Connecting boost and buck converter in series and relocating switch S_2 to the return path yield the circuit shown in Figure 5.34a. In the circuit, the two diodes D_1 and D_2 share a common N node, and they can be replaced with a grafted diode, as shown in Figure 5.34b. When the active switches S_1 and S_2 are turned on, the voltages imposed on both D_1 and D_2 are equal to V_c; that is, $V_1 = V_2$. Thus, the blocking diodes D_{B1} and D_{B2} can be shorted, as shown in Figure 5.34c, and it can be recognized that switches S_1 and S_2 become paralleled, and they can be replaced with a single one, S_{12}, with higher current rating, as shown in Figure 5.34d. Moving capacitor C_1 from the return path to the forward path yields the boost-buck (Ćuk) converter, as shown in Figure 5.34e.

5.3.3.2 Grafting Half-Bridge Inverter on Interleaved Boost Converters in DCM

Two interleaved boost converters connected with a half-bridge inverter in series are shown in Figure 5.35a, in which the two boost converters are operated in DCM and one diode is located in the return path for easily integrating switches. In the circuit, switches S_1 and S_3 can be operated in unison, and they share a common

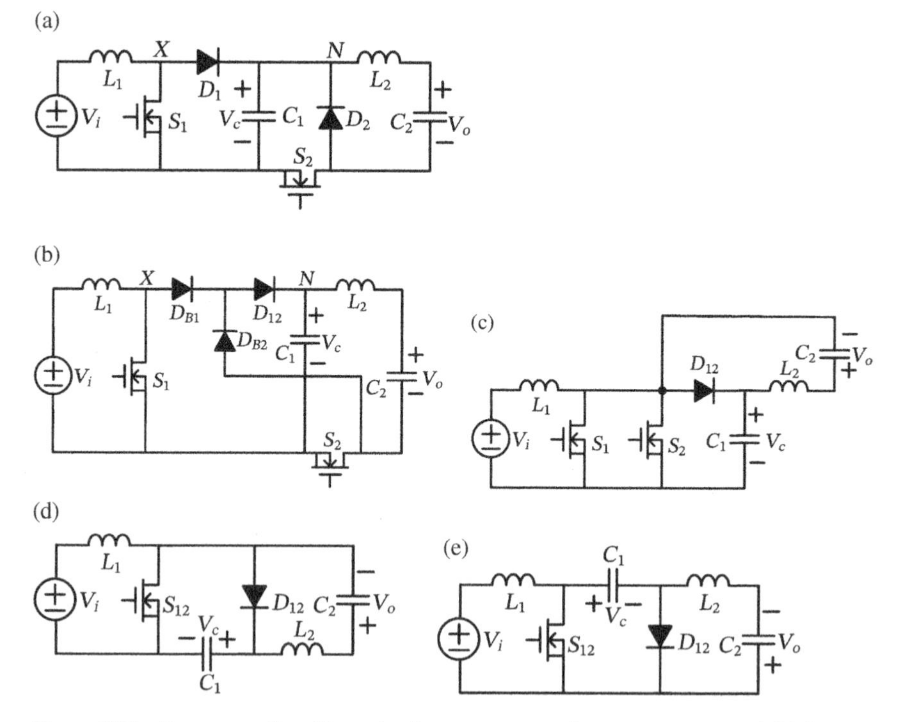

Figure 5.34 Processes of grafting a buck converter on a boost converter to yield the boost–buck (Ćuk) converter by using the grafted diode technique.

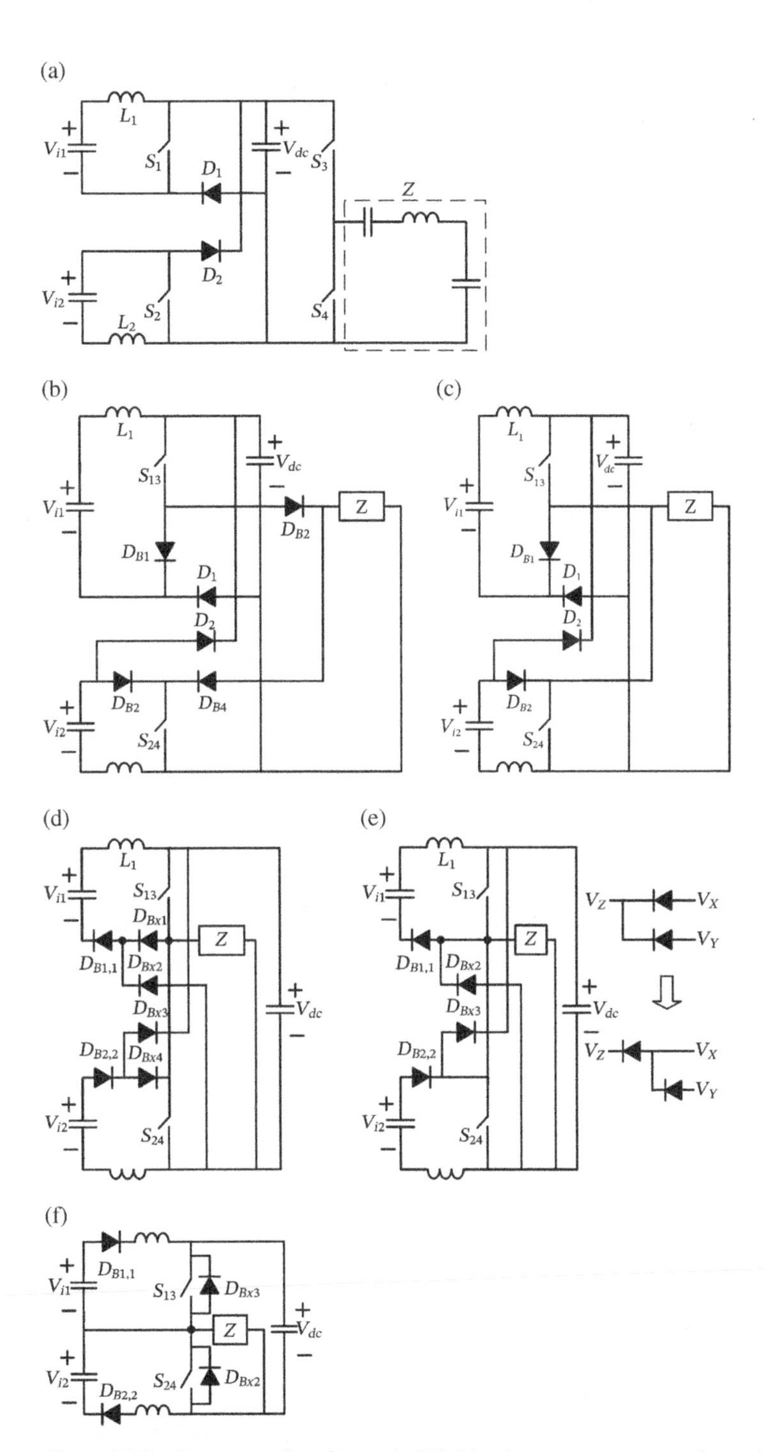

Figure 5.35 Processes of grafting a half-bridge inverter on two interleaved boost converters in DCM.

D–D node; thus, they can be replaced with a TGS. Similarly, switches S_2 and S_4 can be operated in unison, and they share a common S–S node; thus, they can be replaced with another TGS, as shown in Figure 5.35b. Investigating the operational principle of the DCM boost and half-bridge inverter reveals that the voltage imposed on switch S_3 or S_4 is V_{dc}, but the one on switch S_1 can be either V_{dc} in CCM or V_{i1} in DCM, and the one on switch S_2 can be either V_{dc} in CCM or V_{i2} in DCM. Since a boost converter has its output voltage higher than its input, voltage V_{dc} is always higher than V_{i1} and V_{i2}. Thus, blocking diodes D_{B3} and D_{B4} can be removed from the circuit, but diodes D_{B1} and D_{B2} need to be kept in the circuit, as shown in Figure 5.35c, to block the voltage difference between V_{dc} and V_{i1} or V_{i2}. Now, diodes D_1 and D_{B1} share a common N node, while diodes D_2 and D_{B2} share a common P node. They can be replaced with grafted diodes, respectively, as shown in Figure 5.35d. It can be proved that voltages $V_{X1} > V_{Y1}$ when switch S_{13} is turned on and voltages $V_{X2} < V_{Y2}$ when switch S_{24} is turned on. Thus, blocking diodes D_{Bx1} and D_{Bx4} are no longer needed and can be shorted in the circuit, as shown in Figure 5.35e. Now, diode D_{Bx3} is antiparalleled with switch S_{13}, and it can be just its body diode. Similarly, diode D_{Bx2} is antiparalleled with switch S_{24}, and it can be just its body diode, too. The overall circuit is shown in Figure 5.35f in which diode $D_{B1,1}$ is moved to the forward path, but diode $D_{B2,2}$ is moved to the return path. In the applications, input voltage V_{i1} is set to be equal to voltage V_{i2}. With the grafted switch and diode techniques, we can save two diodes.

5.3.3.3 Grafting *N*-Converters with TGS

Figure 5.36a shows that N-converters are represented in Thevenin equivalent circuits. When switches are turned off, their voltage stresses are $V_1, V_2, \ldots, V_n$, respectively, in which $D_1, D_2, \ldots, D_n$ are used to conduct bidirectional current. The active switches are with a common S–S node, and when operating in unison, they can be integrated into a grafted switch, as shown in Figure 5.36b. In the circuit, all of the blocking diodes are kept since the relationships among the voltages, $V_1, V_2, \ldots, V_n$, are unknown and diodes $D_{F1}, D_{F2}, \ldots, D_{Fn}$ are used to conduct bidirectional currents.

5.3.3.4 Grafting *N*-Converters with ΠGS

Figure 5.37a shows that N-converters are represented in Norton equivalent circuits in which diodes $D_1, D_2, \ldots, D_n$ are used to conduct bidirectional currents. The switches with a common S–S can be grafted, so can be those with a common D–D, and then they become with a common D–S or S–D node and can be grafted to be the one shown in Figure 5.37b. In the circuit, diodes D_{FF1}–D_{FFn} can circulate current difference among converters.

In principle, even though there are more than two converters to be grafted, we only conduct two at a time. In fact, we graft two switches at a time and repeat the process until all of the active switches have been grafted.

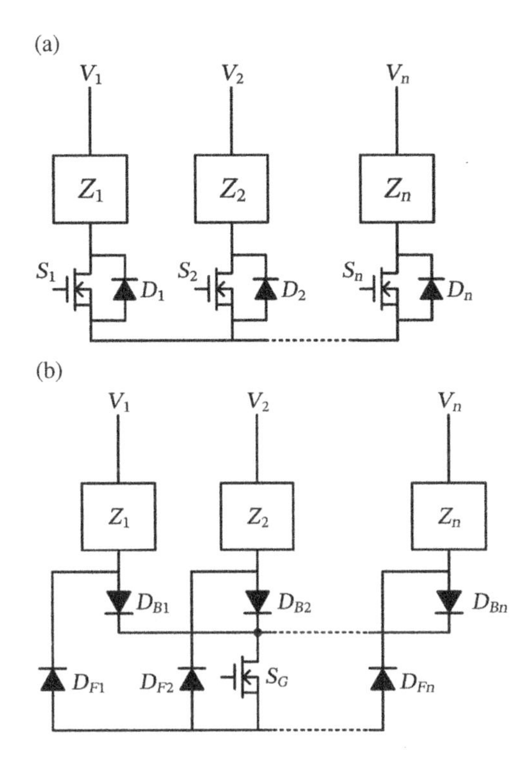

Figure 5.36 Processes of grafting *N*-converters with a TGS.

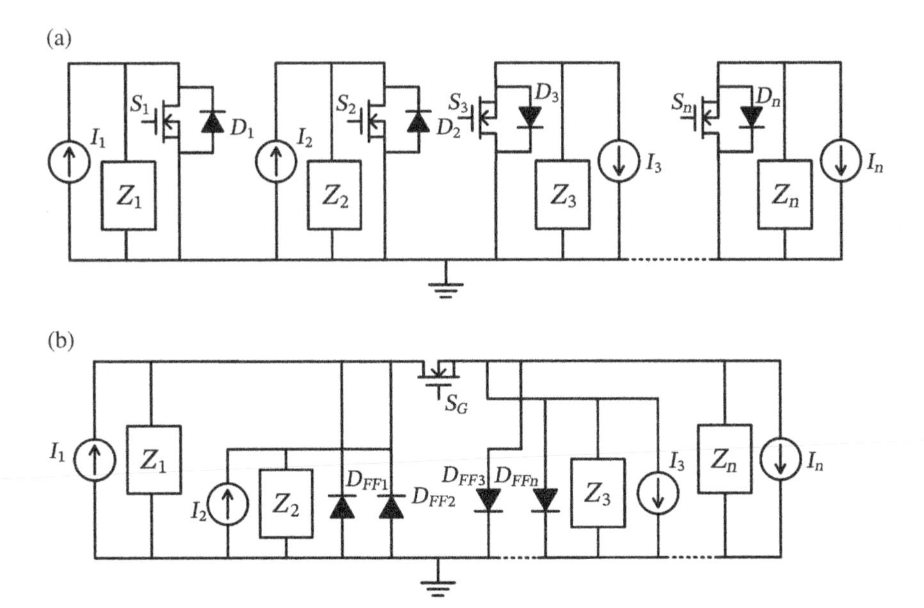

Figure 5.37 Processes of grafting *N*-converters with a ΠGS.

Further Reading

Berkovich, Y., Shenkman, A., Ioinovici, A., and Axelrod, B. (2006). Algebraic representation of DC-DC converters and symbolic method of their analysis. *Proceedings of the IEEE Convention. Electrical and Electronics Engineers*, IEEE, pp. 47–51.

Khan, F.H., Tolbert, L.M., and Peng, F.Z. (2006). Deriving new topologies of DC-DC converters featuring basic switching cells. *Proceedings of the IEEE Workshops on Computers in Power Electronics*, IEEE, pp. 328–332.

Maksimovic, D. and Ćuk, S. (1989). General properties and synthesis of PWM DC-to-DC converters. *Proceedings of the IEEE Power Electronics Specialists Conference*, IEEE, pp. 515–525.

Severns, R.P. and Bloom, G.E. (1985). *Modern DC-to-DC Switch Mode Power Converter Circuits*. New York: Van Nostrand Reinhold Co.

Tao, F.-F. and Lee, F.C. (2000). An interleaved single-stage power-factor-correction electronic ballast. *Proceedings of the IEEE Applied Power Electronics Conference*, IEEE, pp. 617–623.

Wu, T.-F. (2016). Decoding and synthesizing transformerless PWM converters. *IEEE Trans. Power Electron.* 30 (9): 6293–6304.

Wu, T.-F. and Chen, Y.-K. (1996). A systematic and unified approach to modeling PWM DC/DC converters using the layer scheme. *Proceedings of the IEEE Power Electronics Specialists Conference*, IEEE, pp. 575–580.

Wu, T.-F. and Chen, Y.-K. (1998a). A systematic and unified approach to modeling PWM DC/DC converter based on the graft scheme. *IEEE Trans. Ind. Electron.* 45 (1): 88–98.

Wu, T.-F. and Chen, Y.-K. (1998b). Modeling PWM DC/DC converter out of basic converter units. *IEEE Trans. Power Electron.* 13 (5): 870–881.

Wu, T.-F. and Yu, T.-H. (1998). Unified approach to developing single-stage power converters. *IEEE Trans. Aerosp. Electron. Syst.* 34 (1): 221–223.

Wu, T.-F., Chen, Y.-K., and Liang, S.-A. (1999). A structural approach to synthesizing, analyzing and modeling quasi-resonant converters, *Proceedings of the IEEE Power Electronics Specialists Conference*, IEEE, pp. 1024–1029.

Wu, T.-F., Liang, S.-A., and Chen, Y.-K. (2003). A structural approach to synthesizing soft switching PWM converters. *IEEE Trans. Power Electron.* 18 (1): 38–43.

6

Synthesizing Process with Layer Scheme

Continuing from Chapter 5, this chapter will focus on synthesizing process with the layer scheme. According to the code configurations presented in Chapter 4, some of them are with feedback path that cannot be synthesized with the graft scheme, but can be done with the layer scheme. The layer scheme was originally used for processing twigs or branches to generate new plants, which is adapted here for processing converters in the feedback path of code configurations.

This chapter first reviews the layer scheme used in generating new plants, which then is adapted to layer converters. Illustrations of the layer scheme for buck family and boost family will be addressed. Other illustrations for synthesizing codes with the layer scheme to become converters will be also presented. Additionally, evolution of the existing converters in different component configurations but with the same transfer code will be discussed.

6.1 Converter Layering Scheme

Layer scheme is originally used for generating new trees from the existing one, as illustrated in Figure 6.1. In Figure 6.1a, step 1 is to peel the bark of a healthy twig or branch, step 2 is to bend it to ground and cover the peeled place with soil, and step 3 is to hold the twig steadily with enough soil and wait until new roots growing out and finally become a new tree, as conceptually shown in Figure 6.1b. The newly growing tree has several same physical characteristics as its origin, such as tree type, fruit shape, and environmental adaptation, but with different properties, such as young, healthy, and fast growing.

Analogously, some converters with the layer scheme have the same component count and the same transfer type, but with different input-to-output transfer codes, and might increase additional transfer types. We treat the input source a

Origin of Power Converters: Decoding, Synthesizing, and Modeling, First Edition.
Tsai-Fu Wu and Yu-Kai Chen.
© 2020 John Wiley & Sons, Inc. Published 2020 by John Wiley & Sons, Inc.

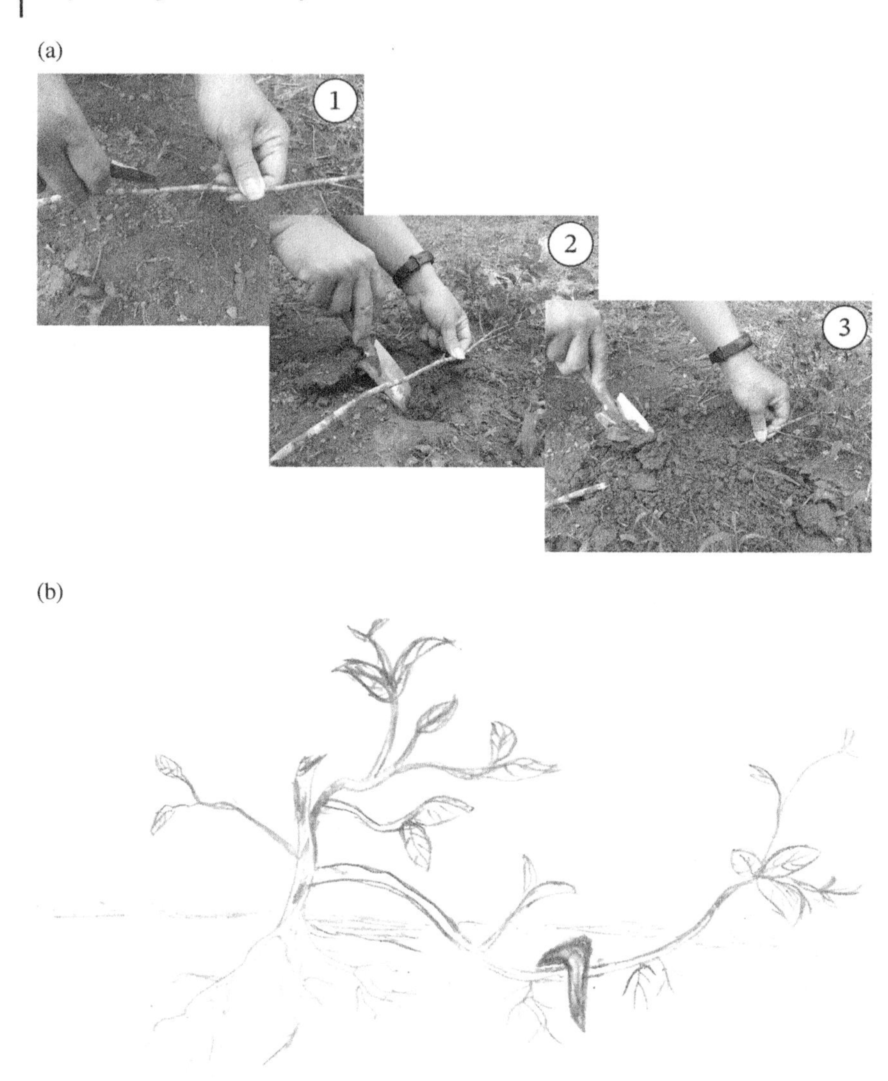

Figure 6.1 (a) Illustration of a layer scheme in three steps and (b) conceptual diagram of the layer scheme.

root and the output a twig. By taking a twig from the output and connecting it to the root (ground), we have the same layer scheme. As shown in Figure 6.2, the buck converter in Figure 6.2a with the layer scheme yields the buck–boost converter shown in Figure 6.2d, which has the same component count (C_f and C_o can be combined into a single one from topological point of view) and the same

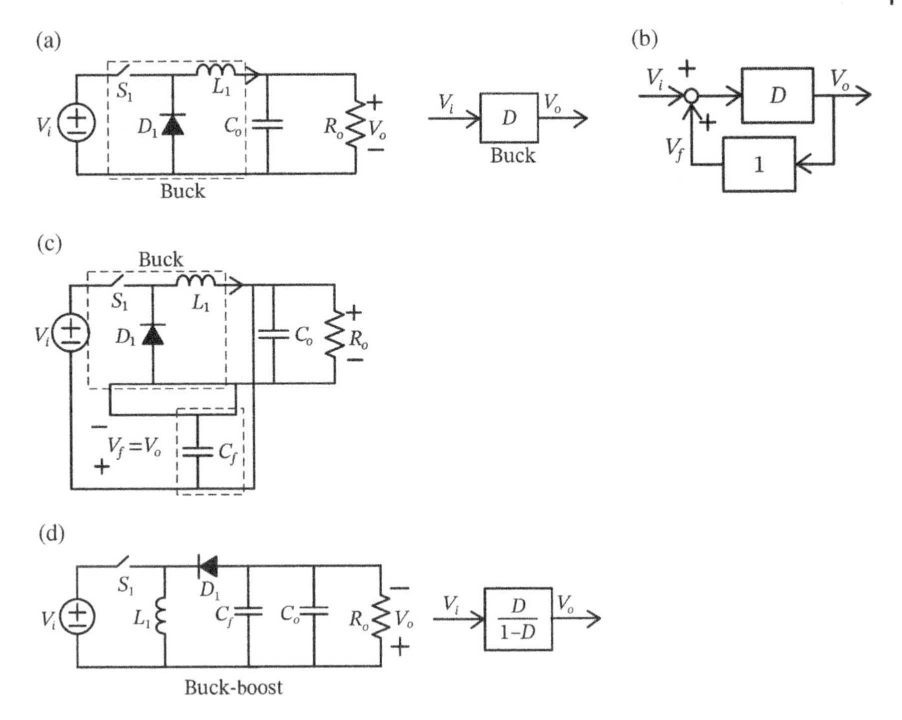

Figure 6.2 Derivation of the buck–boost converter with the converter layering scheme.

step-down type as the buck converter, but with a different transfer code ($D/(1 - D)$), and adds another step-up transfer type. Thus, the buck–boost converter becomes a step-down and step-up type of converter. Note that C_f is not necessary in the feedback path.

6.2 Illustration of Layering Converters

This section presents illustrations of two converter families, buck and boost, synthesized with the layer scheme. The buck family includes the buck, buck–boost, and Zeta (buck–boost–buck) converters, while the boost family includes the boost, boost–buck (Ćuk), and sepic (boost–buck–boost) converters.

6.2.1 Buck Family

Figure 6.2a shows the buck converter with its code D in CCM operation. With a positive unity feedback, the code configuration of $D/(1 - D)$ can be achieved, as shown in Figure 6.2b. This code configuration can be synthesized with a buck converter, a feedback capacitor C_f, and the layer scheme, as shown in Figure 6.2c. Redrawing the component configuration yields the well-known buck–boost converter topology shown in Figure 6.2d where capacitors C_f and C_o can be combined into a single one.

A code configuration shown in Figure 4.11c is redrawn and shown in Figure 6.3a, in which the transfer code TC_{f1} is irregular, and therefore, TC_{f2} and TC_{b1} must be LC filter with its gain as a unity. For illustrating the synthesis of Zeta converter, the code configuration shown in Figure 6.3a is redrawn with new notations, as shown in Figure 6.3b where D_p denotes the pulsating transfer ratio of the buck converter and both F_1 and F_2 are LC filters. Based on this code configuration, we can have converter configuration shown in Figure 6.3c. It can be recognized that $V_p/V_i = D_p$ and D_p with F_1 (L_1C_o filter) yields the code D. Again, D_p with F_2 ($L_f C_f$ filter) forms a smooth DC voltage, which can be combined with the smooth input source voltage V_i.

The transfer code expression for Figure 6.3b is given as

$$\frac{V_o}{V_i} = \frac{D_p F_1}{1 - D_p F_2}. \tag{6.1}$$

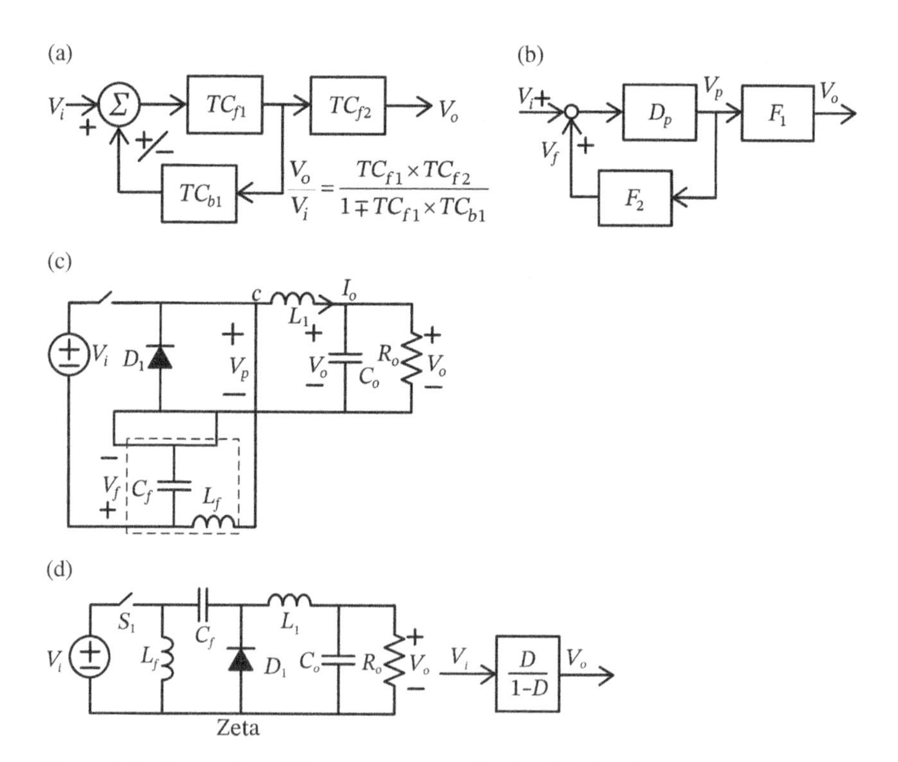

Figure 6.3 Illustration of Zeta converter synthesis with a buck converter and the layer scheme.

Let $F_1 = F_2$. At DC the LC filters act as a unity gain. Thus, we have the following expression:

$$\frac{V_o}{V_i} = \frac{D}{1-D}, \tag{6.2}$$

which is the transfer code of Zeta converter. By redrawing the component configuration shown in Figure 6.3c, we can have the converter configuration shown in Figure 6.3d, in which the output has the same polarity as the input.

In Figure 6.2a, the buck converter has the components of S_1, D_1, L_1, and C_o. Basically, C_o is belonged to the output voltage filter. Thus, a buck converter consists only of S_1, D_1, and L_1. The buck–boost converter shown in Figure 6.2d is derived from the buck converter with the layer scheme. So, it includes a buck converter in the forward path. However, in deriving the Zeta converter, there are only S_1 and D_1 in the forward path, but L_1 and C_o are in the output portion. Thus, there is no buck converter in the Zeta converter. If we say the converters belonged to the same family should have the same deoxyribonucleic acid (DNA), how can we say that the Zeta converter is belonged to the buck family because there is no buck converter in the derived Zeta converter? With a close observation on the circuit shown in Figure 6.3c, inductors L_1 and L_f share a common node. Thus, we can add another inductor L_x to the common node "X," as shown in Figure 6.4. Now, we have three inductors sharing a common node, and one of them is a dependent inductor. Thus, the dynamic degree is still kept two and its DC transfer gain is unity. This additional inductor L_x does not change the degree of dynamics and DC transfer code. Therefore, the circuit shown in Figure 6.4 can be used to derive the Zeta converter, and it includes a buck converter in the derived Zeta converter. Now, we can say that the Zeta converter has the buck DNA and it is belonged to the buck family.

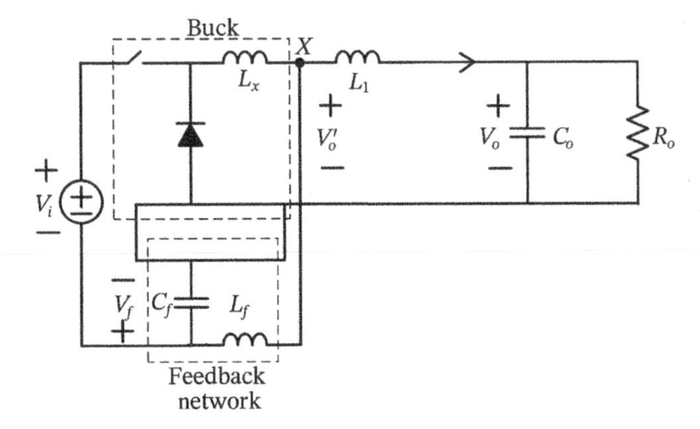

Figure 6.4 Universal converter configuration of the buck family.

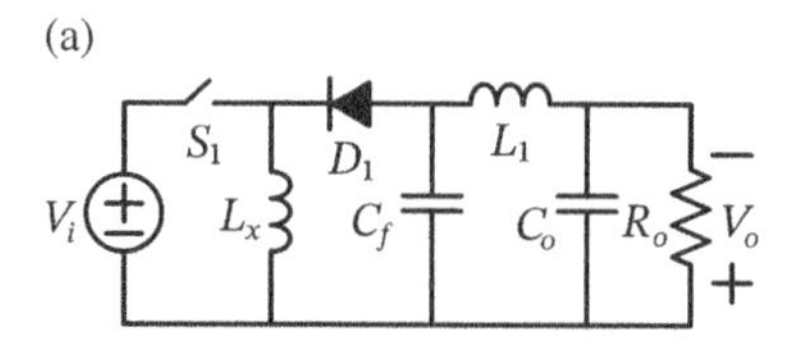

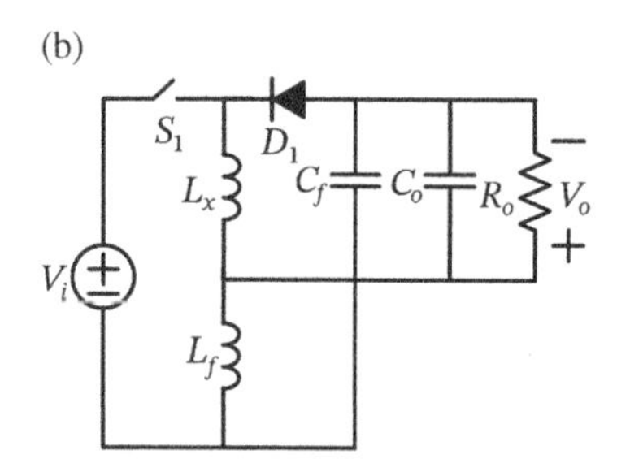

Figure 6.5 Two additional converter topologies derived from the universal form of the buck family: (a) buck-boost converter with an additional *LC* filter, and (b) Eta converter.

It is worthy of noting that when adding the additional inductor L_x to the converter shown in Figure 6.3c to form the one shown in Figure 6.4, we have the converter with three inductors sharing a common node. Thus, one of the inductors becomes a dependent one and can be removed from the converter shown in Figure 6.4 at a time. When removing inductor L_x from the converter, we have the one as the well-known Zeta converter. When removing inductor L_f from the converter, we have the one shown in Figure 6.5a. However, when removing inductor L_1, we have the one shown in Figure 6.5b. For the converter shown in Figure 6.5a, it is a buck–boost converter with an additional *LC* filter (L_1C_o), and it has the same transfer code of $D/(1-D)$. Basically, it is not a new one, but a buck–boost converter with an additional *LC* filter. For the converter, namely, Eta, shown in Figure 6.5b, its transfer code can be derived based on volt-second balance principle, which is again $D/(1-D)$. However, its component configuration is different from those of the buck–boost and Zeta converters. Both of the converters shown in Figure 6.5 are belonged to the buck family because they have the buck DNA.

When removing inductors L_1 and L_f from the converter configuration shown in Figure 6.4, we have the buck–boost converter. When removing L_1 and the feedback circuit ($L_f C_f$), we have the buck converter. Thus, the converter configuration shown in Figure 6.4 can be treated as the universal form of a buck family.

6.2.2 Boost Family

Boost family includes boost, boost–buck (Ćuk), and boost–buck–boost (sepic), and they contain the DNA of a boost converter. Before synthesizing the converters, let us see a property of the buck and boost family. Boost is the dual of the buck. Therefore, when deriving the buck family, we use input-to-output voltage transfer code, while when deriving the boost family, we use the input-to-output current transfer code, as shown in Figure 6.6a. It should be noted that a buck converter consists of an active switch, a diode, and an inductor connected in T shape.

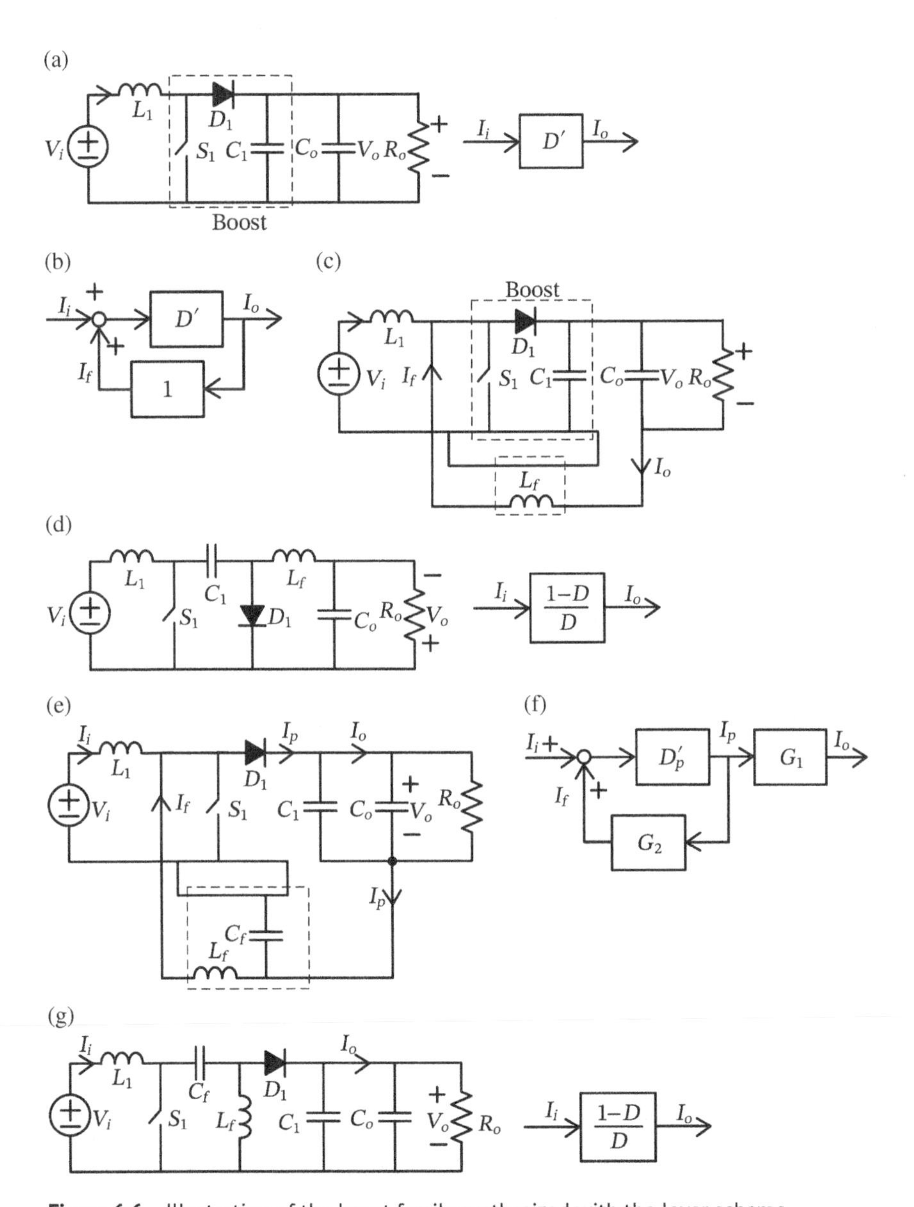

Figure 6.6 Illustration of the boost family synthesized with the layer scheme.

On the other hand, a boost converter consists of an active switch, a diode, and a capacitor connected in Π shape. Therefore, in Figure 6.6a, the output capacitor is split into two capacitors (C_1 and C_o), in which C_1 is associated with switch S_1 and D_1 to form a Π shape, and C_o is belonged to the output filter. Basically, the input current is a smooth one, and we need capacitor C_1 to smooth the pulsating current through diode D_1.

The input-to-output current transfer code of the boost converter is $I_o/I_i = D'$, where $D' = 1 - D$. With the layer scheme, the output current is fed back and added to the input current source, as shown in Figure 6.6b, which yields the following transfer code:

$$\frac{I_o}{I_i} = \frac{D'}{(1-D')} = \frac{(1-D)}{D}. \tag{6.3}$$

This is the transfer code of a boost converter. By following the code configuration shown in Figure 6.6b, we can have its corresponding converter configuration shown in Figure 6.6c, in which the feedback circuit is an inductor L_f. This configuration yields the Ćuk converter, as shown in Figure 6.6d with its transfer code $I_o/I_i = (1 - D)/D$.

Similar to the pulsating voltage feedback in deriving the Zeta converter, we have a pulsating current feedback to derive the sepic converter, as shown in Figure 6.6e, in which the feedback circuit is a CL filter type. The input-to-output current transfer ratio expression is shown in Figure 6.6f, in which D'_p denotes a pulsating current transfer code and filters $G_1 = C_1//C_oL_f$ and $G_2 = C_fL_f$. From Figure 6.6f, we can have

$$\frac{I_o}{I_i} = \frac{D'_pG_1}{(1-D'_pG_2)}, \tag{6.4}$$

in which based on the boost filter operation, $D'_pG_1 = D'$. If letting $G_1 = G_2$, we can have the following expression:

$$\frac{I_o}{I_i} = \frac{D'}{(1-D')} = \frac{(1-D)}{D}. \tag{6.5}$$

By power conservation,

$$V_iI_i = V_oI_o, \tag{6.6}$$

we have

$$\frac{V_o}{V_i} = \frac{D}{(1-D)}. \tag{6.7}$$

Rearranging the component configuration shown in Figure 6.6f yields the well-known sepic converter shown in Figure 6.6g.

In the derivation of the sepic converter, we use pulsating current feedback as shown in Figure 6.6e. If we add another capacitor C_x to the converter shown in Figure 6.6e, capacitors C_1, C_f, and C_x form a loop, as shown in Figure 6.7, one of which is a dependent capacitor, without increasing the dynamic degree. Capacitor C_x acts as a unity DC gain. Thus, the converter configuration shown in Figure 6.7 can be used to derive the sepic converter.

Again, once capacitor C_x is added to the converter shown in Figure 6.7, it forms a loop with the other two capacitors, $C_1//C_o$ and C_f. It cannot be differentiated from one another. Therefore, we can remove one capacitor at a time. When removing capacitor C_x, we have the sepic converter. When removing capacitor C_f, we have the converter shown in Figure 6.8a, and when removing capacitor $C_1//C_o$, we have the one shown in Figure 6.8b.

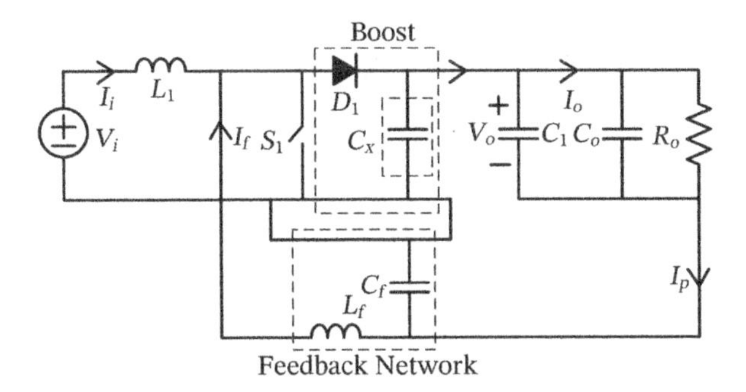

Figure 6.7 Universal form of the boost family.

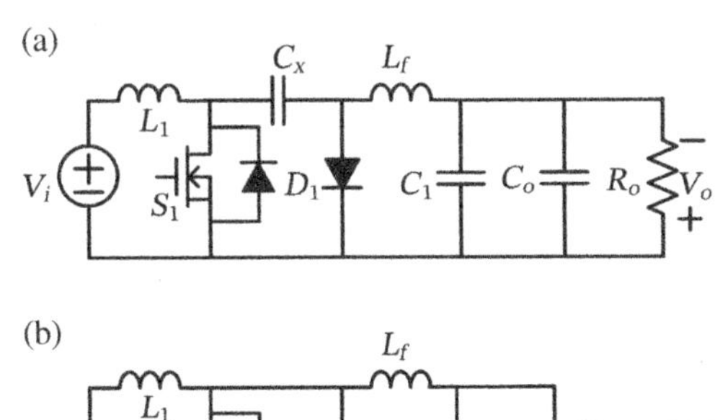

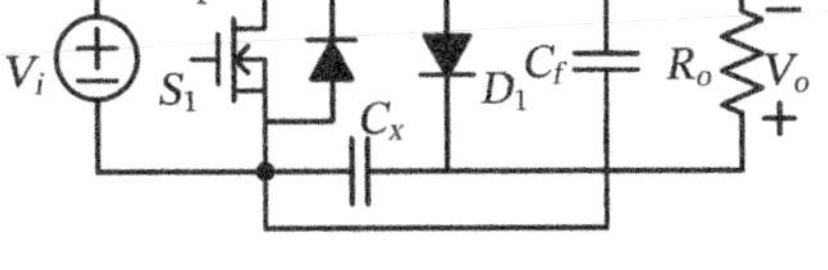

Figure 6.8 Two additional converter topologies derived from the universal form of the boost family: (a) Ćuk converter, and (b) Theta converter.

The converter topology shown in Figure 6.8a is identical to the Ćuk converter, and the one shown in Figure 6.8b is a new converter, namely, Theta. Based on volt-second balance principle, its transfer code can be derived as $D/(1-D)$.

6.2.3 Other Converter Examples

DC PWM converters are always with unidirectional current flow and fixed voltage polarity. They can be layered from their outputs. In the following, two examples are illustrated with the layer scheme. Figure. 6.9a shows a Zeta converter with a positive unity output feedback that yields the transfer code configuration shown in Figure 6.9b. With further processing, we have the final converter configuration with its transfer code shown in Figure 6.9c.

Similarly, Figure 6.10a shows the sepic converter with a positive unity feedback that yields the transfer code configuration shown in Figure 6.10b. With further processing, we have the final converter configuration with its transfer code shown in Figure 6.10c.

The layer scheme is a useful technique for deriving other PWM converter topologies based on the converter itself. If there is no extra component added, the converter is just changed with its component connection to yield a new converter topology. The number of converter topologies is quite limited, because further feedback will come back to its original converter. This argument can be illustrated by the converters depicted in Figure 6.9c and Figure 6.10c, which are processed

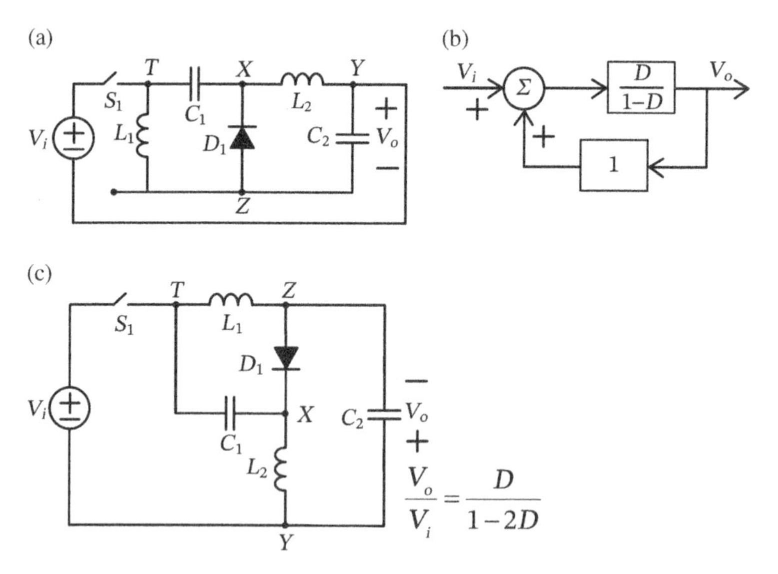

Figure 6.9 Zeta converter with a positive unity feedback and the layer scheme yielding the transfer code of $D/(1-2D)$.

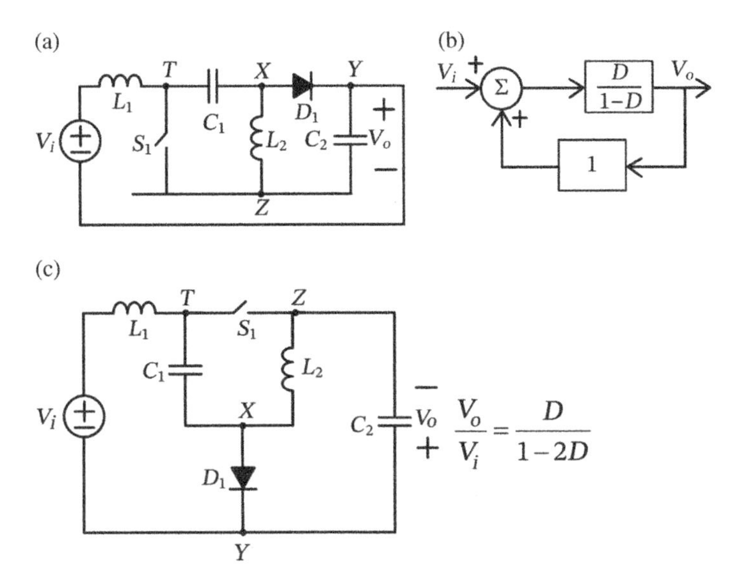

Figure 6.10 Sepic converter with a positive unity feedback and the layer scheme yielding the transfer code of $D/(1-2D)$.

further with the layer scheme. Note that the converters shown in Figures 6.9c and 6.10c cannot be fed back with positive unity because their positive polarities are connected to the source negative polarity. Figure 6.11a shows the converter depicted in Figure 6.9c with a negative unity feedback, and its transfer code configuration is shown in Figure 6.11b. With further processing, we have the converter configuration and its transfer code shown in Figure 6.11c, which is exactly the same as Zeta converter. Similarly, Figure 6.12 shows those of the converter depicted in Figure 6.10c with a negative unity feedback, which comes out the identical sepic converter.

However, if there are components added to the converter, basically, we can derive more converter topologies. This will be illustrated in Chapter 7.

As shown in the derivation of a boost family, the feedback is a kind of current type. If the feedback is using a voltage type, we can have the following converter topologies, as illustrated in Figures 6.13 and 6.14. Figure 6.13a shows a boost converter with a positive unity voltage feedback, and its transfer code configuration is shown in Figure 6.13b. With further processing, we can have the converter topology and its transfer code shown in Figure 6.13c, which is considered a new converter topology. However, diode D_1 should be a controllable switch with dc-blocking capability.

Figure 6.14a shows the Ćuk converter with a negative unity voltage feedback, and its transfer code configuration is shown in Figure 6.14b. With further processing,

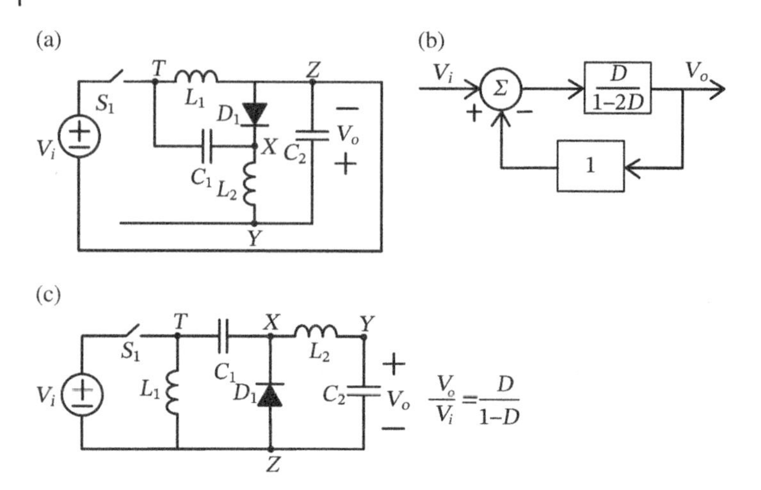

Figure 6.11 The converter depicted in Figure 6.9c with a negative unity feedback yielding the Zeta converter.

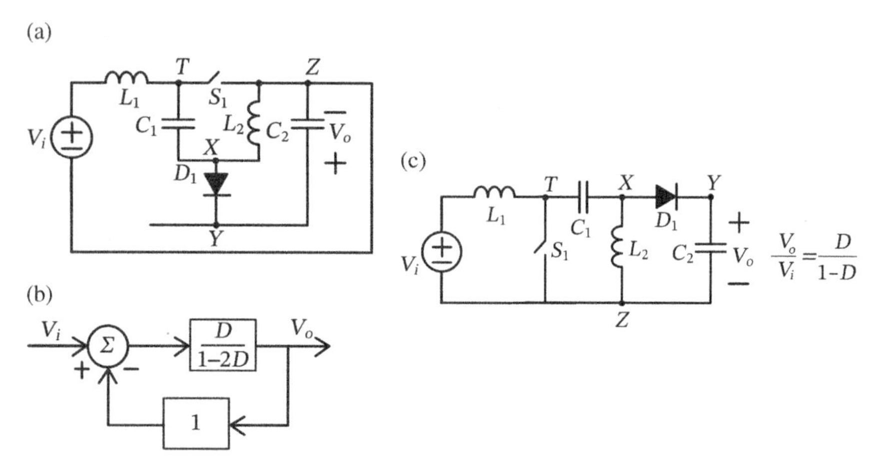

Figure 6.12 The converter depicted in Figure 6.10c with a negative unity feedback yielding the sepic converter.

we have the converter topology and its transfer code shown in Figure 6.14c. Since its transfer code is D, which is equivalent to that of a buck converter, it is supposed to have the same converter topology as the buck converter. Basically, with the fundamental of inductor splitting presented in Chapter 3, we can do the reverse process by combining the two inductors shown in Figure 6.14c. First, moving inductor L_1 from the forward path to the return path yields the one shown in Figure 6.14d, and then inductors L_1 and L_2 are combined into a single one (L_{12}), as shown in Figure 6.14e. Finally, moving inductor L_{12} to the forward path yields the

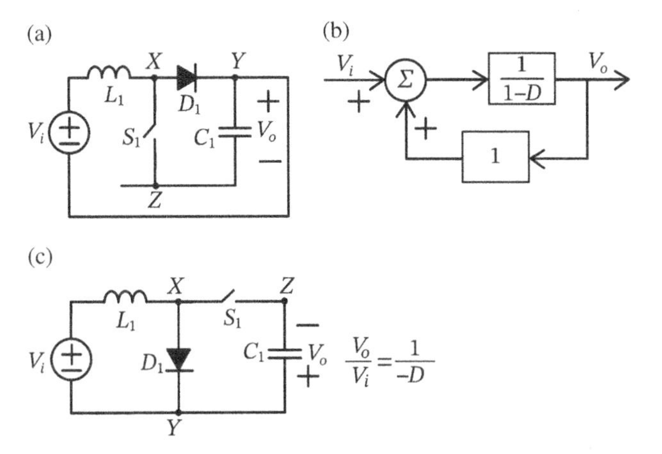

Figure 6.13 Illustration of a boost converter with a positive unity voltage feedback yielding the converter with transfer code of $1/-D$.

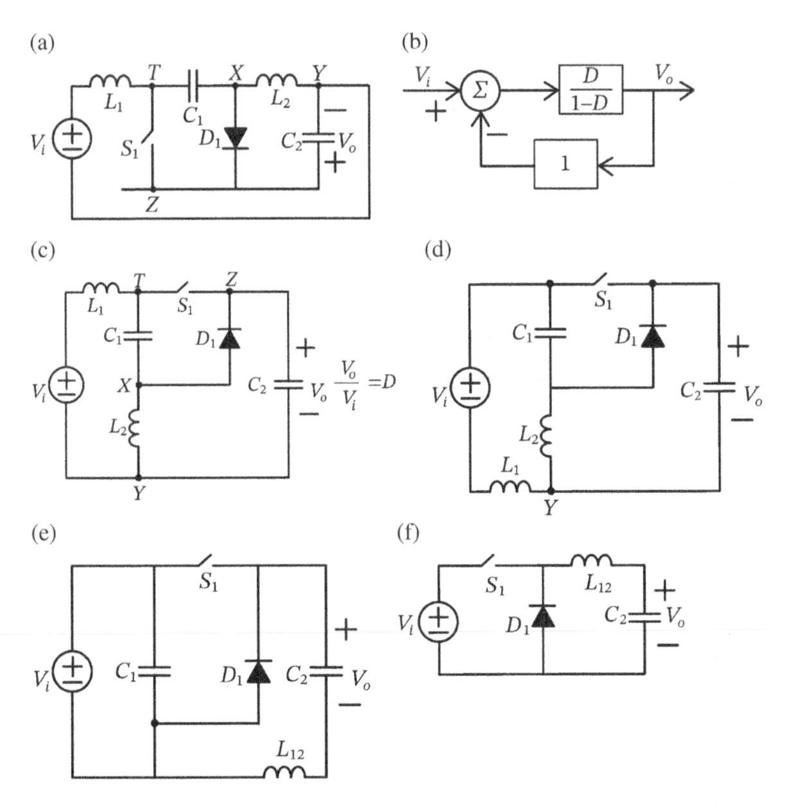

Figure 6.14 Illustration of Ćuk converter with a negative unity voltage feedback yielding the buck converter.

buck converter shown in Figure 6.14f, in which capacitor C_1 is just in parallel with input voltage source V_i and can be merged with V_i. This proves that the same transfer code should have the same converter topology.

6.3 Discussion

As shown in Figure 6.14, Ćuk converter with a negative unity voltage feedback yields the buck converter. Similarly, a buck–boost converter with a negative unity voltage feedback will also yield the buck converter, as illustrated in Figure 6.15. Figure 6.15a shows a buck–boost converter with a negative unity voltage feedback, and Figure 6.15b shows its transfer code configuration. With further processing, we can have the buck converter with its transfer code D, as shown in Figure 6.15c.

The above derivation proves that the same transfer code will yield the same converter topology and might have different capacitor DC offset voltages or with extra LC filters. It should be noted that the capacitor having different DC offset voltages will not change its input-to-output voltage transfer code. In the following, several converters with the same transfer code will be proved to be identical in converter topology.

6.3.1 Deduction from Ćuk to Buck–Boost

Ćuk and buck–boost converters have the same transfer code and have the same output voltage polarity. However, Ćuk converter has more component counts than those of the buck–boost converter. From DC input-to-output voltage transfer code point of view, these two converters should be identical. Figure 6.16 illustrates the

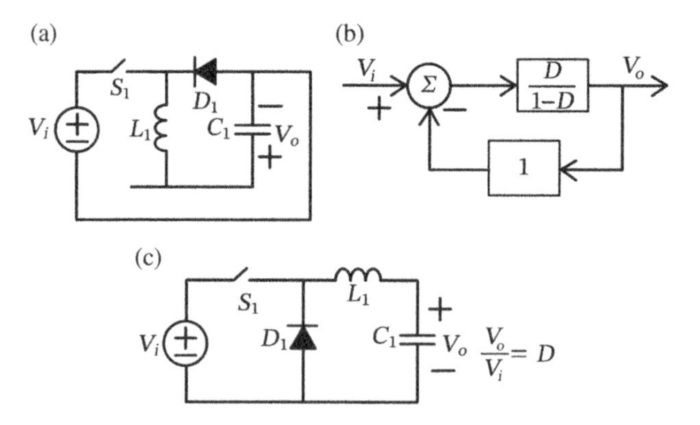

Figure 6.15 Illustration of a buck–boost converter with a negative unity voltage feedback yielding the buck converter with transfer code D.

deduction from the Ćuk converter to the buck–boost converter. Figure 6.16a shows the Ćuk converter, and Figure 6.16b shows the one with inductor L_1 left in the return path. With DC voltage offsetting scheme, presented in Chapter 3, the left terminal of capacitor C_1 can move from the positive polarity of input voltage source V_i to its negative polarity. In other words, capacitor C_1 has voltage $-V_i$ offset. The circuit becomes the one shown in Figure 6.16c. Again, capacitor C_1 has another V_o voltage offset, which means that its right terminal moves from negative polarity of V_o to the positive polarity, as shown in Figure 6.16d. With further processing on the component connection, we have a buck–boost converter with an extra LC filter.

It can be proved that the voltage across capacitor C_1 in the Ćuk converter is $V_i + V_o$. After having the voltage offsets of $-V_i$ at the positive node of C_1 and V_o at the negative node of C_1, the voltage across capacitor C_1 in the buck–boost converter becomes 0 V, which can be proved from the circuit of L_1, L_2, and C_1 forming a loop, as shown in Figure 6.16e. In fact, capacitor C_1 in the Ćuk converter acts as a buffer that does not affect the input-to-output voltage transfer code. Since the voltage across capacitor C_1 is 0 V, it acts as an AC filter, and in DC sense, it is a shorted circuit, and the two inductors L_1 and L_2 are equivalently connected in parallel. Thus, from topological point of view, the converter shown in Figure 6.16e is the same as a buck–boost converter. Or it can be simply said that the buck–boost converter is equipped with either an L_1C_1 or C_1L_2 extra filter.

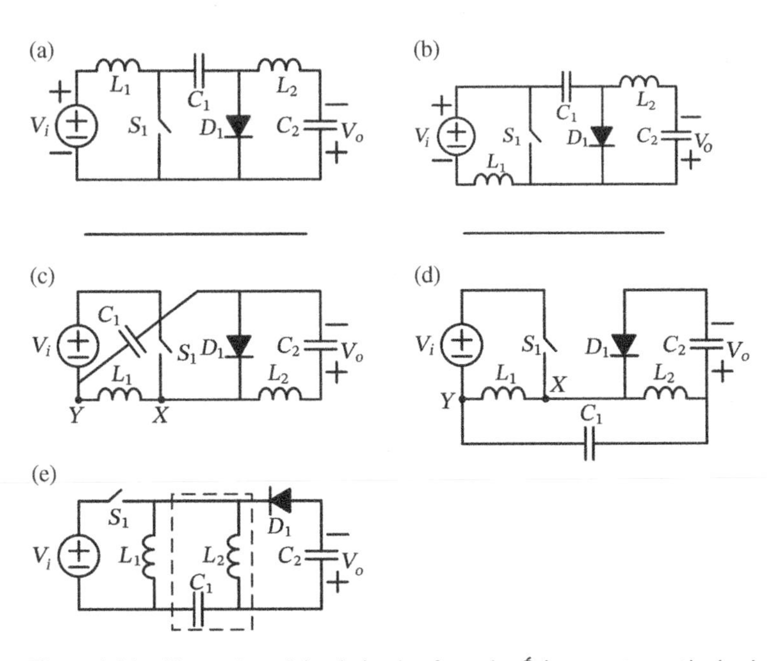

Figure 6.16 Illustration of the deduction from the Ćuk converter to the buck–boost one.

6.3.2 Deduction from Sepic to Buck–Boost

Sepic and buck–boost converters have the same voltage transfer code, but have different output voltage polarities. Additionally, they are belonged to different families in which the sepic belongs to the boost family, while the buck–boost belongs to the buck one. However, since their voltage transfer codes are identical, they will have the same converter topology, which can be illustrated by Figure 6.17. Again, the DC voltage offsetting scheme is adopted. Figure 6.17a shows the sepic converter, and Figure 6.17b shows the converter with its inductor L_1 in return path and its capacitor C_1 with $-V_i$ voltage offset. With further processing on component connection, we have the one shown in Figure 6.17c, and then redrawing it in the form similar to the buck–boost converter with diode D_1 in the return path, as shown in Figure 6.17d. Moving diode D_1 from the return path to the forward path yields the buck–boost converter with an extra LC filter, as shown in Figure 6.17e.

The voltage across capacitor C_1 in the sepic converter is V_i, while with $-V_i$ voltage offset, the voltage across capacitor C_1 in the buck–boost converter shown in Figure 6.17e is 0 V. Again, it acts as an AC filter. From the above derivation, even though belonged to different families, they have identical voltage transfer code and finally come out the same converter topology.

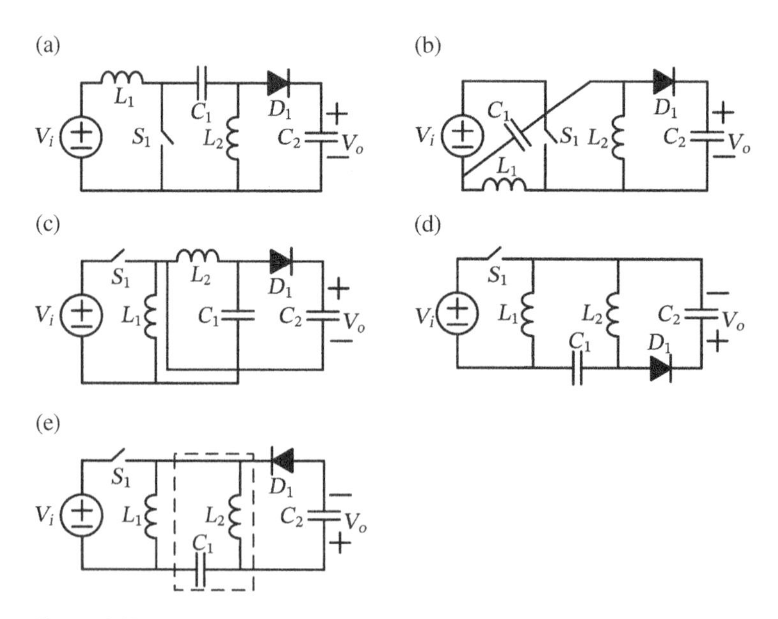

Figure 6.17 Illustration of the deduction from the sepic converter to the buck–boost one.

6.3.3 Deduction from Zeta to Buck–Boost

Zeta and buck–boost converters are belonged to the buck family, but they have different output voltage polarities. However, they have the same voltage transfer code, and they are supposed to have the same converter topology. This can be illustrated by Figure 6.18. Figure 6.18a shows a Zeta converter, and Figure 6.18b shows the one with inductor L_2 in return path. Capacitor C_1 with $-V_o$ voltage offset yields the converter shown in Figure 6.18c. With further processing, we have the one with diode D_1 in return path, as shown in Figure 6.18d. Finally, moving diode D_1 to the forward path yields the buck–boost converter with an extra *LC* filter, as shown in Figure 6.18e.

The voltage across capacitor C_1 can be proved to be V_o, while with $-V_o$ offset, the one shown in Figure 6.18e is 0 V. Capacitor C_1 acts as an AC filter, whose gain is unity. Therefore, the converter shown in Figure 6.18e has the transfer code of $D/(1-D)$, which is identical to that of a buck–boost converter.

From Sections 6.3.1 to 6.3.3, we can observe that even with different output voltage polarities, belonging to different families, with different component counts, once their voltage transfer codes are identical, they will finally yield the same converter topology but might be with an extra *LC* filter and different DC offset voltages in the buffer capacitor, C_1. If the two converters have originally the same component counts, they will yield the same converter topology but with different offset voltages. This will be explained in Section 6.3.4.

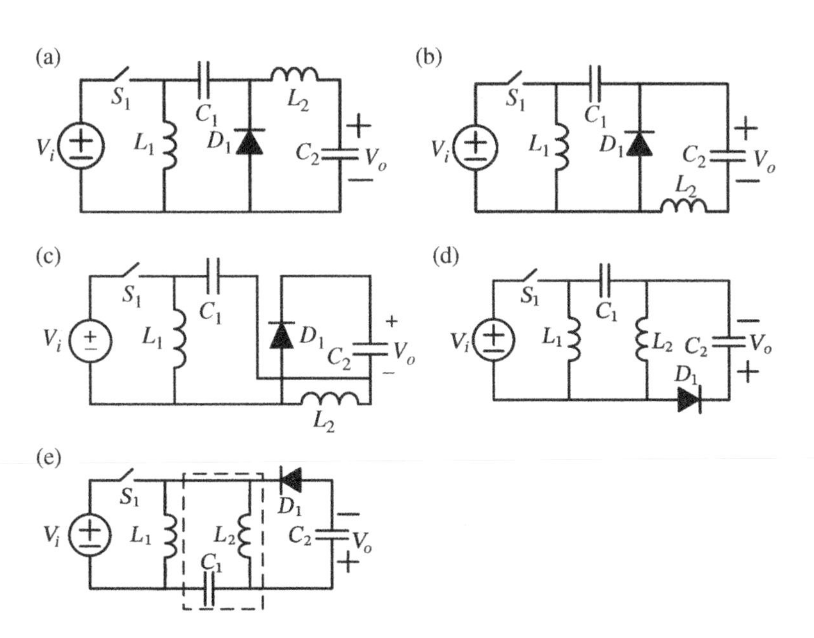

Figure 6.18 Illustration of the deduction from the Zeta converter to the buck–boost one.

6.3.4 Deduction from Sepic to Zeta

Sepic and Zeta converters belong to different families, but with the same component counts and the same transfer code. Additionally, their output voltage polarities are also identical. How come they have different component configurations? Their buffer capacitors might have different offset voltages. Figure 6.19 shows the illustration of the deduction from the sepic converter to the Zeta converter. Figure 6.19a shows the sepic converter, and Figure 6.19b shows the inductor L_1 in the return path. When capacitor C_1 is with $-V_i$ voltage offset at its left terminal and with $-V_o$ voltage offset at its right terminal, we have the converter shown in Figure 6.19c. Reconfiguring the components while not changing its operational principle, we have the converter shown in Figure 6.19d. Moving capacitor C_1 from the return path to the forward path yields the well-known Zeta converter, as shown in Figure 6.19e.

The voltage across capacitor C_1 in sepic converter can be proved to be V_i, while that in Zeta converter is V_o. Again, the DC voltage offset does not change it transfer code. Thus, these two converters have different capacitor voltage offsets, but they have the same transfer code.

From all of the above deductions, we can recognize that the buck–boost, Ćuk, Zeta, and sepic converters have the same transfer code, but they have different voltage

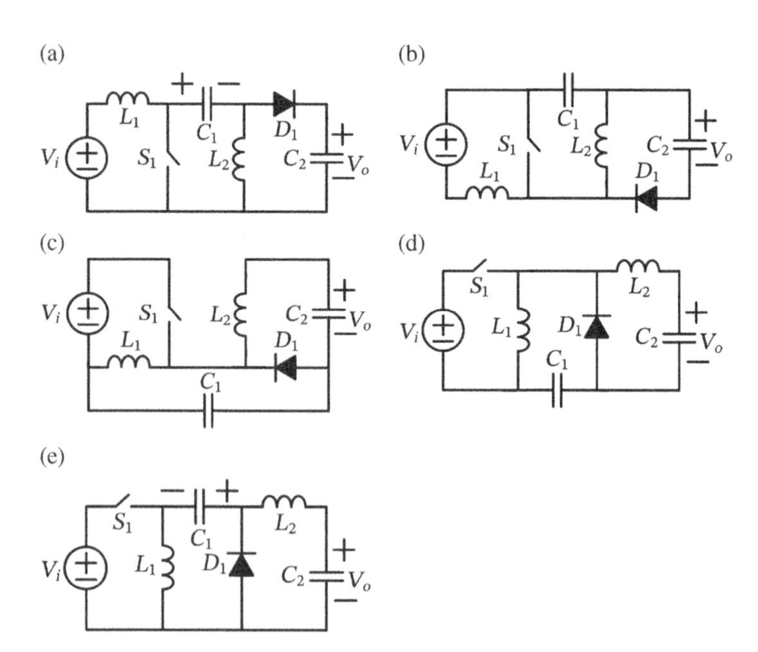

Figure 6.19 Illustration of the deduction from the sepic converter to the Zeta one.

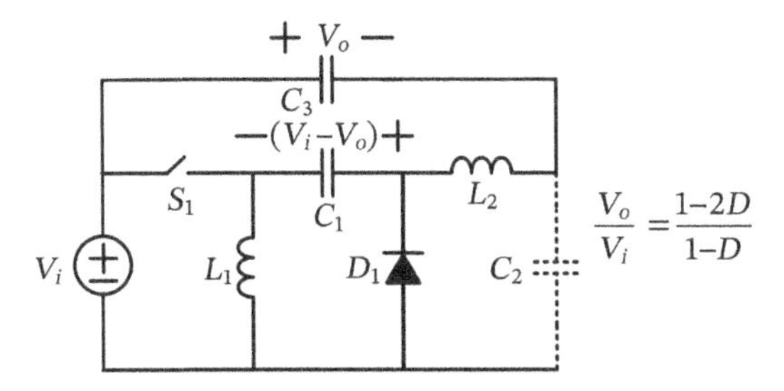

Figure 6.20 Taking output from the other port of the Zeta converter to illustrate the buffer capacitor with DC offset voltage of (V_i - V_o).

offsets across the buffer capacitor and add an extra *LC* filter to the buck–boost converter. In Ćuk, sepic, and Zeta converters, the buffer capacitors have the DC offset voltages ($V_i + V_o$), V_i, and V_o, respectively. In fact, the extra *LC* filter in the buck–boost converter can be treated as the buffer capacitor with 0 V DC offset. When moving the buffer capacitor from the forward path to the return path, we have the DC offset voltages $-(V_i + V_o)$, $-V_i$, and $-V_o$, respectively. There is a question coming to our mind that if there exists ($V_i - V_o$) or ($V_o - V_i$) DC offset voltage across the buffer capacitor. Figure 6.20 shows a converter configuration in which the buffer capacitor C_1 has the DC offset voltage of ($V_i - V_o$). Essentially, the converter is obtained by taking the output from the other port (capacitor C_3) of the Zeta converter, instead of the conventional approach with the output from capacitor C_2, from which the input-to-output voltage transfer code can be derived as $V_o/V_i = (1-2D)/(1-D)$.

Basically, with the same voltage transfer ratio of $D/(1-D)$, the buck–boost converter has the least component count, while the rest of them have included another *LC* filter, which can improve their input or output dynamic characteristics. This argument will be discussed in Part II.

Further Reading

Axelrod, B., Borkovich, Y., and Ioinovici, A. (2008). Switched-capacitor/switched-inductor structures for getting transformerless hybrid dc–dc PWM converters. *IEEE Trans. Circuits Syst. I* 55 (2): 687–696.

Cain, M.L., Damman, H., Lue, R.A., and Yoon, C.K. (2000). *Discover Biology*. Sinauer Association, Inc. and W. W. Norton & Company.

Erickson, R.W. (1983). Synthesis of switched-mode converters. *Proceedings of the IEEE Power Electronics Specialists Conference*, IEEE, pp. 9–22.

Makowski, M.S. (1993). On topological assumptions on PWM converters: a re-examination. *Proceedings of the IEEE Power Electronics Specialists Conference*, IEEE, pp. 141–147.

Maksimovic, D. and Ćuk, S. (1989). General properties and synthesis of PWM DC-to-DC converters. *Proceedings of the IEEE Power Electronics Specialists Conference*, IEEE, pp. 515–525.

Williams, B.W. (2008). Basic DC-to-DC converters. *IEEE Trans. Power Electron.* 23 (1): 387–401.

Wu, T.-F. (2016). Decoding and synthesizing transformerless PWM converters. *IEEE Trans. Power Electron.* 30 (9): 6293–6304.

Wu, T.-F. and Chen, Y.-K. (1996). A systematic and unified approach to modeling PWM DC/DC converters using the layer scheme. *Proceedings of the IEEE Power Electronics Specialists Conference*, IEEE, pp. 575–580.

Wu, T.-F. and Yu, T.-H. (1998). Unified approach to developing single-stage power converters. *IEEE Trans. Aerosp. Electron. Syst.* 34 (1): 221–223.

Wu, T.-F., Liang, S.-A., and Chen, Y.-K. (2003). A structural approach to synthesizing soft switching PWM converters. *IEEE Trans. Power Electron.* 18 (1): 38–43.

7

Converter Derivation with the Fundamentals

As described in Chapter 3, there are many fundamentals, including graft and layer schemes, which can be used to synthesize converter topologies. In Chapters 5 and 6, the graft and layer schemes alone have been adopted to synthesize converters, respectively. In this chapter, we will exploit all of the fundamentals discussed in Chapter 3 and use the decoding and synthesizing processes to derive converters according to given transfer codes.

This chapter is organized as follows. First, we start with a single code figuration for synthesizing the origin converter, buck converter. Next, several well-known transfer codes for synthesizing z-source converters and switched-capacitor/switched-inductor converters are reviewed. Finally, new converter topologies are developed as examples.

7.1 Derivation of Buck Converter

Buck converter is the original converter that can be used to derive other converters. On the other hand, from other converters, we can also derive the buck converter. Basically, the buck converter has already had a very simple transfer code D, and it is no need to be configured to another complex code configuration. However, to illustrate the derivation of converters with the fundamentals, decoding and synthesizing processes, we use the buck converter as an example. Through decoding process, Figure 7.1 shows a code configuration for yielding the voltage transfer code D. There are four converters, buck–boost, Ćuk, sepic, and Zeta converters, with the transfer code of $D/(1-D)$, but only can buck–boost and Ćuk converters have negative unity feedbacks, which will be used to synthesize the code configuration shown in Figure 7.1.

Origin of Power Converters: Decoding, Synthesizing, and Modeling, First Edition.
Tsai-Fu Wu and Yu-Kai Chen.
© 2020 John Wiley & Sons, Inc. Published 2020 by John Wiley & Sons, Inc.

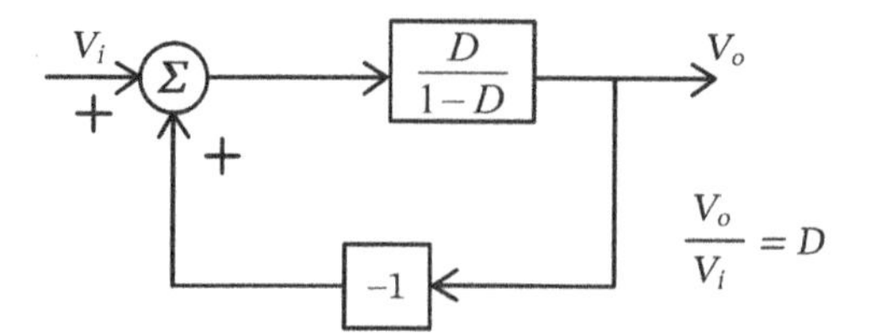

Figure 7.1 Code configuration for synthesizing the buck converter.

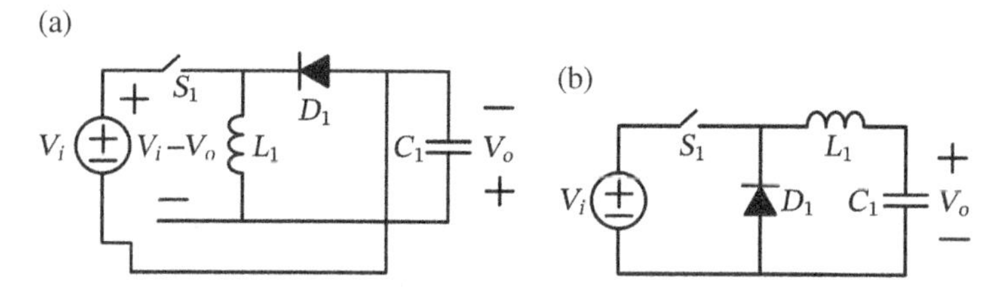

Figure 7.2 Buck converter synthesized with the buck–boost converter and a negative unity feedback.

7.1.1 Synthesizing with Buck–Boost Converter

In Figure 7.1, the voltage transfer code $D/(1-D)$ with a negative unity feedback can be synthesized with the buck–boost converter, as shown in Figure 7.2a. Rearranging its component configuration will yield the buck converter, as shown in Figure 7.2b.

7.1.2 Synthesizing with Ćuk Converter

Again, the voltage transfer code $D/(1-D)$ with a negative unity feedback shown in Figure 7.1 can be synthesized with the Ćuk converter, as shown in Figure 7.3a. Rearranging the component configuration without changing its operational principle will yield the one shown in Figure 7.3b. With inductor splitting, inductor L_2 is split into two inductors, L_{21} and L_{22}, and connected to the return paths of input and output, respectively, as shown in Figure 7.3c. Then, inductors L_1 and L_{21} are in series and can be combined into a single one L_1', and after rearranging the overall component configuration, the buck converter with an extra LC filter is shown in Figure 7.3d.

7.2 Derivation of z-Source Converters

Z-Source converters are famous for their DC/DC and DC/AC applications with low component count and simple configuration. There are three types of z-source converters, including voltage-fed, current-fed, and quasi-z-source converters.

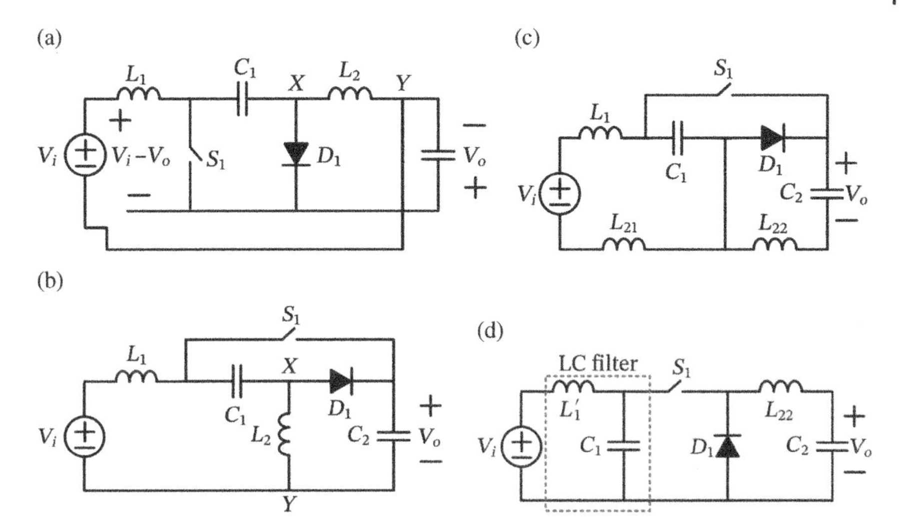

Figure 7.3 Buck converter synthesized with the Ćuk converter and a negative unity feedback.

However, there is no systematic approach to deriving these converters. Basically, there are PWM converters with certain transfer codes, and they are supposed that they can be derived through decoding and synthesizing processes. In the following, we derive these three types of z-source converters with the processes and fundamentals in detail.

7.2.1 Voltage-Fed z-Source Converters

A voltage-fed z-source converter has the transfer code, $(1-D)/(1-2D)$. First of all, through a long division, we have

$$\frac{V_o}{V_i} = \frac{(1-D)}{(1-2D)} = 1 + \frac{D}{(1-2D)} = 1 + TC_r(D), \tag{7.1}$$

where $TC_r(D)$ is the residual transfer code and can be further defined as

$$\frac{V_o'}{V_i} = TC_r(D) = \frac{D}{(1-2D)}, \tag{7.2}$$

where V_o' is a sub-output that will be combined with the unity gain to yield output voltage V_o. This transfer code is not obviously close to the well-known fundamental codes. Thus, we need to use cross multiplication approach to deriving a relation that can be readily recognized in a feedback configuration.

With a cross multiplication for (7.2), we have

$$V_o'(1-2D)=V_iD, \tag{7.3}$$

which is also equal to the following expression:

$$V_o'(1-D)=(V_i+V_o')D. \tag{7.4}$$

That is,

$$V_o'=\frac{(V_i+V_o')D}{1-D}. \tag{7.5}$$

Equation (7.5) can be put into a code feedback configuration, as shown in Figure 7.4a, in which $D/(1-D)$ is a fundamental code. Alternately, the expression in (7.3) can be expressed in the following two simultaneous equations:

$$V_o''=\frac{V_o''D+V_i}{1-D}. \tag{7.6}$$

and

$$V_o'=DV_o'. \tag{7.7}$$

The above two equations can be configured into the code feedback configuration shown in Figure 7.4b, in which the two forward codes, $1/(1-D)$ and D, and a feedback code, D, are all fundamental codes. Combining the unity gain in (7.1) with (7.5) or (7.6) and (7.7) can yield the transfer code: $(1-D)/(1-2D)$, and its code configurations are shown in Figure 7.5a and b, which are corresponding to Figure 7.4a and b, respectively. The code configuration shown in Figure 7.4a can be synthesized with either sepic or Zeta converter, which is presented in Sections 7.2.1.1 and 7.2.1.2.

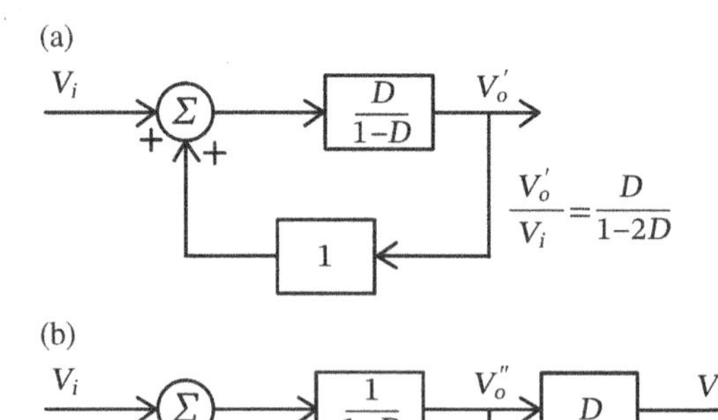

Figure 7.4 Two possible configurations of $D/(1-2D)$: (a) feedback configuration and (b) feedback with forward configuration.

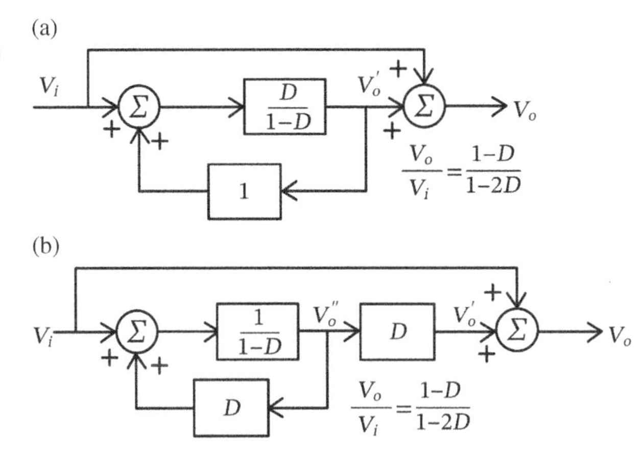

Figure 7.5 The two configurations shown in Figure 7.4 combined with a unity-gain feedforward.

7.2.1.1 Synthesizing with Sepic Converter

The code configuration shown in Figure 7.4a is synthesized with sepic converter and a positive unity feedback, as shown in Figure 7.6a. After rearranging the component location, we have the one shown in Figure 7.6b. To realize a unity input voltage V_i feedforward, as shown in Figure 7.5a, we take the output from the upper port, in which $V_o = V_i + V_o'$. When pulling output V_o to the right-hand side of the converter, we have the one shown in Figure 7.6c. Changing the DC offset of capacitor $C_2' \left(= C_2 // C_f\right)$ from negative V_i to positive V_i yields the converter configuration shown in Figure 7.6d. Then, rearranging the component location will yield a voltage-fed z-source like converter, as shown in Figure 7.6e, which has the same operational principle as that of the well-known voltage-fed z-source converter, as shown in Figure 7.6f. Essentially, when split inductor L_1 shown in Figure 7.6e into L_{11} and L_{12}, we have the exact voltage-fed z-source converter, as shown in Figure 7.6f. It can be determined with the volt-second balance principle that the input-to-output voltage transfer code is $V_o/V_i = (1-D)/(1-2D)$.

For the code configurations shown in Figures 7.4a and 7.5b, since there exist code D feedback and code D feedforward that will be synthesized with buck converters, it might need more active and passive switches for synthesis, as illustrated in Figure 7.7. Figure 7.7a shows one boost converter, for synthesizing the code of $1/(1-D)$, connected to two buck converters, in which one buck is in feedback path, the other is in forward path. Through switch grafting, switches S_2 and S_3 can be replaced with a grafted switch S_{23}, and then the two diodes D_2 and D_3 are in parallel and can be replaced with a single one D_{23}, as shown in Figure 7.7b. When combining inductors L_2 and L_3 into a single one L_{23}, we have the converter shown in Figure 7.7c, and rearranging the component location will yield the one shown in Figure 7.7d. Adding a positive unity feedforward to the converter shown in Figure 7.7d and pulling the output $V_o = V_i + V_o'$ to the

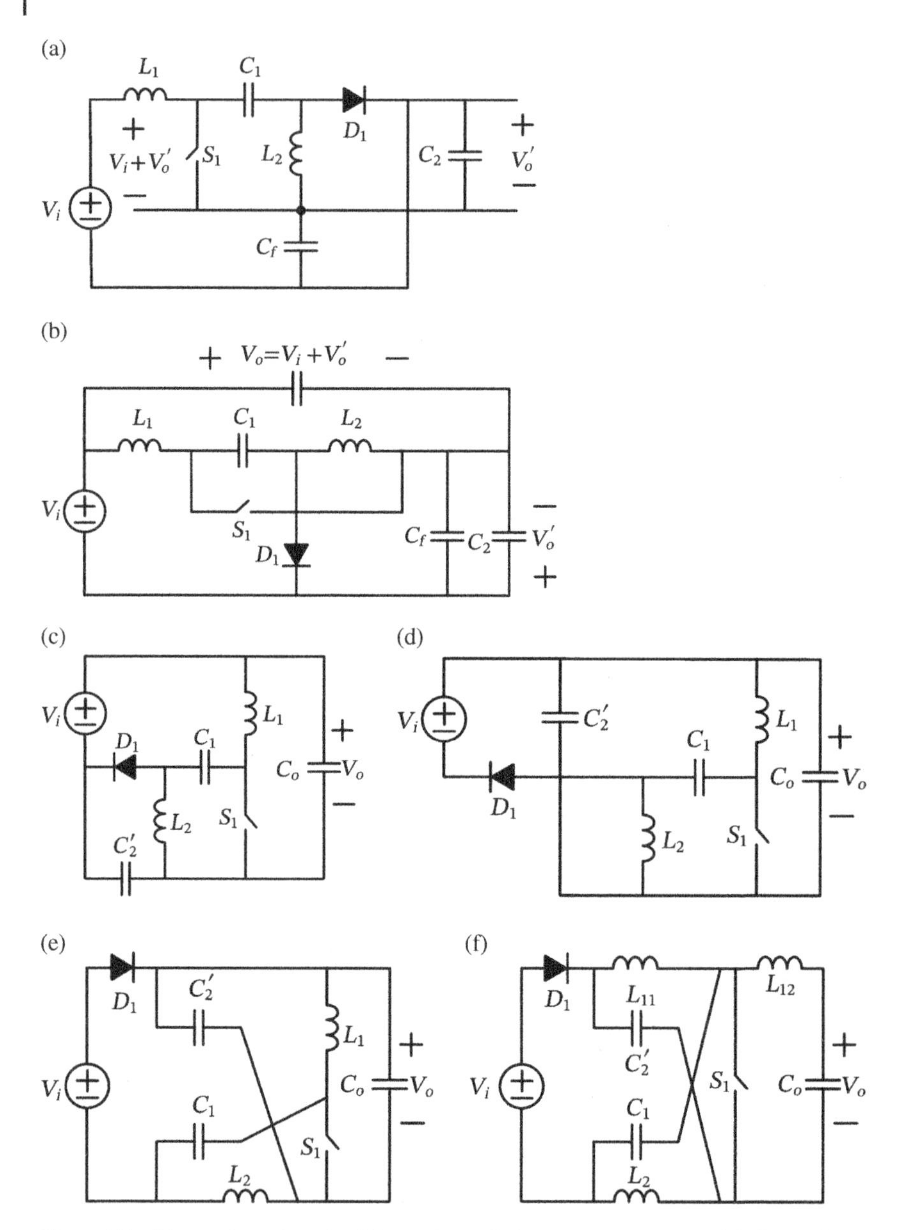

Figure 7.6 Illustration of synthesizing voltage-fed z-source converter with the sepic converter and using the code configuration shown in Figure 7.5a.

right-hand side of the converter will yield the one shown in Figure 7.7e. It can be seen that there are two active switches and two diodes. Since there is no common node between switches S_1 and S_{23}, the two switches cannot be grafted to become a single one.

It is interesting to note that with the code configuration shown in Figure 7.5a, there is no converter code in the feedback or feedforward path but with a unity feedback only. Thus, there is no need of using switch grafting and still keeping with a single active and passive switches. However, with the code configuration shown in Figure 7.5b, two buck codes D are in the paths, and we need to use switch grafting technique to graft active switches. It might result in multiple active switches in the converter. In fact, the buck converter grafted on the boost converter will yield the Ćuk converter, which has a negative output voltage polarity, different from the positive polarity at the output of a sepic converter. This might be the reason that the synthesis of the code configuration shown in Figure 7.5b yields the converter with multiple active and passive switches. This issue needs further study.

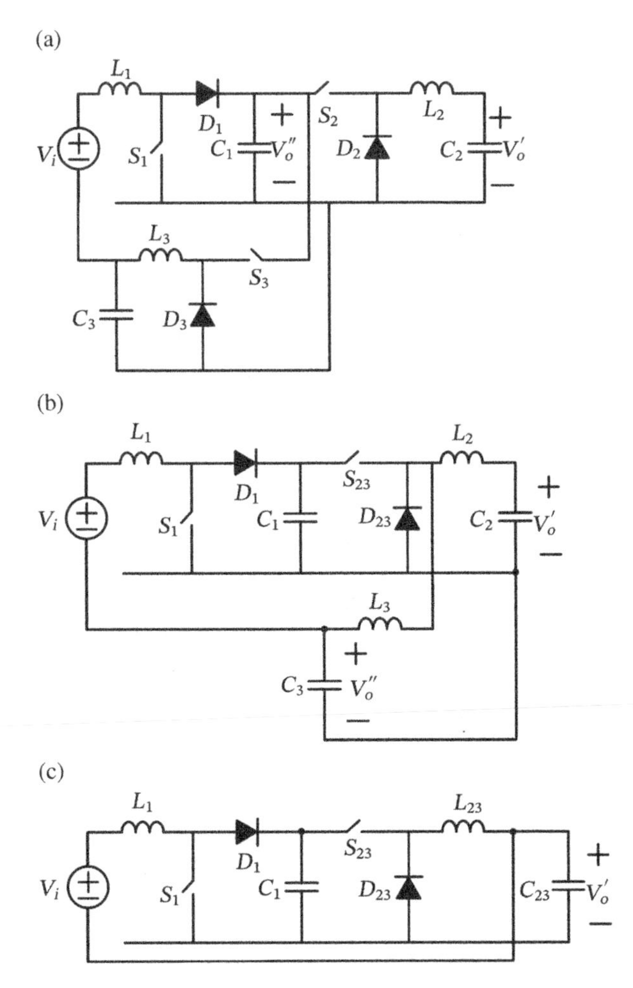

Figure 7.7 Illustration of synthesizing voltage-fed z-source converter with the sepic converter and using the code configuration shown in Figure 7.5b.

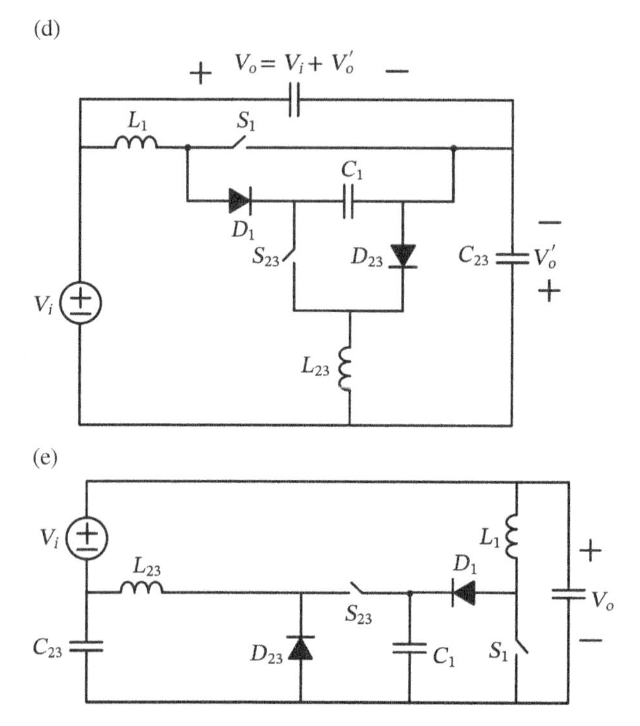

Figure 7.7 (Continued)

7.2.1.2 Synthesizing with Zeta Converter

The code configuration shown in Figure 7.4a can be synthesized with a Zeta converter, as shown in Figure 7.8a. Rearranging the component location and adding a positive unity feedback yield the converter shown in Figure 7.8b. It should be noted here that C_f has been merged to C_2. By pulling the output of $V_o = V_i + V_o'$ to the right-hand side of the converter, we have the one shown in Figure 7.8c, in which C_2 has been changed with its DC offset from $-V_i$ to $+V_i$. Moving inductor L_2 to the forward path and changing the DC voltage offset of capacitor C_1 will yield the converter shown in Figure 7.8d. Again, moving the diode from the return path to the forward path and rearranging the component location will yield the converter shown in Figure 7.8e. Basically, the converter has the same operational principle as that of the voltage-fed z-source converter. To have the well-known configuration of a voltage-fed z-source converter, we split inductor L_1 into L_{11} and L_{12}, and move L_{12} from the return path to the forward one, as shown in Figure 7.8f.

It is worth noting that the sepic and Zeta converters have the same transfer code and have the same output voltage polarity, and therefore, they can be used to synthesize the code of $D/(1-D)$ with a positive unity feedback. Basically, the converters shown in Figures 7.6e and 7.8e are good enough to fulfill the voltage-fed

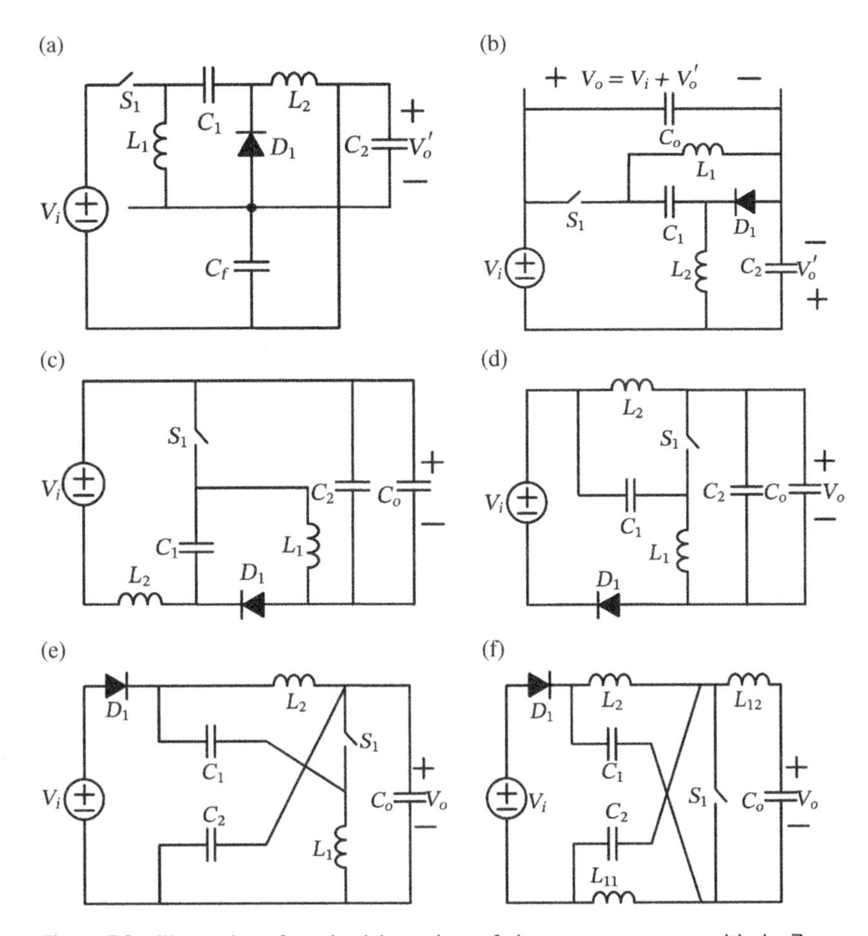

Figure 7.8 Illustration of synthesizing voltage-fed z-source converter with the Zeta converter and using the code configuration shown in Figure 7.5a.

z-source converter function. However, they split inductor L_1 into two inductors L_{11} and L_{12} without changing their operational principle. It needs an extra inductor but may improve dynamics, which is worthy of future study.

7.2.2 Current-Fed z-Source Converters

Current-fed z-source converter has the following transfer code:

$$\frac{V_o}{V_i} = \frac{(2D-1)}{1-D}. \tag{7.8}$$

With long division, Eq. (7.8) can be expressed as follows:

$$\frac{V_o}{V_i} = \frac{(2D-1)}{1-D} = \frac{D}{(1-D)} - 1. \tag{7.9}$$

The transfer code of (7.9) can be put into a code configuration shown in Figure 7.9. buck–boost, Ćuk, sepic, and Zeta converters can synthesize transfer code $D/(1-D)$. However, only can either sepic or Zeta converter combined with a negative unity feedforward synthesize the code configuration shown in Figure 7.9.

7.2.2.1 Synthesizing with SEPIC Converter

Figure 7.10a shows a sepic converter with a negative unity feedforward, which is equivalent to $V_o' - V_i$, pulling V_o to the right-hand side of the converter, as shown in Figure 7.10b. Changing the DC voltage offset of C_2 by $-V_o$ will yield the one shown in Figure 7.10c. Then, splitting inductor L_1 into L_{11} and L_{12} will yield the current-fed z-source converter, as shown in Figure 7.10d.

7.2.2.2 Synthesizing with Zeta Converter

The code configuration shown in Figure 7.9 can be synthesized with a Zeta converter and a negative unity feedforward, as shown in Figure 7.11a, which yields the transfer code $V_o/V_i = (2D-1)/(1-D)$. Pulling the output V_o to the right-hand side of the converter will yield the one shown in Figure 7.11b. Then, changing the DC offset of capacitor C_2, moving inductor L_2 from the return path to the forward one, and changing the DC offset of capacitor C_1 will have the converter configuration shown in Figure 7.11c. Splitting inductor L_1 into L_{11} and L_{12} and relocating the components will yield the final version of current-fed z-source converter, as shown in Figure 7.11d.

7.2.3 Quasi-z-Source Converter

A quasi-z-source converter has the transfer code

$$\frac{V_o}{V_i} = \frac{D}{(1-2D)}. \tag{7.10}$$

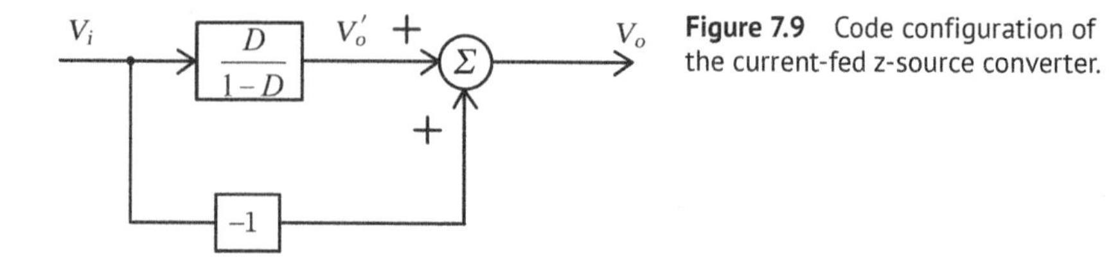

Figure 7.9 Code configuration of the current-fed z-source converter.

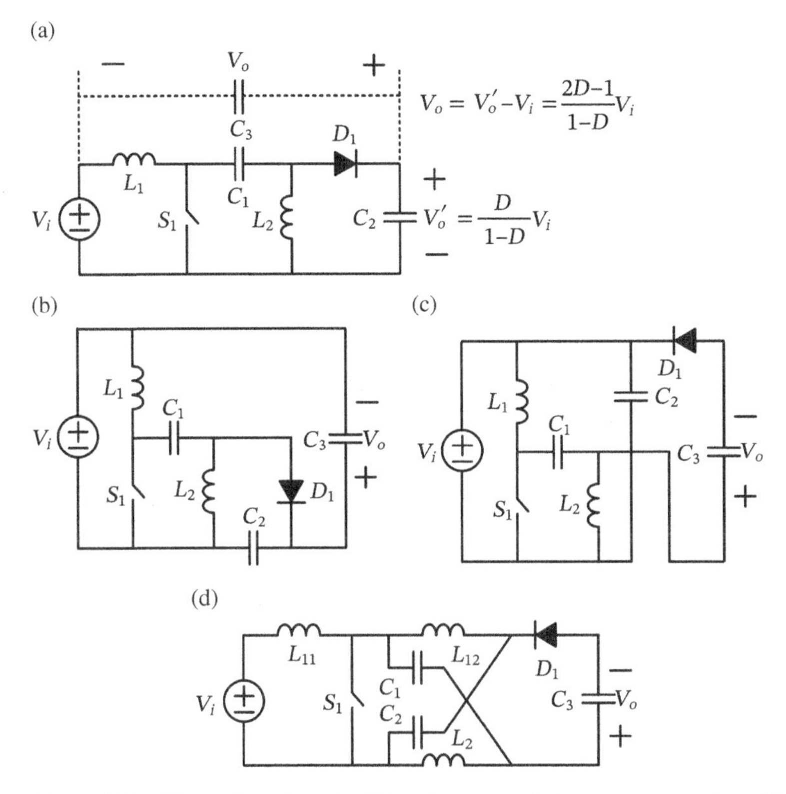

Figure 7.10 Illustration of synthesizing the current-fed z-source converter with the sepic converter and a negative unity feedforward.

With a cross multiplication, we have the following expression:

$$V_o - 2DV_o = V_i D \tag{7.11}$$

or

$$V_o(1-D) = (V_i + V_o)D. \tag{7.12}$$

Equation (7.12) can be further processed to become

$$V_o = (V_i + V_o)\frac{D}{(1-D)}, \tag{7.13}$$

in which output V_o is equal to input V_i plus output V_o multiplied by $D/(1-D)$. They can be represented in the code configuration shown in Figure 7.12.

To synthesize the code configuration shown in Figure 7.12 with transfer ratio $D/(1-D)$ and a positive unity feedback, there are two converters, sepic and Zeta.

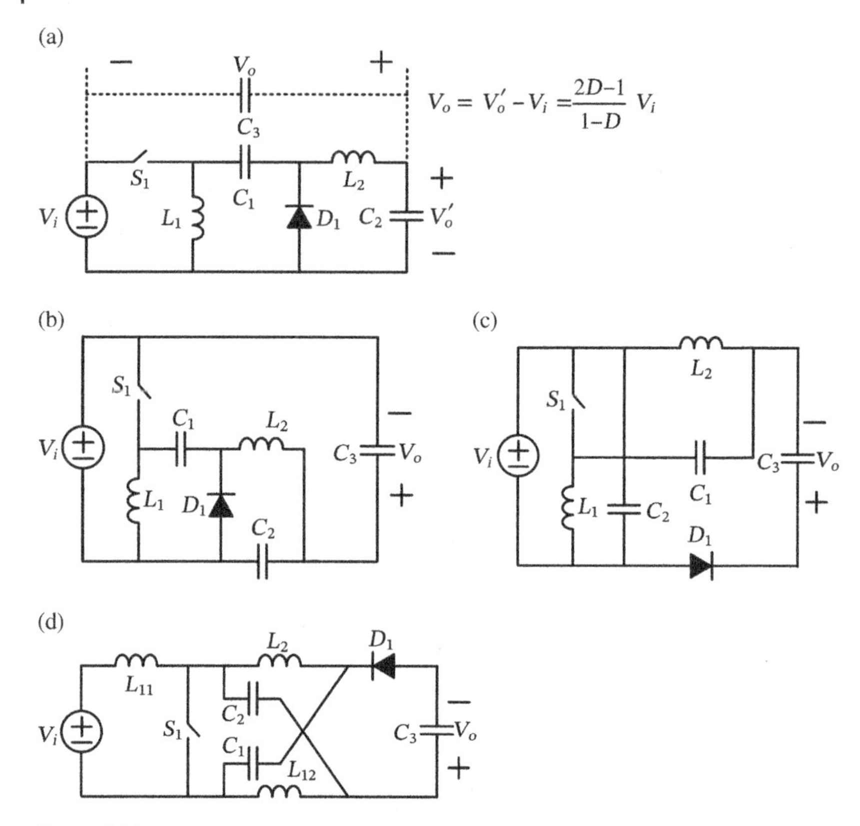

Figure 7.11 Illustration of synthesizing the current-fed z-source converter with a Zeta converter and a negative unity feedforward.

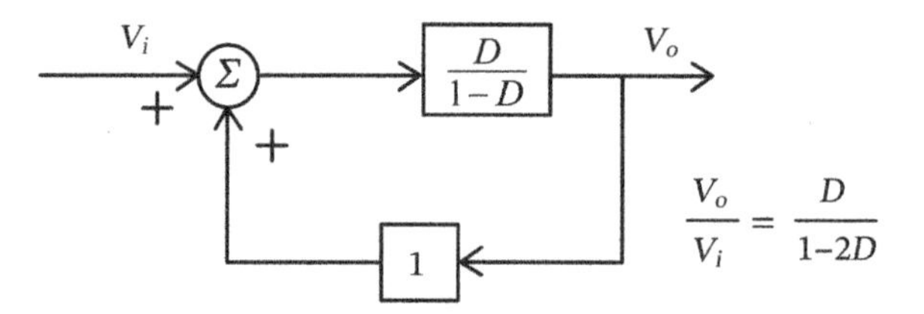

Figure 7.12 Code configuration of the quasi-z-source converter.

7.2.3.1 Synthesizing with Sepic Converter

Figure 7.13a shows a sepic converter with a positive unity feedback. Relocating the components without feedback path yields the one shown in Figure 7.13b, in which its output positive voltage polarity is connected to the negative polarity of the input source. Moving inductor L_1 from the forward path to the return path and

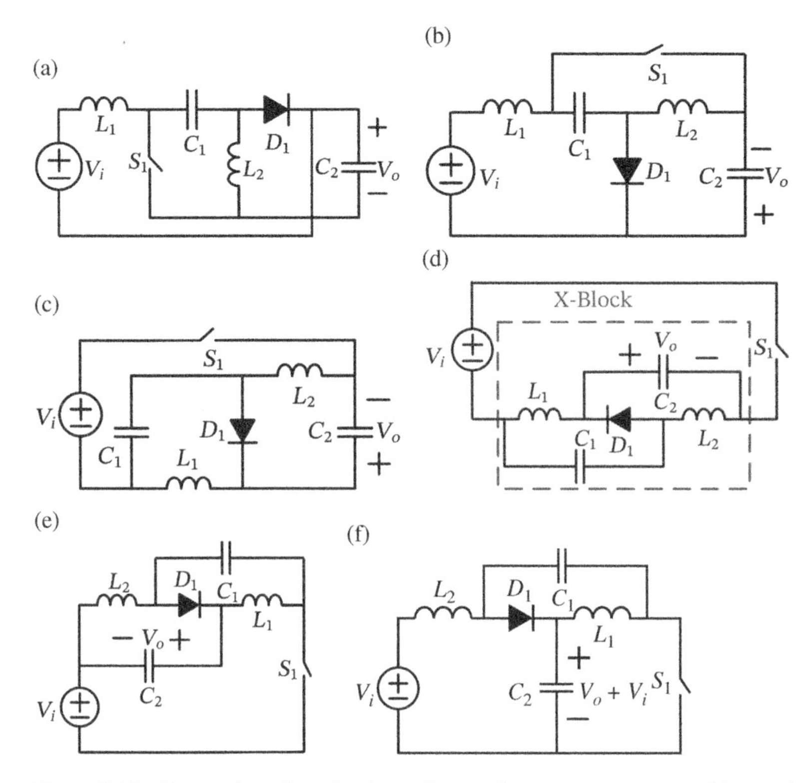

Figure 7.13 Illustration of synthesizing the quasi-z-source converter with a sepic converter and a positive unity feedback.

changing the DC offset of capacitor C_1 will yield the one shown in Figure 7.13c. Then, pulling the active switch to the right-hand side of the converter will become the one shown in Figure 7.13d, in which the rest of the components are marked with X-Block. Moving the X-Block from the return path to the forward one will yield the converter configuration shown in Figure 7.13e, which is a quasi-z-source converter. However, the well-known topology is the one shown in Figure 7.13f, in which the DC offset voltage of capacitor C_2 is changed from V_o to $V_o + V_i$. Basically, both of the two converter topologies shown in Figure 7.13e and f have the same operational principle and will yield the same transfer code.

7.2.3.2 Synthesizing with Zeta Converter

A Zeta converter with a positive unity feedback to synthesize the code configuration shown in Figure 7.12 is shown in Figure 7.14a. Rearranging the components without feedback path yields the one shown in Figure 7.14b. Pulling the active switch to the right-hand side of the converter will yield the one

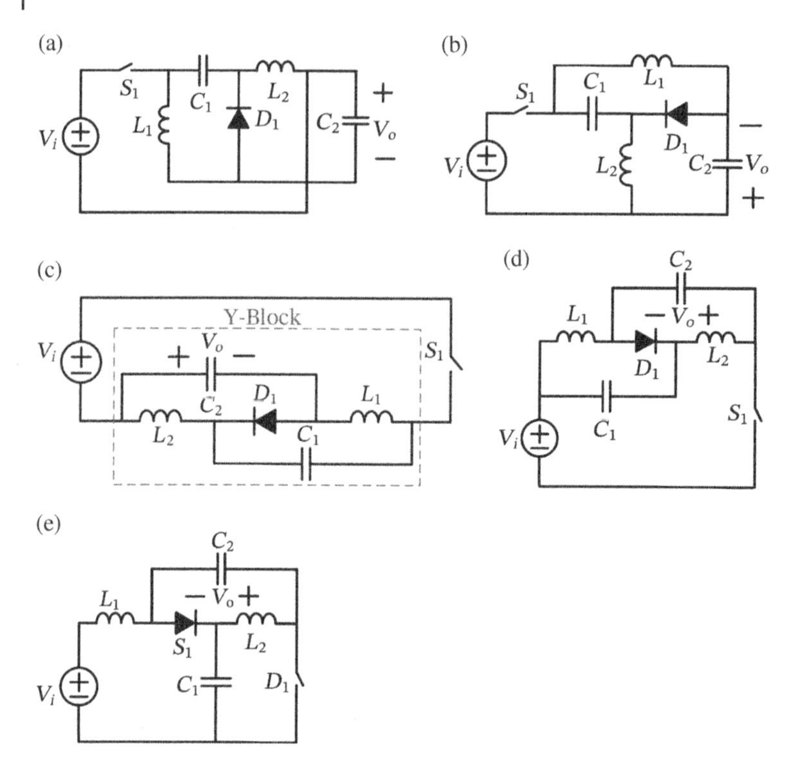

Figure 7.14 Illustration of synthesizing the quasi-z-source converter with a Zeta converter and a positive unity feedback.

shown in Figure 7.14c, in which the Y-Block includes the rest of the components. Moving the Y-Block from the return path to the forward one yields the one shown in Figure 7.14d. Then, changing the DC offset voltage of capacitor C_1 will yield the well-known quasi-z-source converter. Again, the converters shown in Figure 7.14d and e have the same operational principle and transfer code. Therefore, they can be treated as the same quasi-z-source converter.

7.3 Derivation of Converters with Switched Inductor or Switched Capacitor

Before talking about the derivation of converters with switched capacitor or switched inductor, we have to introduce three inverse converters, inverse buck (I-buck), inverse boost (I-boost), and inverse buck–boost (I-buck–boost), with the transfer codes, $1/D$, $1-D$, and $(1-D)/D$, respectively, as shown in Figure 7.15.

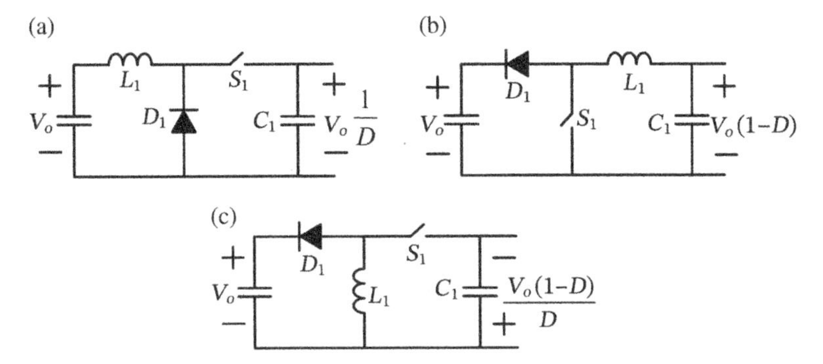

Figure 7.15 Three inverse converters including (a) I-buck, (b) I-boost, and (c) I-buck–boost converters.

Essentially, they are the same as their original converters but with the outputs located in the left-hand side and the inputs referred to the output V_o. Thus, their inputs are marked with $V_o(1/D)$, $V_o(1-D)$, and $V_o(1-D)/D$ for I-buck, I-boost, and I-buck–boost, respectively, where D is still defined as the duty ratio of the active switch S_1.

The purpose to define the three inverse converters is that when using the graft scheme, instead of connecting the input to the other converter, we connect the output to the converter. In the following sections, derivations of switched-inductor converters and switched-capacitor converters are addressed.

7.3.1 Switched-Inductor Converters

Conventionally, there are six PWM converters with regular step-down and step-up transfer codes. There are high step-down and high step-up transfer codes with switched-inductor or switched-capacitor structures. In the following, derivation of the several examples is presented.

7.3.1.1 High Step-Down Converter with Transfer Code *D*/(2 − *D*)

A high step-down switched-inductor converter with the transfer code $D/(2-D)$ is first decoded and synthesized as follows. Through cross multiplication, $V_o/V_i = D/(2-D)$ can be expressed as

$$V_o(2-D) = V_i D. \tag{7.14}$$

or

$$V_i D - V_o(1-D) = V_o. \tag{7.15}$$

Equation (7.15) can be put into a feedback and feedforward form as

$$\left[V_i - \frac{V_o(1-D)}{D}\right]D = V_o. \tag{7.16}$$

Equation (7.16) can be configured as shown in Figure 7.16a. One more possibility to decode transfer code $D/(2-D)$ is shown as follows:

$$\frac{V_o}{V_i} = \frac{D}{(2-D)} = D \cdot \frac{1}{(2-D)}. \tag{7.17}$$

Define $V_o'/V_i = 1/(2-D)$ and we have

$$\frac{V_o}{V_i} = D \cdot \frac{V_o'}{V_i}. \tag{7.18}$$

Through cross multiplication, V_o'/V_i can be expressed as

$$\frac{V_o'}{V_i} = \frac{1}{(2-D)} \tag{7.19}$$

$$V_o'(2-D) = V_i. \tag{7.20}$$

or

$$V_i - V_o'(1-D) = V_o'. \tag{7.21}$$

Combining the feedback expression of (7.21) with a forward term D yields the code configuration shown in Figure 7.16b.

To synthesize the code configuration shown in Figure 7.16a, we use a buck converter as a forward path and use I-buck–boost as a feedback path, as shown in Figure 7.17a. To establish a common node for switches S_1 and S_2, switch S_1 is moved to the return path, as shown in Figure 7.17b. It can be identified that switches S_1 and S_2 have a common D–S node and they can be replaced with a Π-type grafted switch, as shown in Figure 7.17c. Next step is to find the

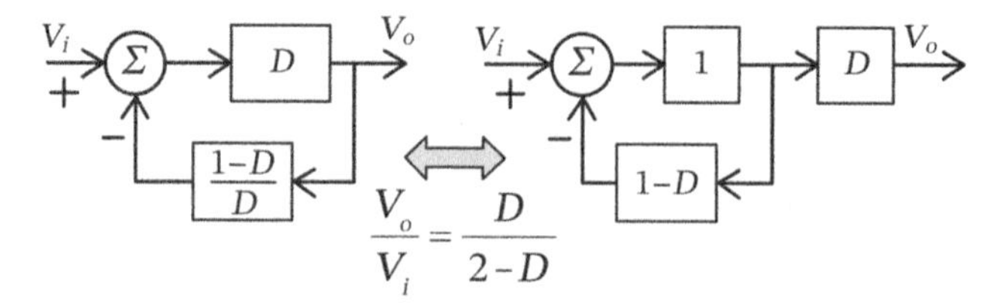

Figure 7.16 Code configurations of the high step-down transfer code $D/(2-D)$.

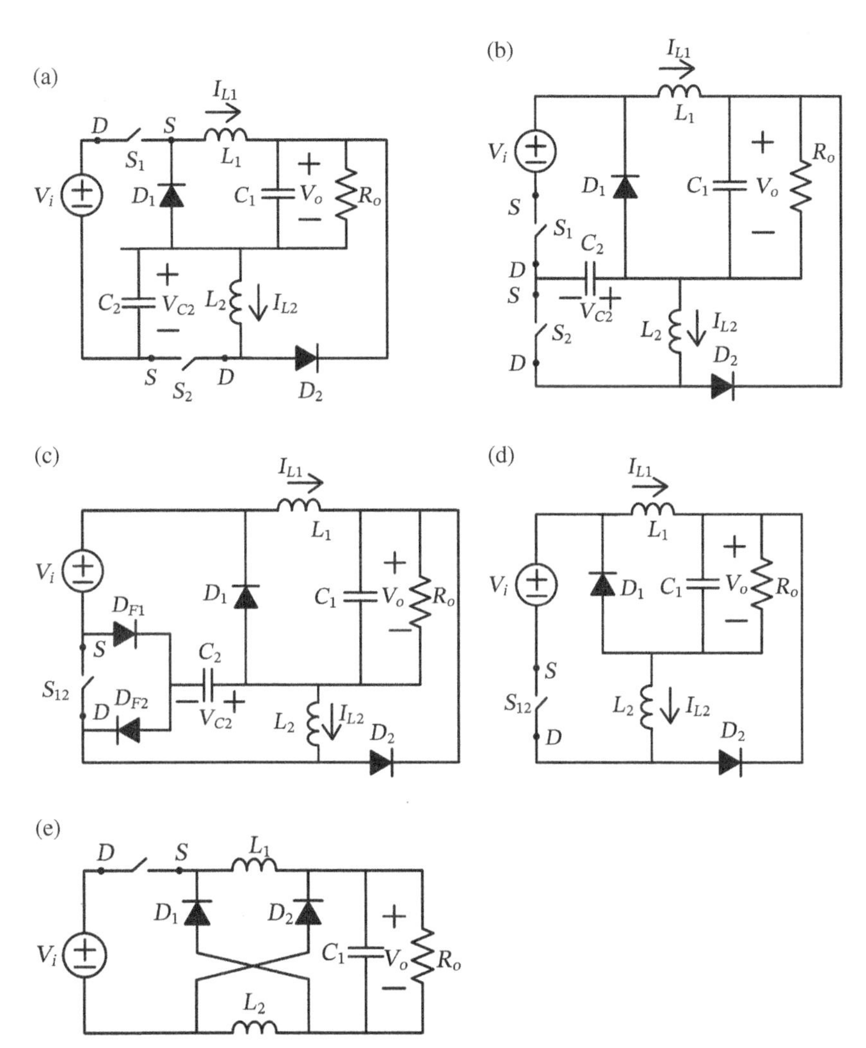

Figure 7.17 Illustration of synthesizing the high step-down converter with a buck converter as a forward path and an I-buck–boost converter as a feedback path.

relationship between the currents through switches S_1 and S_2. When both switches are turned on simultaneously, the voltage across inductor L_1 is $V_i(1-D)/(2-D)$. Basically, the current through switch S_1 in the buck converter is equal to that of inductor L_1, and that through switch S_2 in the I-buck-boost can be proved from that the voltage across inductor L_2 is also equal to $V_i(1-D)/(2-D)$. Under the condition of inductors $L_1 = L_2$, we have $I_{L1} = I_{L2}$, and we can conclude that the currents through switches S_1 and S_2 are identical. In fact, capacitor C_2 has the possibility to conduct current only when active switch S_2 is in the on state. To satisfy with ampere-second balance, there is no average current through capacitor C_2 in the steady state. Therefore, the currents through switches S_1 and S_2 are identical, diodes D_{F1} and D_{F2} are no longer needed, and capacitor C_2 can be removed from the converter, too. The converter becomes the one shown in Figure 7.17d, in which there is only switch S_{12} left. Rearranging the components in a special form will yield the high step-down converter as shown in Figure 7.17e.

For the code configuration shown in Figure 7.16b, although there is an I-boost converter that can serve as the feedback path, there is no converter to synthesize the unity when its input is combined with the negative feedback path. Thus, even though we have a correct code configuration, it is not necessary to have converters to synthesize the code configuration because there is no isolation transformer in the synthesis.

Code configuration is not unique for a given transfer code. Except to the configuration shown in Figure 7.16, transfer code $D/(2-D)$ can be expressed alternately as follows, and they can be configured into different code configurations:

$$\frac{V_o}{V_i} = \frac{D}{(2-D)}. \tag{7.22}$$

With cross multiplication, we have

$$V_o(2-D) = V_i D, \tag{7.23}$$

or

$$V_o(1-D) = V_i D - V_o = \left(V_i - \frac{V_o}{D}\right)D. \tag{7.24}$$

or

$$V_o = \frac{\left(V_i - \frac{V_o}{D}\right)D}{(1-D)}. \tag{7.25}$$

A code configuration to realize (7.25) is shown in Figure 7.18a, in which V_o has a feedback gain $-1/D$, combined with input V_i, which is fed to a forward gain $D/(1-D)$. With similar processes shown in (7.17)–(7.21), we can modify the code configuration shown in Figure 7.18a to the one shown in Figure 7.18b.

To synthesize the code configuration shown in Figure 7.18a, we select a buck–boost converter for the forward transfer code $D/(1-D)$ and an I-buck converter for the

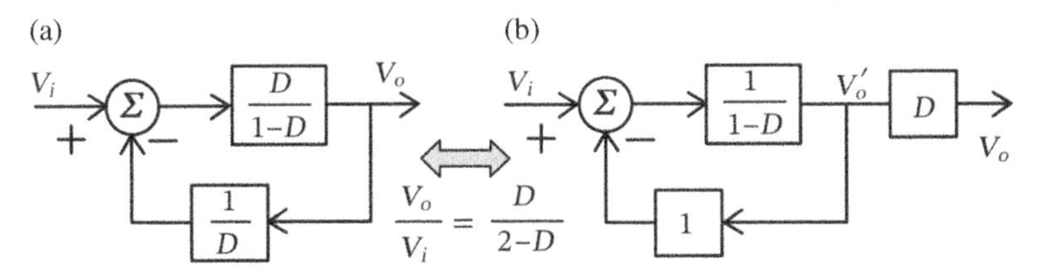

Figure 7.18 Alternate code configurations of the high step-down transfer code $D/(2 - D)$.

feedback transfer code $1/D$. The converter configuration is shown in Figure 7.19a. Moving switch S_1 from the forward path to the return path will establish a common D–S node for switches S_1 and S_2, as shown in Figure 7.19b. Then, if they are operated in unison, the two active switches can be replaced with a Π-type grafted switch S_{12} and with two diodes D_{F1} and D_{F2}, as shown in Figure 7.19c. When both switches S_1 and S_2 are in the on state, the voltage across inductors L_1 and L_2 are identical ($V_i(1-D)/(2-D)$) and yield identical inductor current, $I_{L1} = I_{L2}$. In fact, there is no average current through capacitor C_2 in the steady state. Therefore, we have the same current through switches S_1 and S_2 under their on states, and diodes D_{F1} and D_{F2} conduct no current, so does capacitor C_2, and they can be removed from the converter, as shown in Figure 7.19d. After rearranging the component positions, we have the final high step-down switched-inductor converter, as shown in Figure 7.19e.

For the code configuration shown in Figure 7.18b, since the forward path has a transfer code of $1/(1-D)$, it will be synthesized with a boost converter. However, a boost converter has no negative feedback. Thus, there is no converter to synthesize the code configuration shown in Figure 7.18b. Similar to that of Figure 7.16b, not all of the valid code configurations can be synthesized with converters. For instance, other code configurations for achieving the transfer code $D/(2- D)$ are shown in Figure 7.20. However, there is no converter to synthesize the transfer code $D/2$ shown in Figure 7.20a, and there is no negative feedback for an I-buck converter with the transfer code $1/D$, either, as shown in the inner loop of Figure 7.20b.

Comparing the code configurations shown in Figures 7.16b, 7.18b, and 7.20 with those shown in Figures 7.16a and 7.18a will lead to the following observations: (i) Less items of forward path in the code configurations will lead to more chance to be synthesized with converters; (ii) even though there are valid code configurations, it is not necessary to have converters with valid feedbacks to synthesize them; and (iii) combining the transfer codes as much as possible but still keeping with fundamental codes will have more chance to yield converters with a single active switch. These observations will be the rules of thumb and help derive code configurations that can be synthesized with fundamental converters and

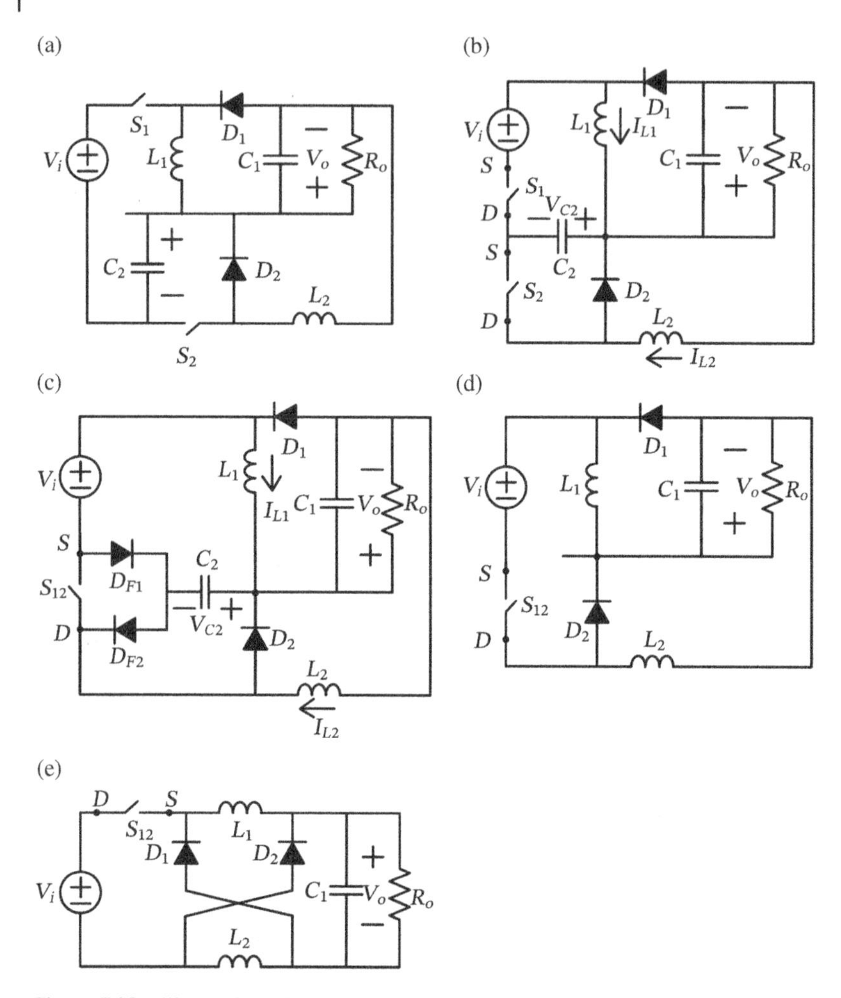

Figure 7.19 Illustration of synthesizing the high step-down converter with a buck-boost converter as a forward path and an I-buck converter as a feedback path.

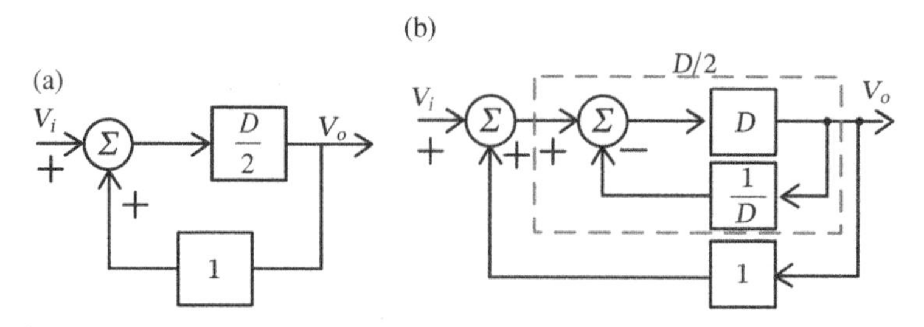

Figure 7.20 Other code configurations for achieving the transfer code $D/(2-D)$.

yield the converters with only a single active switch. In the later derivation of converters, we will follow the same rules.

7.3.1.2 High Step-Down Converter with Transfer Code $D/(2(1-D))$

The second example of high step-down switched-inductor converter with the transfer code $D/(2(1-D))$ is decoded as follows. Through cross multiplication, we have

$$\frac{V_o}{V_i} = \frac{D}{2(1-D)} \tag{7.26}$$

or

$$2V_o(1-D) = V_i D. \tag{7.27}$$

Equation (7.27) can be expressed as follows:

$$\frac{V_o(1-D)}{D} = V_i - \frac{V_o(1-D)}{D} \tag{7.28}$$

or

$$V_o = \frac{\left[V_i - \frac{V_o(1-D)}{D}\right]D}{(1-D)}. \tag{7.29}$$

Based on (7.29), a code configuration to realize the expression is shown in Figure 7.21. There are four possible converter topologies, buck–boost, Ćuk, sepic, and Zeta, to synthesize the transfer code $D/(1-D)$ and four possible inverse converter topologies, I-buck–boost, I-Ćuk, I-sepic, and I-Zeta, to synthesize $(1-D)/D$, too. However, since the feedback converter input has the same positive voltage polarity as that of the forward converter output and the feedback output has a negative voltage polarity connected to input voltage source V_i, there are only four possible combinations, sepic and I-Ćuk, Zeta and I-buck–boost, sepic and I-buck–boost, and Zeta and I-Ćuk, among the aforementioned converters and inverse converters. In the following, their syntheses are presented one by one in detail.

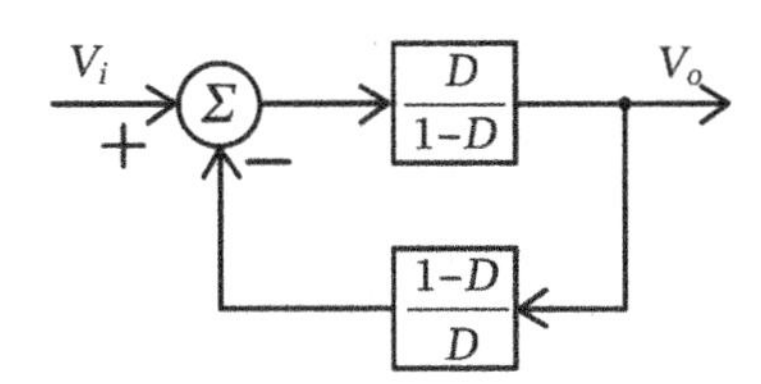

Figure 7.21 A code configuration to realize the expression shown in Eq. (7.29).

7.3.1.2.1 Synthesis with SEPIC and I-Ćuk Figure 7.22a shows a sepic converter in the forward path and an I-Ćuk converter in the feedback path, which can synthesize the code configuration shown in Figure 7.21. The two active switches S_1 and S_2 in the figure can be identified to have a common S–D node, and they can be replaced with a Π-type grafted switch S_{12} and two diodes D_{F1} and D_{F2}, as shown in Figure 7.22b. Based on the converter shown in Figure 7.22a, it can be proved that both voltages across capacitors C_1 and C_3 are $V_i/2$, and all voltages across inductors L_1–L_4 are $V_i/2$ under the active switch on state. On the other hand, all voltages across the inductors are $-V_o$ under the active switch off state. If we let all of the inductors have the same inductance, the current through switches S_1 and S_2

(a)

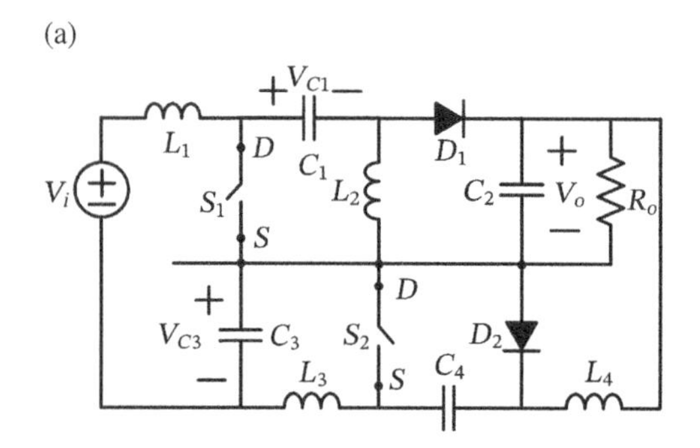

(b)

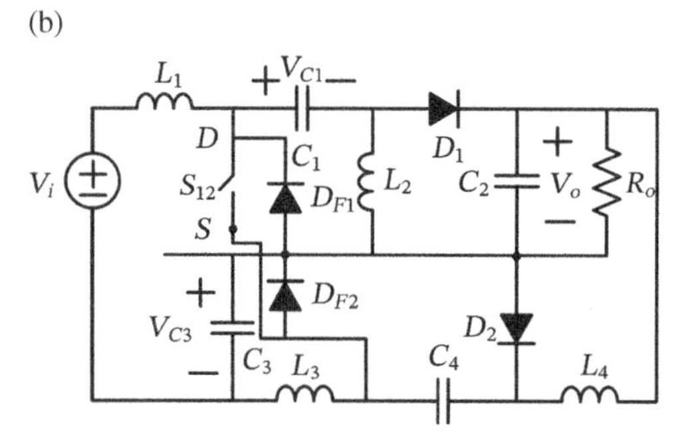

(c)

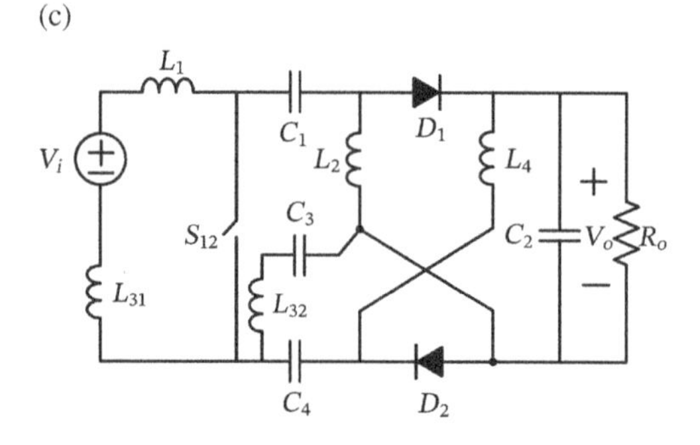

Figure 7.22 Illustration of synthesizing the high step-down converter with a sepic converter as a forward path and an I-Ćuk converter as a feedback path.

Figure 7.22 (Continued)

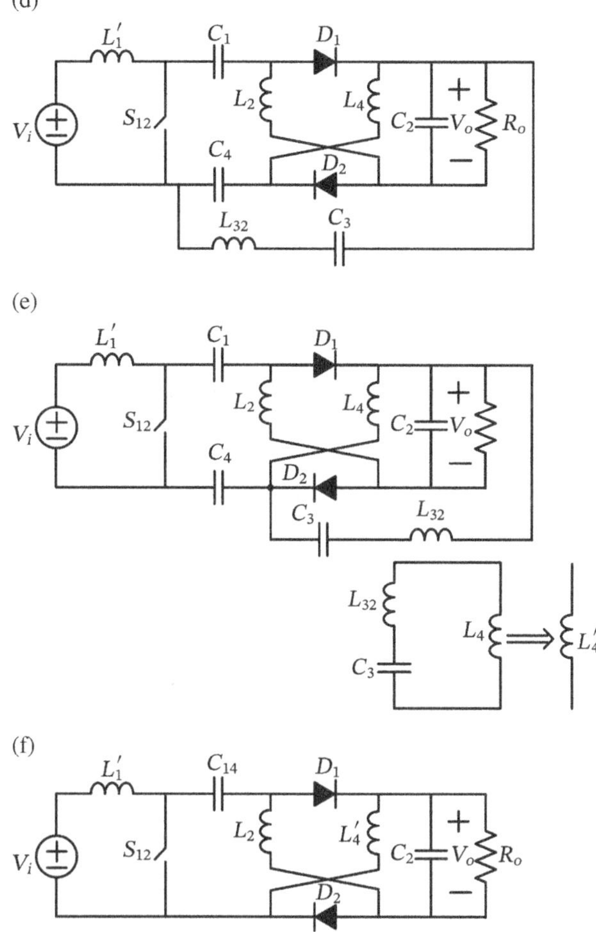

will be identical. Again, there is no average current through capacitor C_3 in the steady state. Thus, the two diodes D_{F1} and D_{F2} are no longer needed and can be removed from the converter shown in Figure 7.22b. The overall converter becomes the one shown in Figure 7.22c, in which inductor L_3 is split into two inductors L_{31} and L_{32}. Combining inductor L_{31} with inductor L_1 becomes L_1' and changing the DC offset voltage of capacitor C_3 by increasing V_o will yield the converter shown in Figure 7.22d. Changing the DC offset voltage of capacitor C_3 by decreasing $(V_{C3} + V_o)$, the voltage across capacitor C_4, yields the one shown in Figure 7.22e. Since the series connected circuit C_3 and L_{32} in parallel with inductor L_4 acts as an energy buffer circuit and the average voltage across C_3 is 0 V, they can be equivalent to a single inductor L_4' from topological point of view. Finally, the high step-down switched-inductor converter is shown in Figure 7.22f, in which C_4 is moved to the forward path and combined with C_1 to form C_{14}. By the way, since there is no current through capacitor C_3 and it can be removed from the circuit shown in Figure 7.22b. Thus, inductor L_3 can be combined with L_1 to form L_1', and the overall circuit becomes the one shown in Figure 7.22f after combining C_1 and C_4.

7.3.1.2.2 Synthesis with Zeta and I-Buck–Boost The second combination to synthesize the code configuration shown in Figure 7.21 is the converter shown in Figure 7.23a in which the Zeta converter is located in forward path and I-buck–boost converter is in the feedback path. Relocating switch S_1 to the return path will have a common D–S node with switch S_2, as shown in Figure 7.23b. They can be replaced

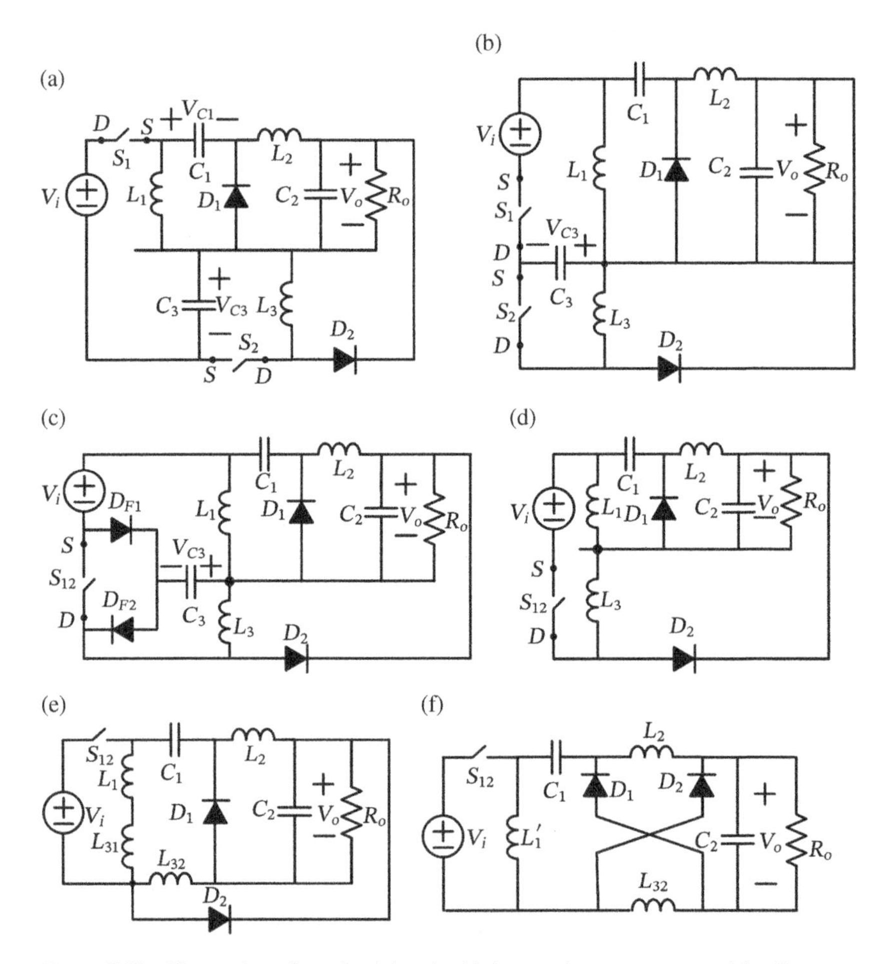

Figure 7.23 Illustration of synthesizing the high step-down converter with a Zeta converter as a forward path and an I-buck–boost converter as a feedback path.

with a Π-type grafted switch S_{12} with two diodes D_{F1} and D_{F2}, as shown in Figure 7.23c. In the switch on state, the currents through inductors L_1 and L_2 will flow through inductor L_3 since there is no average current through capacitor C_3. Thus, the two diodes are no longer needed and can be removed from the converter, as shown in Figure 7.23d. With inductor splitting, inductor L_3 is split into L_{31} and L_{32}, as shown in Figure 7.23e. Combining L_1 with L_{31} to become L_1' and relocating the components will yield the high step-down switched-inductor converter, as shown in Figure 7.23f.

7.3.1.2.3 Synthesis with SEPIC and I-Buck–Boost The third combination to synthesize the code configuration shown in Figure 7.21 is the converter shown in Figure 7.24, in which the sepic converter is located in the forward path and the I-buck–boost converter is in the feedback path. Since there is no common node between switches S_1 and S_2, we cannot graft the two switches. However, this converter configuration still can achieve the high step-down transfer code $D/(2(1 - D))$ although it has two active switches.

7.3.1.2.4 Synthesis with Zeta and I-Ćuk The fourth combination to synthesize the code configuration shown in Figure 7.21 is the converter shown in Figure 7.25, in which Zeta converter is located in the forward path and I-Ćuk converter is in the feedback path. Again, since there is no common node between switches S_1 and S_2, we cannot graft the two switches. However, this converter configuration still can achieve high step-down transfer code $D/(2(1 - D))$ although it has two active switches.

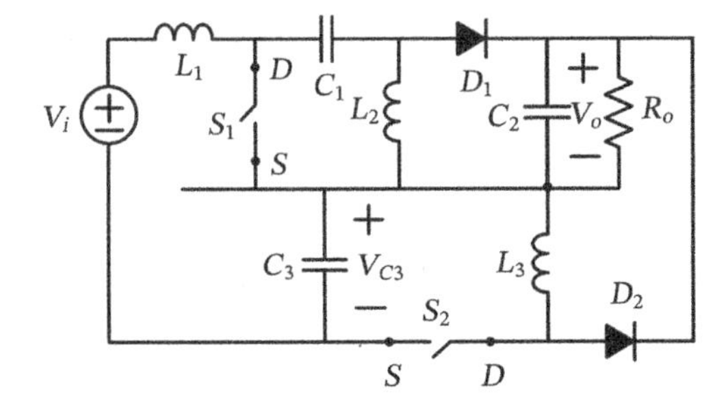

Figure 7.24 Illustration of a sepic converter as a forward path and an I-buck–boost converter as a feedback path to achieve the high step-down converter but with two active switches.

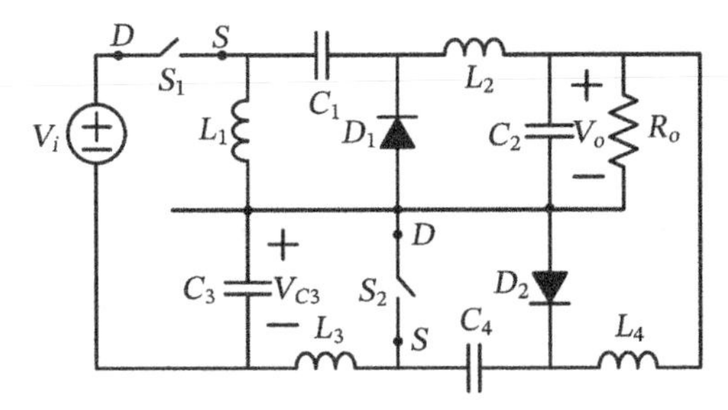

Figure 7.25 Illustration of a Zeta converter as a forward path and an I-Ćuk converter as a feedback path to achieve the high step-down converter but with two active switches.

In the third and fourth combinations, sepic and I-buck–boost and Zeta and I-Ćuk, each of which has two converters, but they belong to different families, such as sepic belongs to the boost family, while I-buck–boost belongs to the buck family. Thus, in the converter, there is no common node between the two active switches, and they cannot be grafted. This is the only unique feature of such combination, but not a general feature of grafting converters from different families. For instance, we can graft a buck on a boost to form a Ćuk converter, while it has only a single active switch.

7.3.2 Switched-Capacitor Converters

In literature, there are high step-up transfer codes with switched-capacitor structures. In the following, derivation of several examples is presented.

7.3.2.1 High Step-Up Converter with Transfer Code (1 + *D*)/(1 − *D*)

A high step-up switched-capacitor converter with the transfer code $(1+D)/(1-D)$ is first decoded and synthesized as follows. Through decoding process, $V_o/V_i=(1+D)/(1-D)$ can be expressed as

$$\frac{V_o}{V_i}=\frac{(1+D)}{(1-D)}=\frac{1}{(1-D)}+\frac{D}{(1-D)}. \tag{7.30}$$

To synthesize (7.30), one can choose a boost converter for the transfer code $1/(1-D)$, select either Ćuk or buck–boost converter for $D/(1-D)$, and then they can be combined to yield $V_o/V_i=(1+D)/(1-D)$. First, let's start with the synthesis of a boost converter plus a Ćuk converter, as shown in Figure 7.26a, in which

$$V_o=V_{o1}+V_{o2}, \tag{7.31}$$

where $V_{01}=V_i/(1-D)$ and $V_{02}=V_iD/(1-D)$. From Figure 7.26a, we can identify a common *S*–*S* node between active switches S_1 and S_2, and they have the same voltage stress, $V_i/(1-D)$, under switch off state. Thus, these two active switches can be replaced with a T-type grafted switch S_{12}, as shown in Figure 7.26b, and therefore the two inductors L_1 and L_2 can be combined into a single one, L_{12}. Relocating the components will yield the converter shown in Figure 7.26c, from which it can be observed that inductor L_3 is located in the return path and capacitor C_3 is across diode D_2 and inductor L_3. Changing the DC offset voltage of capacitor C_3 by increasing V_{01} and moving inductor L_3 from the return path to the forward one will yield the high step-up switched-capacitor converter, as shown in Figure 7.26d.

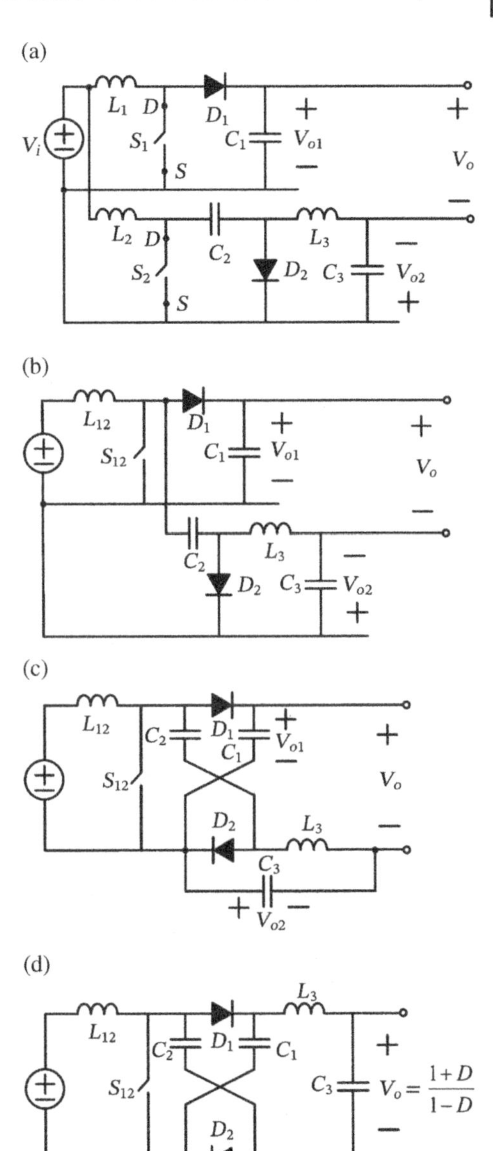

Figure 7.26 Illustration of synthesizing the high step-up converter with a boost converter and a Ćuk converter to achieve the transfer code $(1+D)/(1-D)$.

Equation (7.30) can be expressed alternately as

$$\frac{V_o}{V_i} = \frac{D}{(1-D)} + \frac{1}{(1-D)}, \tag{7.32}$$

which can be synthesized by a Ćuk converter plus a boost converter, as shown in Figure 7.27a. Again, it can be identified that the two active switches sharing a

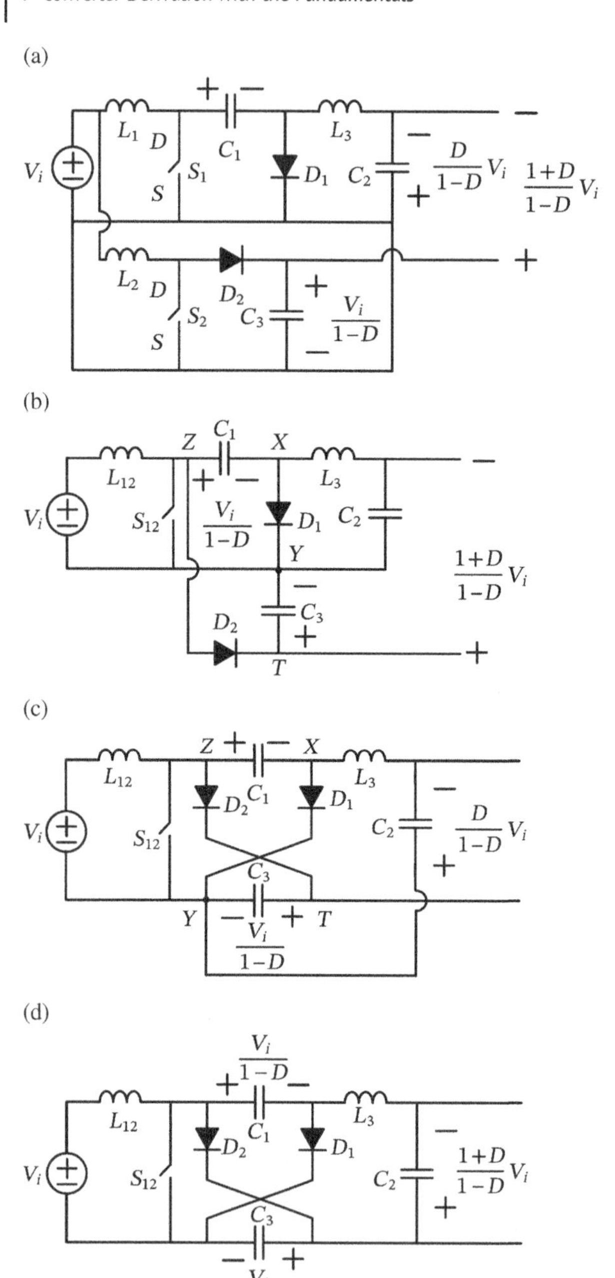

Figure 7.27 Illustration of synthesizing the high step-up converter with a Ćuk converter and a boost converter to achieve the transfer code $(1+D)/(1-D)$.

common S–S node have the same voltage stress under off state and they can be replaced with a T-type grafted switch S_{12}, as shown in Figure 7.27b, where inductors L_1 and L_2 are in parallel and can be replaced with a single one, L_{12}. Relocating the components will yield the one shown in Figure 7.27c. Changing the DC offset voltage of capacitor C_2 by increasing $V_i/(1-D)$ will yield the high step-up switched-capacitor converter with the transfer code $(1+D)/(1-D)$, as shown in Figure 7.27d.

It is interesting to note that with commutative property of addition, we exchange the order of $1/(1-D)$ and $D/(1-D)$ and synthesize them with the corresponding orders of boost and Ćuk converters. Then, they come out different converter topologies, as shown in Figures 7.26 and 7.27. Basically, it can be proved that they are identical but with inductor L_3 changed from return to forward path at the output side. This feature further proves that identical transfer code ratio should yield identical converter topology.

7.3.2.2 High Step-Up Converter with Transfer Code $2D/(1-D)$

The second example has the transfer code $2D/(1-D)$, which is the one of a high step-up switched-capacitor converter. The transfer code can be decoded as

$$\frac{V_o}{V_i} = \frac{2D}{(1-D)} = \frac{D}{(1-D)} + \frac{D}{(1-D)}. \tag{7.33}$$

The above expression can be synthesized with a Zeta converter plus a buck–boost converter as shown in Figure 7.28a. The two active switches share a common D–D node and can operate in unison; thus, they can be replaced with a T-type grafted switch S_{12}. After we graft the two active switches, inductors L_1 and L_3 become paralleled, and they can be replaced with a single one L_{13}. The converter with the grafted switch and the paralleled inductors is shown in Figure 7.28b. Relocating the components will yield the one shown in Figure 7.28c. Changing the DC offset voltage of capacitor C_2 by increasing $V_{o2} = V_i D/(1-D)$ will have the output $V_o = V_i\, 2D/(1-D)$, as shown in Figure 7.28d, which is the transfer code of a high step-up switched-capacitor converter.

Swapping Zeta converter and buck–boost converter will yield the one shown in Figure 7.29a, and its output voltage polarity will become negative. Grafting the two active switches and integrating inductors L_1 and L_2 yield the one shown in Figure 7.29b. Relocating the components will have the one shown in Figure 7.29c. Changing the DC offset voltage of capacitor C_2 by increasing V_{o1} will yield the high step-up switched-capacitor converter, as shown in Figure 7.29d.

In fact, the converters shown in Figures 7.28 and 7.29 are identical, except that their output voltage polarities are opposite. However, this difference is due primarily to the connection of components.

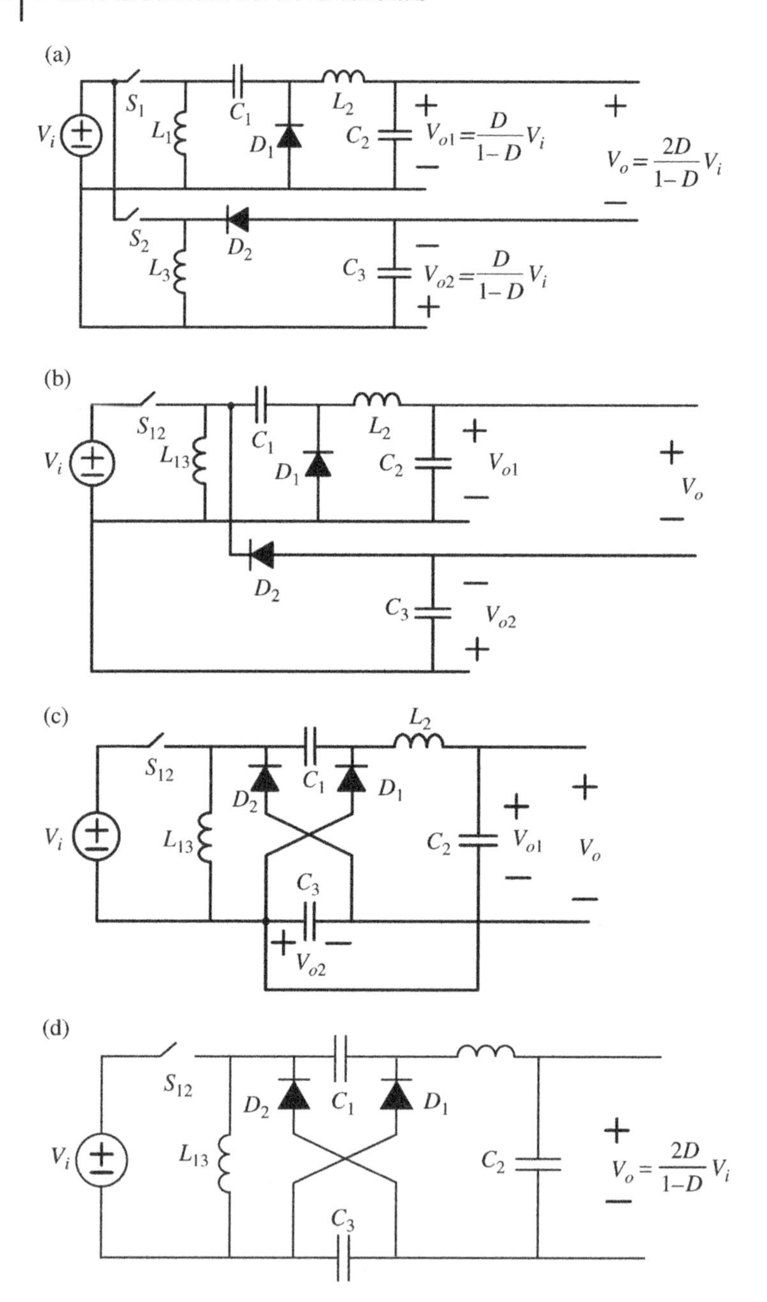

Figure 7.28 Illustration of synthesizing the high step-up converter with a Zeta converter and a buck–boost converter to achieve the transfer code $2D/(1-D)$.

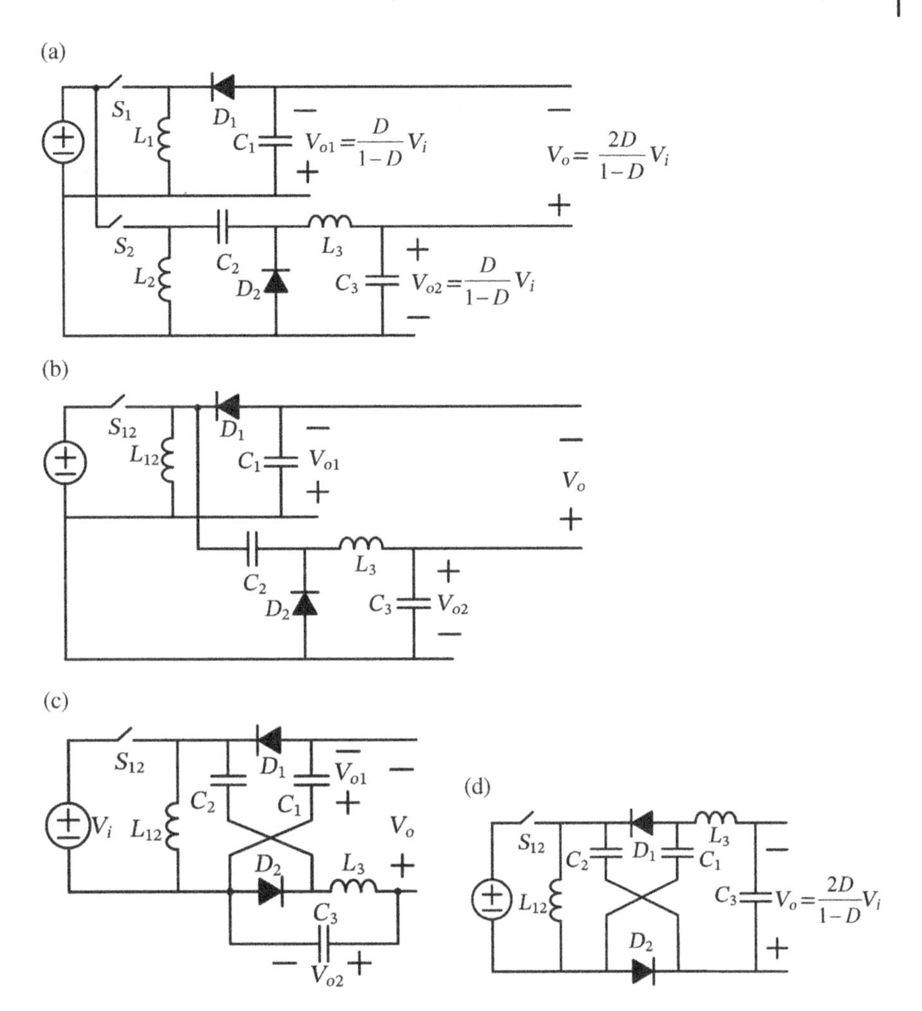

Figure 7.29 Illustration of synthesizing the high step-up converter with a buck–boost converter and a Zeta converter to achieve the transfer code $2D/(1-D)$.

Another choice of converter combination to achieve transfer code $2D/(1-D)$ is the sepic converter plus Ćuk converter, as shown in Figure 7.30a. Since the two active switches S_1 and S_2 have a common S–S node and they can operate in unison, they can be replaced with a grafted switch S_{12}, and then inductors L_1 and L_3 become paralleled. With grafted switch S_{12} and paralleled inductor L_{13}, we have the one shown in Figure 7.30b. Relocating the components without changing their operational principle yields the one shown in Figure 7.30c. Changing the DC offset voltages of capacitors C_3 and C_4 by decreasing V_{C1} and increasing V_{o1}, respectively, will yield the high step-up switched-capacitor converter, as shown in Figure 7.30d.

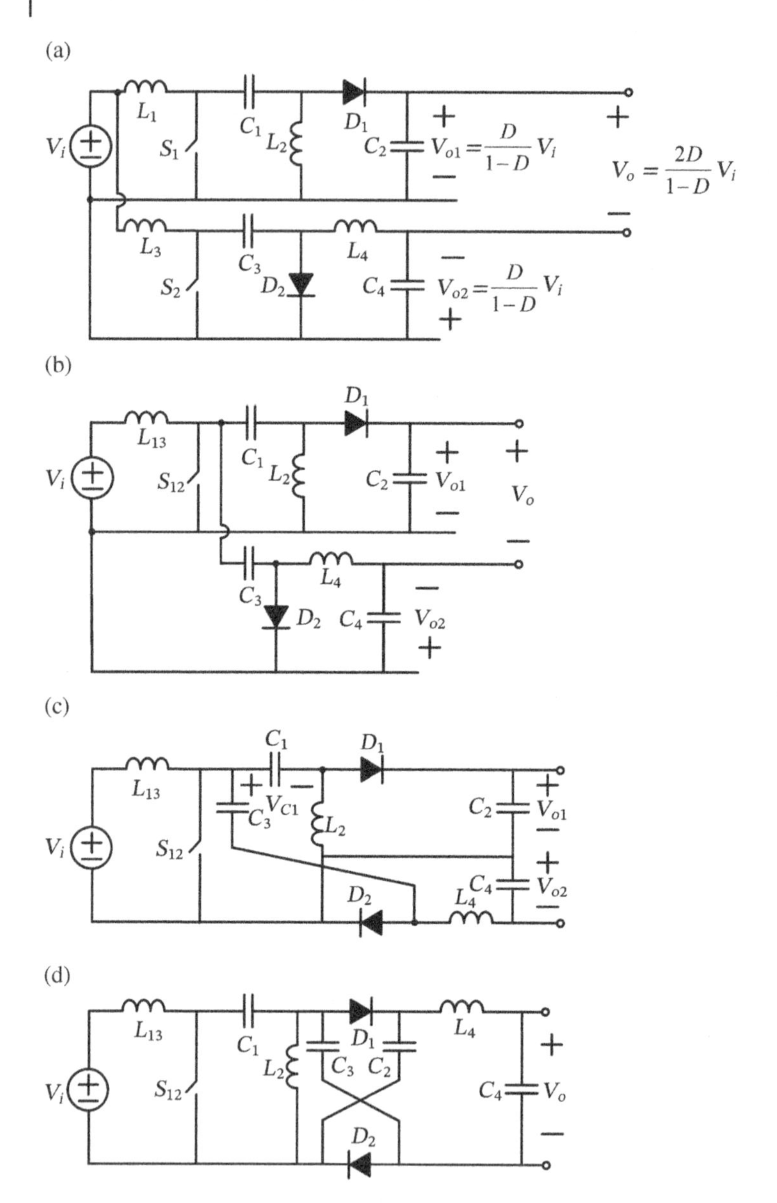

Figure 7.30 Illustration of synthesizing the high step-up converter with a sepic converter and a Ćuk converter to achieve the transfer code $2D/(1-D)$.

7.3.2.3 High Step-Up Converter with Transfer Code $D/(1-2D)$

Another high step-up transfer code $D/(1-2D)$ can be synthesized with regular converters, but not switched-capacitor converters. The decoding process of $D/(1-2D)$ is shown as follows. From the transfer code and with cross multiplication, we have

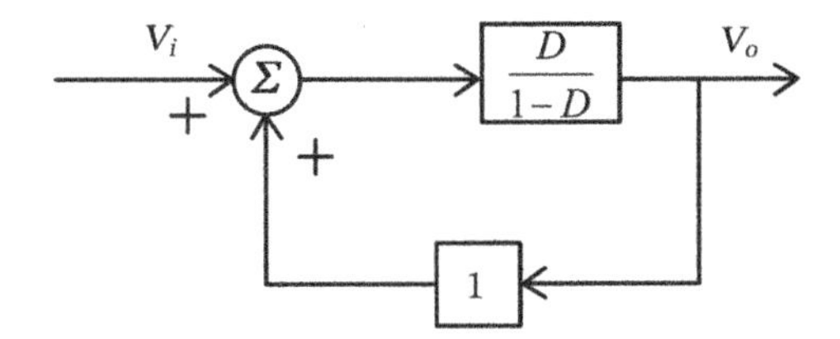

Figure 7.31 A code configuration of the transfer code $D/(1-2D)$.

$$\frac{V_o}{V_i}=\frac{D}{(1-2D)} \tag{7.34}$$

or

$$V_o(1-2D)=V_iD. \tag{7.35}$$

$$\Rightarrow V_o(1-D)=(V_i+V_o)D$$

$$\begin{aligned}&\Rightarrow V_o(1-D)=(V_i+V_o)D\\&\Rightarrow V_o=\frac{(V_i+V_o)D}{(1-D)}.\end{aligned} \tag{7.36}$$

From (7.36), we can have the code configuration shown in Figure 7.31, in which the positive unity feedback is added to the input and then fed forwardly to the converter with the transfer code $D/(1-D)$.

To synthesize the code configuration shown in Figure 7.31, we first choose a sepic converter with a positive unity feedback, as shown in Figure 7.32a. Relocating its component connection will yield a high step-up converter with the transfer code $D/(1-2D)$, as shown in Figure 7.32b.

Another choice of synthesizing the code configuration shown Figure 7.31 is using a Zeta converter with a positive unity feedback, as illustrated in Figure 7.33.

In Section 7.3, what have been synthesized is the existing converters. In the following, several novel synthesizing examples will be illustrated.

7.4 Syntheses of Desired Transfer Codes

For certain applications, we might need special transfer codes, such as for high step-up, high step-down, or high step-up/step-down. Thus, we will obtain transfer codes first and then conduct decoding and synthesizing processes.

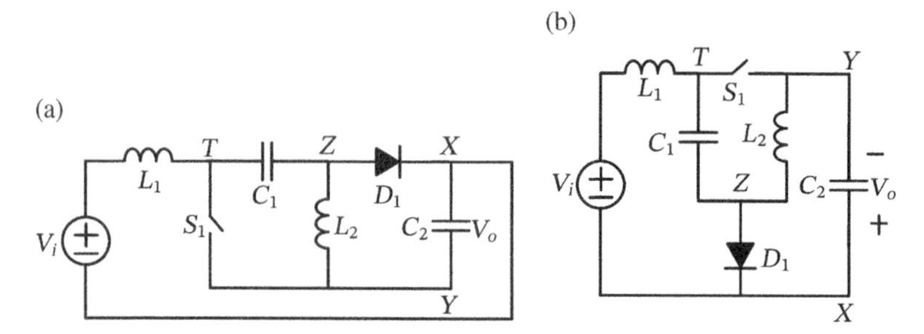

Figure 7.32 Illustration of synthesizing the code configuration shown in Figure 7.31 with a sepic converter plus a positive unity feedback to achieve the transfer code $D/(1-2D)$.

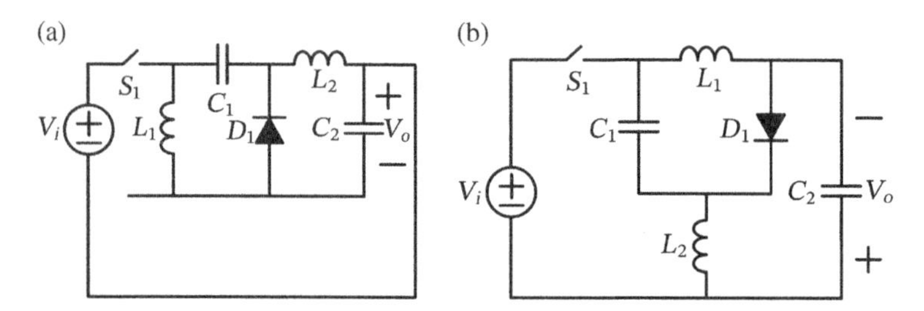

Figure 7.33 Illustration of synthesizing the code configuration shown in Figure 7.31 with a Zeta converter plus a positive unity feedback to achieve the transfer code $D/(1-2D)$.

7.4.1 Synthesis of Transfer Code: $D^2/(D^2-3D+2)$

Given a transfer code as

$$\frac{V_o}{V_i}=\frac{D^2}{\left(D^2-3D+2\right)},\tag{7.37}$$

we first conduct a decoding process. Through factorization of (7.37), we have

$$\frac{V_o}{V_i}=\frac{D}{(1-D)}\cdot\frac{D}{(2-D)}.\tag{7.38}$$

The above equation can be configured with two transfer codes in cascade, as shown in Figure 7.34, in which the two transfer codes can be synthesized with fundamental converters. Transfer code $D/(1-D)$ can be synthesized with buck–boost, Ćuk, sepic, or Zeta converter, while $D/(2-D)$ can be synthesized with the high step-down switched-inductor converter shown in Figure 7.19e.

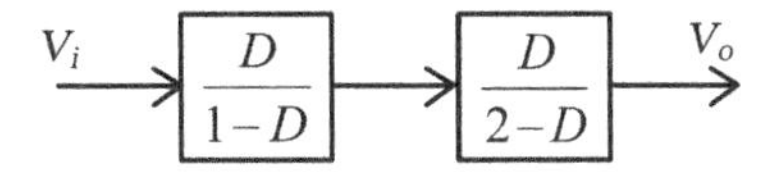

Figure 7.34 Cascade code configuration of transfer code $D^2/(D^2-3D+2)$.

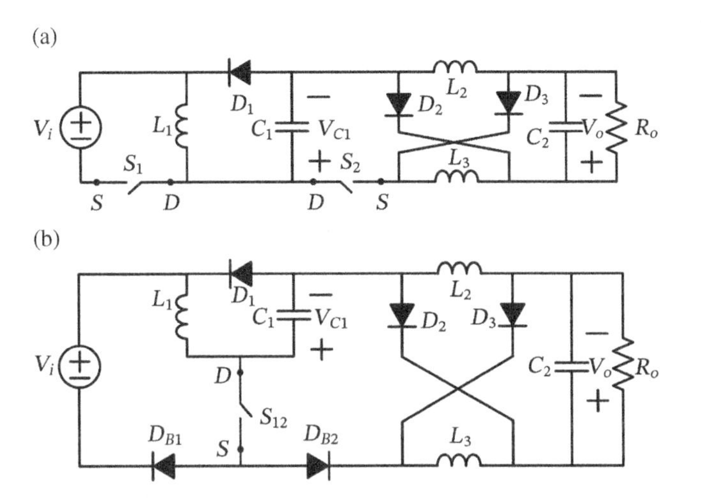

Figure 7.35 Illustration of synthesizing the code configuration shown in Figure 7.34 with a buck–boost converter and the converter shown in Figure 7.19e.

7.4.1.1 Synthesizing with Buck–Boost Converter

A buck–boost converter is used to synthesize the transfer code $D/(1-D)$, which is cascaded with the high step-down converter shown in Figure 7.19e. The overall converter is shown in Figure 7.35a, in which switch S_1 is located in the return path. It can be identified that the two active switches S_1 and S_2 share a common D–D node and when operating in unison, they can be replaced with an inverted T-type switch, as shown in Figure 7.35b. Next, we have to identify the relationship between the voltages across the two switches under their off states. For switch S_1, the voltage stress is

$$V_1 = V_i + V_{C1} = V_i\left(1+\frac{D}{(1-D)}\right) = V_i\left(\frac{1}{(1-D)}\right). \tag{7.39}$$

For switch S_2, it is

$$\begin{aligned} V_2 &= V_{C1} + V_o \\ &= V_i\left(\frac{D}{(1-D)} + \frac{D^2}{(1-D)(2-D)}\right) \\ &= V_i\left(\frac{2D}{(1-D)(2-D)}\right). \end{aligned} \tag{7.40}$$

Then,

$$V_1 - V_2 = V_i\left(\frac{(2-3D)}{(1-D)(2-D)}\right). \tag{7.41}$$

Since (7.41) can be either positive when $D < 2/3$ or negative when $D > 2/3$, diodes D_{B1} and D_{B2} need to be kept in the converter. Therefore, the final converter topology becomes the one shown in Figure 7.35b.

7.4.1.2 Synthesizing with Zeta Converter

The transfer code of $D/(1-D)$ can be synthesized with a Zeta converter. When it is combined with the converter shown in Figure 7.19e, the overall converter is shown in Figure 7.36a. With both active switches in the return path, we can have the one shown in Figure 7.36b. It can be identified that switches S_1 and S_2 share a common D–S node and when operating in unison, they can be replaced by a Π-type grafted

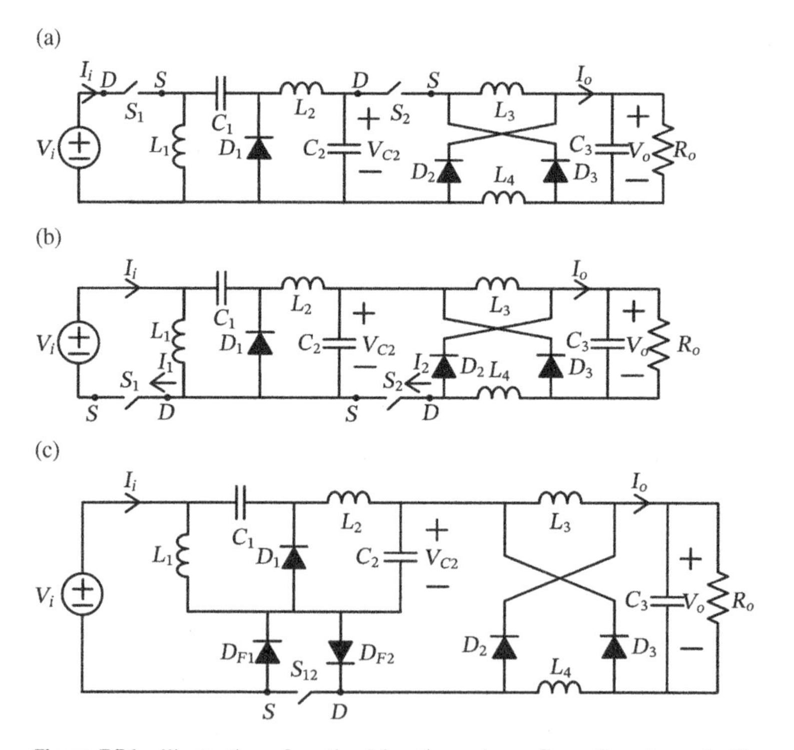

Figure 7.36 Illustration of synthesizing the code configuration shown in Figure 7.34 with a Zeta converter and the converter shown in Figure 7.19e.

switch, as shown in Figure 7.36c. Next, we have to identify the relationship of the currents through the switches under their on states. Based on power conservation, we have $V_i I_i = V_o I_o$, where I_i and I_o are the average input and output currents, respectively. The current I_1 through switch S_1 in the on state can be determined as

$$I_1 = \frac{I_i}{D}, \tag{7.42}$$

where D is the duty ratio of the active switches and the current I_2 through switch S_2 can be determined as

$$I_2 = I_o. \tag{7.43}$$

Since

$$V_o = \frac{D^2 V_i}{\left[(1-D)(2-D)\right]}, \tag{7.44}$$

we have

$$I_2 = I_o = \frac{V_i I_i}{V_o} = \frac{I_i(1-D)(2-D)}{D^2}. \tag{7.45}$$

Thus,

$$I_1 - I_2 = \frac{I_i\left(4D - D^2 - 2\right)}{D^2}. \tag{7.46}$$

From (7.46), it can prove that I_1 could be greater or less than I_2 when D varies from 0 to 1. Therefore, the two diodes D_{F1} and D_{F2} cannot be removed from the converter shown in Figure 7.36c, which will be the final converter topology to synthesize the transfer code shown in (7.37).

7.4.1.3 Synthesizing with Ćuk Converter

The transfer code $D/(1-D)$ can be synthesized with a Ćuk converter. When it is combined with the one shown in Figure 7.19e, the overall converter becomes the one shown in Figure 7.37a. It can be identified that the two active switches share a common S–D node and they can be replaced with an inverse Π-type grafted switch, as shown in Figure 7.37b. Next, we have to find out the relationship between the currents through switches S_1 and S_2. Based on power balance principle, we have

$$V_i I_i = I_2 V_{C2} = V_o I_o. \tag{7.47}$$

(a)

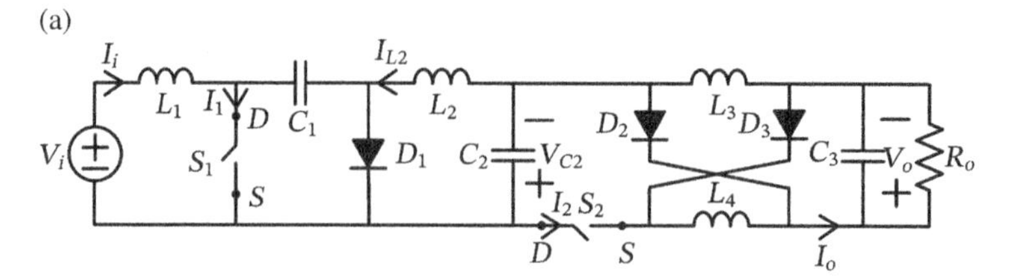

(b)

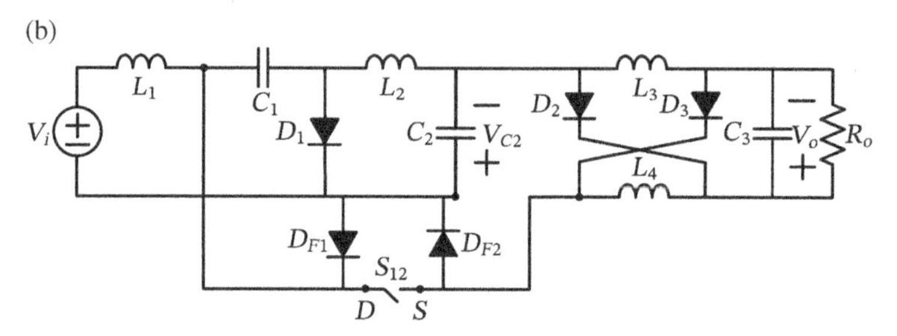

Figure 7.37 Illustration of synthesizing the code configuration shown in Figure 7.34 with a Ćuk converter and the converter shown in Figure 7.19e.

From Figure 7.37a, we have

$$V_{C2} = \frac{V_i D}{(1-D)}, \tag{7.48}$$

$$I_1 = I_i + I_{L2} = I_i + \frac{I_i(1-D)}{D}, \tag{7.49}$$

and

$$I_2 D V_{C2} = I_o V_o \tag{7.50}$$

or

$$I_2 = \frac{I_o}{(2-D)} = \frac{I_i(1-D)}{D^2}. \tag{7.51}$$

Then,

$$I_1 - I_2 = \frac{I_i(2D-1)}{D^2}. \tag{7.52}$$

From above equation, one can see that when $D < 0.5$, $I_1 < I_2$, while when $D > 0.5$, $I_1 > I_2$. For an overall range of duty ratio, the two diodes, D_{F1} and D_{F2}, in Figure 7.37b need to be kept. That is, the converter topology shown in Figure 7.37b is the final converter version.

Basically, the transfer code $D/(1-D)$ can be synthesized with a sepic converter. It is left for the readers to practice themselves.

In principle, there are many transfer codes that can be synthesized with the proposed approaches to yielding a big number of converters, depending on desired applications. The converters might consist of more than one active and passive switches. Usually, a simple code may result in single active and passive switch pair, as the conventional well-known PWM converters.

7.4.2 Synthesizing Converters with the Fundamentals

In Chapter 3, we have discussed seven types of fundamentals that can be used to develop new converter topologies. They are shown in the following:

1) DC voltage and DC current offsetting
2) Inductor and capacitor splitting
3) DC voltage blocking and filtering
4) Magnetic coupling
5) DC transformer
6) Switch and diode grafting
7) Layer technique

These fundamental techniques must be associated with given converters. Based on the derived converters shown in previous sections, we can apply the fundamentals to further develop new converter topologies. In the following, we use the Zeta converter as a base and apply the fundamentals one by one to see if there are new converters generated.

7.4.2.1 DC Voltage and DC Current Offsetting

A Zeta converter is shown in Figure 7.38. First of all, moving switch S_1 from the forward path to the return one, we have the one shown in Figure 7.39a. Then, we can identify that capacitor C_1 is connected to input voltage source V_i, and it can be changed with its DC voltage offsetting from V_o to $V_i + V_o$, as shown in Figure 7.39b. Redrawing the circuit and putting the components in proper positions will become the one shown in Figure 7.39c. Finally, moving capacitor C_1 and inductor L_2 to the forward path will become the Ćuk converter, as shown in Figure 7.39d.

Figure 7.38 Zeta converter topology.

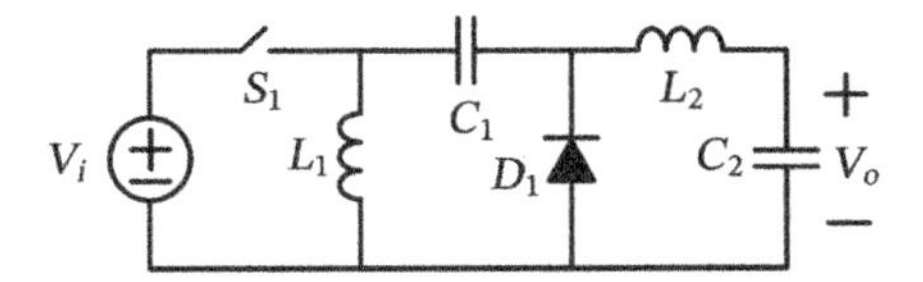

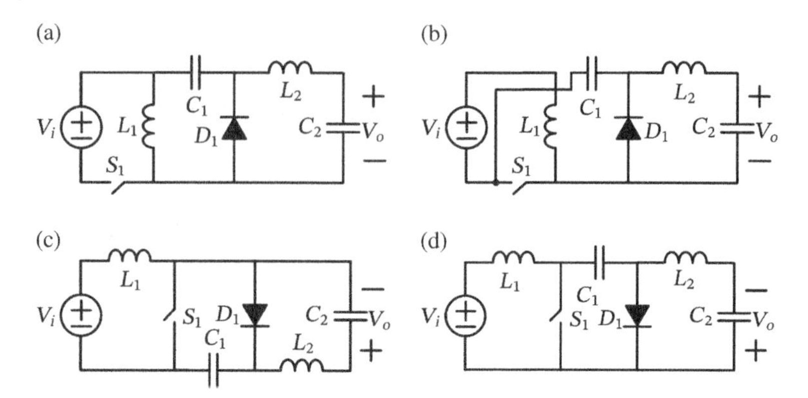

Figure 7.39 Illustration of the Zeta converter with a DC voltage offsetting.

However, since there is no current source in the converter, we cannot use the DC current offsetting fundamental to derive a new converter topology.

7.4.2.2 Inductor and Capacitor Splitting

Examining the Zeta converter shown in Figure 7.38 reveals that there is no more than a single loop including only linear components. Thus, inductor and capacitor splitting fundamentals cannot be applied to the Zeta converter for developing a new converter topology.

7.4.2.3 DC Voltage Blocking and Filtering

For illustrating the DC voltage blocking and filtering, the Zeta converter is re-depicted and shown in Figure 7.40a in which an L_fC_b filter is connected in parallel with active switch S_1. Since the voltage across switch S_1 is $(V_i + V_o)$ when the switch is operated in the off state, its duration time is $(1-D)T_s$ where T_s is the switching period. Inductor L_f acts as a filter, but it needs to satisfy volt-second balance in the steady state, and this is done by the blocking capacitor C_b that blocks the average voltage as

$$\begin{aligned} V_{o2} &= (V_i + V_o)(1-D) \\ &= \left(V_i + \frac{DV_i}{1-D}\right)(1-D) \\ &= V_i. \end{aligned} \tag{7.53}$$

It is worth noting that the pulsating voltage across diode D_1 has been filtered and DC blocked by the L_2 and C_2, respectively, which yields output voltage $V_{o1} = DV_i/(1-D)$.

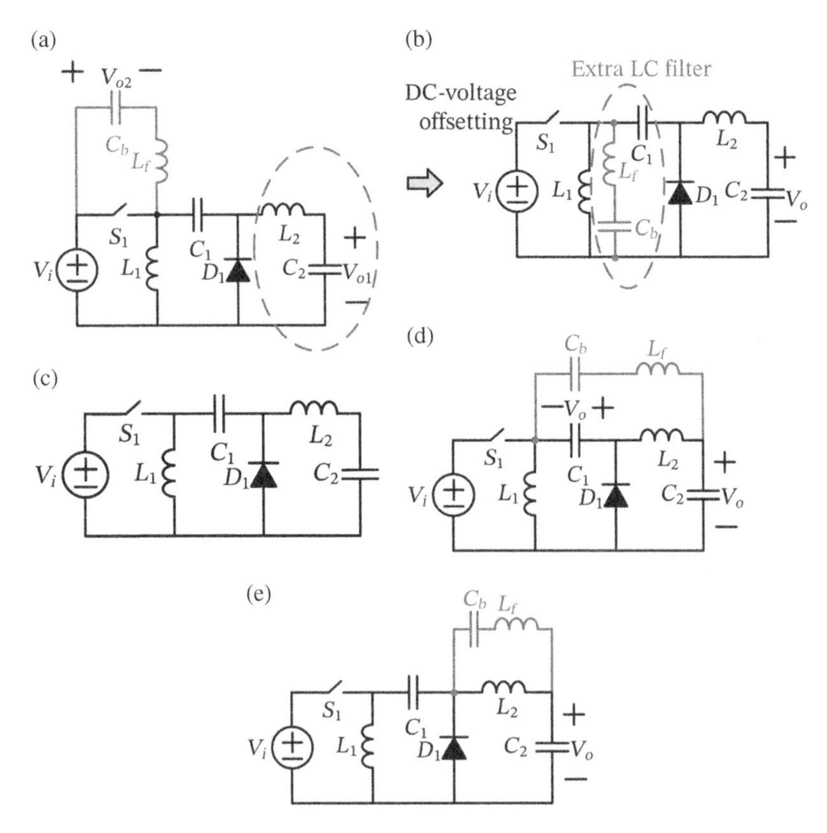

Figure 7.40 Illustration of the Zeta converter with a DC voltage blocking and filtering.

Essentially, the C_b shown in Figure 7.40a can be changed with its DC voltage offsetting from V_i to 0, as shown in Figure 7.40b, which forms as an extra *LC* filter to inductor L_1. Thus, finally, it comes back to the Zeta converter, as shown in Figure 7.40c. Or the L_fC_b can be changed with its DC voltage offsetting to $-V_o$ or to 0 and in parallel to inductor L_2. However, they finally will be degenerated to the Zeta converter. Therefore, Zeta converter is the one with the minimum component count in a fourth-order converter with the input-to-output voltage transfer code of $D/(1-D)$.

7.4.2.4 Magnetic Coupling

First of all, it can be proved that the two inductors in the Zeta converter can be wound on the same core because their voltages across the inductors are V_i and $-V_o$ during switch on and off states, respectively. The converter can be shown in Figure 7.41a in which the two inductors are coupled on the same core.

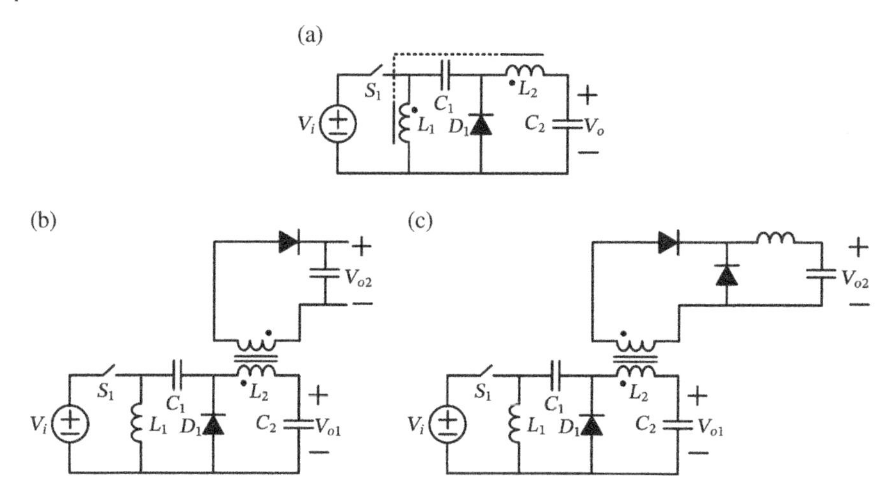

Figure 7.41 Illustration of the Zeta converter with a magnetic coupling.

Since all of the magnetic components must satisfy volt-second balance in a steady state, we can couple any extra winding on the core with magnetic components, as shown in Figure 7.41b and c. Figure 7.41b shows a flyback type of magnetic coupling circuit at the secondary side, while Figure 7.41c shows that with a forward type of secondary circuit. Of course, one can also couple the magnetic circuits on the other inductor (L_1).

7.4.2.5 DC Transformer

A DC transformer can be inserted into any PWM converter to derive a new converter topology. For instance, a Zeta converter is shown in Figure 7.42a marked with a location *b*–*b*′. By inserting a DC transformer into location *b*–*b*′, we have the one shown in Figure 7.42b, in which switch S_1 operates with duty ratio d_1, switches S_{D1} and S_{D2} operate with identical duty ratio d_2 ($=d_1$) to convert DC voltage into AC pulse voltages, and diodes D_{D1} and D_{D2} rectify the AC voltages to a DC voltage. Note that the switching frequency of S_1 should be doubled that of switches S_{D1} and S_{D2}. When switch S_1 is turned off, inductor current I_{L1} is flowing through the body diode D_{b1} and D_{b2} of switches S_{D1} and S_{D2}, respectively. In addition, diode D_1 in Figure 7.42b can be removed, because inductor current I_{L2} can be freewheeling through diodes D_{D1} and D_{D2}. The final version of the converter is shown in Figure 7.42c.

Basically, a DC transformer can be inserted into any location between input voltage source V_i and output capacitor C_2. And it may be degenerated by combining the original switch or diode in the converter with the switches and diodes in the DC transformer.

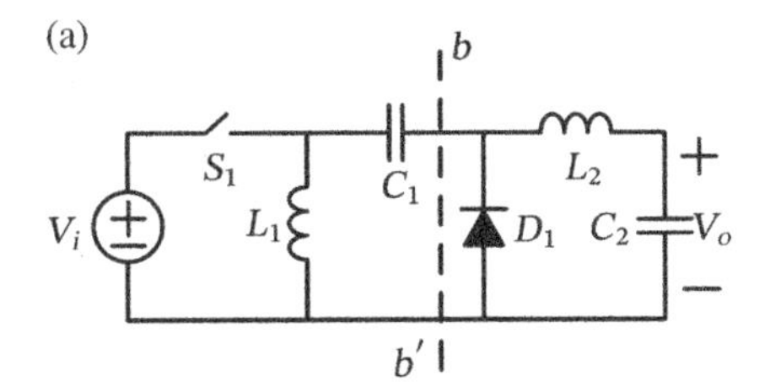

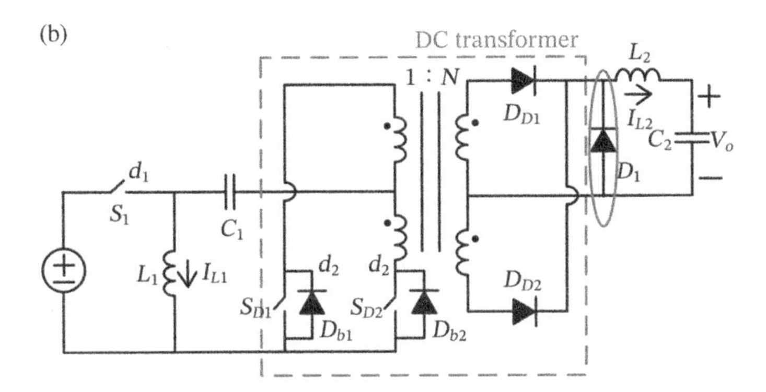

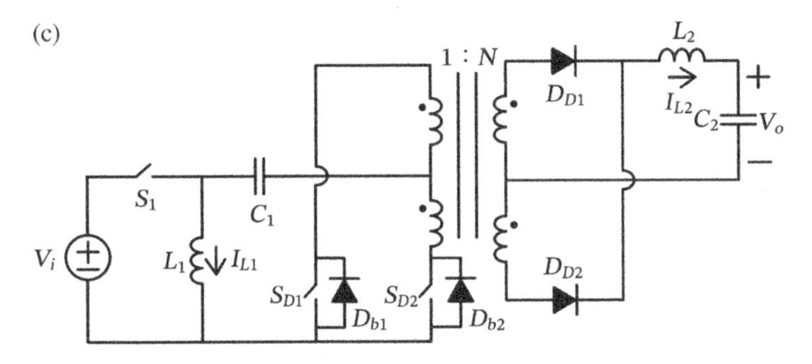

Figure 7.42 Illustration of the Zeta converter with a DC transformer.

7.4.2.6 Switch and Diode Grafting

There are no two active switches or no two diodes in the Zeta converter; thus, switch and diode grafting cannot be applied in this case.

7.4.2.7 Layer Technique

With a layer technique, the Zeta converter can have the one with output voltage feedback, as shown in Figure 7.43a. With a proper component connection without changing its operational principle, we have the converter shown in Figure 7.43b, in which its voltage transfer code is

$$\frac{V_o}{V_i} = \frac{\frac{D}{1-D}}{\left[1 - \frac{D}{1-D}\right]} = \frac{D}{1-2D}. \tag{7.54}$$

We can also feedback the pulsating voltage from node N, which is the voltage across diode D_1 with an L_fC_f filter, as shown in Figure 7.43c. Essentially, the role of L_fC_f is the same as that of L_2C_2; thus, the voltage transfer code will be identical to that of the one shown in Figure 7.43b. With a slight component relocation, we can yield the one shown in Figure 7.43d.

Figure 7.44 shows the Zeta converter with the layer technique and a buck converter feedback. Since the two active switches have no common node, they cannot be grafted. The voltage transfer code becomes $V_o/V_i = D/(1-D-D^2)$.

There is one special case that when the transfer code is $V_o/V_i = -1$, which can be decoded into the code configuration shown in Figure 7.45a, it can be synthesized with a Zeta converter and an inverse buck converter, as shown in Figure 7.45b. Moving switch S_1 from the forward path to the return path can identify that the two active switches S_1 and S_2 share a common D–D node and they can be replaced with a T-type grafted switch, as shown in Figure 7.45c. Examining the voltages across S_1 and S_2 yields that they are both having V_o/D and the two blocking diode D_{B1} and D_{B2} can be shorted. The final version of the converter is shown in Figure 7.45d. It should be noted that since capacitor C_3 is connected in series to switch S_{12}, it cannot satisfy ampere-second balance, and it requires an external voltage source V_o/D.

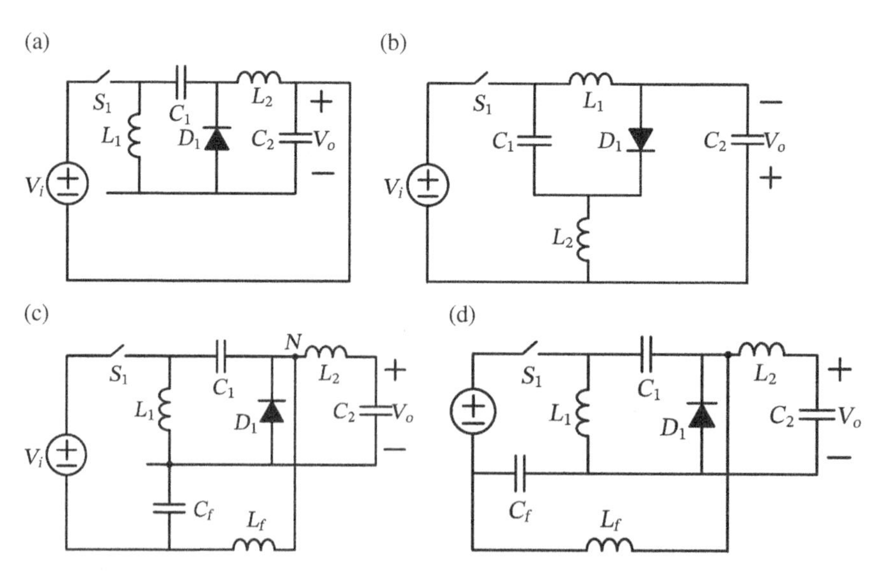

Figure 7.43 Illustration of the Zeta converter with the layer technique and a positive output voltage feedback.

Figure 7.44 Illustration of the Zeta converter with the layer technique and a buck converter feedback.

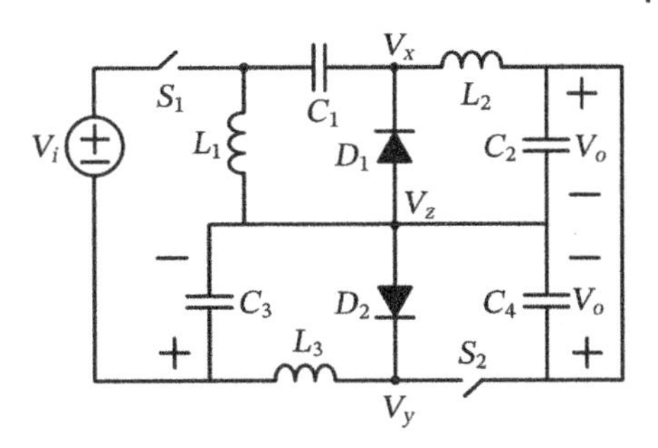

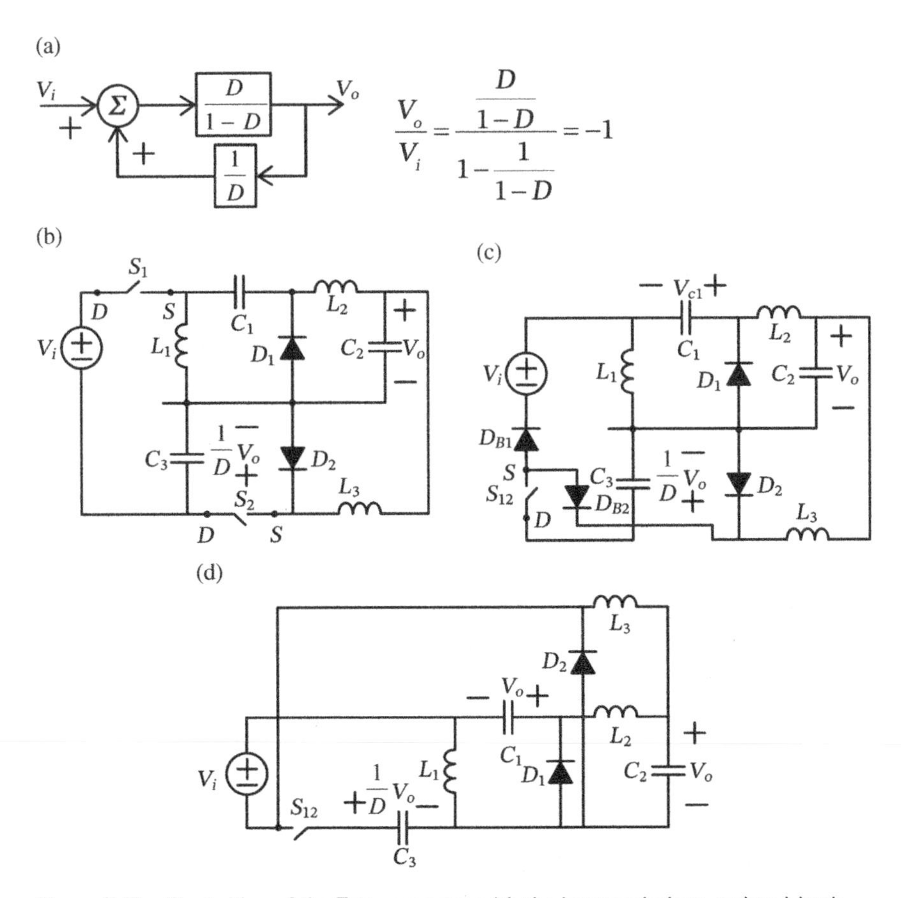

Figure 7.45 Illustration of the Zeta converter with the layer technique and an I-buck converter feedback.

Further Reading

Anderson, J. and Peng, F.Z. (2008). A class of quasi-Z source inverters. *Proceedings of the IEEE Industrial Application Meeting*, IEEE, pp. 1–7.

Axelrod, B., Borkovich, Y., and Ioinovici, A. (2008). Switched-capacitor/switched-inductor structures for getting transformerless hybrid DC–DC PWM converters. *IEEE Trans. Circuits Syst. I* 55 (2): 687–696.

Cao, D. and Peng, F.Z. (2009). A family of Z source and quasi-Z source DC–DC converters, *Proceedings of the IEEE Applied Power Electronics Conference*, IEEE, pp. 1097–1101.

Ćuk, S. (1979). General topological properties of switching structures. *Proceedings of the IEEE Power Electronics Specialists Conference*, IEEE, pp. 109–130.

Peng, F.Z. (2003). Z source inverter. *IEEE Trans. Ind. Appl.* 39 (2): 504–510.

Severns, R.P. and Bloom, G.E. (1985). *Modern DC-to-DC Switch Mode Power Converter Circuits*. New York: Van Nostrand Reinhold Co.

Williams, W. (2014). Generation and analysis of canonical switching cell DC-to-DC converters. *IEEE Trans. Ind. Electron.* 61: 329–346.

Wu, T.-F. (2016). Decoding and synthesizing transformerless PWM converters. *IEEE Trans. Power Electron.* 30 (9): 6293–6304.

Wu, T.-F. and Chen, Y.-K. (1996). A systematic and unified approach to modeling PWM DC/DC converters using the layer scheme. *Proceedings of the IEEE Power Electronics Specialists Conference*, IEEE, pp. 575–580.

Wu, T.-F. and Chen, Y.-K. (1998a). A systematic and unified approach to modeling PWM DC/DC converter based on the graft scheme. *IEEE Trans. Ind. Electron.* 45 (1): 88–98.

Wu, T.-F. and Chen, Y.-K. (1998b). Modeling PWM DC/DC converter out of basic converter units. *IEEE Trans. Power Electron.* 13 (5): 870–881.

Wu, T.-F. and Yu, T.-H. (1998). Unified approach to developing single-stage power converters. *IEEE Trans. Aerosp. Electron. Syst.* 34 (1): 221–223.

Wu, T.-F., Chen, Y.-K., and Liang, S.-A. (1999). A structural approach to synthesizing, analyzing and modeling quasi-resonant converters. *Proceedings of the IEEE Power Electronics Specialists Conference*, IEEE, pp. 1024–1029.

Wu, T.-F., Liang, S.-A., and Chen, Y.-K. (2003). A structural approach to synthesizing soft switching PWM converters. *IEEE Trans. Power Electron.* 18 (1): 38–43.

8

Synthesis of Multistage and Multilevel Converters

Based on the original converter, buck converter, many single-stage PWM converters can be synthesized, as shown in Chapters 5–7. However, there are existing many multilevel and multistage PWM converters, such as single-phase converters, three-phase converters, flywheeling capacitor converters (FCC), Vienna rectifier, neutral-point clamped converters (NPC), modular multi-level converters (MMC), etc., which are useful for generating AC output voltages or currents. How to synthesize the converters has not been discussed yet. In this chapter, syntheses of the multilevel and multistage converters are presented.

8.1 Review of the Original Converter and Its Variations of Transfer Code

There are many multilevel and multistage converters that can generate AC output voltages or currents, including single-phase converters, three-phase converters, FCC, Vienna rectifier, NPC, MMC, etc. Basically, they are all derived from the original converter, buck converter, with some modifications. Thus, this section reviews the buck converter and its variation versions.

Figure 8.1a shows a half-bridge type of buck converter with bidirectional switches. For DC output V_{AO}, active switch S_2 can be replaced with a single diode D_2, and diode D_1 can be removed, which becomes a conventional buck converter, as shown in Figure 8.1b. It is worth noting that when diode D_2 in wheeling, it also clamps the voltage across active switch S_1 to input voltage V_{dc}. However, for AC output, we need to combine two DC-type buck converters, in which one deals with the positive half cycle, while the other deals with the negative half cycle, as shown in Figure 8.1c. How to combine the positive half cycle and the negative

Origin of Power Converters: Decoding, Synthesizing, and Modeling, First Edition.
Tsai-Fu Wu and Yu-Kai Chen.
© 2020 John Wiley & Sons, Inc. Published 2020 by John Wiley & Sons, Inc.

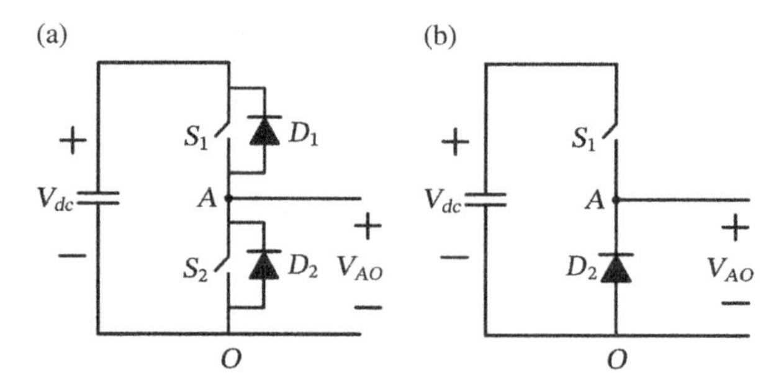

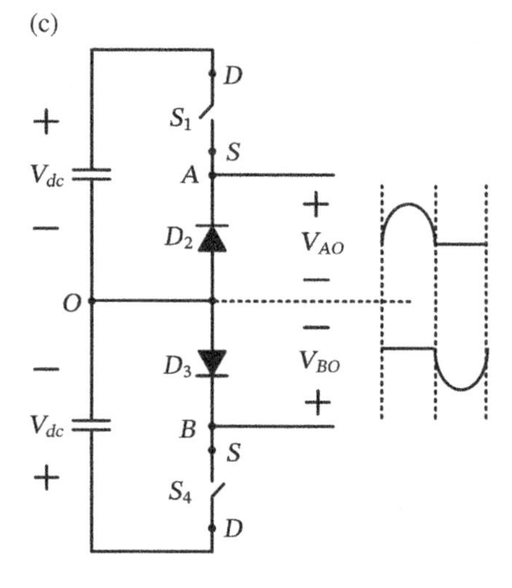

Figure 8.1 Variations of the original converter: (a) a general type, (b) DC type, and (c) two DC-type buck converters to generate AC outputs.

half cycle circuits shown in Figure 8.1c to form practical converters will be presented in Section 8.2.

When operating reversely, from the output to the input, we can have a boost-type converter for rectification in offline applications. Essentially, the buck converter can be operated bidirectionally if the switches are replaced with bidirectional ones.

Output V_{AO} is a pulsating voltage, and it needs a filter to smooth out, such as an *LC* or *LCL* filter. The gain is defined as

$$\frac{V_{AO}}{V_{dc}} = f(d), \tag{8.1}$$

where d is the duty ratio of active switch S_1 or S_2. It can be manipulated further, such as

$$F(d) = f_1(d) + f_2(d) + \cdots + f_n(d), \tag{8.2}$$

$$F(d)=f_1(d)-f_2(d), \tag{8.3}$$

$$F(d)=ST\left(f_1(d),f_2(d),f_3(d),\right) \tag{8.4}$$

and

$$F(d)=f(f_1(f_2(\cdots f_n(d)\cdots))), \tag{8.5}$$

where ST is defined as a function of subtraction between two arguments. For instance, the one shown in (8.4) has the set of six possibilities:

$$ST\left(f_1(d),f_2(d),f_3(d)\right)=\begin{Bmatrix} f_1(d)-f_2(d), f_2(d)-f_1(d), f_1(d)-f_3(d), \\ f_3(d)-f_1(d), f_2(d)-f_3(d), f_3(d)-f_2(d) \end{Bmatrix}. \tag{8.6}$$

Based on the above equations, we can synthesize various kinds of converter topologies for generating AC output voltages or currents.

8.2 Syntheses of Single-Phase Converters

As shown in Figure 8.1c, the two buck converters are used to deal with the positive and the negative half cycles, respectively. Although the converters can generate AC outputs, we need an extra switch to switch between outputs V_{AO} and V_{BO} and two isolated DC sources V_{dc}. If we combine the two DC sources, the converter will become the one shown in Figure 8.2a in which both outputs V_{AO} and V_{BO} become positive. However, if exchanging the positions of switch S_4 and diode D_3 and taking output from the differential of V_{AO} and V_{BO}, we will have both positive and negative outputs, as shown in Figure 8.2b, in which all of the switches are replaced with bidirectional ones to allow AC currents to flow. Since we flip over the output of V_{BO} and take a differential output, the diagonal switches (S_1&S_4 or S_2&S_3) need to be turned on or turned off in unison. Its transfer code becomes

$$\frac{V_{AB}}{V_{dc}}=f_A(d_1)-f_B(d_2). \tag{8.7}$$

Equation (8.7) can yield the result either positive or negative; thus, the converter shown in Figure 8.2b can have AC outputs. The converter is known as a full-bridge single-phase converter. It should be noted that when $d_1 = d_2$ and the two diagonal switch pairs take turn conducting, the full bridge is operated in bipolar mode.

Transfer code $f_A(d_1)$ or $f_B(d_2)$ is ranging from 0 to 1. If we replace $f_B(d_2)$ with 1/2, which in turn biases the output voltage to $V_{dc}/2$, we have the V_{AO}/V_{dc} as follows:

$$\frac{V_{AO}}{V_{dc}}=f_A(d_1)-\frac{1}{2}. \tag{8.8}$$

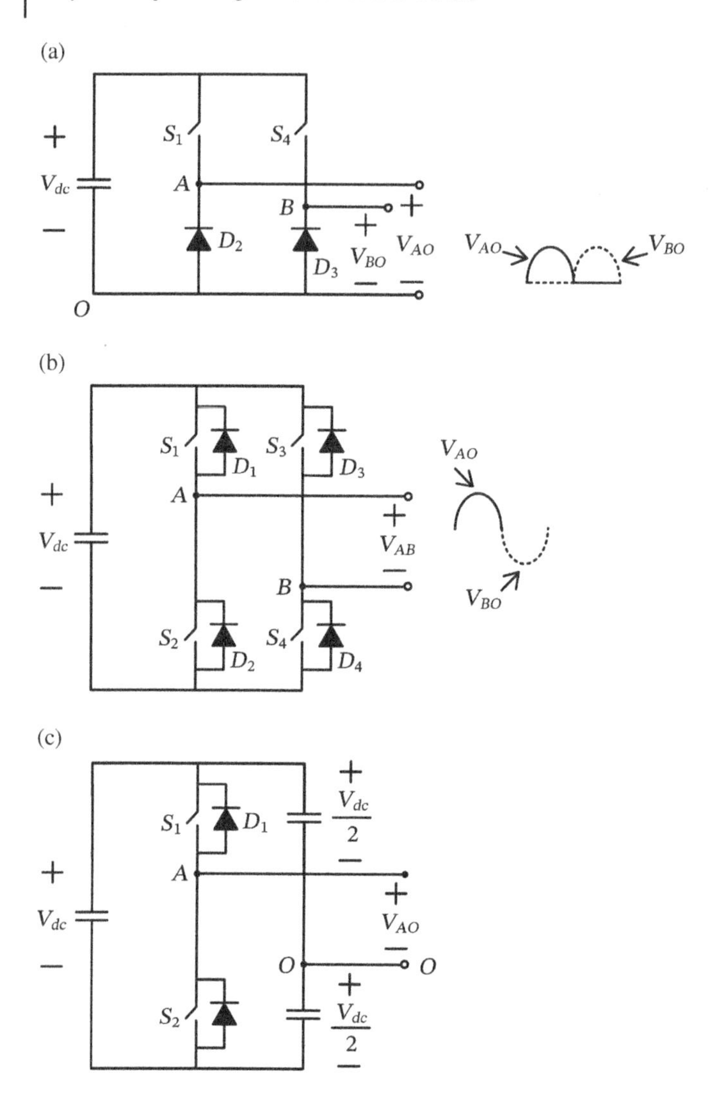

Figure 8.2 Syntheses of single-phase converters: (a) combined two DC sources and two DC-type buck converters in parallel, (b) combined single AC output to form a full-bridge converter, and (c) a half-bridge converter.

The above equation can be synthesized with the converter shown in Figure 8.2c, which is known as a half-bridge single-phase converter. In the converter, switch S_1 controls the positive half line cycle, while switch S_2 controls the negative one.

Another variation of Figure 8.1c is by flipping over the voltage polarities of the lower arm V_{dc}, as shown in Figure 8.3a where both diode D_3 and switch S_4

are correspondingly reversed to accommodate the power flow. Switch S_1 and diode D_2 take care of the positive half cycle to yield V_{AO}, while switch S_4 and diode D_3 take care of the negative one to yield $-V_{BO}$, as illustrated in Figure 8.3a. To combine V_{AO} and V_{BO}, and to avoid shorting the voltage source V_{dc}, it needs two additional switches, S^+ and S^-, to switch between positive and negative half cycles, as shown in Figure 8.3b. This is known as a single-phase NPC converter. It is worth noting that when switch S_1 or S_4 turns off, its voltage stress is clamped to V_{dc} by diode D_2 or D_3, respectively, while the converter increases conduction loss because the current always flows through two active switches, or one active switch and one diode.

A similar type of NPC converter is called T-type converter, as shown in Figure 8.3c. It is formed with two buck converters and configured as a half-bridge type, in which switches S_1, S_{b1} and diode D_{b2} are responsible for the positive half cycle, while switches S_2, S_{b2} and diode D_{b1} are for the negative half one. To combine two buck converters to generate AC outputs, we need two bidirectional switches S_{b1} and S_{b2} to serve as freewheeling paths for output filter inductors, either an LC or LCL filter.

8.3 Syntheses of Three-Phase Converters

A full-bridge single-phase converter has two legs, as shown in Figure 8.2b, which can have two outputs V_A and V_B. When taking their differential, we have a single-phase output V_{AB}. When including three legs, we have three outputs V_A, V_B, and V_C, as shown in Figure 8.4a, which is known as a full-bridge three-phase three-wire converter. Based on (8.4), for positive-sequence three-phase outputs, we have V_{AB}, V_{BC}, and V_{CA}, while for negative-sequence outputs, we have V_{AC}, V_{CB}, and V_{BA}. With proper modulation on the six switches, such as space-vector PWM (SVPWM), we can have three-phase sinusoidal output waveforms when the three outputs are connected to LC or LCL filters.

Triplicating the half-bridge single-phase converter shown in Figure 8.2c will generate a half-bridge three-phase four-wire converter, as shown in Figure 8.4b, in which the fourth wire connects the neutral point of the two DC sources to the common node (ground) of the three outputs. Similarly, three sets of the single-phase NPC converter are connected to the same DC voltage sources that will form a three-phase NPC converter, as shown in Figure 8.4c.

A buck-type converter is used for power injection to grid. On the other hand, the reverse operation of the buck-type converter is a boost-type converter, which acts as an active rectifier. When two boost converters are used to actively rectify AC grid voltage, in which one deals with the positive half cycle, while the other deals with the negative one, we have the converter shown in Figure 8.5a. In Figure 8.5a,

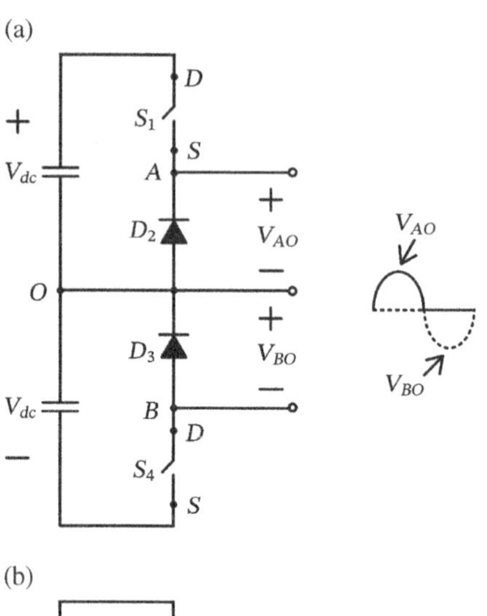

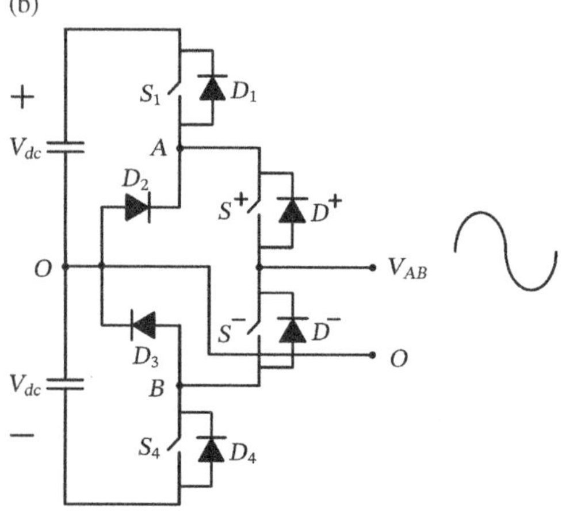

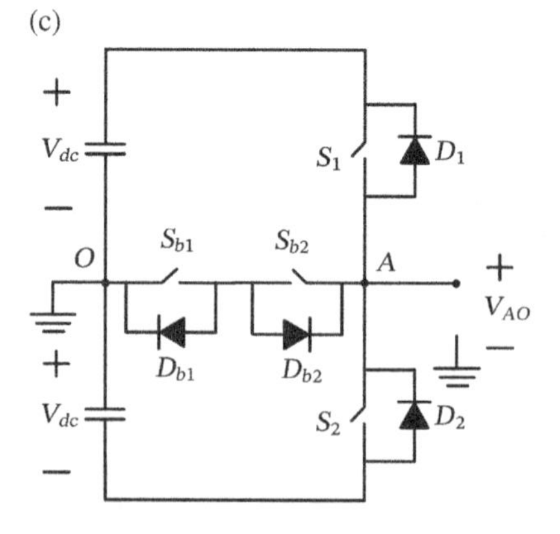

Figure 8.3 Syntheses of other single-phase converters: (a) combined two DC sources with different voltage polarities, (b) using two additional switches S^+ and S^- to combine the positive and negative half cycles to form a single-phase NPC converter, and (c) a T-type NPC converter.

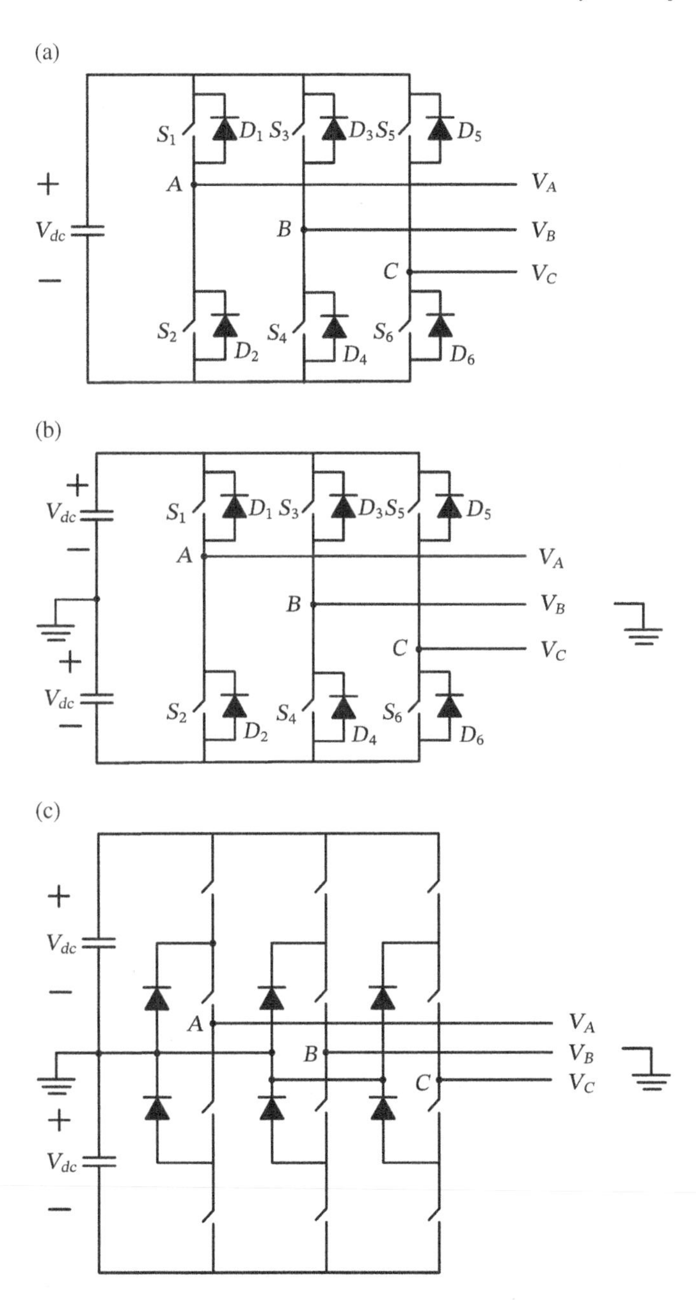

Figure 8.4 Syntheses of three-phase converters: (a) full-bridge three-phase three-wire converter, (b) half-bridge three-phase four-wire converter, and (c) three-phase NPC.

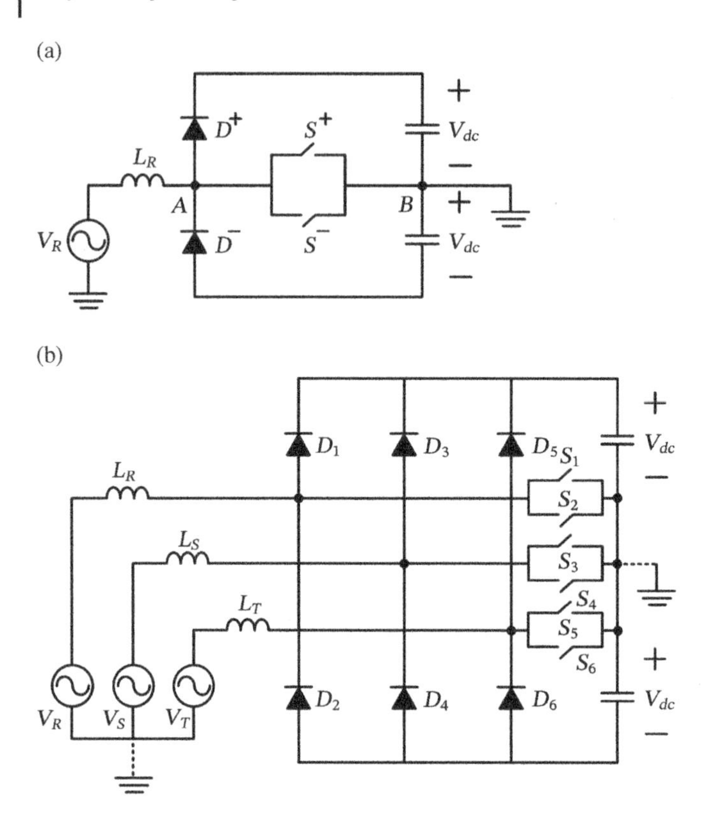

Figure 8.5 Synthesis of Vienna rectifier: (a) single phase and (b) three phase.

switches S^+ and D^+ are used to process the positive half cycle, while switch S^- and diode D^- are used to process the negative one. Like a boost converter, when the active switch turns on, input source V_R magnetizes inductor L_R, and when it turns off, the inductor current freewheels through the passive diode and output V_{dc}. Thus, it needs to connect the neutral of the two DC sources to the neutral of the AC source.

For a three-phase converter system, three single-phase converters are connected to the same DC sources, while at the AC side, each is connected to one phase, as shown in Figure 8.5b. Under undistorted and balanced three-phase currents, the total current through the neutral wire is 0. Therefore, the neutral wire can be removed from the converter topology, as indicated with dashed line in Figure 8.5b. This is the well-known Vienna rectifier.

In Vienna rectifier, since active switches S^+ and S^-, as shown in Figure 8.6a, are not operated simultaneously, they can be replaced with an active switch with four diodes, as shown in Figure 8.6b. It can save one active switch.

However, it increases conduction loss to two diodes. With diode grafting, diodes D^+ and D_1 sharing a common P node and D^- and D_3 sharing a common N node can be relocated to the positions shown in Figure 8.6c. In this type of switch configuration, during inductor current freewheeling through D_1 and D^+, it increases conduction loss, while it can reduce ringing at S_1 turn-off transition. Similarly, it applies to diodes D_3 and D^-.

8.4 Syntheses of Multilevel Converters

Multilevel converters have many types of configurations. Typically, they are derived from the buck-type converter with the summation of transfer codes, as shown in (8.2), in which each transfer code $f_i(d)$ is the transfer code of a buck converter or its variation where subscript i is 1, 2, ..., n. A three-phase modular multilevel converter (MMC) is shown in Figure 8.7a, in which each cell can be either a half-bridge type or a full-bridge type of converter, as shown in Figure 8.7b and c, respectively. For a

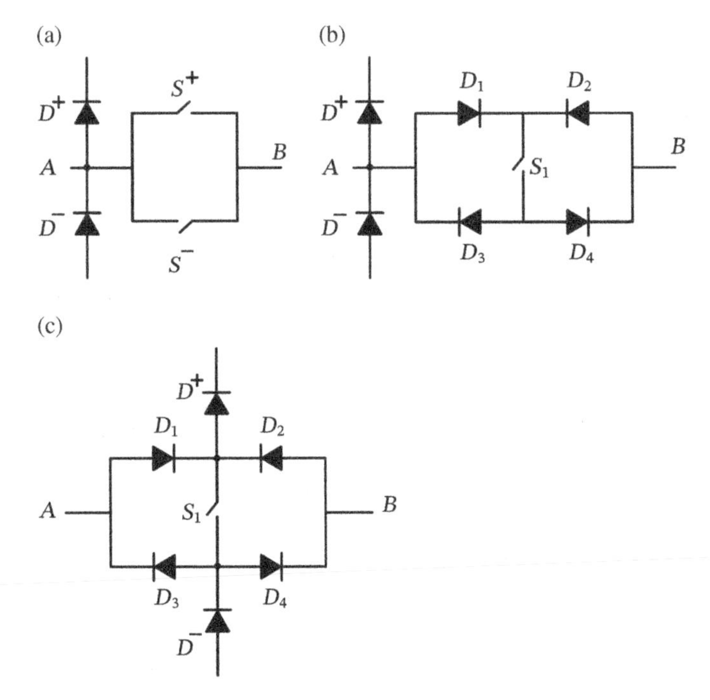

Figure 8.6 A bidirectional switch realized with (a) two active switches, (b) an active switch and four diodes, and (c) the one shown in (b) with diode grafting.

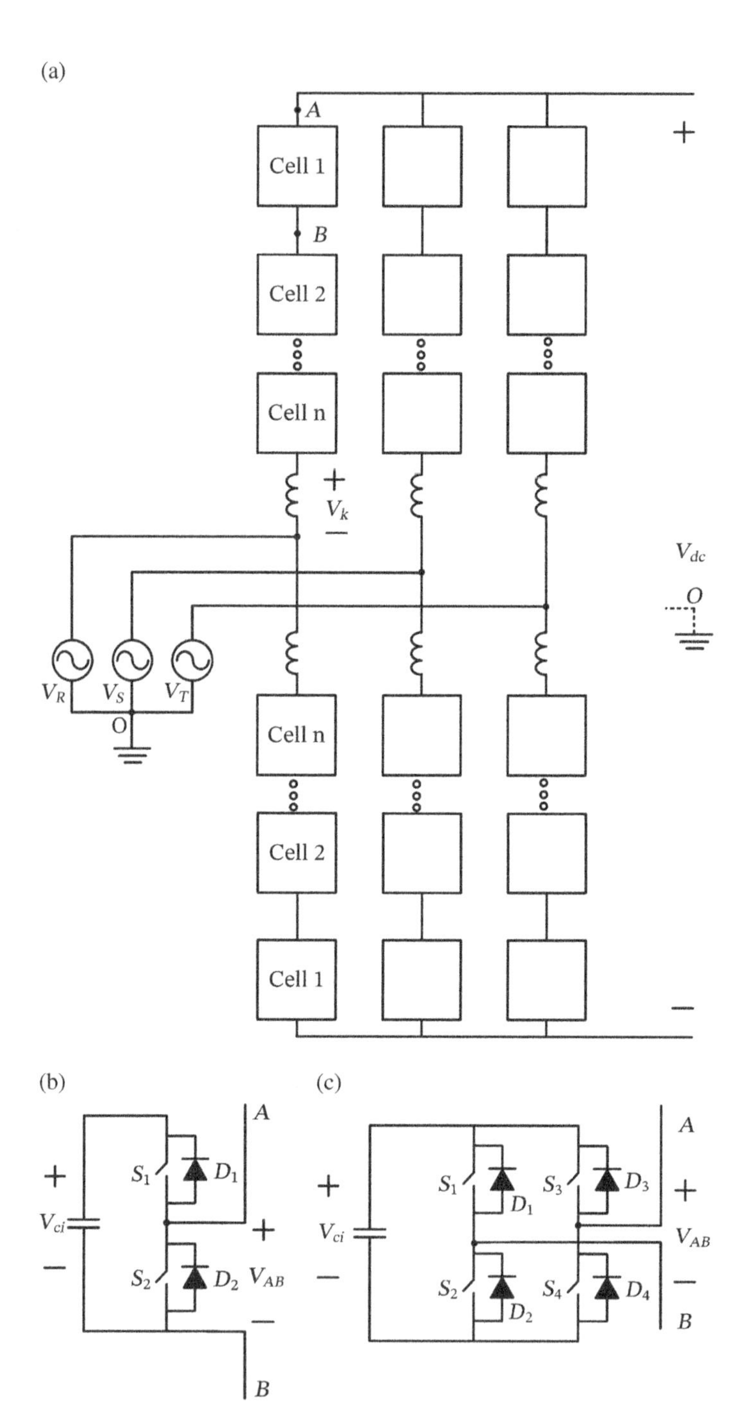

Figure 8.7 Configurations of (a) an MMC, (b) a half-bridge cell, and (c) a full-bridge cell.

half-bridge cell, average voltage V_{AB} can be represented as $d_i V_{Ci}$ where d_i is the duty ratio of the upper-arm switch of the ith cell, while for a full-bridge cell, V_{AB} can be $\pm d_i V_{Ci}$ where d_i is the duty ratio of switches S_3 and S_2 in the ith cell.

For an MMC, its transfer ratio can be related to the equation shown in (8.2) because each cell blocks a $d_i V_{Ci}$ voltage, like a buck converter shown in Figure 8.1b blocking dV_{dc}, and the total blocking voltage V_B becomes

$$V_B = d_1 V_{C1} + d_2 V_{C2} + \cdots + d_n V_{Cn}. \tag{8.9}$$

Thus, the voltage across the filter inductor is V_k:

$$V_k = \left(\left(\frac{V_{dc}}{2} - V_B \right) - V_R \right), \tag{8.10}$$

which is equivalent to that, $((dV_{dc}) - V_o)$, of a buck converter. It can be proved that the $V_{dc}/2$ in (8.10) is the equivalent voltage between "+" polarity and virtual ground "O," as indicated in Figure 8.7a.

For an FCC or a capacitor-clamped converter (CCC), it can yield either DC or AC output. For a DC output, it has a transfer code as shown in (8.5) and repeated as

$$\frac{V_o}{V_{dc}} = f_1\left(f_2\left(\cdots\left(f_n\left(d\right)\right)\right)\right). \tag{8.11}$$

A three-level FCC is shown in Figure 8.8a in which capacitor C_1 and switches S_2 and S_3 form a buck converter as shown in Figure 8.8b and capacitor C_{dc} and switches S_1–S_4 form another buck converter. The overall converter has the following transfer code:

$$\frac{V_o}{V_{dc}} = f_1\left(f_2\left(d\right)\right). \tag{8.12}$$

One of the possible transfer codes is shown as follows:

$$\frac{V_o}{V_{dc}} = \min\left(f_1\left(d_1\right), f_2\left(d_2\right)\right) + f_2\left(d_2\right). \tag{8.13}$$

where d_1 is the duty ratio of switch S_1 and d_2 is that of switch S_2. Typically, voltage V_C is controlled to be $V_{dc}/2$. For AC outputs, there are many types of modulations. However, capacitor voltage V_C needs to be kept to $V_{dc}/2$ to achieve symmetrical AC output waveforms.

In principle, more than three-level FCC can be generated. In practice, however, voltage balance can be a big problem when the level is higher than five.

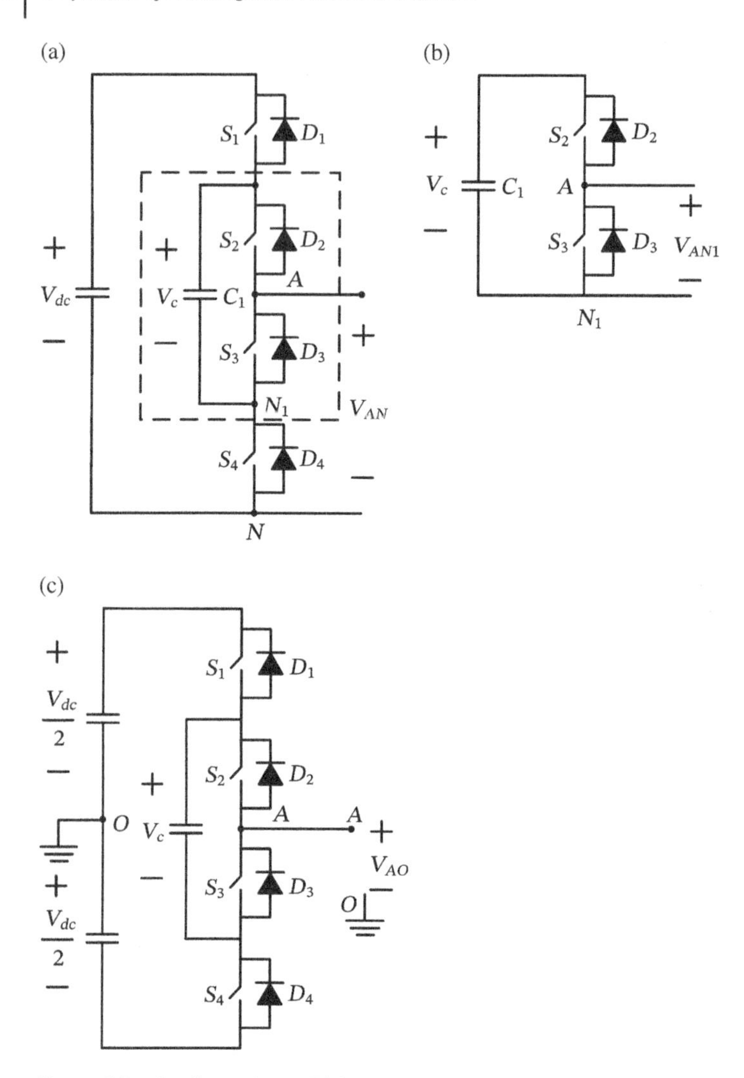

Figure 8.8 Configurations of (a) a full-bridge three-level FCC for generating DC outputs, (b) an inner buck cell, and (c) a half-bridge three-level FCC for generating AC outputs.

8.5 *L–C* Networks

In the aforementioned half-bridge and full-bridge inverter configurations, such as the ones shown in Figures 8.2b, c, 8.3b, c, and 8.4, their outputs can be connected with resonant networks, such as *LC*, *LCL*, *LLC*, *LCC*, etc. If the natural frequencies of the networks are far below the switching frequencies, the networks act as current and voltage ripple filters. Their output gains still follow the PWM voltage

(a)

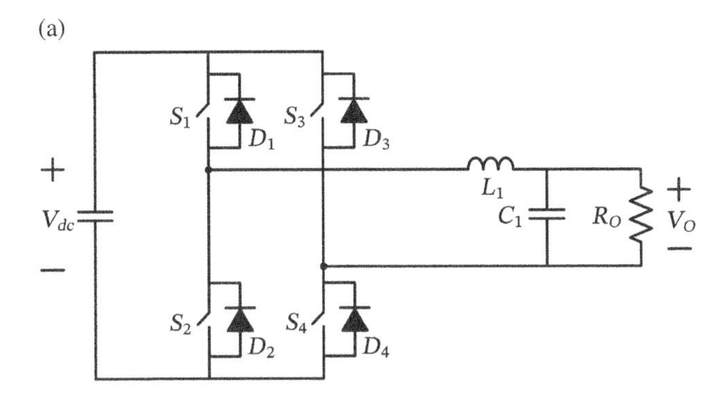

(b)

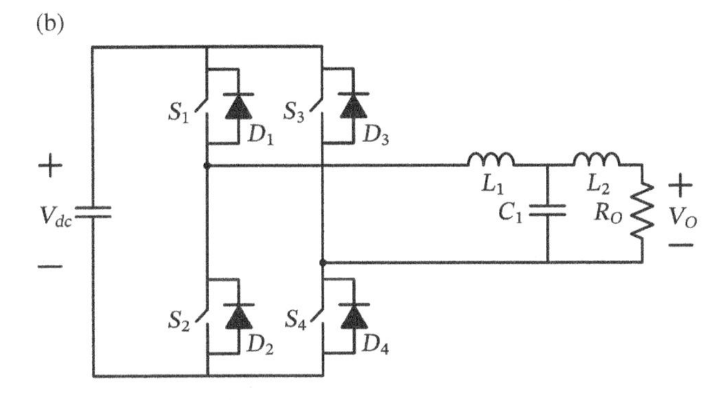

(c)

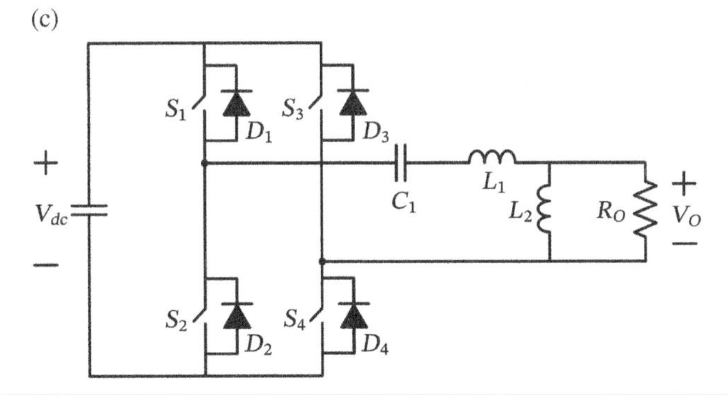

Figure 8.9 Full-bridge resonant converters with (a) *LC*, (b) *LCL*, and (c) *LLC* resonant tanks.

transfer gains. However, if their natural frequencies are close to the switching frequencies, they become resonant converters, and their output gains function of switching frequencies, phase-shift angle, resonant frequencies, and load conditions. The analyses of the resonant converters are much more complicated than those of PWM converters and are beyond the scope of this book.

Therefore, synthesis of resonant converters is identical to that of single-phase PWM converters shown in Section 8.2. We have to design the resonant networks (tanks) carefully to fit to the desired input-to-output voltage transfer gains. Figure 8.9 shows the full-bridge converters with *LC*, *LCL*, and *LLC* resonant tanks. Since the output voltage V_o is a symmetrical AC, we can insert a transformer at the output to become an isolated resonant converter.

Further Reading

Behrouzian, E., Bongiorno, M., and De La Parra, H.Z. (2013). An overview of multilevel converter topologies for grid connected applications. *Proceedings of 15th EPE Applications Conference*, IEEE, pp. 1–10.

Carpaneto, M., Marchesoni, M., and Vaccaro, L. (2007). A new cascaded multilevel converter based on NPC cells. *Proceedings of the IEEE International Symposium on Industrial Electronics*, IEEE, pp. 1033–1038.

Holmes, D.G. and Lipo, T.A. (2003). *Pulse Width Modulation for Power Converters: Principles and Practice*. Hoboken: Wiley.

Kolar, J.W. and Ertl, H. (1999). Status of the techniques of three-phase rectifier systems with low effects on the mains. *Proceedings of the IEEE International Telecommunications Energy Conference*, IEEE, session 14-1.

Kolar, J.W. and Zach, F.C. (1994). A novel three-phase utility interface minimizing line current harmonics of high-power telecommunications rectifier modules. *Proceedings of the IEEE International Telecommunications Energy Conference*, pp. 367–374.

Kolar, J.W., Ertl, H., and Zach, F.C. (1996). Design and experimental investigation of a three phase high power density high efficiency unity power factor PWM (Vienna) rectifier employing a novel integrated power semiconductor module. *Proceedings of the IEEE Applied Power Electronics Conference*, P.514–523, vol. 2.

Ladoux, P., Serbia, N., Rubino, L., and Marino, P. (2014). Comparative study of variant topologies for MMC. *Proceedings of the International Symposium on Power Electronics, Electrical Drives, Automation and Motion (SPEEDAM)*, IEEE, Ischia, Italy, pp. 659–664.

Lai, J.S. and Peng, F.Z. (1995). Multilevel converters – new breed of power converters. *Proceedings of the IEEE Industry Applications Society Annual Meeting*, pp. 2348–2356.

Meynard, T., Fadel, M., and Aouda, N. (1997). Modeling of multilevel converters. *IEEE Trans. Ind. Electron.* 44 (3): 356–364.

Mohan, N., Undeland, T.M., and Robbins, W.P. (1989). *Power Electronics*, 2e. Wiley.

Nabae, A., Takahashi, I., and Akagi, H. (1981). A new neutral-point-clamped PWM inverter. *IEEE Trans. Ind. Appl.* 17 (5): 518–523.

Shu, Z., Liu, M., Zhao, L. et al. (2016). Predictive harmonic control and its optimal digital implementation for MMC-based active power filter. *IEEE Trans. Ind. Electron.* 63 (8): 5244–5254.

Wilkinson, R., de Mouton, H., and Meynard, T. (2006). Natural balance of multicell converters: the two-cell case. *IEEE Trans. Power Electron.* 21 (6): 1649–1657.

Yuang, X., Stemmler, H., and Barbi, I. (2001). Self-balancing of the clamping-capacitor-voltages in the multilevel capacitor-clamping-inverter under sub-harmonic PWM modulation. *IEEE Trans. Power Electron.* 16 (2): 256–263.

9

Synthesis of Soft-Switching PWM Converters

In previous chapters, we have addressed how to synthesize PWM power converters, which are all operated with hard switching, except the resonant converters. For the half-bridge and full-bridge resonant converters, they are derived with the switch pairs and resonant tanks and operated with variable switching frequencies and/or phase shift to achieve zero-voltage or zero-current switching. Their operational principles are quite different from those of conventional PWM converters. However, from converter topological point of view, they are identical to the PWM converters.

In literature, there are many hard-switching PWM converters associated with soft-switching cells to form soft-switching PWM converters. Basically, the soft-switching cells just deal with the switching transitions during switch turn-on and turn-off, without changing the main operational principle of their PWM converter counterparts. Thus, the transfer ratios are almost identical to those of their original hard-switching PWM converters. Soft-switching cells are just cells. They cannot exist independently. Without the PWM converters, they are meaningless and do not have input-to-output voltage or current transfer ratios. In this chapter, we will first present how to generate passive and active soft-switching cells to achieve near-zero-voltage switching (NZVS) and near-zero-current switching (NZCS). Then, by combining the graft and the layer schemes, we will show some examples to illustrate how to synthesize soft-switching PWM converters.

9.1 Soft-Switching Cells

Soft-switching PWM converters need soft-switching cells to take care of switch turn-on and turn-off transitions to achieve either NZVS or NZCS. Soft-switching cells include passive and active ones, in which only are lossless soft-switching cells discussed in this section.

Origin of Power Converters: Decoding, Synthesizing, and Modeling, First Edition.
Tsai-Fu Wu and Yu-Kai Chen.
© 2020 John Wiley & Sons, Inc. Published 2020 by John Wiley & Sons, Inc.

9.1.1 Passive Lossless Soft-Switching Cells

Both active and passive lossless cells can be employed to achieve soft switching. Several kinds of passive lossless cells have been presented to reduce switching losses. They need only passive switches and reactive components, which are usually with simple structures and high efficiency. They reduce switching loss by lowering the rates of di/dt and dv/dt of active switches to achieve NZVS or NZCS. Furthermore, by controlling the rate of di/dt of the active switch, the change rate of reverse recovery currents of the diodes is also controlled. The only loss that is not recovered from the passive cell is the energy stored in the internal parasitic capacitance of the power switch. However, this loss is usually much smaller than the others.

This subsection describes the basic nature of how the zero-current turn-on and zero-voltage turn-off passive soft-switching cells are added to the PWM converters and derives general topological and electrical properties common to all passive lossless soft-switching PWM converters.

9.1.1.1 Near-Zero-Current Switching Mechanism

Figure 9.1 shows the waveforms of switch gate signal V_G, switch current I_{DS}, and switch voltage V_{DS}, illustrating a zero-current switching mechanism. However, according to Figure 9.1, the main requirement of NZCS turn-on or turn-off is to slow down the current rising time or falling time, respectively, during switching transition. This NZCS feature can be achieved by placing a snubber inductor L_1 in series with the loop containing the active switch or freewheeling diode so that the switch current is limited to rise or fall slowly at switching transition. Figure 9.2 shows an equivalent circuit of a PWM converter with possible locations of the snubber inductor L_1. Thus, the rate of di/dt of the switch current is restricted, and the switching transition is close to zero-current transition, namely, NZCS or ZCT.

Snubber inductor L_1 will store energy that needs to be recovered ultimately and delivered to the output load or the input voltage source to maintain lossless operation. In this case, the snubber cell must provide a conduction path when a freewheeling diode or a switch turns off. If there is no path provided, either device might be destroyed by a voltage spike. Thus, as shown in Figure 9.3, two freewheeling diodes and one buffer capacitor are added to recover the absorbed energy and transfer it to the output. The energy absorbed in snubber inductor L_1 is transferred to buffer capacitor C_1. Then, energy recovery is achieved by discharging buffer capacitor C_1 and transferring the energy to the output load.

In Figure 9.3, if we consider the equivalent circuit as a PWM boost converter with zero-current turn-on snubber, the operational principle of energy recovery is presented as follows: during the turn-on process of S_M, the freewheeling diode D_M conducts reversely to charge its junction capacitor. The growth rate of the reverse

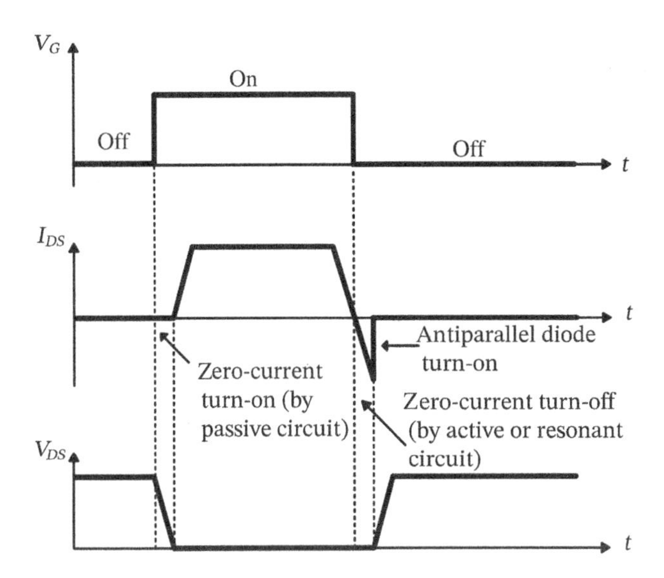

Figure 9.1 Illustration of a zero-current switching.

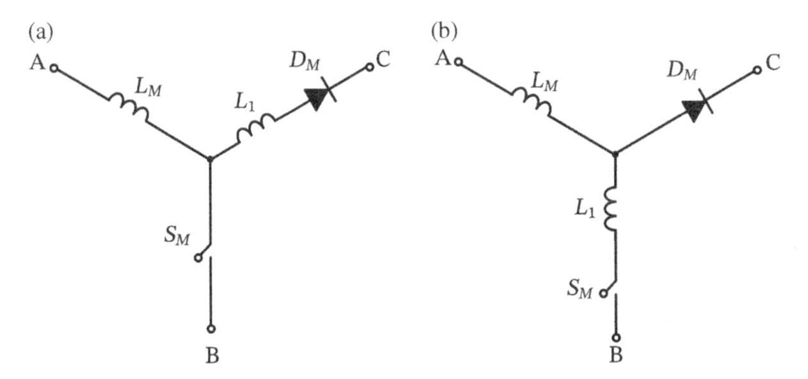

Figure 9.2 Proper locations of snubber inductor L_1 for PWM converters to achieve NZCS: (a) connected to a freewheeling diode and (b) connected to an active switch.

recovery current is restricted by snubber inductor L_1, which is placed in series with diode D_M, as shown in Figure 9.3a. After the reverse recovery of D_M, it is turned off, and the first resonance path is formed by L_1, D_1, and C_1. The stored energy of L_1 in current form is transferred to buffer capacitor C_1 through freewheeling diode D_1. After power switch S_M is turned off and the first resonance is stopped, the recovered energy is delivered to the output through the second resonance path L_M–L_1–C_1–D_2. Operational principles of the snubber can be extended to other topologies. Complete buck and boost PWM converters with the zero-current turn-on snubber are shown in Figures 9.4 and 9.5.

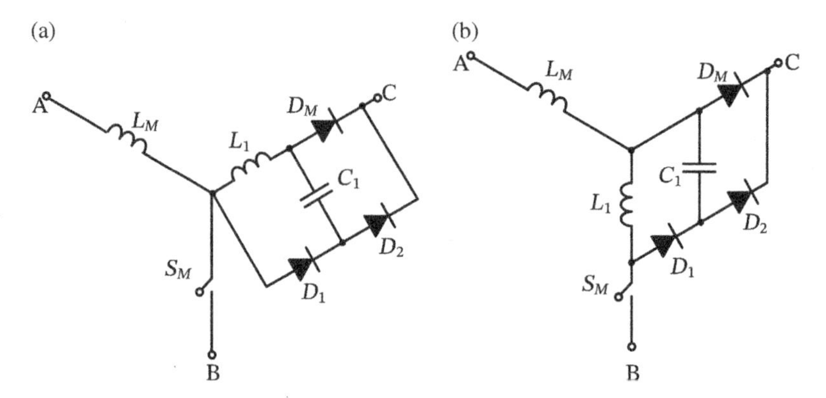

Figure 9.3 NZCS lossless cells with energy recovery: (a) associated with the freewheeling diode and (b) associated with the active switch.

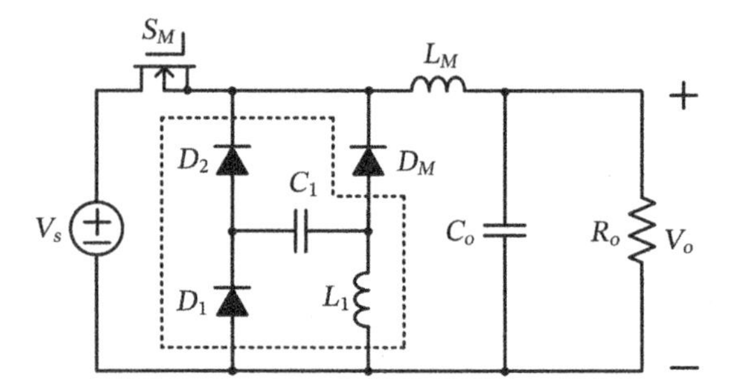

Figure 9.4 Passive soft-switching buck converter with NZCS, in which the soft switching cell is associated with main diode D_M.

9.1.1.2 Near-Zero-Voltage Switching Mechanism

According to Figure 9.6, a main task of NZVS or zero-voltage transition (ZVT) is to slow the rising time of the switch voltage. This is done by inserting a snubber capacitor in parallel with the switch to limit dv/dt of the drain-source voltage. Figure 9.7 shows equivalent circuits of PWM converters with possible locations of a snubber capacitor C_1. Thus, the near-zero-voltage turn-off can be considered NZVS. In the circuit, diode D_1 must be in the direction to conduct the switch current when main switch S_M is turned off. This diode is added to isolate the power switch from the snubber capacitor. It can also prevent the energy in C_1 from dissipating to the switch when it is turned on.

It is worth noting that capacitor C_1 in Figure 9.7 can be connected to positions A, B, and C, because it can have different DC voltage offsets while still keeping resonance with other snubber components. The positions A, B, and C are corresponding to the input, ground, and output, respectively.

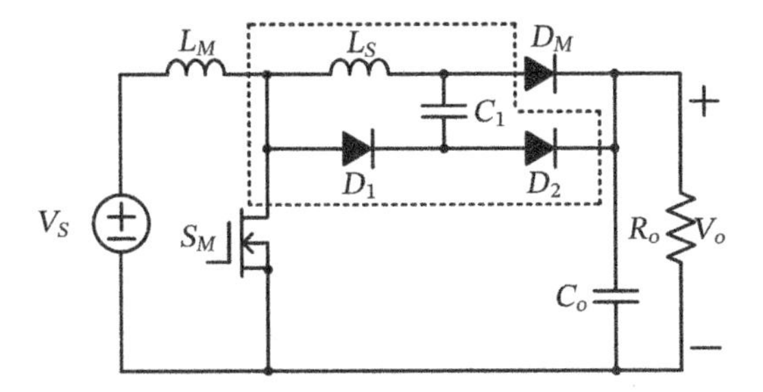

Figure 9.5 Passive soft-switching boost converter with NZCS, in which the soft-switching cell is associated with main diode D_M.

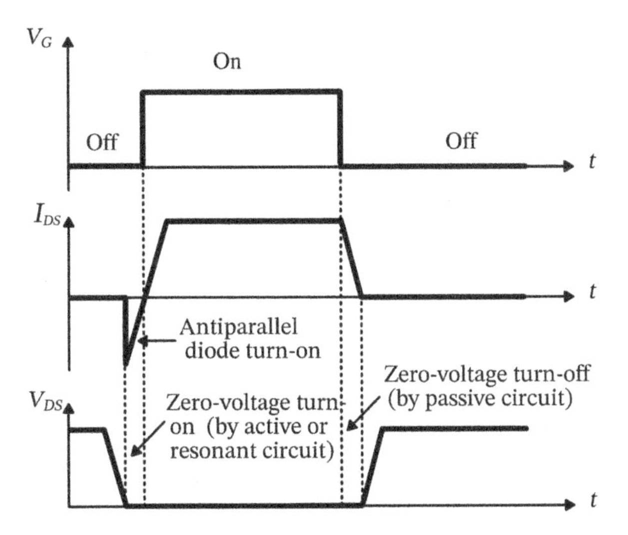

Figure 9.6 Illustration of a zero-voltage switching.

Snubber capacitor C_1 will accumulate energy that needs to be transferred to the output or input voltage source before next turn-off interval to ensure lossless operation. Using LC resonance is the most effective way to efficiently transfer this energy. This is implemented by a snubber inductor L_1 and a buffer capacitor C_2, as shown in Figure 9.8. At turn-on process of switch, S_M, C_1, C_2, and L_1 resonate through S_M and D_2, in which the energy stored in snubber capacitor C_1 is transferred to capacitor C_2. At the turn-off process, C_2 and L_1 resonate through inductor L_M, in which the energy stored in capacitor C_2 is transferred to the output. Thus, this kind of snubber circuit is lossless, and the conversion efficiency is relatively high.

Figures 9.9 and 9.10 show the buck and boost soft-switching converters, respectively, and each of which provides both NZCS and NZVS for the main switch S_M.

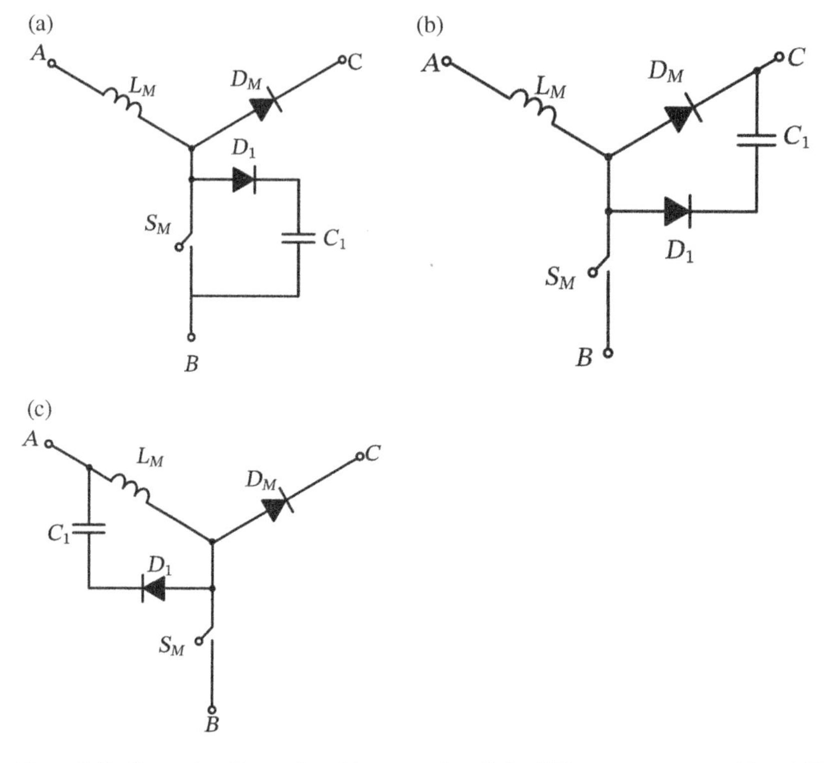

Figure 9.7 Proper locations of snubber capacitor C_1 for PWM converters to achieve NZVS.

Zero-voltage turn-on and turn-off of the freewheeling diode D_M are also obtained in these converters. All energy absorbed in inductor L_1 and capacitor C_1 is transferred to capacitor C_2. Energy recovery is therefore achieved by discharging buffer capacitor C_2 to the output. Switching loss and EMI level can be effectively reduced. Thus, efficiency of the converters is high, and switching frequency can be increased greatly, which would decrease the size of inductors and capacitors significantly.

9.1.2 Active Lossless Soft-Switching Cells

Generally, the passive lossless soft-switching snubbers are effective in reducing the turn-on and turn-off switching losses to achieve NZCS and NZVS. A NZCS snubber limits the rising rate of the switch current and thereby suppresses the current surge due to reverse recovery of a freewheeling diode at switch turn-on transition. On the other hand, a NZVS snubber reduces turn-off loss and the voltage spike by reducing increase rate of the switch voltage. The voltage spike due to wire and circuit parasitic inductance occurs at switch turn-off transition.

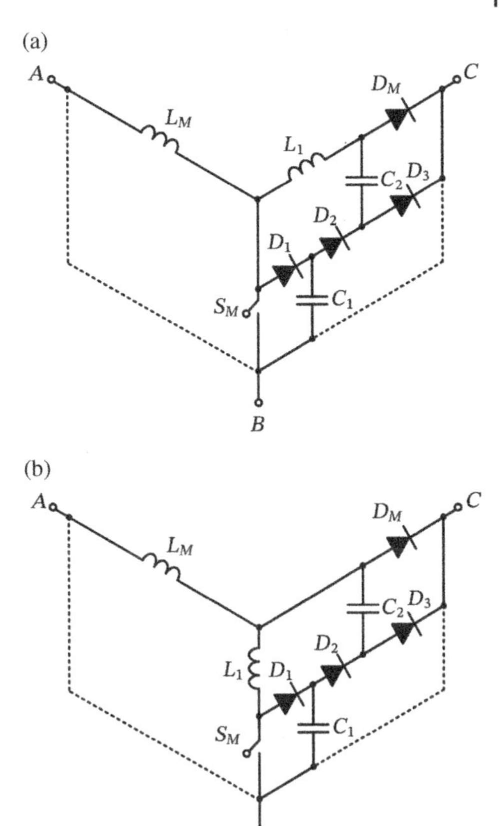

Figure 9.8 Near-zero-voltage turn-off lossless snubbers with energy recovery.

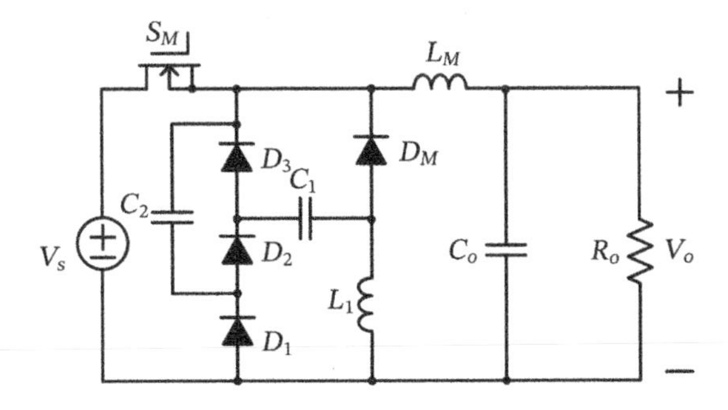

Figure 9.9 Passive soft-switching buck converter with NZCS and NZVS.

Passive lossless snubbers, however, cannot achieve ZVS and ZCS features. Essentially, ZVS eliminates the capacitive turn-on loss, and ZCS reduces switch turn-off loss. Active soft-switching converters have been proposed to achieve either ZVS or ZCS. Generally, the active lossless snubber employs an auxiliary

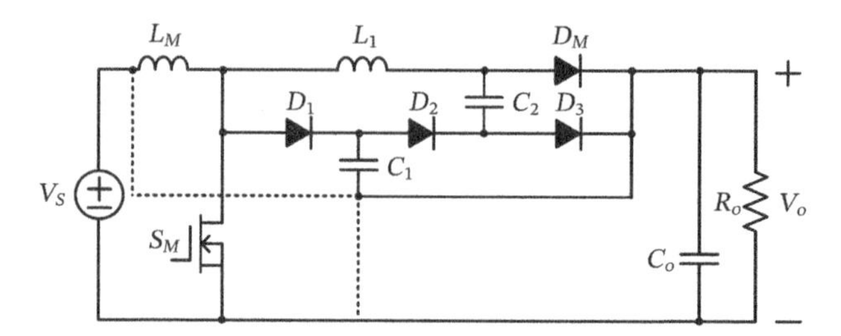

Figure 9.10 Passive soft-switching boost converter with NZCS and NZVS.

active switch with several passive components such as inductors, capacitors, and rectifiers.

In this section, the concept of fundamental ZVS or ZCS cells was adopted to generate active lossless soft-switching PWM converters. These cells can be extended to other converter topologies. Generation of the PWM soft-switching cells is described as follows:

9.1.2.1 Zero-Voltage Switching Mechanism

Mechanism of a ZVS feature is illustrated in Figure 9.6. The energy stored in the MOSFET output parasitic capacitance is not dissipated within the device when it is turned on. This will increase the system efficiency. A ZVS feature is achieved by directing the current in the snubber inductor to discharge the output parasitic capacitance of the switch. The power switch is then turned on when the antiparallel diode of the active switch (MOSFET) has been conducting.

In the literature, a number of active soft-switching PWM converters with ZVS have been proposed. Each of them employs an auxiliary active switch with passive components to form an active lossless snubber, which is used only to create condition for achieving ZVS at the switch turn-on transition. Thus, these ZVS converters still possess the same steady-state characteristics as those of their PWM converter counterparts. However, the soft-switching PWM converters require auxiliary power switch, increase control complexity in a certain level, and need to establish a soft-switching feature for wide line and load ranges.

Figure 9.11 shows an equivalent PWM converter with a ZVS soft-switching cell. This topology differs from a conventional PWM converter by an additional auxiliary switch S_1, a resonant inductor L_1, a resonant capacitor C_1, which includes the output parasitic capacitance of the power switch S_M, and a freewheeling diode D_1. This resonant branch L_1–S_1–C_1 is active only during a short switching transition time to create a ZVS condition for main switch S_M. A ZVS feature can be always performed for one switching direction by utilizing the current source to charge and discharge the resonant capacitor placed across the switch. During the switching transition, the resonant network is activated to create a partial resonance to

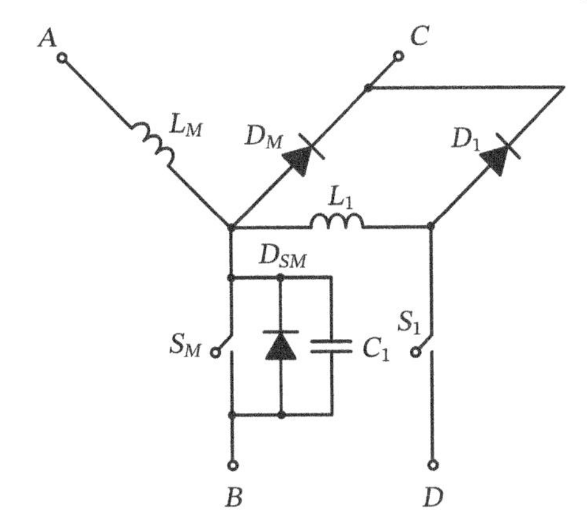

Figure 9.11 Zero-voltage turn-on lossless snubbers with energy recovery.

achieve ZVS. When switching transition is over, the circuit simply acts as a conventional PWM converter. As a consequence, the converter can achieve soft switching while preserving the advantages of the PWM converters.

It should be mentioned that ZVS can be implemented in many ways. From the circuit topological point of view, each active soft-switching converter with ZVS can be viewed as a variation of the equivalent circuit shown in Figure 9.11. By incorporating this type of resonant network, it creates a resonance to achieve ZVS. The ZVS concept can be extended to generate different types of soft-switching converters. Figures 9.12–9.14 show typical ZVS-PWM topologies. These active ZVS-PWM converters can achieve ZVS on the main switch without significantly increasing voltage and current stresses of the switch and can achieve soft switching on the auxiliary switch.

Figure 9.12a shows a circuit diagram of the ZVS-PWM soft-switching boost converter. The additional resonant circuitry of the converter is composed of a resonant inductor L_1, an auxiliary switch S_1, two diodes D_1, D_2, and one resonant capacitor C_1. The converter operates at a fixed frequency and is controlled by a PWM scheme.

From Figure 9.12b, it can be seen that the current and voltage waveforms of the switches in the converter are essentially square wave except at the turn-on and turn-off switching intervals. Both the active switch and the rectifier diode are imposed by minimum voltage and current stresses. At t_0, auxiliary switch S_1 is turned on. The current of resonant inductor i_{L1} is linearly built up until it reaches the input current level I_{LM}, and the current through diode D_M declines to zero. Since inductor L_1 limits the increasing rate of di/dt, switch S_1 is turned on with ZCT. At t_1, inductor current i_{L1} equals I_{LM}, and diode D_M is naturally turned off with ZCT. Thus, S_1 can be turned on with ZCT, and D_M is with ZCT turn-off. During the time interval t_1–t_2, capacitor C_1 resonates with inductor L_1

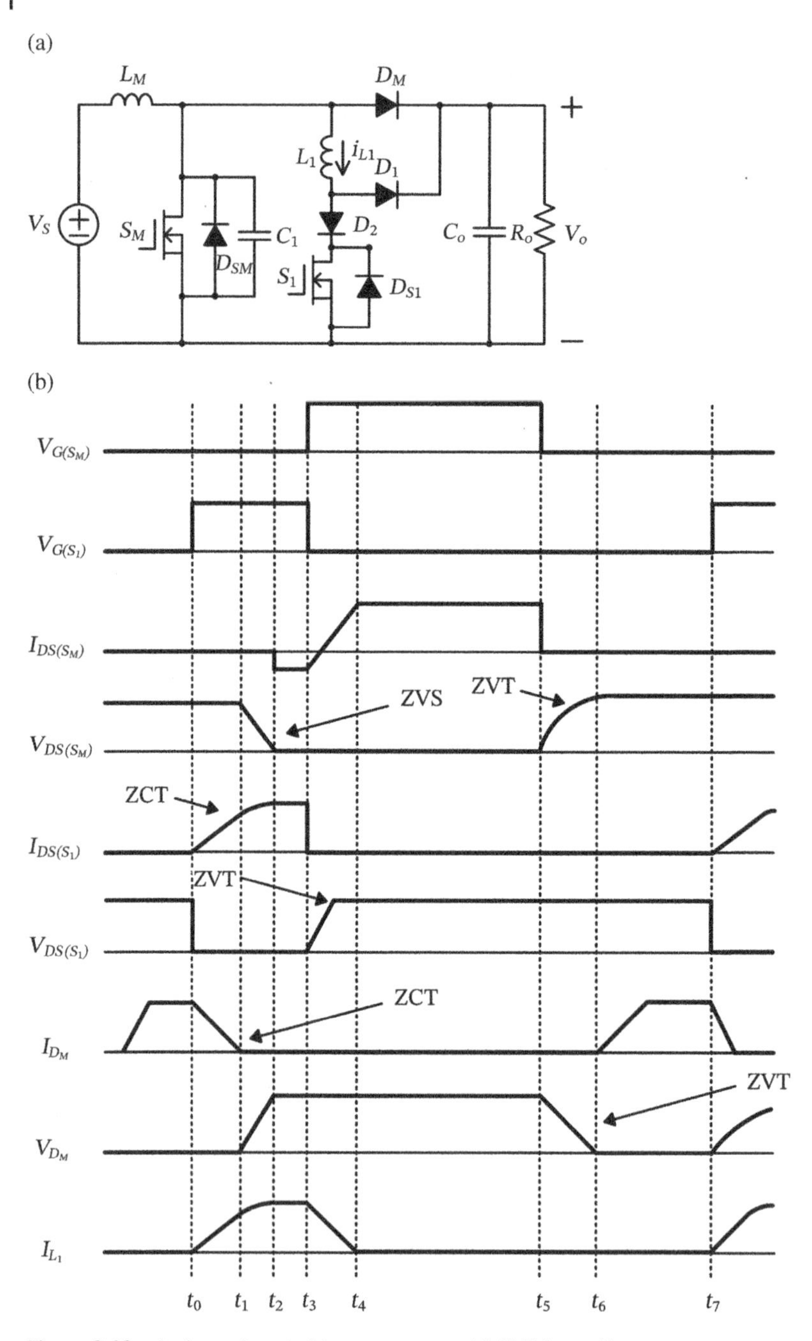

Figure 9.12 Active soft-switching converter with ZVS (type 1).

(c)

Switches	Turn-on conditions	Turn-off conditions
S_M	ZVS	ZVT
S_1	ZCT	ZVT
D_M	ZVT	ZCT

Figure 9.12 (Continued)

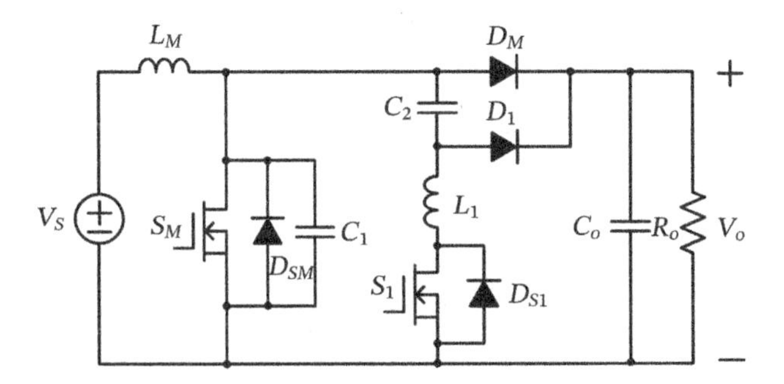

Figure 9.13 Active soft-switching converter with ZVS (type 2).

to discharge the energy stored in C_1. This mode is not terminated until capacitor voltage V_{C1} drops to zero. Once V_{C1} drops to zero, the body diode D_{SM} of main switch S_M will conduct; thus, a ZVS operational opportunity is created. At t_3, main switch S_M is turned on with ZVS and its current ramp up when i_{L1} drops below I_{LM}. At the same time, switch S_1 is turned off, and the energy stored in the resonant inductor L_1 will be dumped to the output by way of D_1. The current of inductor L_1 decreases linearly until it reaches zero at t_4. During t_4–t_5 interval, the main switch keeps conducting. This operation mode is the same as that in a conventional boost PWM converter. At t_5, main switch S_M is turned off, and resonant capacitor C_1 starts to charge. Since C_1 will limit the increasing rate of dv/dt, main switch S_M has a ZVT turn-off. When V_{C1} equals V_0, diode D_M conducts, and power is transferred from the source to the load. Thus, D_M has a ZVT turn-on. A complete switching cycle is ended at t_7, at which auxiliary switch S_1 is turned on again. The switching conditions of the power devices are shown in Figure 9.12c.

Figure 9.13 shows another active soft-switching PWM converter with ZVS. One important difference from the previous converter is that the extra resonant capacitor C_2 is in series with the resonant inductor L_1. The capacitor allows for an LC resonance between L_1 and C_2 to return the inductor current to zero. Unlike the previous one, this resonance does not require the auxiliary switch to turn off to change the direction of the resonant inductor current.

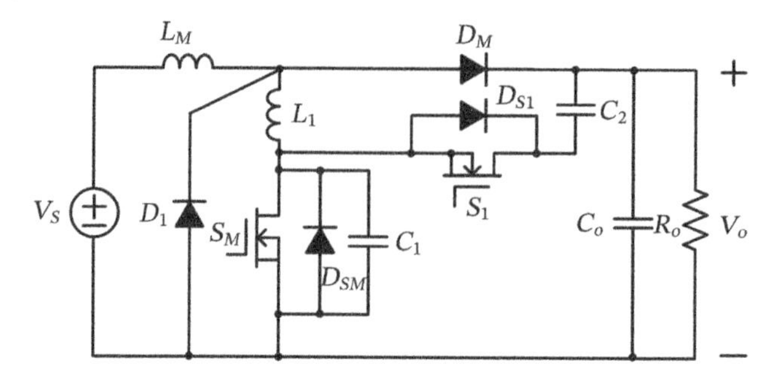

Figure 9.14 Active soft-switching converter with ZVS (type 3).

The other circuit diagram of the boost converter that employs the ZVS soft-switching technique is shown in Figure 9.14. The circuit uses the auxiliary switch S_1 and resonant capacitor C_2 connected in series to discharge the energy stored in resonant inductor L_1 to the output load after switch S_1 is turned off. Diode D_1 is employed to eliminate the parasitic ringing between the junction capacitance of diode D_M and resonant inductor L_1 by clamping the anode of D_M to ground.

Three types of ZVS-PWM soft-switching converters with active snubbers are presented, and the one shown in Figure 9.12 is analyzed as an example by using the concept of *LC* resonance. These converters present ZVS on the main switch and can also achieve almost no switching loss at turn-on/turn-off for the auxiliary switch and freewheeling diodes. The same technique can be extended to generate new types of ZVS-PWM soft-switching converters.

9.1.2.2 Zero-Current Switching Mechanism

In applications with high voltage and high values of root-mean-square currents through the power semiconductor devices, IGBTs are preferred as the power switches, because they present lower conduction loss than MOSFETs. Nevertheless, the presence of the "tail current," during their turn-off process, induces significant switching losses in the semiconductor switches. Thus, a ZCS feature is mainly required when using IGBTs as switches.

The switch turn-off loss, which is usually the dominant switching loss in high-power applications, cannot be alleviated effectively with the ZVS technique. From Figure 9.1, it can be seen that the ZCS technique can significantly reduce the switch turn-off loss by forcing the outgoing switch current to zero prior to its turn-off. Figure 9.15 shows the equivalent PWM converter with zero-current turn-off soft-switching cell that can represent the basic active soft-switching PWM converter with a ZCS function. It differs from a conventional PWM converter by including additionally an auxiliary switch S_1, a resonant inductor L_1, a resonant

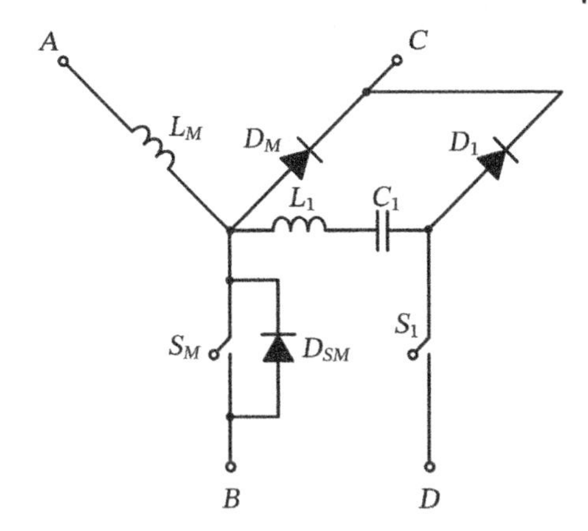

Figure 9.15 Zero-current turn-off lossless snubber with energy recovery.

capacitor C_1, and a freewheeling diode D_1. The resonant branch is only active during a relatively short switching time to create a ZCS condition for the main PWM switch without substantially increasing voltage or current stresses.

The circuit of ZCS operates with a resonant network and an auxiliary switch to create a zero-current condition for the main switch S_M during turn-off. The auxiliary switch S_1 is turned on before the main switch S_M is turned off so that the resonant tank current can rise sinusoidally and allow zero-current switching for the main switch S_M. Thus, to implement this ZCS mechanism, an L–C network is needed to add around the switch so that the switch current may be kept at zero during switching transition. When the soft-switching transition is over, the converter simply comes back to the basic PWM operating mode. According to the basic concept of Figure 9.15, the ZCS-PWM soft-switching converters can be possibly implemented in a number of ways. From the circuit topological point of view, one should keep it in mind that every ZCS converter can be viewed as a variation of the equivalent circuit shown in Figure 9.15.

In this section, two ZCS schemes are proposed and investigated to further improve the ZCS technique. With modified control and topology, the main switches are switched off under ZCS conditions; thus, the switching losses and stresses are reduced significantly. Figures 9.16 and 9.17 show two typical ZCS-PWM topologies. These converters have the following features:

1) ZCS for the main switch
2) Low voltage/current stresses on the main switch and rectifier diode
3) Minimal circulating energy
4) Wide line and load ranges for ZCS
5) Constant frequency operation

In the following, we use Figure 9.16 as an example to illustrate the operational principle for a ZCS-PWM converter. Referring now to Figure 9.16a and b, prior

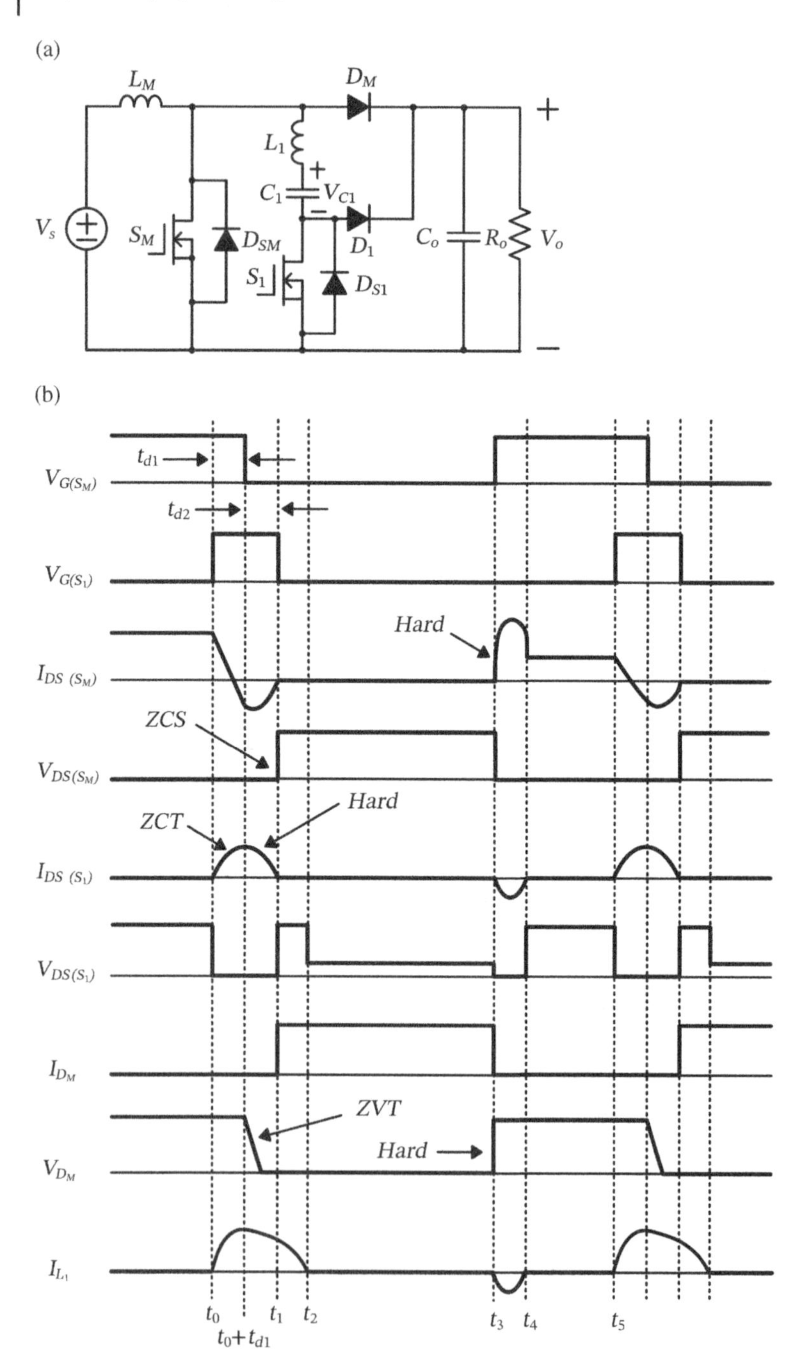

Figure 9.16 Active soft-switching converter with ZCS (type1).

(c)

Switches	Turn-on conditions	Turn-off conditions
S_M	Hard	ZCS
S_1	ZCT	Hard
D_M	ZVT	Hard

Figure 9.16 (Continued)

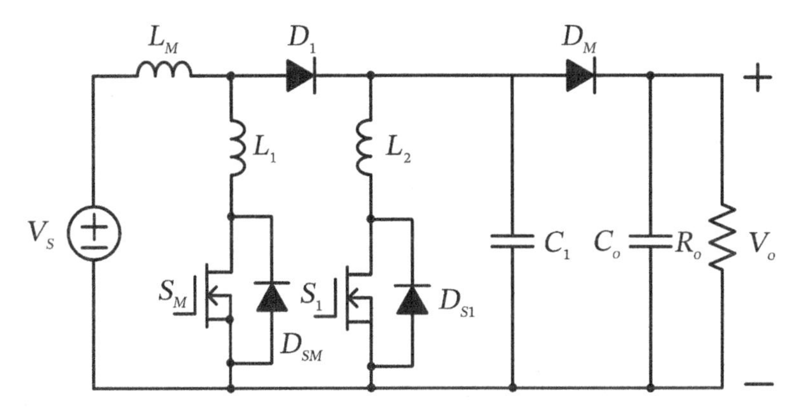

Figure 9.17 Active soft-switching converter with ZCS (type2).

to $t = t_0$, the main switch S_M is conducting, and the resonant capacitor C_1 is charged with certain negative voltage $-V_{C1}$. At $t = t_0$, the auxiliary switch S_1 is turned on, starting a resonance between resonant inductor L_1 and resonant capacitor C_1. This resonance forces the switch current through the main switch S_M to decrease in a sinusoidal fashion. Since inductor L_1 limits the increasing rate of di/dt, auxiliary switch S_1 can be achieved with ZCT at transition interval. After a quarter of the resonant period t_{d1}, the voltage across the resonant capacitor C_1 reaches zero, and the current through resonant inductor L_1 reaches its maximum. After the current through the main switch S_M has been reduced to zero and a small current flows through the antiparallel diode, the gate signal of switch S_M is disabled at $t = t_0 + t_{d1}$. It can be seen that the main switch S_M can be achieved with ZCS at this switching interval. After S_M is turned off, auxiliary switch S_1 is turned off shortly. Both freewheeling diode D_M and D_1 will start to conduct, and L_1 and C_1 continue resonating until the resonant inductor current drops to zero at $t = t_2$. At $t = t_2$, L_1 and C_1 complete the half-cycle resonance, and diode D_1 is reverse biased. The operating stage, t_2–t_3, is identical to the freewheeling stage of the PWM boost converter. At $t = t_3$, main switch S_M is turned on with hard switching, and the freewheeling diode D_M becomes reverse biased with hard switching. Meanwhile,

L_1 and C_1 form a half-cycle resonance through switch S_M and the antiparallel diode of switch S_1, which reverse the polarity of the resonant capacitor C_1 voltage. The operation during this t_4–t_5 interval is similar to a hard-switching PWM boost converter. At time t_5, the auxiliary switch S_1 turns on, and the cycle repeats.

In this circuit, since the resonant transition time is very short with respect to the switching cycle, the resonant inductance is very small as compared with the boost inductance. Thus, the circulating energy of the ZCS-PWM soft-switching converter is quite small.

According to Figure 9.17, the active soft-switching PWM converter is proposed with a ZCS function. The soft-switching cell is composed of two power switches (main switch S_M and auxiliary switch S_1), two diodes D_1 and D_M, two resonant inductors L_1 and L_2, and one resonant capacitor C_1. The main switch S_M has the major responsibility for the power transfer to the load. The function of the auxiliary switch S_1 is to provide conditions for the main switch to achieve a ZCS under constant frequency.

Therefore, with an elimination of switching losses, operation of the ZCS-PWM converter at high frequency will be possible, which allows a size reduction and consequently an increase in its power density.

9.2 Synthesis of Soft-Switching PWM Converters with Graft Scheme

When synthesizing hard-switching PWM converters, we use graft scheme to integrate two synchronous switches in the converters to become a single-stage one. The scheme is adopted here again to integrate two soft-switching PWM converters with either passive or active soft-switching cells. In the following, they are presented separately.

9.2.1 Generation of Passive Soft-Switching PWM Converters

As shown in Figure 9.18, the passive soft-switching buck and boost converters are recognized as the two fundamental converters, and the passive soft-switching cells enclosed in the dashed box are formed with only passive switches and reactive elements. From the two converters, another possible passive soft-switching buck–boost, Zeta, Ćuk, sepic, etc. can be generated by using the graft scheme. With the passive PWM soft-switching-cell, the converter operation is not changed except at the switching transition. For the rest of the switching period, these converters operate as regular PWM converters.

In the following, only is the soft-switching buck–boost PWM converter illustrated.

To obtain the passive soft-switching single-stage converters (SSCs), the cascaded/cascoded converters are grafted through switch integration. For instance, grafting passive soft-switching PWM boost converter on passive soft-switching PWM buck converter yields a soft-switching buck–boost SSC. Derivation of this SSC is illustrated in Figure 9.19. Figure 9.19a shows the two converters in cascade connection. First of all, we observe that the link network, L_R, C_5, an L_L, of the two converters is equivalent to a current buffer I_L, which can be represented by Figure 9.19b. Relocating switch S_R, without changing its operating principle, results in an equivalent converter configuration, as shown in Figure 9.19c. This figure reveals that active switches S_R and S_L are in the Π-type configuration (with a D–S common node); thus, with the graft scheme, S_R and S_L are replaced with a grafted switch S_{RL} and diodes D_{F1} and D_{F2}. The derived circuit is depicted in Figure 9.19d. It can be further simplified by removing diodes D_{F1} and D_{F2} to

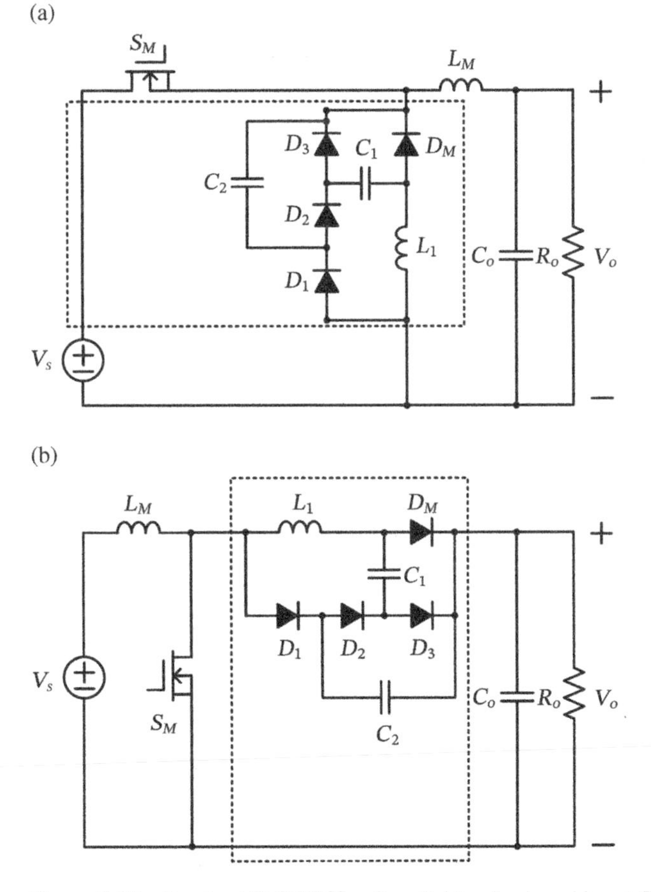

Figure 9.18 Passive NZVS/NZCS soft-switching buck and boost PWM converters: (a) buck converter and (b) boost converter.

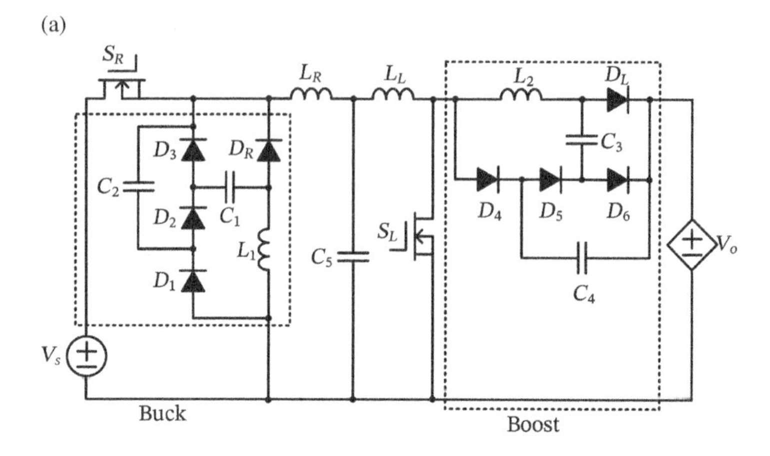

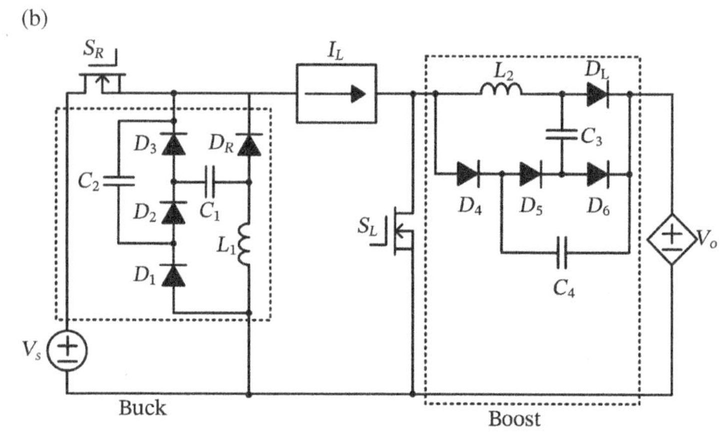

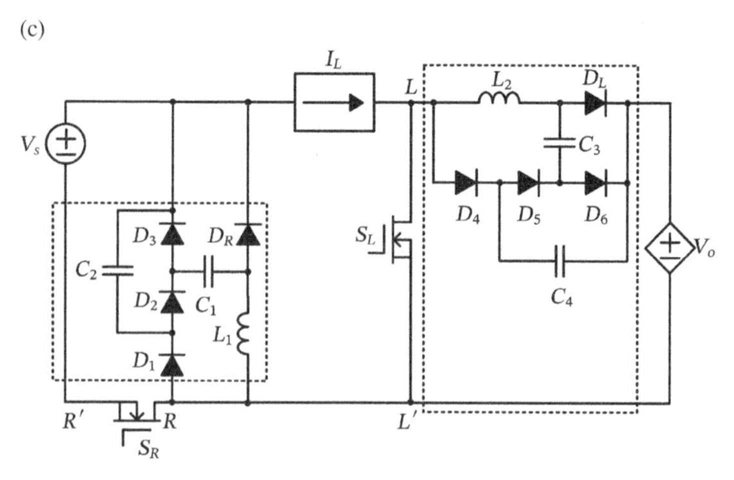

Figure 9.19 Derivation of the passive soft-switching PWM buck–boost PWM converter.

(d)

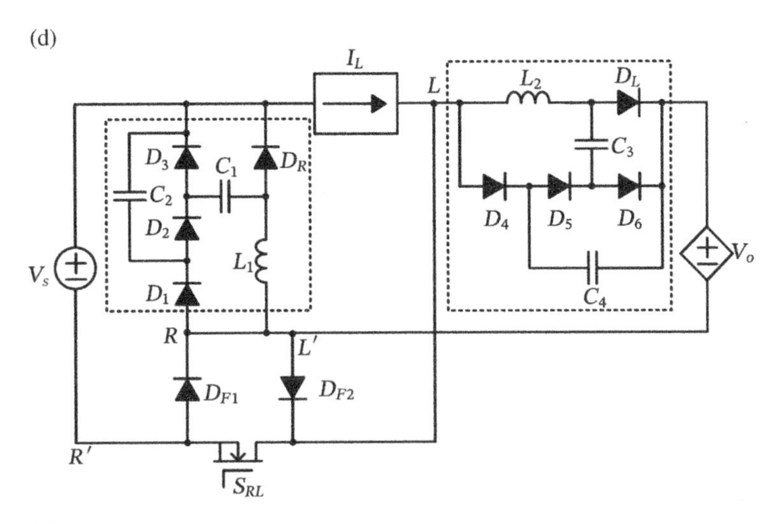

(e)

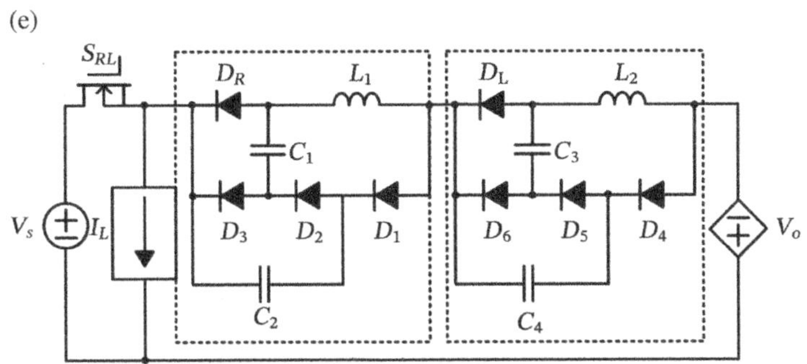

(f)

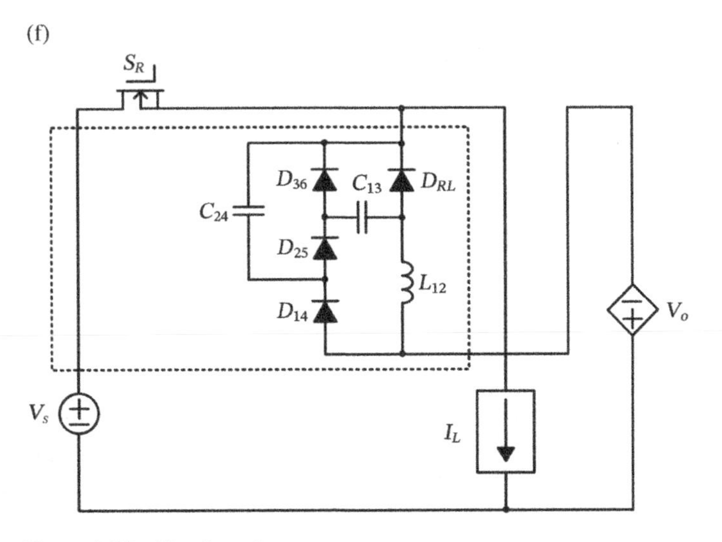

Figure 9.19 (Continued)

become the one shown in Figure 9.19e, because the currents through S_1 and S_2 are identical during the steady on state, and no differential current circulates through D_{F1} and D_{F2}. It is apparent that, in Figure 9.19e, the two soft-switching cells with the same configuration are in series; thus, they can be replaced with a single soft-switching cell due to the networks with passive devices and unidirectional current flow. By properly relocating switch S_{RL} and rearranging the overall circuit configuration, the passive soft-switching PWM buck–boost converter is obtained and depicted in Figure 9.19f. Note that although the configuration of the soft-switching cell depicted in Figure 9.19f is the same as that shown in Figure 9.19e, the components of the cells might be with different ratings.

Additionally, it can be noted that since the two soft-switching snubbers are finally in series and can be merged into a single one, we can just integrate a soft-switching PWM converter with a hard-switching one, which will result in a soft-switching PWM converter. This is an interesting point when integrating a soft-switching PWM converter with a hard-switching one, we can obtain a soft-switching PWM converter. From a genetic point of view, the soft-switching PWM converter has a dominance feature. When it is combined with a hard-switching one, the final converter will become a soft-switching PWM converter.

9.2.2 Generation of Active Soft-Switching PWM Converters

According to the active soft-switching buck converter and the active soft-switching boost converter, as shown in Figure 9.20a, properly configuring these two converters and operating their active switches in unison can yield an SSC to fulfill both step-down and step-up conversions with ZVS. First, replacing the LCL filter with a current buffer I_L yields the one shown in Figure 9.20b. Since switches S_R and S_L operate in unison and share a common node D–S, they can be replaced by a Π-type grafted switch, as shown in Figure 9.20c. The currents through both switches S_R and S_L are identical, diodes D_{F1} and D_{F2} are no longer needed, and they can be removed from the circuit, as shown in Figure 9.20d. By separating the voltage sources of the main power stages from the active soft-switching snubbers, we have the one shown in Figure 9.20e. Note that in Figure 9.20e, the voltage sources V_s and V_0 are assumed to be independent of the soft-switching cells for convenience of analysis. Redrawing the component connections will yield the one shown in Figure 9.20f.

It is apparent that in Figure 9.20f, two soft-switching cells are in series so that the auxiliary active switches S_1 and S_2 can be combined by the graft scheme, as illustrated in Figure 9.20g–i. Since inductor L_1 and switch S_2 only operate at switching transitions, they can be relocated to the return paths of the active snubbers, as shown in Figure 9.20g. Investigation of Figure 9.20g reveals that the auxiliary active switches are in Π-type configuration (with a D–S common node); thus, switches S_1 and S_2 are replaced with a Π-type grafted switch, which is depicted in Figure 9.20h. During the off states of

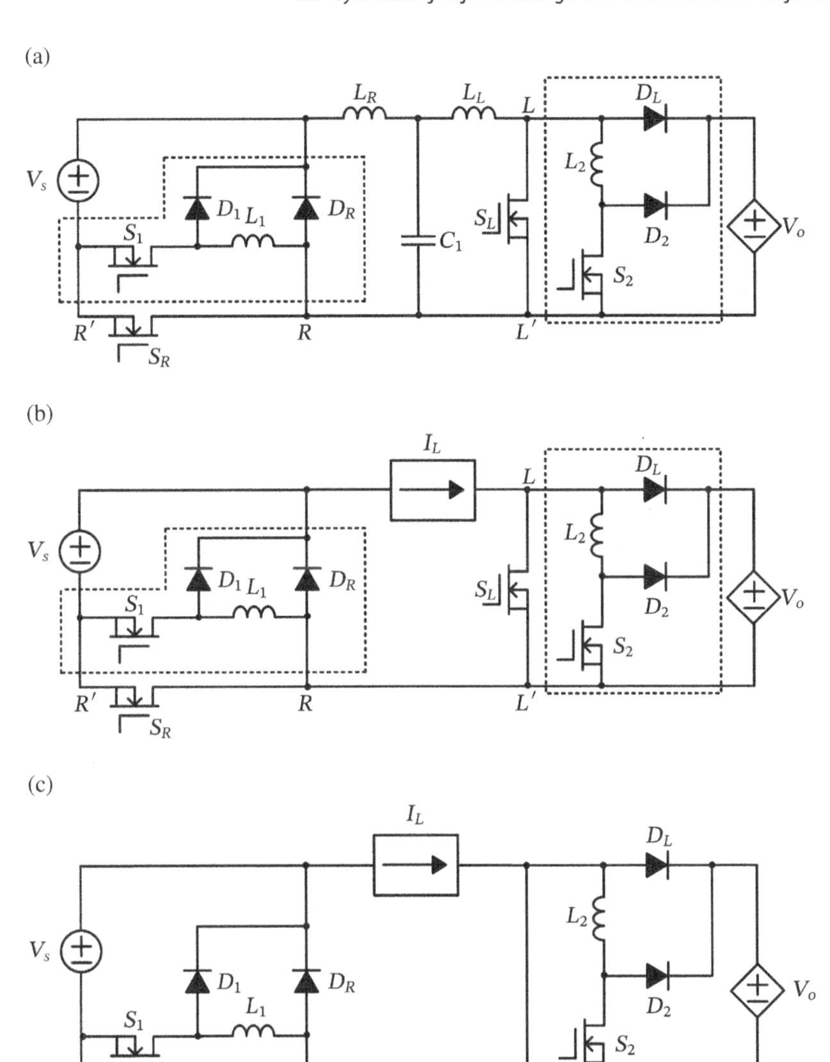

Figure 9.20 Derivation of the active soft-switching PWM buck–boost SSC with ZVS.

(d)

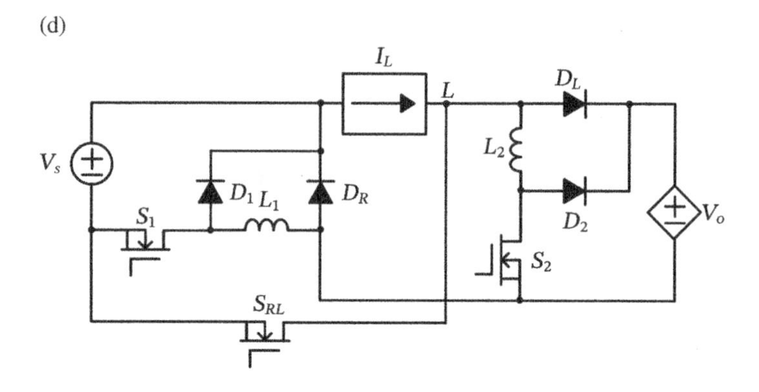

(e)

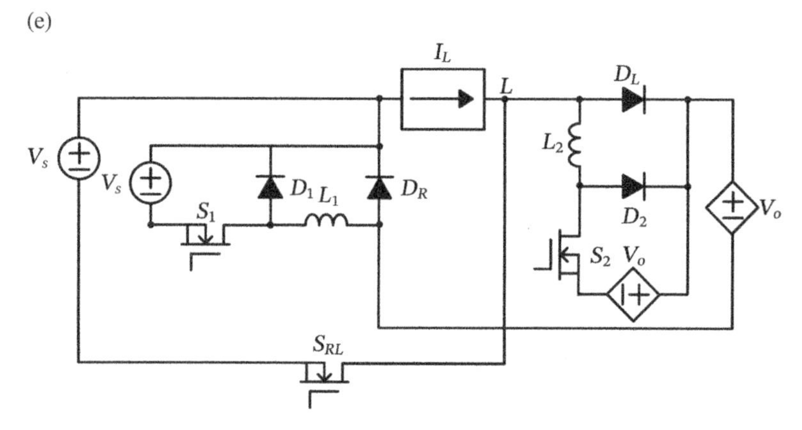

(f)

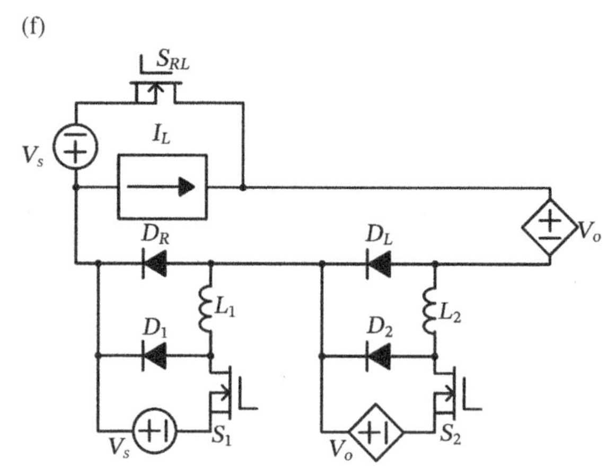

Figure 9.20 (Continued)

(g)

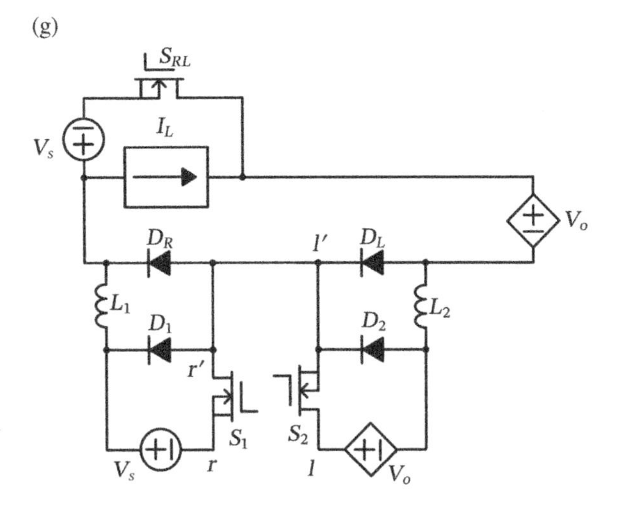

(h)

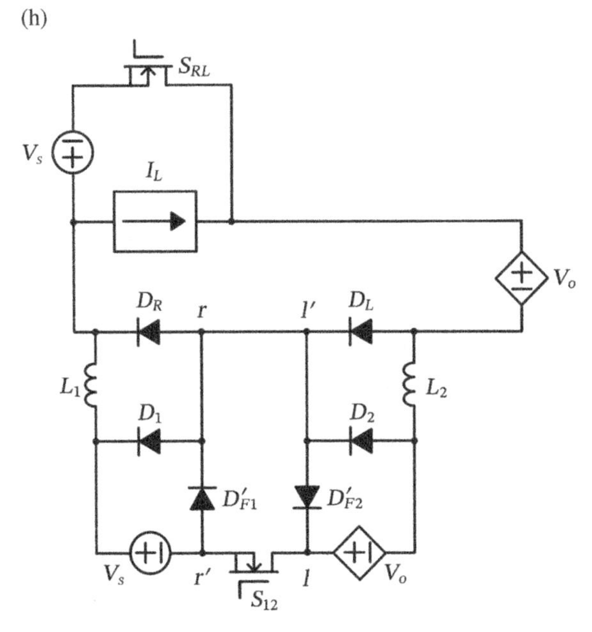

Figure 9.20 (Continued)

(i)

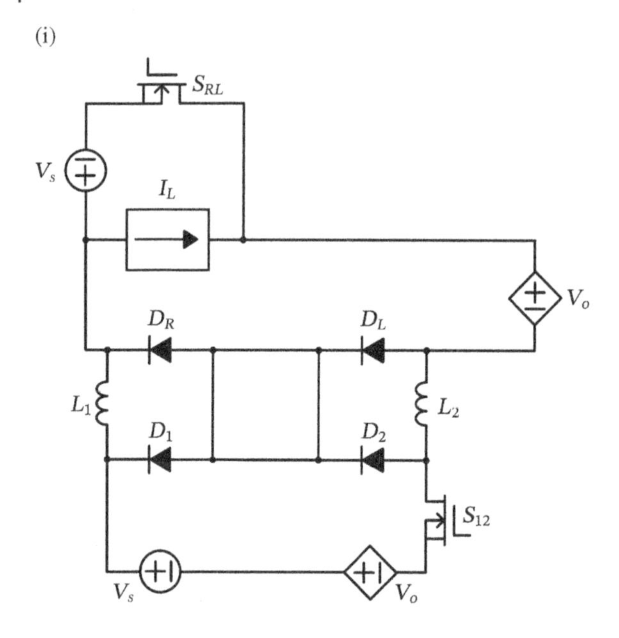

(j)

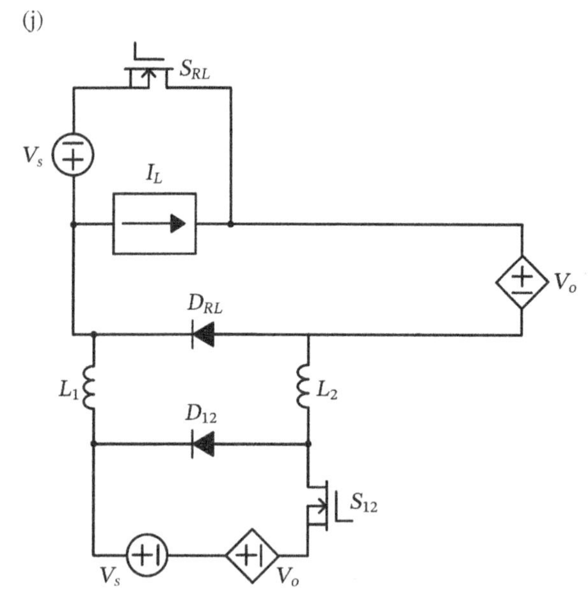

Figure 9.20 (Continued)

(k)

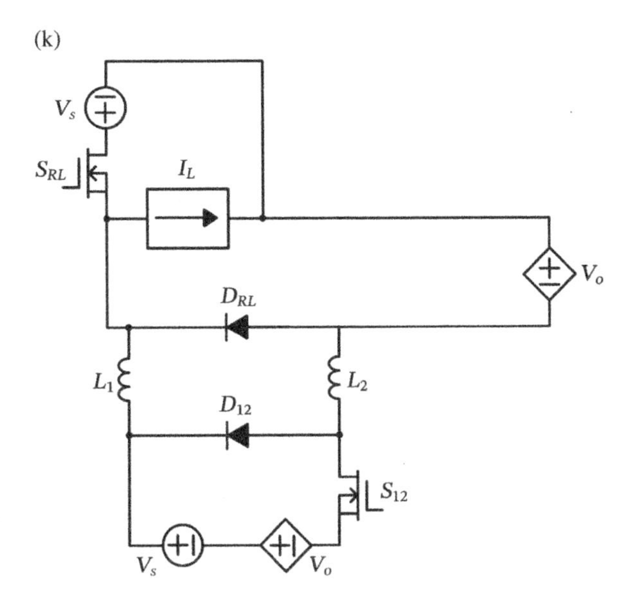

(l)

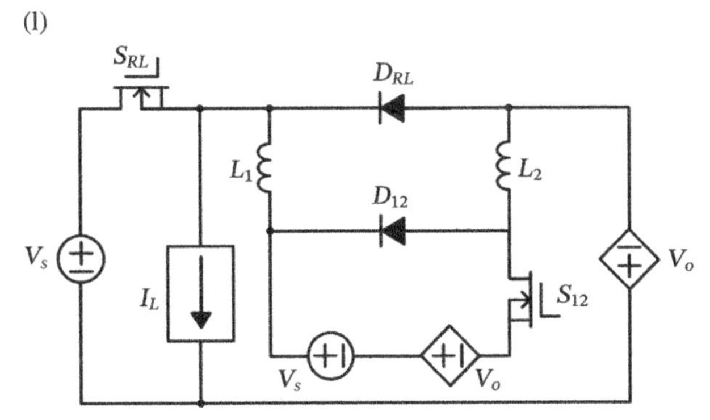

(m)

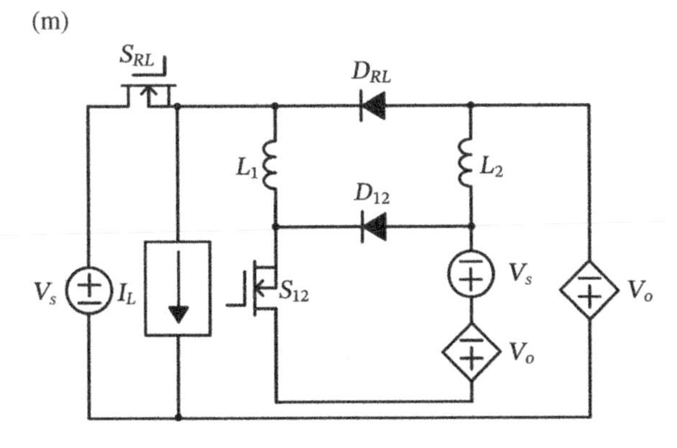

Figure 9.20 (Continued)

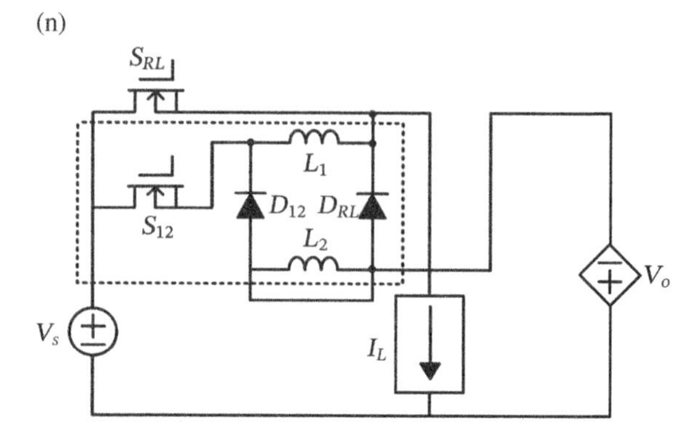

Figure 9.20 (Continued)

switches S_{RL} and S_{12}, the currents flow through diodes D_R and D_L, and D_1 and D_2 are equal; thus, the circuit can be simplified as depicted in Figure 9.20i. To reduce the component counts, let the current of D_1 equal to that of D_2; that is, diodes D_1 and D_2 can be replaced with a single one but with a higher voltage rating, and the current of D_R must be equal to that of D_L. Thus, the network can be simplified to D_{RL} and D_{12}, as shown in Figure 9.20j. By properly relocating switches S_{RL}, S_{12} and voltage sources $V_s + V_o$ and then reconfiguring the circuit, the active soft-switching PWM buck–boost SSC with ZVS can be obtained through the circuit configurations depicted in Figure 9.20k–n, in which inductor L_2 is equivalent to be shorted, degenerated into a single-inductor snubber.

Similarly, by applying the graft scheme, the other converters with active soft switching can also be derived, including Ćuk, Zeta, and sepic converters. Moreover, with a ZCS configuration, the active soft-switching PWM SSC can be derived by following the same procedure described above.

9.3 Synthesis of Soft-Switching PWM Converters with Layer Scheme

The layer scheme has been used to synthesize hard-switching PWM converters. Again, it can be used to synthesize soft-switching ones.

9.3.1 Generation of Passive Soft-Switching PWM Converters

Many passive soft-switching configurations have been proposed to reduce switching losses at high-frequency operation. Passive configurations are formed with only passive switches and resonant elements to achieve soft switching. Typical passive soft-switching PWM converters consist of buck, buck–boost, Zeta, boost, Ćuk, and sepic, as shown in Figure 9.21. These converters can be configured into either a buck or a boost

(a)

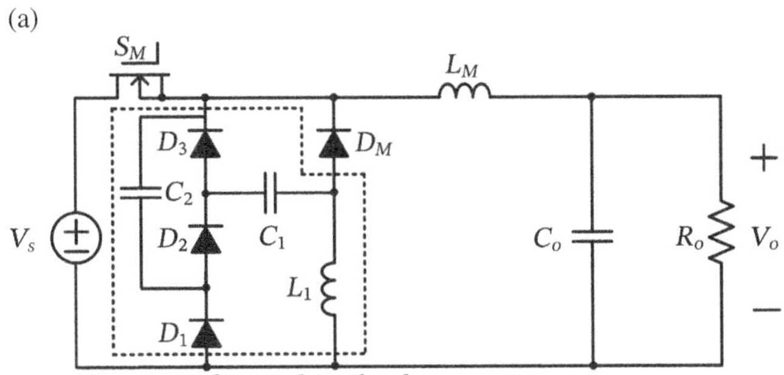

(b)

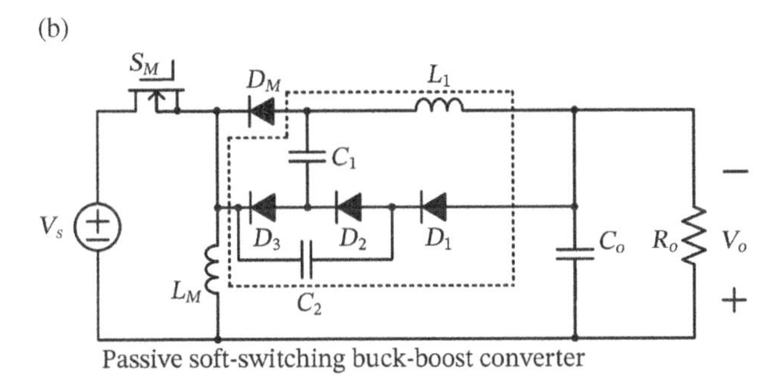

(c)

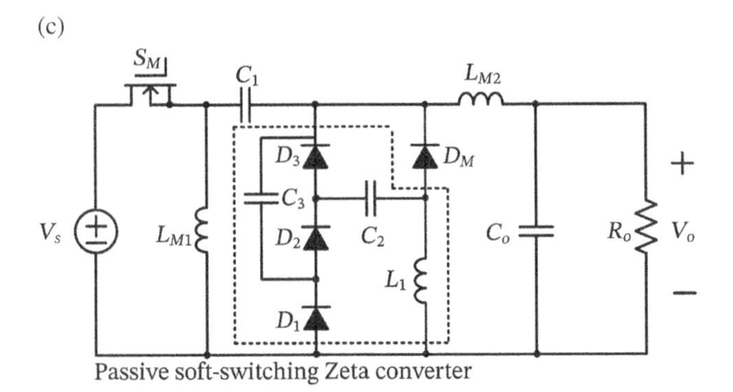

(d)

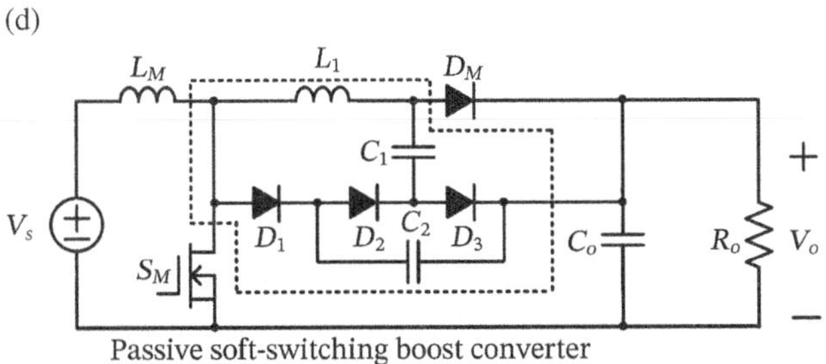

Figure 9.21 Passive soft-switching PWM DC/DC converters derived from the six well-known PWM hard-switching converters.

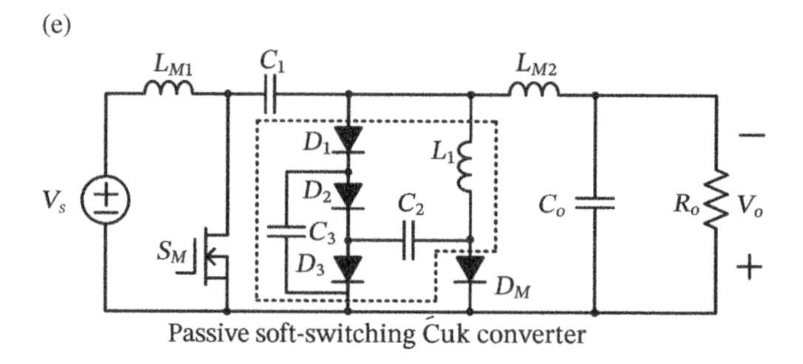

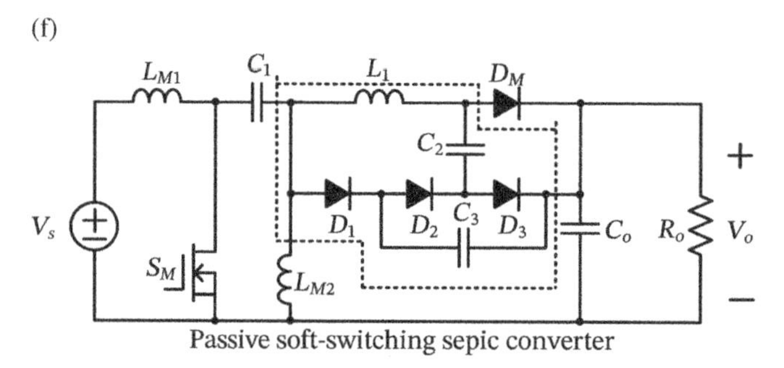

Figure 9.21 (Continued)

converter, plus linear devices, inductors, and capacitors, as illustrated in Figures 9.22 and 9.23. It should be pointed out that inductor L_x is added to the circuit shown in Figure 9.22c without varying its operating principle and without increasing its dynamic order because inductors L_x, L_f, and L_M exclusively share a common node. With the auxiliary inductor L_x, the buck basic converter unit (BCU) can be obviously recognized. By duality, an additional capacitor C_x is supplemented to the sepic converter so as the boost BCU can appear explicitly. Thus, the buck and boost converters are recognized as the two BCUs, and the passive soft-switching PWM converters are classified into buck family and boost family.

The converters shown in Figures 9.22 and 9.23 then can be conceptually represented with a configuration shown in Figure 9.24, in which the source can be either a voltage or a current one, the BCU can be either a buck BCU or a boost one, the load can be either an inductive or a capacitive type, and the feedback network can be either a capacitor, an inductor, or a combination of them. With this configuration, small signal modeling of the converters can be readily performed. This makes the reconfiguration of the passive soft-switching converters become uniquely valuable. It should be further pointed out that modeling of the converters shown in Figures 9.22c and 9.23c can also effectively derive the dynamics of

(a)

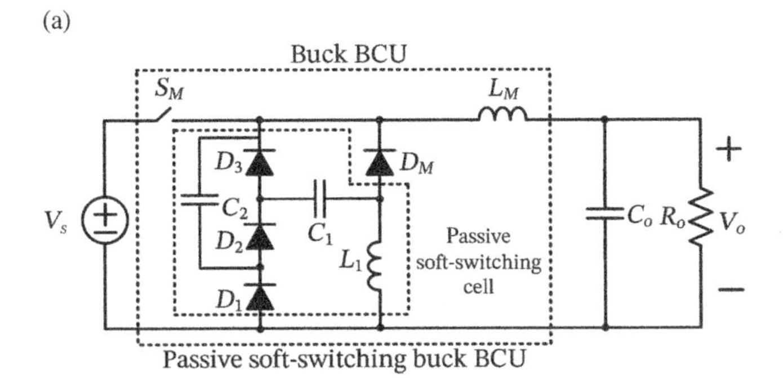

(b)

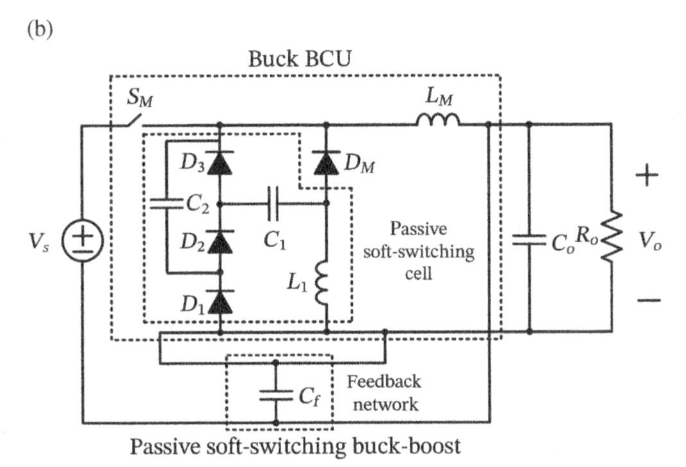

(c)

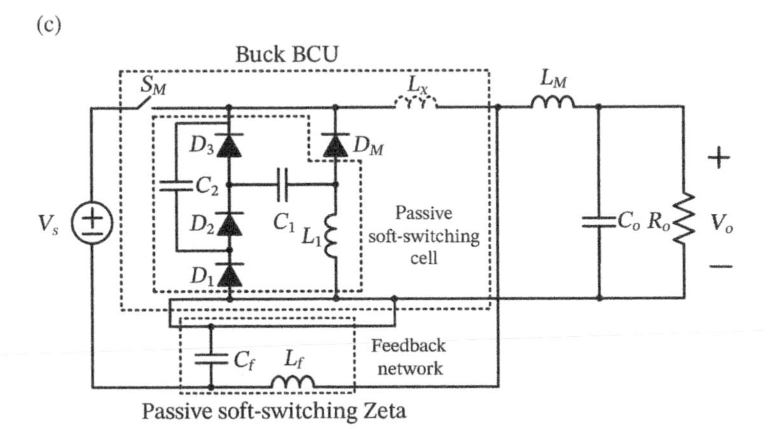

Figure 9.22 Illustration of buck–boost and Zeta passive soft-switching converters derived from the soft-switching buck BCU using the layer scheme.

(a)

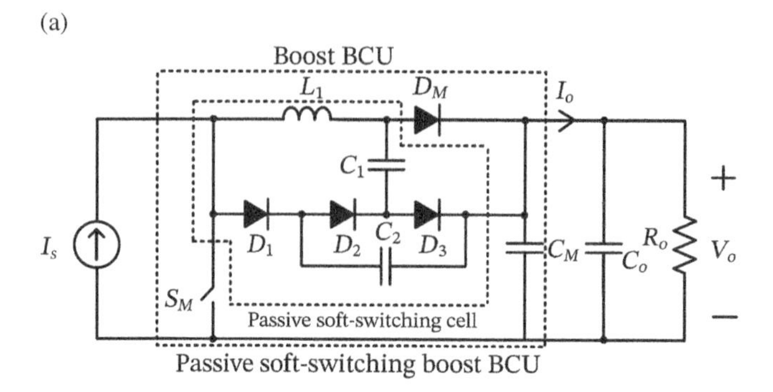

(b)

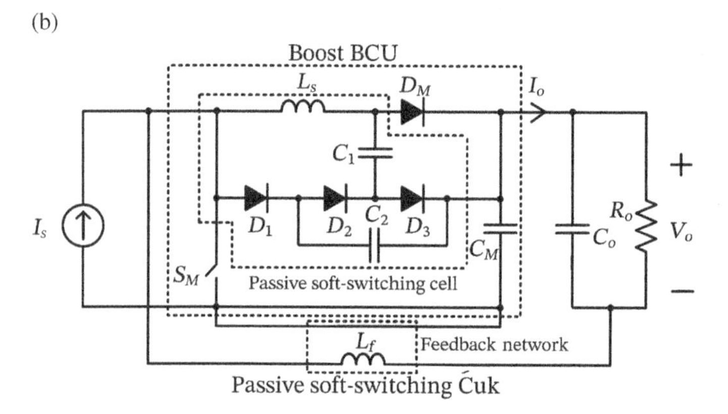

(c)

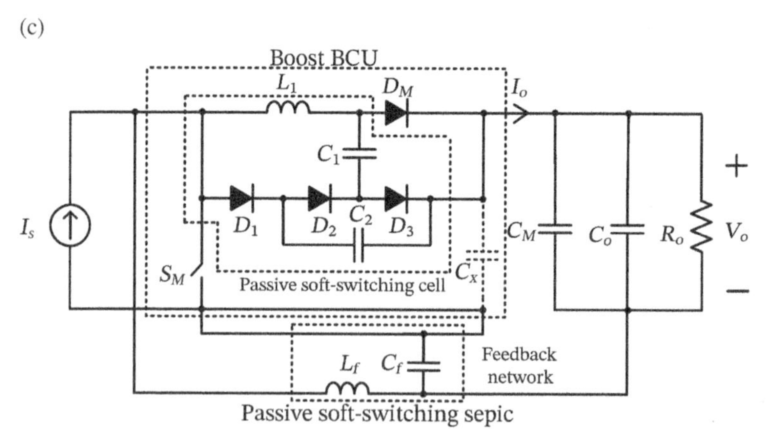

Figure 9.23 Illustration of Ćuk and sepic passive soft-switching converters derived from the soft-switching boost BCU using the layer scheme.

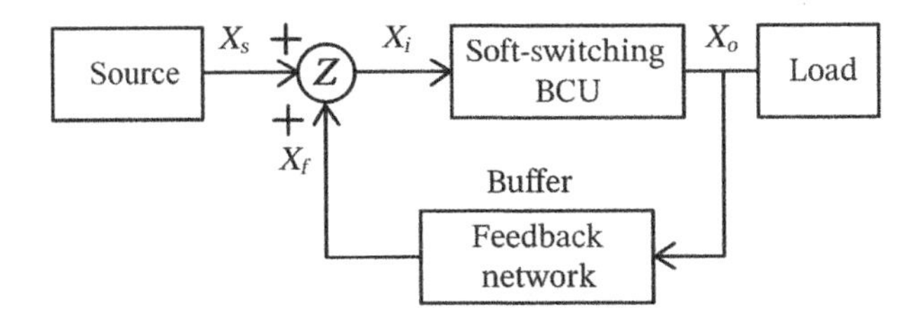

Figure 9.24 General configuration of a soft-switching BCU with a feedback network.

the rest. For instance, once the small signal dynamics of the Zeta converter shown in Figure 9.22c is derived, that of the buck–boost converter shown in Figure 9.22b can be obtained by letting $L_f = 0$ and $L_M = 0$, and that of the buck converter shown in Figure 9.22a can be also obtained by letting $L_x = 0$ and removing the L_f–C_f feedback network. Similarly, the dynamics of boost family can be derived from that of sepic converter shown in Figure 9.23c.

It is worth noting that from the general Zeta converter configuration shown in Figure 9.22c, a useful soft-switching converter can be degenerated by removing L_M, which is depicted in Figure 9.25a. Likewise, when the output capacitors C_M and C_o of the general sepic converter configuration shown in Figure 9.23c are removed, a useful soft-switching converter can be degenerated, as depicted in Figure 9.25b.

9.3.2 Generation of Active Soft-Switching PWM Converters

Except for passive soft-switching cells, active ones also have been proposed by using additional active and passive switches and resonant elements to reduce switching losses. With active cells, the converters can easily achieve soft switching for active switches (main and auxiliary switches) and passive switch (rectifier diodes) while minimizing their voltage and current stresses. However, using active cells increases the complexity of both the power circuit and control circuit, which, in turn, increases the circuit cost and decreases reliability. Even though the soft-switching cells are classified as either passive or active in nature, each of which has its own benefits and drawbacks in the aspects of switching loss, electrical stress, efficiency, cost, and circuit complexity, depending on application requirements. For instance, to achieve ZVS for the active switches, Duarte and Barbi employed the six fundamental active soft-switching cells to generate 36 topologies of ZVS PWM converters. These converters also can be configured into either buck or boost converter, plus six different active soft-switching cells, namely, buck, buck–boost, Zeta, boost, Ćuk, and sepic. Therefore, the buck and boost converters with soft-switching cells are recognized as the two BCUs, and the active soft-switching PWM converters are classified into buck family and boost family, as shown in Figures 9.26

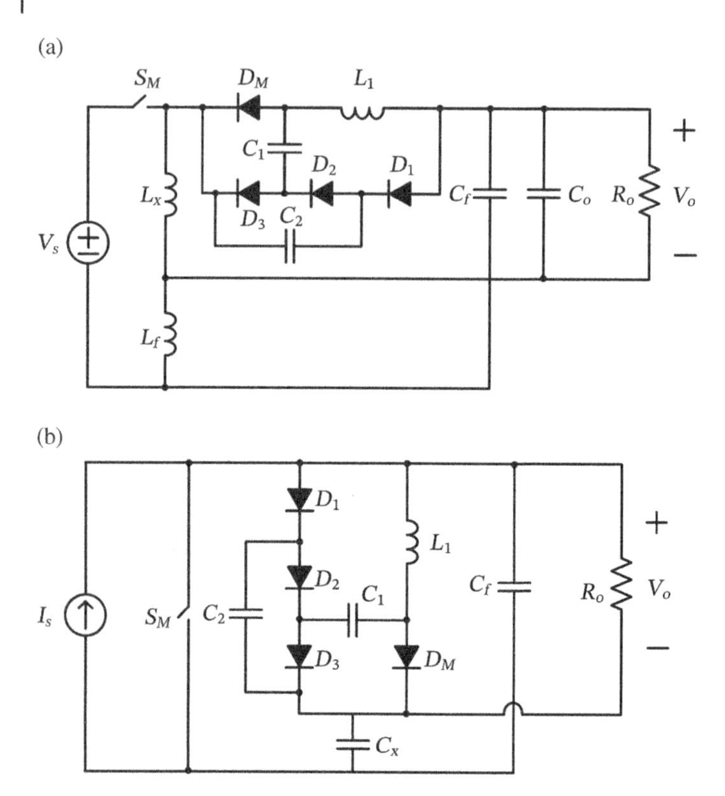

Figure 9.25 Two derived passive soft-switching converters from the general converter forms of the buck and boost families.

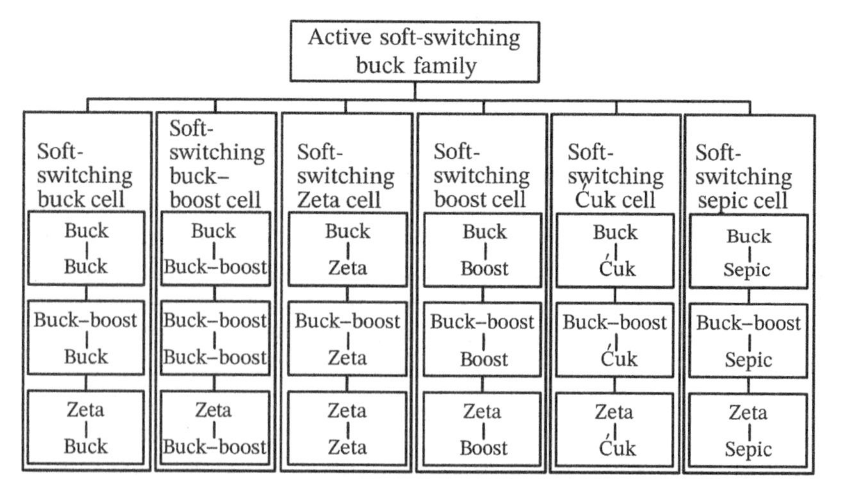

Figure 9.26 A list of active soft-switching buck family with six basic soft-switching cells.

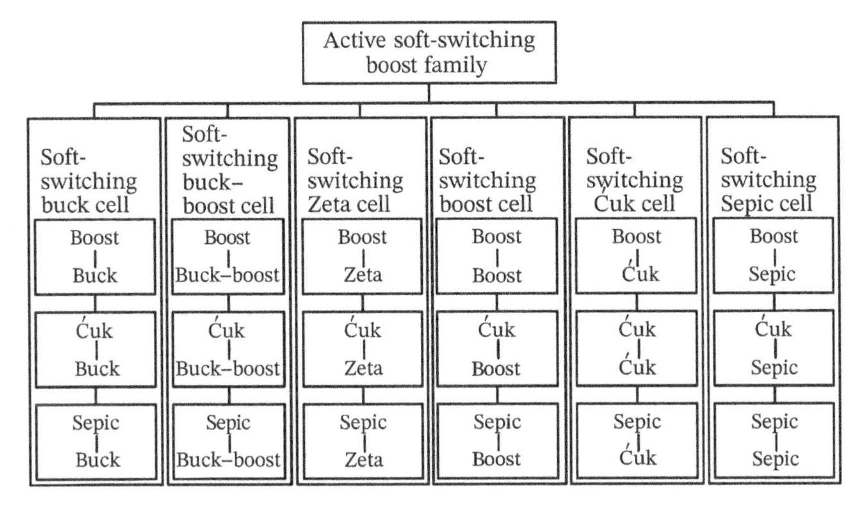

Figure 9.27 A list of active soft-switching boost family with six basic soft-switching cells.

and 9.27. For convenience of illustration, the active soft-switching buck cell and linear devices are used to synthesize the buck–boost–buck and Zeta-buck soft-switching PWM converters, as shown in Figures 9.28 and 9.29; thus, following the same procedure, the other buck family topologies can also be achieved with different soft-switching cells. Similarly, Ćuk–buck and sepic–buck can be synthesized from a boost–buck BCU, as drawn in Figures 9.30 and 9.31. The same discussion can be applied to derive the rest of the soft-switching converters.

9.4 Discussion

According to a proper reconfiguration, the general passive/active soft-switching PWM converter representations for the buck family, including soft-switching buck, buck–boost, and Zeta converters, and the boost family, including soft-switching boost, Ćuk, and sepic converters, can be readily identified. Moreover, with the proposed structural approach, synthesis and analysis can be performed systematically and effectively. Since each soft-switching PWM converter is partitioned into four portions in this book, including source, load, BCU, and feedback network, in developing the dual topology of a soft-switching PWM converter, we can begin in establishing the dual of each portion or subnetwork and then changing the configuration of the BCU and feedback network. This

(a)

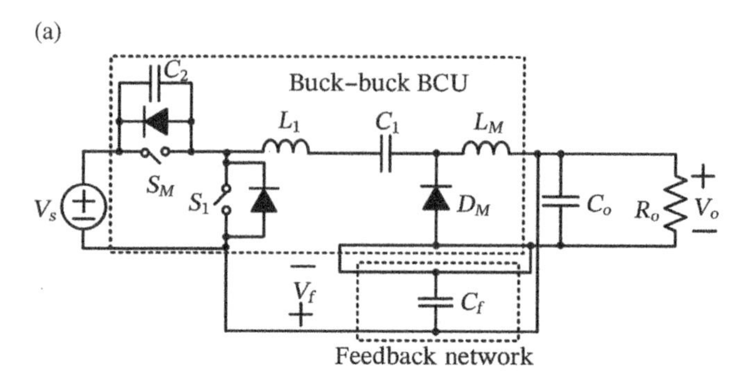

(b)

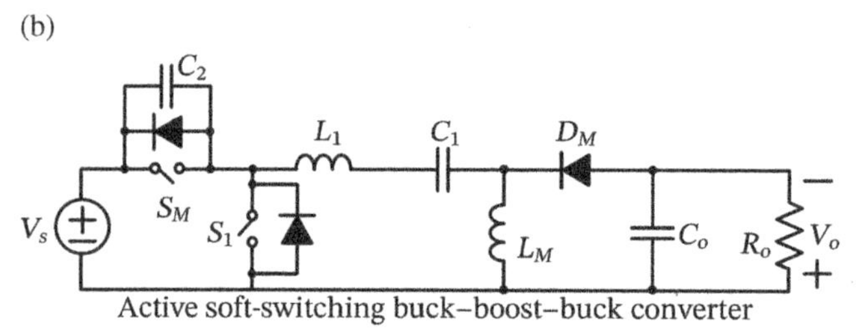

Figure 9.28 Illustration of the active soft-switching buck–boost–buck converter derived from the soft-switching buck–buck BCU.

(a)

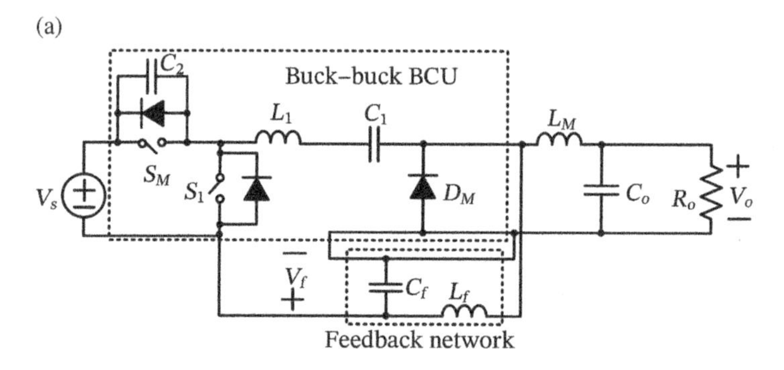

(b)

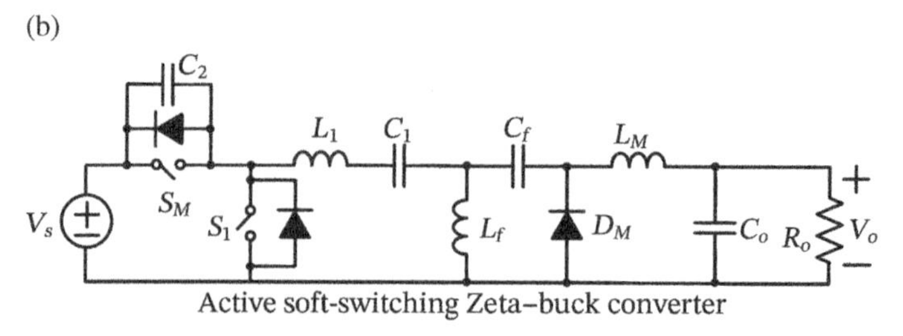

Figure 9.29 Illustration of the active soft-switching Zeta-buck converter derived from the soft-switching buck–buck BCU.

(a)

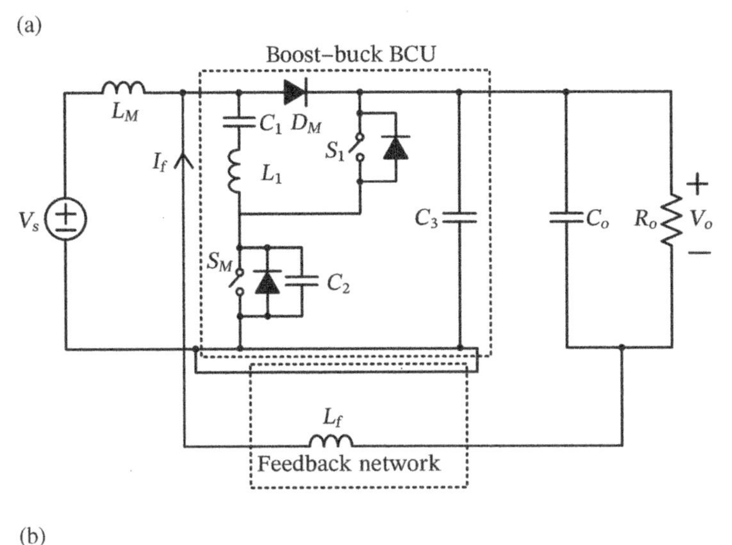

(b)

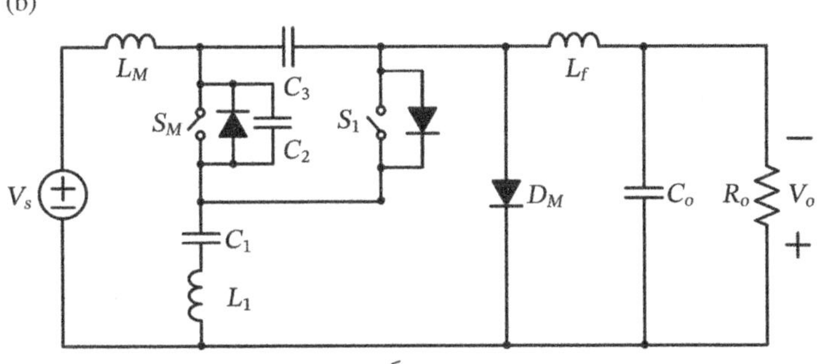

Figure 9.30 Illustration of the active soft-switching Ćuk–buck converter derived from the soft-switching boost–buck BCU.

description can be demonstrated by comparing Figure 9.22b with Figure 9.23b, Figure 9.22c with Figure 9.23c, Figure 9.27a with Figure 9.29a, and Figure 9.28a with Figure 9.30a. The dualities of the soft-switching PWM converters in buck and boost families are collected in Table 9.1.

A special feature of the structural approach is that it is capable of classifying the soft-switching PWM converters into families. Furthermore, as shown in literature, more physical insights into the converters in a family and more relationships among converters can be explored. These merits have made the proposed approach become uniquely valuable.

(a)

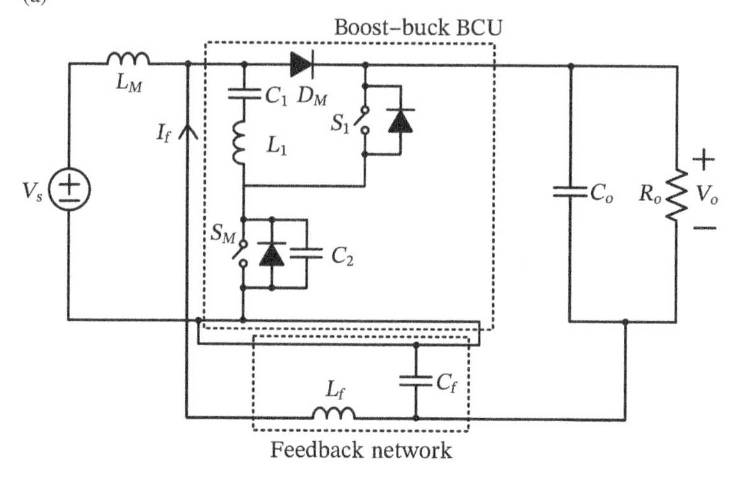

(b)

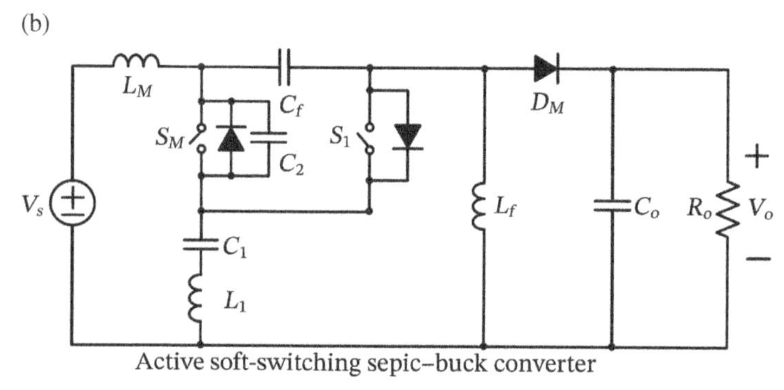

Figure 9.31 Illustration of the active soft-switching sepic–buck converter derived from the soft-switching boost–buck BCU.

Table 9.1 Duality of the soft-switching PWM buck and boost converter families.

	Soft-switching Buck family (buck, buck–boost, Zeta)	**Soft-switching Boost family (boost, Ćuk, sepic)**
Source	Voltage	Current
Load	Inductive	Capacitive
BCU	Soft-switching buck	Soft-switching boost
Feedback network	C_f C_f L_f	L_f L_f C_f
Configuration	Series–shunt	Shunt–series

Further Reading

Batarsech, I. and Lee, C.Q. (1991). Steady-state analysis of the parallel resonant converter with LLCC type commutation network. *IEEE Trans. Power Electron.* 6 (3): 526–538.

Ben-Yaakov, S. and Ivensky, G. (1997). Passive lossless snubbers for high frequency PWM converters. *Tutorial of the IEEE Power Electronics Specialists Conference*, IEEE.

Canesin, C.A. and Barbi, I. (1997). Novel zero-current-switching PWM converters. *IEEE Trans. Ind. Electron.* 44 (3): 372–381.

Carsten, B. (1987). A hybrid series-parallel resonant converter topologies. *Proceedings of High Frequency Power Conversion Conference*, IEEE, pp. 41–47.

Ćuk, S. (1997). General topological properties of switching converters. *Proceedings of the IEEE Power Electronics Specialists Conference*, IEEE, pp. 109–130.

Duarte, C.M.C. and Barbi, I. (1997). A family of ZVS-PWM active-clamping DC-to-DC converters: synthesis, analysis, design, and experimentation. *IEEE Trans. Circuits Syst. I* 44 (8): 698–704.

Elasser, A. and Torry, D.A. (1996). Soft switching active snubbers for DC/DC converters. *IEEE Trans. Power Electron.* 11 (5): 710–722.

Farrington, R.W. (1994). Reduced voltage/zero current transition boost power converter. US Patent 5,313,382.

Henze, C.P., Martin, H.C., and Parsley, D.W. (1988). Zero-voltage-switched in high frequency power converters using pulse width modulation. *Proceedings of the IEEE Applied Power Electronics Conference*, IEEE, pp. 33–40.

Hua, G. and Lee, F.C. (1991). A new class of zero-voltage-switched PWM converters. *Proceedings of the IEEE High Frequency Power Conversion Conference*, IEEE, pp. 244–251.

Hua, G. and Lee, F.C. (1995). Soft-switching techniques in PWM converters. *IEEE Trans. Ind. Electron.* 42 (6): 595–603.

Hua, G., Leu, C.S., Jiang, Y., and Lee, F.C. (1994a). Novel zero-voltage-transition PWM converters. *IEEE Trans. Power Electron.* 9 (2): 213–219.

Hua, G., Yang, E.X., Jiang, Y., and Lee, F.C. (1994b). Novel zero-current-transition PWM converters. *IEEE Trans. Power Electron.* 9 (6): 601–606.

Imbertson, P. and Mohan, N. (1991). Asymmetrical duty cycle permits zero switching loss in PWM circuits with no conduction loss penalty. *IEEE Trans. Ind. Electron.* 29 (1): 121–125.

Jitaru, I.D. (1991). Constant frequency, forward converter with resonant transition. *Proceedings of the IEEE High Frequency Power Conversion Conference*, IEEE, pp. 282–292.

Jitaru, I.D. (1992). A new high frequency, zero-voltage switched, PWM converter. *Proceedings of the IEEE Applied Power Electronics Conference*, IEEE, pp. 657–664.

Jitaru, I.D. and Cocian, G. (1994). High efficiency DC–DC converter. *Proceedings of the IEEE Applied Power Electronics Conference*, IEEE, pp. 638–644.

Jovanovic, M.M., Tsang, D.M.C., and Lee, F.C. (1994). Reduction of voltage stress in integrated high-quality rectifier-regulators by variable-frequency control. *Proceedings of the IEEE Applied Power Electronics Conference*, IEEE, pp. 569–575.

Kornetzky, P., Wei, H., and Batarseh, I. (1997). A novel one-stage power factor correction converter. *Proceedings of the IEEE Applied Power Electronics Conference*, IEEE, pp. 251–258.

Lee, Y.-S., Siu, K.-W., and Lin, B.-T. (1996). Single-switch fast-response switching regulators with unity power factor. *Proceedings of the IEEE Applied Power Electronics Conference*, IEEE, pp. 791–796.

Lee, Y.-S., Siu, K.-W., and Lin, B.-T. (1997). Novel single-stage isolated power-factor-corrected power supplies with regenerative clamping. *Proceedings of the IEE Applied Power Electronics Conference*, IEEE, pp. 259–265.

Lee, F.C., Tabisz, W.A., and Jovanovic, M.M. (1989). Recent developments in high-frequency quasi-resonant and multi-resonant converter topologies. *Proceedings of the third European Conference on Power Electronics and Applications*, pp. 401–410.

Liu, R. and Lee, C.Q. (1984). Series resonant converter with third order commutation network. *IEEE Trans. Ind. Electron.* 7 (3): 181–191.

Liu, K.H. and Lee, F.C. (1988). Topological constraints on basic PWM converters. *Proceedings of the IEEE Power Electronics Specialists Conference*, IEEE, pp. 164–172.

Liu, K.H. and Lee, F.C. (1990). Zero-voltage switching technique in DC/DC converter. *IEEE Trans. Power Electron.* 5 (1): 293–304.

Liu, K.H., Oruganti, R., and Lee, F.C. (1985). Resonant switches-topologies and characteristics. *Proceedings of the IEEE Power Electronics Specialists Conference*, IEEE, pp. 62–67.

Liu, K.H., Oruganti, R., and Lee, F.C. (1987). Quasi-resonant converters topologies and characteristics. *IEEE Trans. Power Electron.* 2: 62–74.

Maksimovic, D. and Ćuk, S. (1989a). A general approach to synthesis and analysis of quasi-resonant converters. *Proceedings of the IEEE Power Electronics Specialists Conference*, IEEE, pp. 713–727.

Maksimovic, D. and Ćuk, S. (1989b). General properties and synthesis of PWM DC-to-DC converters. *Proceedings of the IEEE Power Electronics Specialists Conference*, IEEE, pp. 515–525.

Maksimovic, D. and Ćuk, S. (1991). Switching converters with wide DC conversion range. *IEEE Trans. Power Electron.* 6 (1): 151–157.

Ninomiya, T., Matsumoto, N., Nakahara, M., and Harada, K. (1991). State and dynamic and analysis of ZVS half-bridge converter with PWM control. *Proceedings of the IEEE Power Electronics Specialists Conference*, IEEE, pp. 230–237.

Redl, R. and Balogh, L. (1995). Design considerations for single-stage isolated power supplies with fast regulation of the output voltage. *Proceedings of the IEEE Applied Power Electronics Conference*, IEEE, pp. 454–458.

Redl, R., Balogh, L., and Sokal, N.O. (1994). A new family of single-stage isolated power-factor correctors with fast regulation of the output voltage. *Proceedings of the IEEE Power Electronics Specialists Conference*, IEEE, pp. 1137–1144.

Smith, K.M. Jr. and Smedley, K.M. (1997a). Properties and synthesis of lossless, passive soft switching converters. *Proceedings of the 1st International Congress in Israel on Energy, IEEE, Power & Motion Control*, pp. 112–119.

Smith, K.M. Jr. and Smedley, K.M. (1997b). Lossless passive soft switching methods for inverters and amplifiers. *Proceedings of the IEEE Power Electronics Specialists Conference*, IEEE, pp. 1431–1439.

Smith, K.M. Jr. and Smedly, K.M. (1997c). A comparison of voltage-mode soft-switching methods for PWM converters. *IEEE Trans. Power Electron.* 12 (2): 376–386.

Tabisz, W.A. and Lee, F.C. (1988). Zero-voltage-switching multi-resonant technique – A novel approach to improve performance of high frequency quasi-resonant converters. *Proceedings of the IEEE Power Electronics Specialists Conference*, IEEE, pp. 9–17.

Tseng, C.-J. and Chen, C.-L. (1998). Passive lossless snubbers for DC/DC converters, *Proceedings of the IEEE Applied Power Electronics Conference*, IEEE, pp. 1049–1054.

Vorperian, V. (1988). Quasi-square wave converters: topologies and analysis. *IEEE Trans. Power Electron.* 3 (2): 183–191.

Watson, R., Hua, G., and Lee, F.C. (1996). Characterization of an active clamp flyback topology for power factor correction applications. *IEEE Trans. Power Electron.* 11 (1): 191–198.

Wu, T.-F. and Chen, Y.-K. (1996). A systematic and unified approach to modeling PWM DC/DC converters using the layer scheme. *Proceeding of the IEEE Power Electronics Specialists Conference*, IEEE, pp. 575–580.

Wu, T.-F. and Yu, T.-H. (1997). Off-line applications with single-stage converters. *IEEE Trans. Ind. Electron.* 44 (5): 638–647.

Wu, T.-F., Liang, S.-A., and Chen, Y.-K. (1999). High-power-factor single-stage converter with robust controller for universal off-line applications. *IEEE Trans. Power Electron.* 14 (6): 1078–1085.

Wu, T.-F., Yu, T.-H., and Chang, Y.-H. (1995). Generation of power converter with graft technique. *Proceedings of the 15th Symposium on Electrical Power Engineering*, Taiwan, R.O.C., pp. 370–376.

Zheng, T., Chen, D.Y., and Lee, F.C. (1986). Variations of quasi-resonant DC–DC converter topologies. *Proceedings of the IEEE Power Electronics Specialists Conference*, IEEE, pp. 381–392.

10

Determination of Switch-Voltage Stresses

To generate converters, this book starts to decode input-to-output voltage transfer codes into special code configurations and then synthesizes the configurations to form converters. Since we start from the transfer codes, is it possible to identify the voltage stresses imposed on the switches inside the converters from the transfer codes directly? If it is, we can use the voltage stress as a criterion to select a proper code out of several possible codes before synthesizing them into converters. This can save time and efforts in developing converters. As to current stress, it depends on load condition and will not be discussed here.

10.1 Switch-Voltage Stress of the Original Converter

The original converter, buck converter, is shown in Figure 10.1a, in which there are two switches, active switch S_1 and passive switch D_1. When switch S_1 is turned on with a duty ratio D, input voltage V_i is imposed on switch D_1, which in turn becomes a pulsating voltage lasting for DT_s where T_s is the switching period, as illustrated in Figure 10.1b. Through the LC filter, the pulsating voltage across diode D_1 is smoothed out to become output voltage V_o as denoted with the dashed line. When switch S_1 is turned off, diode D_1 will conduct to freewheel current i_L. Thus, the voltage stress imposed on switch S_1 is again input voltage V_i.

If current i_L never drops to zero, the buck converter is operated in continuous conduction mode (CCM). However, if the current drops to zero within a switching period, the converter is operated in discontinuous conduction mode (DCM). When current i_L drops to zero before switch S_1 turns on again, the voltage across switch S_1 is $(V_i - V_o)$, while that across diode D_1 is V_o. For a buck converter, $V_i \geq V_o$. Based on the above discussion, we can conclude that the maximum voltage stress imposed on switch S_1 and diode D_1 is V_i, since even the converter operating in

Origin of Power Converters: Decoding, Synthesizing, and Modeling, First Edition.
Tsai-Fu Wu and Yu-Kai Chen.
© 2020 John Wiley & Sons, Inc. Published 2020 by John Wiley & Sons, Inc.

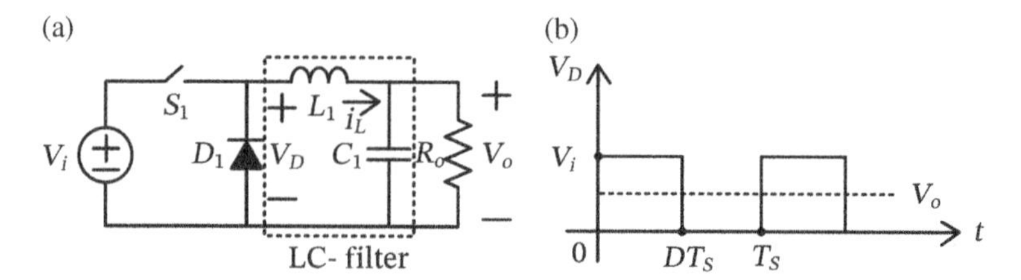

Figure 10.1 (a) Buck converter topology and (b) its associated diode and output voltage waveforms.

DCM, its inductor current i_L will always go through a nonzero value for a while first and then drop to zero. Therefore, the maximum voltage stress is determined by the converter in CCM operation, and in the following, we will discuss the converter operated in CCM only.

For a buck converter operated in CCM, its input-to-output voltage transfer ratio can be expressed as follows:

$$\frac{V_o}{V_i} = D, \tag{10.1}$$

where D is the duty ratio of switch S_1 shown in Figure 10.1a. Or it can be expressed as follows:

$$V_o = DV_i \tag{10.2}$$

$$V_o = \frac{DT_s V_i + (1-D)T_s \cdot 0}{T_s}. \tag{10.3}$$

It should be noted first that in a typical PWM converter, when active switch turns on, its passive diode must be operated in reverse bias. In (10.3), the turn-on time, DT_S, of switch S_1 represents that there exists a switching action, and voltage V_i denotes that the voltage across diode D_1 is V_i, which is the voltage stress imposed on the diode, while $(1-D)T_s$ denotes the conduction time of diode D_1. In other words, the voltage across the switch is V_i. Through an LC filtering action, which averages the right-hand side (RHS) of (10.3), we can obtain a smooth output V_o, as shown in the left-hand side (LHS) of the equation.

In short, if an input-and-output relationship can be expressed in (10.2) or (10.3) in which the RHS of the equation is associated with duty ratio D, while the LHS is with a constant, we can conclude that both voltage stresses of the active switch and the diode are equal to V_i. It should be noted that this is only true when there is only one active switch and one diode in a PWM converter. However, if there are more than one active switch and one diode, it might not be correct.

10.2 Switch-Voltage Stresses of the Fundamental Converters

Fundamental converters include boost, buck–boost, Ćuk, sepic, Zeta, z-source, etc., of which there are only an active switch and a passive one (i.e. diode) in each converter. Following the aforementioned discussions, we can determine the voltage stresses imposed on the switches in the converters.

10.2.1 The Six Well-Known PWM Converters

The six well-known PWM converters include buck (the original), boost, buck–boost, Ćuk, sepic, and Zeta converters. For the buck converter, the switch-voltage stress has been linked to its input-to-output voltage transfer ratio, as shown in (10.2). Following the same process, switch-voltage stresses of the rest of the six converters can be determined from their voltage transfer ratios as well.

10.2.1.1 Boost Converter

Figure 10.2a shows a boost converter, and its switch voltage V_{DS} is shown in Figure 10.2b, in which the pulsating voltage V_{DS} is filtered by filter L_1 and input capacitor C_i to form a smooth V_i as denoted by the dashed line. The input-to-output voltage transfer ratio of the boost converter is

$$\frac{V_o}{V_i} = \frac{1}{(1-D)}. \tag{10.4}$$

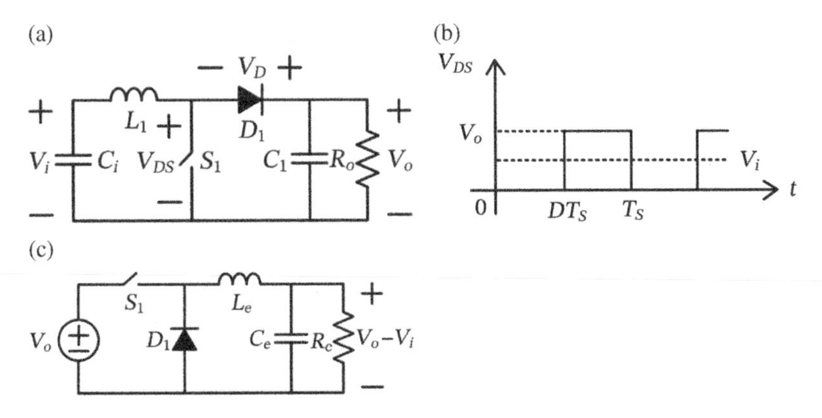

Figure 10.2 (a) Boost converter topology, (b) active switch–voltage waveform, and (c) equivalent buck converter.

Or it can be expressed as

$$V_o - V_i = DV_o. \tag{10.5}$$

To be analogous to the buck converter, Eq. (10.5) is represented by the equivalent converter shown in Figure 10.2c. Thus, we can determine that the voltage stress of both active switch S_1 and passive one D_1 is V_O. This can be proved from the boost converter shown in Figure 10.2a. When switch S_1 is turned on, output voltage V_o is imposed on diode D_1, and when the switch is turned off and diode conducts, voltage V_o is imposed on switch S_1. Therefore, the voltage stresses of both S_1 and D_1 are V_O.

Again, it should be noted that at the RHS of (10.5), duty ratio D denotes a switch action, and DV_o is an average value with LC filtering, and at the LHS of (10.5), it is a constant ($V_o - V_i$), which denotes the averaged output of the equivalent buck converter shown in Figure 10.2c.

10.2.1.2 Buck–Boost Converter

Figure 10.3a shows a buck–boost converter, and Figure 10.3b shows its conceptual diode voltage waveform. The input-to-output voltage transfer ratio of the converter is

$$\frac{V_o}{V_i} = \frac{D}{1-D}. \tag{10.6}$$

Or it can be expressed as follows:

$$V_o = D(V_i + V_o). \tag{10.7}$$

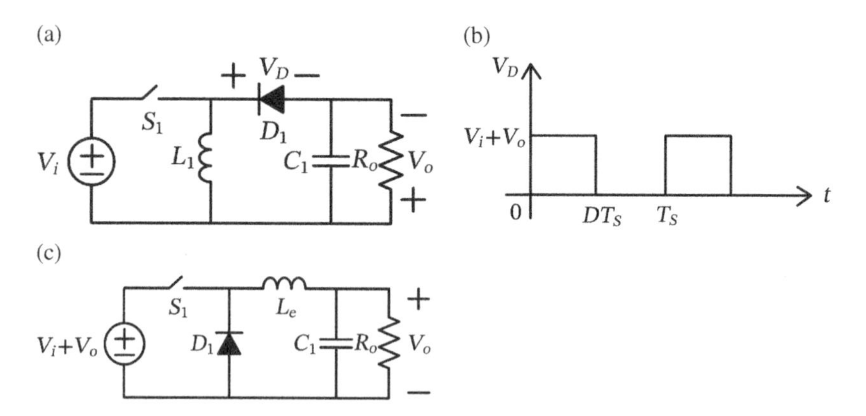

Figure 10.3 (a) Buck–boost converter topology, (b) passive switch–voltage waveform, and (c) equivalent buck converter.

Based on the format of (10.2), Eq. (10.7) can be represented by the equivalent converter shown in Figure 10.3c. From (10.7) or Figure 10.3c, we can determine that the voltage stresses of both active switch S_1 and passive diode D_1 are $(V_i + V_o)$. This can be also proved by investigating the converter operation shown in Figure 10.3a. When switch S_1 is turned on, diode D_1 is imposed by voltage $(V_i + V_o)$, while when it is turned off, the same voltage will impose on switch S_1. It comes out the same voltage stress as that determined from (10.7).

10.2.1.3 Ćuk, Sepic, and Zeta Converters

As to Ćuk, sepic, and Zeta converters, they were all proved to be equivalent to the buck–boost converter with an extra *LC* filter in Chapter 6 and have the same voltage transfer ratio as shown in (10.6). Therefore, they are supposed to all have the same switch-voltage stress, $V_i + V_o$, and the equivalent converter as shown in Figure 10.3c.

Figure 10.4 shows Ćuk, sepic, and Zeta converters. They are all fourth-order converters. When their active switches are turned on, the voltage stress $(V_i + V_o)$ is all imposed on their diodes. When they are turned off, the same voltage stress $(V_i + V_o)$ is imposed on the active switches. This comes out identical to the equivalent converter shown in Figure 10.3c.

As mentioned in Chapter 6, the buck–boost converter can be derived from Ćuk, sepic, and Zeta converters by changing the DC offset voltage of the buffer capacitor C_1 in the converters of Figure 10.4. However, their switch-voltage stress keeps unchanged. This also in turn proves that DC voltage offset on the buffer capacitor does not change the switch-voltage stress.

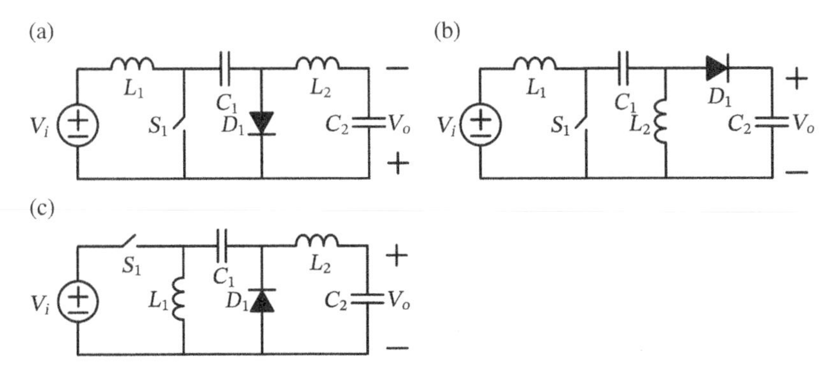

Figure 10.4 (a) Ćuk converter, (b) sepic converter, and (c) Zeta converter.

10.2.2 z-Source Converters

In addition to the well-known six PWM converters, there are three types of z-source converters, voltage-fed z-source, current-fed z-source, and quasi-z-source. They have various voltage transfer ratios, but each converter still has only one active switch and one passive one. When their active switches are turned on, their passive ones are in reverse bias. On the other hand, when their active switches are turned off, their passive ones will conduct. They are the type of fundamental PWM converters, and we can follow the same process as that of the original converter to determine their switch-voltage stresses based on their voltage transfer ratios.

10.2.2.1 Voltage-Fed z-Source Converter

For a voltage-fed z-source converter, as shown in Figure 10.5a, its input-to-output voltage transfer ratio is

$$\frac{V_o}{V_i} = \frac{1-D}{1-2D}. \tag{10.8}$$

With a cross multiplication and some algebraic manipulation, Eq. (10.8) can be expressed as

$$V_o - V_i = D(2V_o - V_i). \tag{10.9}$$

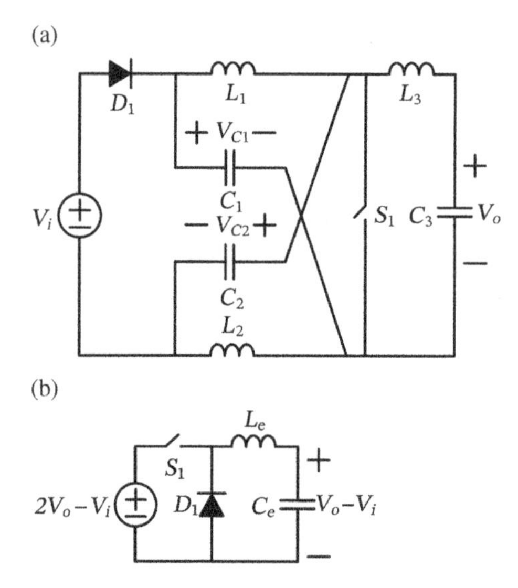

Figure 10.5 (a) Voltage-fed z-source converter and (b) its equivalent buck converter.

Based on (10.9), we can have an equivalent buck converter shown in Figure 10.5b. Obviously, from (10.9), we can read that the voltage stresses of the active switch S_1 and passive diode D_1 are $(2V_o - V_i)$. This can be confirmed from the converter operation. First of all, capacitor voltages V_{c1} and V_{c2} can be determined as V_o. When switch S_1 is turned on, based on KVL, it can be determined that the voltage imposed on diode D_1 is $(2V_o - V_i)$. When switch S_1 is turned off, diode D_1 will be forced to conduct, and the voltage imposed on switch S_1 is again $(2V_o - V_i)$. This voltage stress is identical to that read from (10.9).

10.2.2.2 Current-Fed z-Source Converter

Figure 10.6a shows a current-fed z-source converter with its voltage transfer ratio shown as

$$\frac{V_o}{V_i} = \frac{2D-1}{1-D}. \tag{10.10}$$

Through cross multiplication and some algebraic manipulation, Eq. (10.10) can be expressed as

$$V_o + V_i = D(2V_i + V_o). \tag{10.11}$$

Based on (10.11), we can have an equivalent converter shown in Figure 10.6b. Obviously, from (10.11), we can read that the voltage stresses of the active switch S_1 and passive diode D_1 are $(2V_i + V_o)$. This can be confirmed from the converter

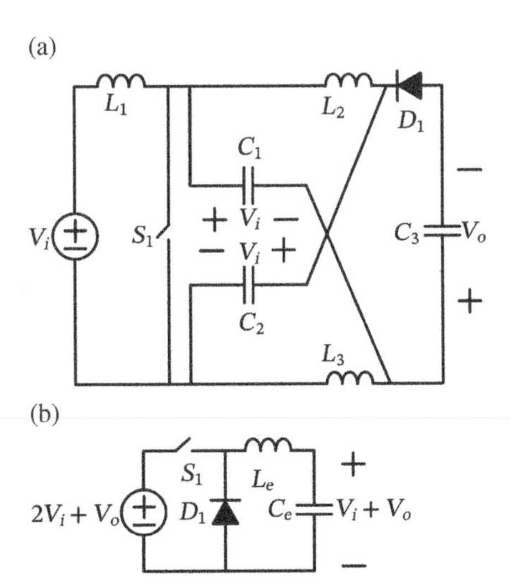

Figure 10.6 (a) Current-fed z-source converter and (b) its equivalent buck converter.

operation. First of all, capacitor voltages V_{c1} and V_{c2} can be determined as V_i. When switch S_1 is turned on, based on KVL, it can be determined that the voltage imposed on diode D_1 is $(2V_i + V_o)$. When switch S_1 is turned off, diode D_1 will be forced to conduct, and the voltage imposed on switch S_1 is again $(2V_i + V_o)$. This voltage stress is identical to that read from (10.11).

10.2.2.3 Quasi-z-Source Converter

A quasi-z-source converter is shown in Figure 10.7a, and its input-to-output voltage transfer ratio is given as follows:

$$\frac{V_o}{V_i} = \frac{D}{1-2D}. \tag{10.12}$$

Or it can be expressed as follows:

$$V_o = D\left(2V_o + V_i\right). \tag{10.13}$$

Based on (10.13), we can have an equivalent converter shown in Figure 10.7b. Obviously, from (10.13), we can read that the voltage stresses of the active switch S_1 and passive diode D_1 are $(2V_o + V_i)$. This can be confirmed from the converter operation. First of all, capacitor C_2 voltage can be determined as $V_i + V_o$. When switch S_1 is turned on, based on KVL, it can be determined that the voltage imposed on diode D_1 is $(2V_o + V_i)$. When switch S_1 is turned off, diode D_1 will be forced to conduct, and the voltage imposed on switch S_1 is again $(2V_o + V_i)$. This voltage stress is identical to that read from (10.13).

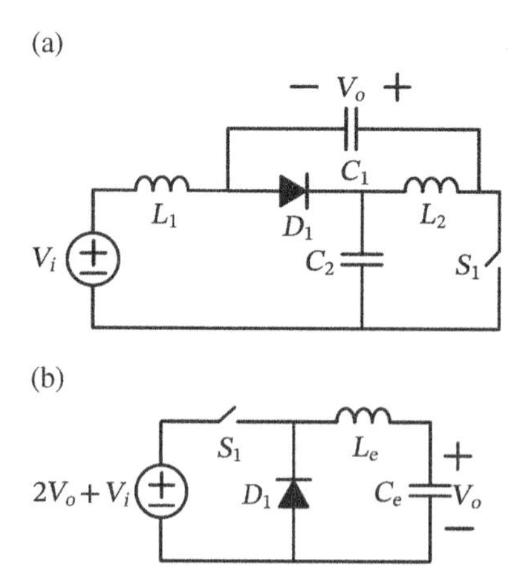

Figure 10.7 (a) Quasi-z-source converter and (b) its equivalent buck converter.

10.3 Switch-Voltage Stresses of Non-Fundamental Converters

For PWM converters, typically, each converter has only one active switch and one passive one (diode). However, there are PWM converters with one single active switch but with multiple diodes, namely, non-fundamental PWM converters, such as high step-up/step-down switched-inductor/switched-capacitor converters, buck–buck converter, boost–boost converter, etc. Can we determine their voltage stresses imposed on switches and diodes from their voltage transfer ratios? To answer this question, we will start from the same discussion presented in Sections 10.1 and 10.2.

10.3.1 High Step-Down Switched-Inductor Converter

A high step-down switched-inductor converter derived from buck and inverse buck–boost converters is shown in Figure 10.8a, in which its input-to-output voltage transfer ratio is given as follows:

$$\frac{V_o}{V_i} = \frac{D}{2-D}. \tag{10.14}$$

(a)

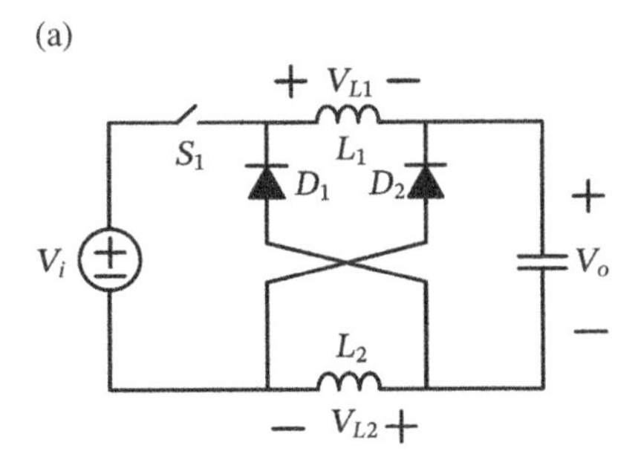

(b)

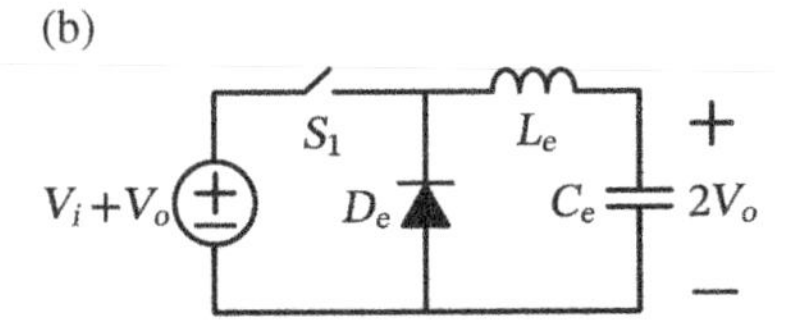

Figure 10.8 (a) High step-down switched-inductor converter with voltage transfer ratio, $D/(2-D)$, and (b) its equivalent buck converter.

Or it can be expressed as follows:

$$2V_o = D\left(V_i + V_o\right). \tag{10.15}$$

Analogously, the voltage stress imposed on the active switch can be read from (10.15) as $(V_i + V_o)$. Since there are two diodes conducting simultaneously when the active switch S_1 is turned off, the voltage stress imposed on each diode is supposed to be one half $(V_i + V_o)$, i.e. $(1/2)(V_i + V_o)$.

The voltage stresses of switch S_1 and diodes D_1 and D_2 can be confirmed from the converter operation. First of all, we have the following assumption that both inductors are identical, and when switch S_1 is turned on, the voltages across inductors L_1 and L_2 are equal to $(1/2)(V_i - V_o)$. Following KVL, we can determine the voltages across both diodes D_1 and D_2 are $(1/2)(V_i + V_o)$. When switch S_1 is turned off, both diodes will be forced to conduct, and we can determine the voltage stress for the switch to be $(V_i + V_o)$, which is identical to that read from (10.15).

It should be noted that although the high step-down switched-inductor converter can achieve a higher step-down output voltage than that of the buck converter, it has higher voltage stress $(V_i + V_o)$ than that (V_i) of the buck converter. Thus, it needs a trade-off between the step-down ratio and voltage stress.

10.3.2 High Step-Down/Step-Up Switched-Inductor Converter

A high step-down/step-up switched-inductor converter synthesized from a Zeta converter and an inverse buck–boost converter is shown in Figure 10.9a, in which its input-to-output voltage transfer ratio is shown as follows:

$$\frac{V_o}{V_i} = \frac{D}{2\left(1-D\right)}. \tag{10.16}$$

Through cross multiplication and some algebraic manipulation, we can express (10.16) as

$$2V_o = D\left(V_i + 2V_o\right). \tag{10.17}$$

Analogously, the voltage stress imposed on the active switch can be read from (10.17) as $(V_i + 2V_o)$. Since there are two diodes conducting simultaneously when the active switch S_1 is turned off, the voltage stress imposed on each diode is supposed to be one half $(V_i + 2V_o)$, i.e. $(1/2)(V_i + 2V_o)$.

The voltage stresses of switch S_1 and diodes D_1 and D_2 can be confirmed from the converter operation. First of all, we have to determine the voltages across capacitor C_1 and inductors L_2 and L_3. Obviously, the voltage V_{C1} across capacitor C_1 can be determined from the loop of L_1, C_1, L_2, C_2, and L_3 based on volt-second

(a)

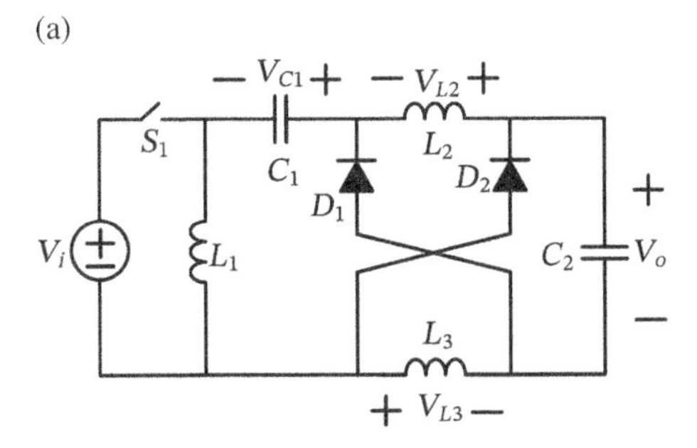

(b)

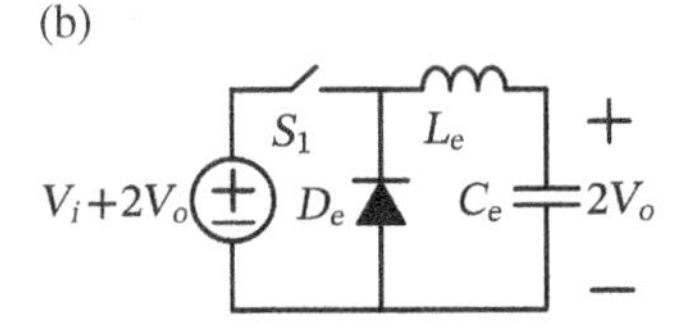

Figure 10.9 (a) High step-down switched-inductor converter with voltage transfer ratio, $D/(2-D)$, and (b) its equivalent buck converter.

balance of the two inductors, which is V_o. Again, we have the following assumption that both inductors are identical. When switch S_1 is turned on, the voltages across inductors L_2 and L_3 are equal to $(V_i/2)$. Following KVL, we can determine the voltages across both diodes D_1 and D_2 are $(1/2)(V_i+2V_o)$. When switch S_1 is turned off, both diodes will be forced to conduct, and we can determine the voltage stress for the switch to be (V_i+2V_o), which is identical to that read from (10.17).

It should be noted that although the high step-down/step-up switched-inductor converter can achieve a higher step-down/step-up output voltage than that of the Zeta converter, it has higher voltage stress (V_i+2V_o) than that (V_i+V_o) of the Zeta converter. Thus, it needs a trade-off between the step-down/step-up ratio and voltage stress.

10.3.3 Compound Step-Down/Step-Up Switched-Capacitor Converter

A compound step-down/step-up switched-inductor converter synthesized from a boost converter and a Ćuk converter is shown in Figure 10.10a, in which its input-to-output voltage transfer ratio is shown as follows:

$$\frac{V_o}{V_i}=\frac{1+D}{1-D}. \tag{10.18}$$

(a)

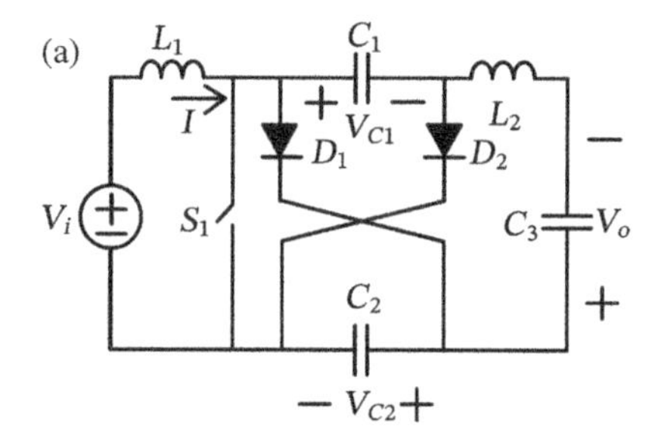

(b)

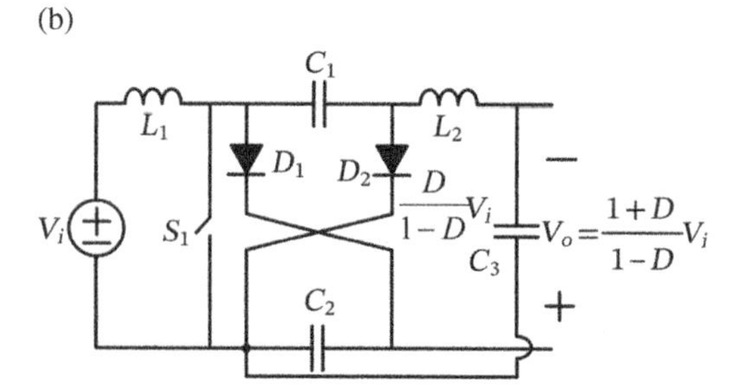

Figure 10.10 (a) Compound step-down/step-up switched capacitor converter with voltage transfer ratio, $1 + D/(1 - D)$, and (b) the grafted converter from the boost and Ćuk converters.

Through cross multiplication and some algebraic manipulation, we can express (10.18) as

$$V_o - V_i = D\left(V_i + V_o\right). \tag{10.19}$$

Analogously, the voltage stress imposed on the active switch can be read from (10.19) as $(V_i + V_o)$. Since there are two diodes conducting simultaneously when the active switch S_1 is turned off, the voltage stress imposed on each diode is supposed to be one half $(V_i + V_o)$, i.e. $(1/2)(V_i + V_o)$.

The voltage stresses of switch S_1 and diodes D_1 and D_2 can be confirmed from the converter operation. First of all, we have to determine the voltages across capacitors C_1 and C_2. Obviously, the voltages V_{C1} and V_{C2} across capacitors C_1 and C_2 can be determined from the loop of V_i, L_1, C_1, L_2, V_o, and C_2 based on volt-second balance of the two inductors, which are $(1/2)(V_i + V_o)$. Again, we have the following assumption that both capacitors are identical. When switch S_1 is turned on, the voltages across inductors L_1 and L_2 are equal to V_i. Following KVL, we can determine the voltages across both diodes D_1 and D_2 are $(1/2)(V_i + V_o)$. When switch S_1 is turned off, both diodes will be forced to conduct, and we can determine the voltage stress for the active switch to be $(1/2)(V_i + V_o)$, which is different from that read from (10.19). In fact, all of the active switch and the passive diodes have the same voltage stress, $(1/2)(V_i + V_o)$.

What is wrong with the rule presented in Sections 10.2, 10.3.1, and 10.3.2? How come it does not apply to this transfer ratio? Let us review the transfer ratio shown in (10.18) which can be expressed as follows:

$$\frac{V_o}{V_i} = \frac{1+D}{1-D} = \frac{1}{1-D} + \frac{D}{1-D}. \tag{10.20}$$

It can be seen obviously that the transfer ratios of $1/(1-D)$ and $D/(1-D)$ can be synthesized with boost and Ćuk converters, respectively. By applying the graft technique to the two active switches shown in the boost and Ćuk converters, we can obtain the grafted converter shown in Figure 10.10b, The voltage stresses imposed on both active switches of the boost and Ćuk converters are equal to $V_i/(1-D)$, which is equal to $(1/2)(V_i + V_o)$. This is the reason why the voltage stress cannot be read directly from (10.19). However, the voltage stress read from the regular rule shown in (10.19) is higher than that derived from the converter operation. It is still not bad. Basically, we can start from reading the voltage stress in the equation represented in the form such as (10.19) and confirm it with the converter operation when it is available.

It should be noted that the converter topology shown in Figure 10.10a can be derived from the one shown in Figure 10.10b by changing the DC voltage offset of its capacitor C_3. And, as described in the previous chapters, a capacitor with DC voltage offset will not change the voltage stress imposed on switches. This point has been double confirmed by the voltage stress shown in the discussed converter.

10.3.4 High Step-Down Converter with Transfer Ratio of D^2

A grafted converter derived by cascading a buck converter with the other buck converter is shown in Figure 10.11a, namely, buck–buck converter, in which there are an active switch and three diodes. The voltage transfer ratio of the buck–buck converter is shown as follows:

$$\frac{V_o}{V_i} = D^2. \tag{10.21}$$

Or it can be expressed as

$$V_o = D\left(D\left(V_i\right)\right). \tag{10.22}$$

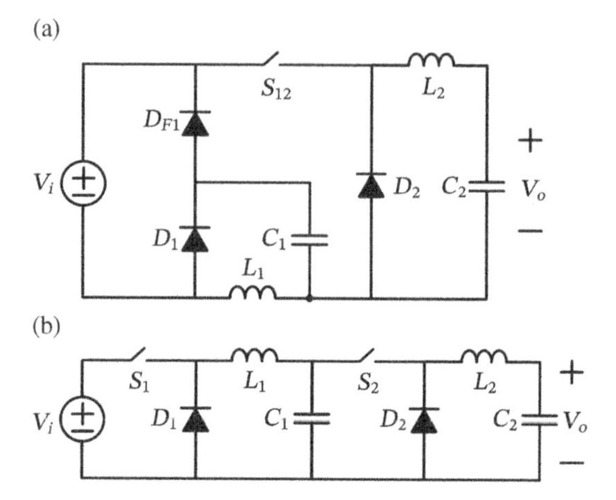

Figure 10.11 (a) Grafted buck–buck converter and (b) two buck converters in cascade.

The voltage stress of switch S_{12} still can be read from (10.22) as V_i. In fact, switch S_{12} is grafted from two active switches shown in Figure 10.11b, in which switch S_1 with the voltage stress of V_i and that of switch S_2 is DV_i. Since V_i is always greater or equal to DV_i, the voltage stress of the grafted switch S_{12} is V_i.

How about the voltage stresses of the three diodes? Similar to those of switches S_1 and S_2, the voltage stresses of diodes D_1 and D_2 are V_i and DV_i, respectively. This can be confirmed from the converter operation. In Figure 10.11a, diodes D_1 and D_2 serve as the freewheeling as the original buck converters. So, diode D_1 has the voltage stress of V_i, while that of diode D_2 is DV_i. Diode D_{F1} in the converter serves to circulate the current difference between switches S_1 and S_2; thus, it conducts when switch S_{12} does. When switch S_{12} is turned off, diodes D_1 and D_2 conduct, and D_{F1} is in reverse bias. The voltage stress of switch S_{12} is V_i, so is diode D_{F1}.

10.3.5 High Step-Up Converter with Transfer Ratio of $1/(1 - D)^2$

A grafted converter derived by cascading a boost converter with the other boost converter is shown in Figure 10.12a, namely, boost–boost converter, in which there are an active switch and three diodes. The voltage transfer ratio of the boost–boost converter is shown as follows:

$$\frac{V_o}{V_i} = \frac{1}{(1-D)^2}. \tag{10.23}$$

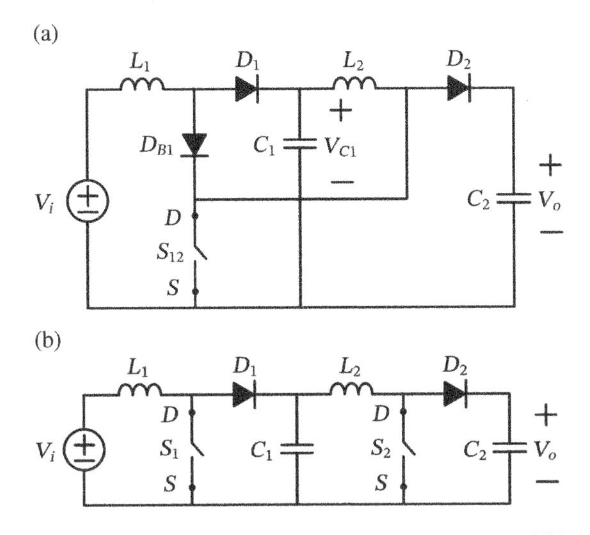

Figure 10.12 (a) Grafted boost–boost converter and (b) two boost converters in cascade.

Or it can be expressed as

$$V_o - V_i = D(2 - D)V_o \tag{10.24}$$

$$V_o - V_i = D(V_o) + D(V_o(1 - D)) \tag{10.25}$$

The voltage stress of switch S_{12} still can be read from (10.25) as V_o. In fact, switch S_{12} is grafted from two active switches shown in Figure 10.12b, in which switch S_1 with the voltage stress of $(1-D)V_o$ and that of switch S_2 is V_o. Since V_o is always greater or equal to $(1-D)V_o$, the voltage stress of the grafted switch S_{12} is V_o.

How about the voltage stresses of the three diodes? Similar to those of switches S_1 and S_2, the voltage stresses of diodes D_1 and D_2 are $(1-D)V_o$ and V_o, respectively. This can be confirmed from the converter operation. In Figure 10.12a, when switch S_{12} is turned on, diodes D_1 and D_2 are reversely biased. So, diode D_1 has the voltage stress of V_{C1}, which is equal to $(1-D)V_o$, while that of diode D_2 is V_o. Diode D_{B1} in the converter serves to block the voltage difference between those of switches S_1 and S_2. When switch S_{12} is turned off, diodes D_1 and D_2 conduct, and D_{B1} is in reverse bias. The voltage stress of switch S_{12} is V_o, and that of diode D_{B1} is DV_o.

It is interesting to be noted that the two terms shown in (10.25) are just corresponding to the two voltage stresses of the two active switches and the two diodes shown in Figure 10.12b, the original two boost converters before grafting.

From Sections 10.1–10.3, it is worth pointing out that only do (10.2) and (10.22) have V_i and V_o separated at the RHS and LHS of the equations. This is one of unique characteristics of a minimum-phase control system. The rest of them are all belonged to nonminimum-phase systems.

Further Reading

Axelrod, B., Borkovich, Y., and Ioinovici, A. (2008). Switched-capacitor/switched-inductor structures for getting transformerless hybrid DC–DC PWM converters. *IEEE Trans. Circuits Syst. I* 55 (2): 687–696.

Cao, D. and Peng, F.Z. (2009). A family of Z source and quasi-Z source DC–DC converters. Proceedings of the IEEE Applied Power Electronics Conference, IEEE, pp. 1097–1101.

Mohan, N., Undeland, T.M., and Robbins, W.P. (1989). *Power Electronics*, 2e. Wiley.

Williams, B.W. (2014). Generation and analysis of canonical switching cell DC-to-DC converters. *IEEE Trans. Ind. Electron.* 61: 329–346.

Wu, T.-F. (2016). Decoding and synthesizing transformerless PWM converters. *IEEE Trans. Power Electron.* 30 (9): 6293–6304.

Wu, T.-F. and Chen, Y.-K. (1998). A systematic and unified approach to modeling PWM DC/DC converter based on the graft scheme. *IEEE Trans. Ind. Electron.* 45 (1): 88–98.

Wu, T.-F. and Yu, T.-H. (1998). Unified approach to developing single-stage power converters. *IEEE Trans. Aerosp. Electron. Syst.* 34 (1): 221–223.

Wu, T.-F., Liang, S.-A., and Chen, Y.-K. (2003). A structural approach to synthesizing soft switching PWM converters. *IEEE Trans. Power Electron.* 18 (1): 38–43.

11

Discussion and Conclusion

From Chapters 1–9, we have addressed the discovery of the original converter, on which fundamentals, graft, and layer schemes were presented for synthesizing hard-switching PWM converters. The synthesis has been extended to resonant converters, multilevel converters, and soft-switching PWM converters. However, there exist several vague points and philosophical issues that need further discussions, such as the following: If identical transfer code will yield the same converter topology? If buck converter is the dual of the boost converter, how come they are not one-to-one correspondence? How to derive the original converter based on the resonant concept? Is it possible to analogize the buck converter to the structure of deoxyribonucleic acid (DNA)? In this chapter, we are going to address these issues and make a conclusion, including possible future study topics.

11.1 Will Identical Transfer Code Yield the Same Converter Topology?

For the most well-known PWM converters, buck–boost, Ćuk, sepic, and Zeta converters, as shown in Figure 11.1, have the same transfer code $D/(1-D)$, but they have different circuit configurations. Except the buck–boost converter, they are all the fourth-order converters. To make a buck–boost converter become a fourth-order one, we insert an extra LC filter to the converter, as shown in Figure 11.2, in which the DC voltage offset of capacitor C_1 is 0V. Now, they have the same fourth-order dynamics, but still with different circuit configurations. How can they yield the same input-to-output voltage transfer ratio? Part of the reasons has been presented in Chapter 6. Here, a more theoretical analysis is presented as follows.

Origin of Power Converters: Decoding, Synthesizing, and Modeling, First Edition.
Tsai-Fu Wu and Yu-Kai Chen.
© 2020 John Wiley & Sons, Inc. Published 2020 by John Wiley & Sons, Inc.

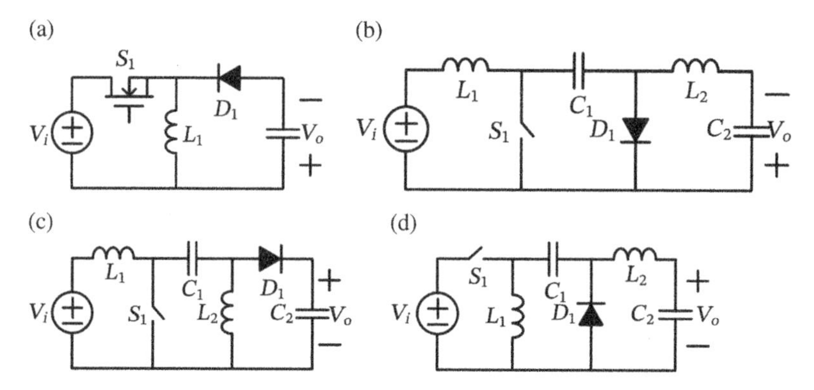

Figure 11.1 (a) The second-order buck–boost converter, (b) Ćuk converter, (c) sepic converter, and (d) Zeta converter.

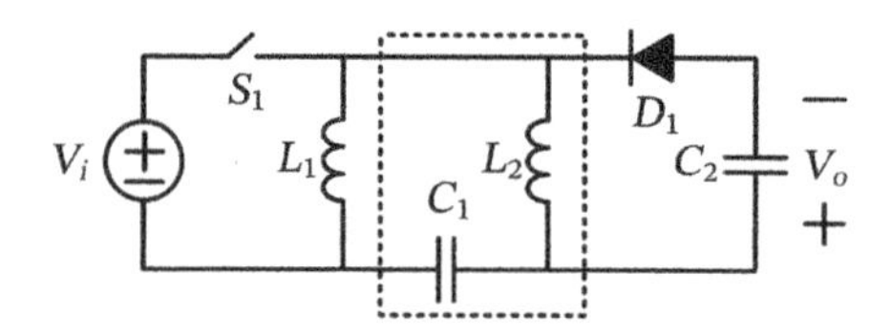

Figure 11.2 A fourth-order buck–boost converter with an extra *LC* filter.

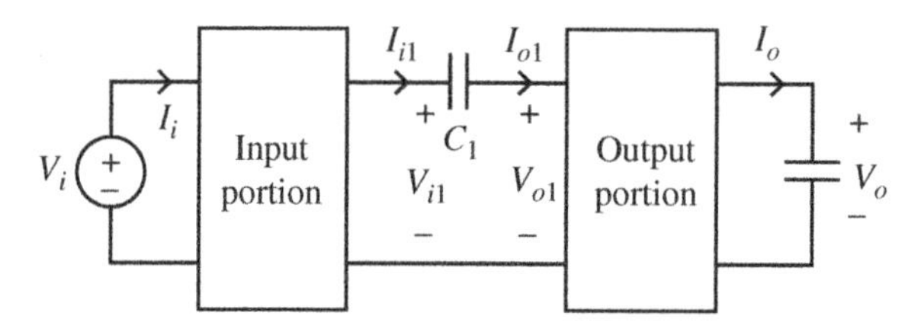

Figure 11.3 Conceptual converter topology with a buffer capacitor located between the input and output portions.

The four converters shown in Figures 11.1b–d and 11.2 can be conceptually represented by the one shown in Figure 11.3, in which a buffer capacitor located between the input and output portions. In the steady state, capacitor C_1 must satisfy with ampere-second balance, and its input power should be equal to its output power. Therefore, we have the following relationship:

$$I_{i1}V_{i1} = I_{o1}V_{o1}. \tag{11.1}$$

Based on power balance, we also have the following relationships:

$$I_i V_i = I_{i1} V_{i1}$$

and

$$I_{o1} V_{o1} = I_o V_o. \tag{11.2}$$

From (11.1) and (11.2), we can have

$$I_i V_i = I_o V_o. \tag{11.3}$$

Basically, the input-to-output power relationship, which in turn can derive the input-to-output voltage transfer ratio, is nothing with the buffer capacitor C_1. It is interesting to note that no matter what is the DC offset voltage of capacitor C_1, it does not change the voltage transfer ratio. The DC offset voltages in capacitor C_1 of the fourth-order buck–boost, Ćuk, sepic, and Zeta are 0, $(V_i + V_o)$, V_i, and V_o, respectively.

So what is the purpose of capacitor C_1? From the converter operation, it can be observed that capacitor C_1 serves to buffer the energy from the input and/or change the output voltage polarity, such as the buck–boost and Ćuk converters. In fact, the transformations among the four converters can be conducted by applying the DC voltage offsetting fundamental to capacitor C_1. Most of the transformations have been presented in Chapter 6. Here, we present the transformation from Zeta converter to Ćuk converter as an illustration example.

Zeta and Ćuk converters belong to different families, but with the same component count and the same transfer ratio. However, their output voltage polarities are different. How come they have different component configurations? Their buffer capacitors might have different offset voltages. Figure 11.4 shows the illustration of the deduction from Zeta converter to Ćuk converter. Figure 11.4a shows the Zeta converter, and Figure 11.4b shows the switch S_1 in its return path. When changing the DC voltage offset for capacitor C_1 from V_o to $V_i + V_o$, we have the converter shown in Figure 11.4c. Reconfiguring the components while not changing its operational principle, we have the converter shown in Figure 11.4d. Moving capacitor C_1 and inductor L_2 from the return path to the forward path yields the well-known Ćuk converter, as shown in Figure 11.4e.

The voltage across capacitor C_1 can be proved to be $V_i + V_o$ in the Ćuk converter, while that in the Zeta converter is V_o. Again, the DC voltage offset does not change its transfer ratio. Thus, these two converters have different capacitor voltage offsets, but they have the same voltage transfer ratio.

Comparing the second-order buck–boost converter shown in Figure 11.1a and the fourth-order one shown in Figure 11.2 reveals that the extra LC filter does not change the static voltage transfer ratio and they are belonged to the same converter

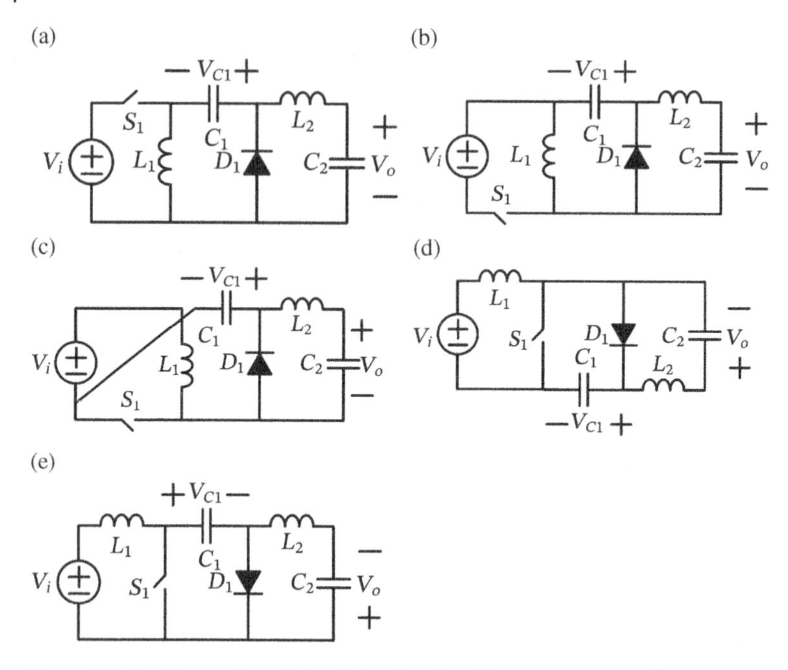

Figure 11.4 Illustration of the deduction from Zeta converter to Ćuk one.

topology. For the fourth-order Ćuk, sepic, and Zeta converters, they have the same voltage transfer ratio, but with different buffer capacitor voltage offsets. They can be considered similar converter topologies. Based on this point of view, can we conclude that one voltage transfer ratio will be uniquely corresponding to one similar converter topology? From our discussions presented in previous chapters, the answer is yes. This can be explained as follows. In a typical PWM converter, there exist only an active switch, a passive diode, and an *LC* filter. To transfer power from input to output, we take turn conducting the active switch and the passive diode. When the active switch is turned on, the passive diode is reversely biased, while when the active switch is turned off, the passive diode will conduct. The pulsating voltage will be smoothed output through the *LC* filter. Thus, the voltage transfer ratio will be highly related to the converter topology. Thus, a transfer ratio will be uniquely associated with one converter topology. However, rigorous proof might be needed to elaborate the above discussion further.

11.2 Topological Duality Versus Circuital Duality

Prof. Ćuk has proposed a duality for the buck and boost converters, and for the buck–boost and Ćuk converters, respectively. First, he represented the buck converter in voltage source, PWM switch (S_1 and D_1) and current sink, as

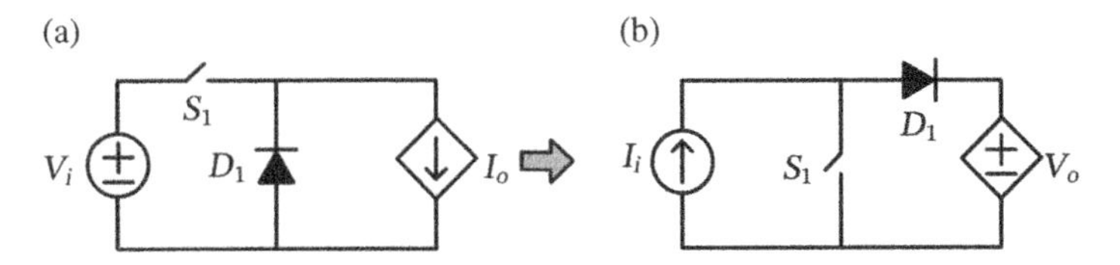

Figure 11.5 Buck and boost converters in topological duality.

shown in Figure 11.5a, and the boost converter in current source, PWM switch and voltage sink, as shown in Figure 11.5b. Since the boost converter is the dual of the buck converter, the voltage source V_i in the buck converter is converted to a current source I_i in the boost converter, the switch S_1 in series with the voltage source is converted to be in parallel with the current source I_i, diode D_1 in parallel with current sink I_o is converted to be in series with voltage sink V_o, and the output voltage sink V_o in the buck converter is converted to a current sink I_o in the boost converter. It can be observed that the components are one-to-one correspondence. However, there is one vague point that how come the input voltage source is in parallel with the input port and the current source in the boost converter is also in parallel with the input port. For duality, the parallel-connected voltage source should be converted to a series-connected current source. Similarly, the parallel-connected output current sink should be converted to a series-connected voltage sink. Maybe, we do not need to care about the input and output ports. At least, they are dual in the sense of one-to-one correspondence.

For the buck and boost converters shown in Figure 11.5, they are represented in voltage source and current sink and current source and voltage sink, respectively. However, in practice, there are only voltage source and voltage sink, but no current source or current sink. There is no circuit component that can realize a current source or current sink. Thus, we name the buck and boost converters shown in Figure 11.5 as topological dual.

To have practical converter topologies, the buck and boost converters are represented in circuit components, as shown in Figure 11.6. The voltage source/sink is realized with a capacitor, while a current source/sink is realized with a capacitor in series with an inductor. How come the voltage source/sink is realized with a capacitor, but the current source/sink cannot be realized with an inductor? There must be something wrong with the circuit configurations of the buck and boost converter topologies.

Before investigating the philosophy behind, we would like to make a note that in fact, a current source/sink cannot be realized with a capacitor and an inductor, as shown in Figure 11.6, but because of PWM switching action, they can be controlled to be close to a current source/sink.

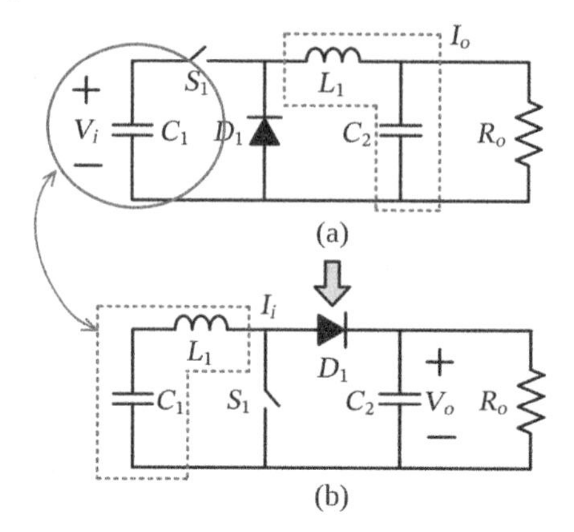

Figure 11.6 Buck and boost converters in circuital forms, but not in circuital duality.

In social science, we talk about "harmony" that is a kind of "give and take" balancing state. People know how to appreciate each other to achieve a peaceful environment. Harmony is the top solicitation of our friendly societies. In our universe, starting from Big Bang, "harmonics" keeps all stars revolving without collapsing each other. Even in our blood circulation system, "resonance" plays an important role in driving blood into organs. All of the aforementioned matters are kept with dynamic balance. How about our converter circuits? Resonance could also be the principle of converter operation or power transfer. Power flow is always an ongoing manner no matter what switch status is.

Let us start with a transmission line model of two-port network, as shown in Figure 11.7a, which are represented in cascaded multiple-stage *LC* components. As we know, power through *LC* components is always in resonant manner and has no loss, which means the total energy stored in the *LC* components is conserved all the time. However, a transmission line cannot control the power flow from input to output. It is just free running. It cannot be a converter, or it can be said as a special converter with the input power equaling to the output power all the time.

If we insert a PWM switch to control the power flow, as shown in Figure 11.7b, we can obtain a converter, namely, new buck converter. The power flow still keeps in resonance when we turn on or off the switch S_1. When switch S_1 is turned on, the power flow is the same as the original transmission line. When switch S_1 is turned off, L_1C_1 are of resonance, so are L_2C_2 through diode D_1. The newly derived buck converter shown in Figure 11.7b has the minimum component count to keep the resonant principle. This converter complies with the universal principle,

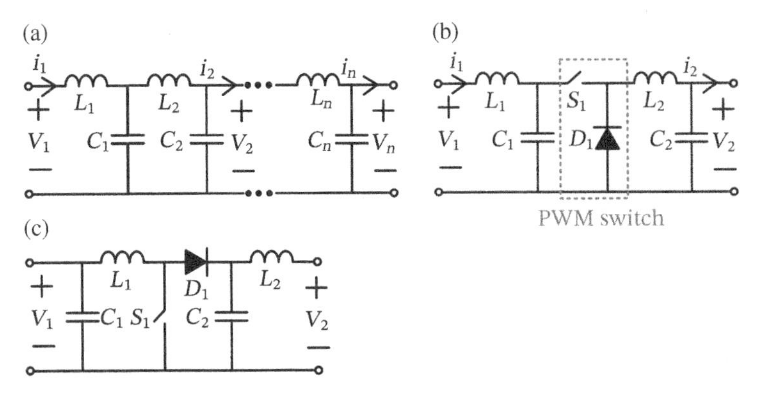

Figure 11.7 (a) A transmission line model of two wires represented in a two-port network, (b) inserting a PWM switch to the transmission line to form a new buck converter with resonance, and (c) the boost converter that is a dual of the buck converter.

harmony, harmonics, or resonance. We do believe this will be a right converter topology for a buck converter.

Through circuital duality theory, we can find a boost dual from the new buck converter shown in Figure 11.7b, and the newly derived boost converter is shown in Figure 11.7c. Comparing the buck converter with the boost converter reveals that the two converters are represented in two-port networks and they have one-to-one correspondence dual components. In the buck converter, the series inductor L_1 is transformed to the parallel capacitor C_1 shown in the boost converter, so is parallel C_1 transformed to series inductor L_1 in the boost converter. Similarly, the PWM switch, inductor L_2, and capacitor C_2 in the buck converter are all transformed to the corresponding components shown in the boost converter. Therefore, we name the buck and boost converters shown in Figure 11.7b and c a "circuital duality" while the ones shown in Figure 11.5a and b a "topological duality."

With the two newly derived buck and boost converters, we can derive the rest of PWM converters based on the graft and layer schemes, which will be presented in next section.

11.3 Graft and Layer Schemes for Synthesizing New Fundamental Converters

In Chapters 5 and 6, we have presented the graft and layer schemes for deriving all of fundamental PWM converters. Here, they are adopted to derive the new buck–boost, Ćuk, sepic, and Zeta converters, as examples, based on the newly derived buck and boost converters.

11.3.1 Synthesis of Buck–Boost Converter

Grafting a new boost converter on a new buck converter yields the buck cascaded with the boost, as shown in Figure 11.8a. It is obvious that capacitors C_2 and C_3 are in parallel, and they can be replaced with a single one, C_{23}, as shown in Figure 11.8b. For steady-state converter operation, *LCL* filter L_2, C_{23}, and L_3 can be simplified to a single inductor, L_{23}, as shown in Figure 11.8c. Moving inductor L_{23} from the forward path to the return path can identify an *S–D* common node for switches S_1 and S_2, and by following the principle of the graft scheme, we can integrate the

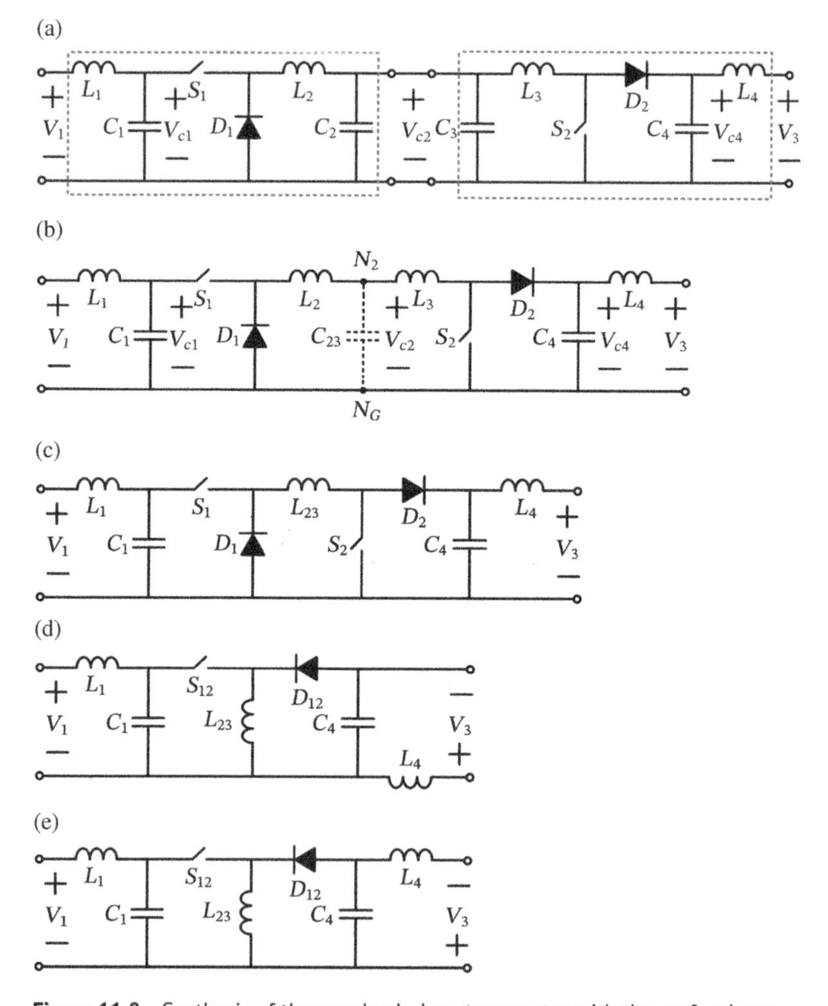

Figure 11.8 Synthesis of the new buck–boost converter with the graft scheme.

two switches into a single one, S_{12}, as shown in Figure 11.8d. Basically, the one shown in Figure 11.8d is a new buck–boost converter. However, we can move inductor L_4 from the return path to the forward path, as shown in Figure 11.8e. This is also a new buck–boost converter topology.

The old buck–boost converter is a second-order topology, while the new one is a fifth-order topology. It increases the order by three. However, it can fulfill the resonance principle during operation. When switch S_{12} is turned on, inductor L_1, capacitor C_1, and inductor L_{23} are resonant, and capacitor C_4 and inductor L_4 can resonate, too. When switch S_{12} is turned off, diode D_{12} is forced to conduct. Again, inductor L_1 and capacitor C_1 are of resonance, so are L_{23}, C_4, and L_4. Through duality principle, we can derive the new Ćuk converter, since from the topological duality the old Ćuk is the dual of the old buck–boost. Or it can be synthesized by grafting a new buck converter on a new boost converter, which will be shown in Section 11.3.2.

11.3.2 Synthesis of Boost–Buck (Ćuk) Converter

Grafting a new buck converter on a new boost converter yields the boost cascaded with the buck, as shown in Figure 11.9a. It is obvious that inductors L_2 and L_3 are in series, and they can be replaced with a single one, L_{23}, as shown in Figure 11.9b. For steady-state converter operation, *CLC* filter, including C_2, L_{23}, and C_3, can be simplified to a single capacitor, C_{23}, as shown in Figure 11.9c. Moving diode D_1 from the forward path to the return path can identify a *D–D* common node for switches S_1 and S_2. By following the principle of the graft scheme, we can replace the two active switches with a T-type grafted switch, S_{12}, and the two diodes D_1 and D_2 become in parallel and can be replaced with a single one, D_{12}, as shown in Figure 11.9d. Basically, the one shown in Figure 11.9d is a new boost–buck converter. However, we can move inductor L_4 from the return path to the forward path, as shown in Figure 11.9e. This is also a new boost–buck converter topology.

The old boost–buck converter is a fourth-order topology, while the new one is a fifth-order topology. It increases the order by one. However, it can fulfill the resonance principle during operation. When switch S_{12} is turned on, inductor L_1 and capacitor C_1 are resonant, and capacitor C_{23}, inductor L_4, and capacitor C_4 can resonate, too. When switch S_{12} is turned off, diode D_{12} is forced to conduct. Again, capacitor C_1, inductor L_1, and capacitor C_{23} are of resonance, so are L_4 and C_4.

It is interesting to note that the old buck–boost converter is a second-order topology, while the old Ćuk converter is a fourth-order one. They cannot be dual with a one-to-one correspondence. However, in the newly synthesized converters, they are all fifth-order converter topologies. Therefore, they can be converted each other by duality principle and, more important, with one-to-one correspondence.

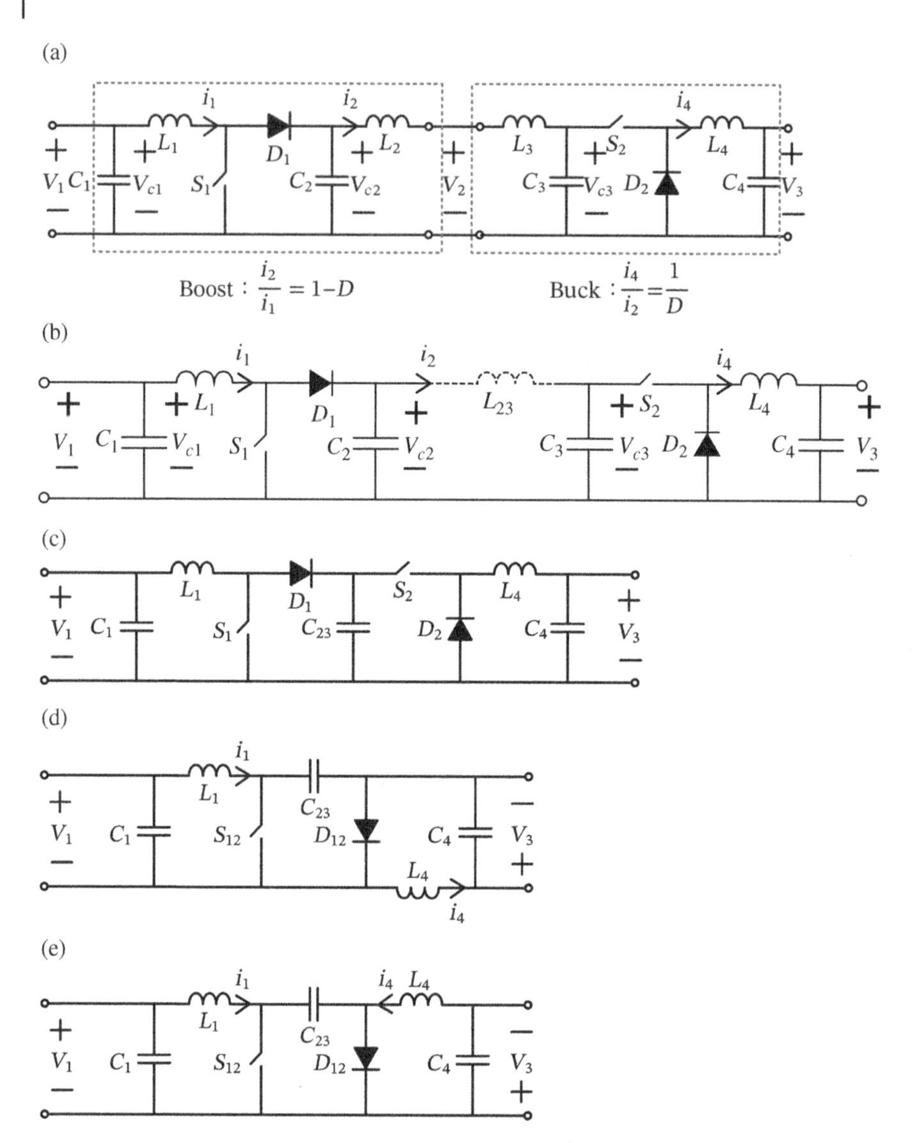

Figure 11.9 Synthesis of the new boost–buck (Ćuk) converter with the graft scheme.

11.3.3 Synthesis of Buck–Boost–Buck (Zeta) Converter

Zeta converter has the same transfer ratio of $D/(1-D)$ as that of a buck–boost converter. However, it requires a new code configuration to derive the transfer ratio of $D/(1-D)$, as shown in Figure 11.10a. A converter configuration to synthesize the code configuration is shown in Figure 11.10b, in which the buck converter

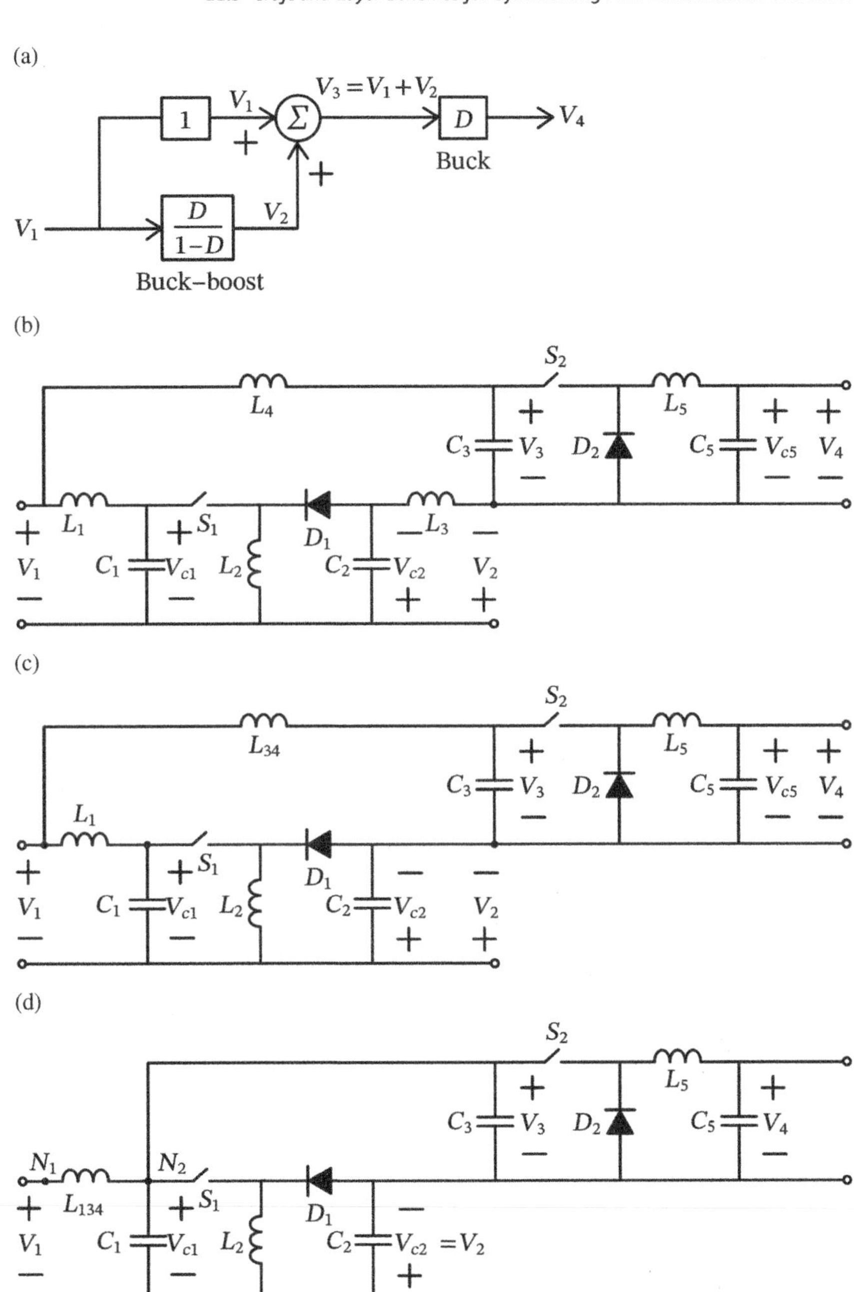

Figure 11.10 (Continued)

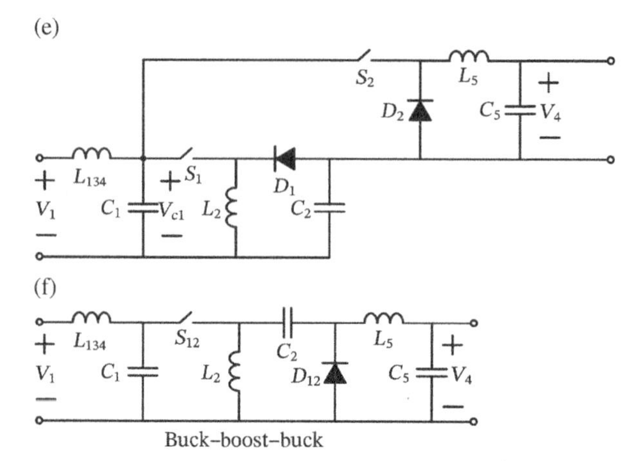

Figure 11.10 Synthesis of the new buck–boost–buck (Zeta) converter with the graft scheme.

is fed by $V_1 + V_2$, equivalent to $V_1(1+ D/(1-D))$. From Figure 11.10b, it can be observed that inductors L_3 and L_4 are in series, and they can be replaced with a single one, L_{34}, as shown in Figure 11.10c. Since the inputs of both buck and buck–boost converters have inductors L_1 and L_{34}, they can be replaced with a single one, L_{134}. Moving the input of the buck converter from node N_1 to node N_2 while still keeping resonance anytime yields the converter circuit shown in Figure 11.10d. In the figure, since capacitors C_1, C_2, and C_3 form a loop, one of them is redundant and can be removed from the circuit. For instance, we remove capacitor C_3 from the circuit, as shown in Figure 11.10e. Now, we can identify a *D–D* common node for switches S_1 and S_2, and following the graft scheme processes to graft the two active switches yields the buck–boost–buck (Zeta) converter shown in Figure 11.10f, which has six orders.

The Zeta converter still keeps resonance no matter what switch turns on or off. When switch S_{12} turns on, all of inductors and capacitors in the converter are of resonance, and when it is turned off, L_{134} and C_1, L_2 and C_2, and L_5 and C_5 are all resonant, respectively. The converter has the same principle of resonance.

11.3.4 Synthesis of Boost–Buck–Boost (Sepic) Converter

Sepic converter has the other name, boost–buck–boost converter, which is the dual of buck–boost–buck (Zeta) converter. In Zeta converter, we use input-to-output voltage transfer ratio, while in sepic converter we use current transfer ratio, as shown in Figure 11.11a, where the transfer ratios of boost–buck and boost are all represented in current type. The converters to synthesize the code configuration

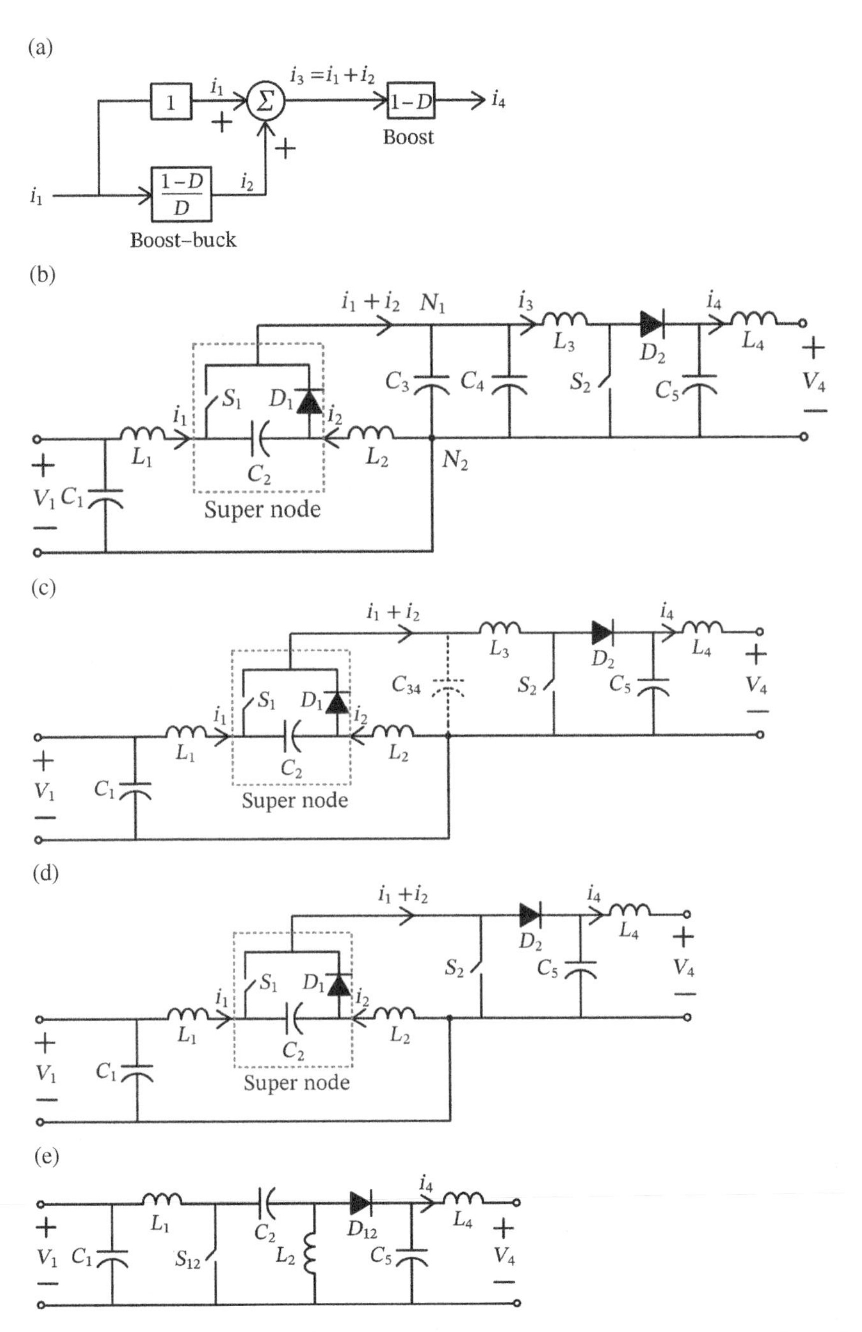

Figure 11.11 Synthesis of the new boost–buck–boost (sepic) converter with the graft scheme.

shown in Figure 11.11a are shown in Figure 11.11b, in which the input and output currents, $i_1 + i_2$, are summed up and fed to the boost converter. To simplify the synthesis, we treat switch S_1, diode D_1, and capacitor C_2 a super node because S_1 and D_1 take turn conduction and the total current $i_1 + i_2$ always flowing through capacitor C_2. It should be noted that the output of the boost–buck converter is supposed to be connected to node N_1, but with DC voltage offset, it is connected to node N_2. Obviously, capacitors C_3 and C_4 are in parallel connection, and they can be replaced with a single one, C_{34}, as shown in Figure 11.11c.

In Figure 11.11c, capacitor C_{34} can be removed because there is no current flow through it in the steady state. Therefore, inductors L_1, L_2, and L_3 share a common node, and one can be removed from the circuit. For instance, we remove inductor L_3, and the converter circuit becomes the one shown in Figure 11.11d. Then, we can identify an *S–D* common node for switches S_1 and S_2. Through the graft scheme processes, the two active switches can be replaced with a grafted switch S_{12}, and the two diodes are replaced with a single D_{12}, as shown in Figure 11.11e.

In the newly derived six PWM converters, their dynamic orders have been increased. The buck and boost converters are increased by two, the buck–boost converter is increased by three, the Ćuk converter is increased by one, while the Zeta and sepic converters are increased by two, respectively. As described previously, buck and boost, buck–boost and boost–buck (Ćuk), and buck–boost–buck (Zeta) and boost–buck–boost (sepic) converters have the same dynamic orders, four, five and six, respectively. They are dual with one-to-one correspondence. In addition, they are always of resonance no matter what switch turns on or off.

11.3.5 Synthesis of Buck-Family Converters with Layer Scheme

Buck-family converters consist of buck, buck–boost, and buck–boost–buck (Zeta) converters. They can be synthesized with the layer scheme. Figure 11.12a shows a new buck converter with a positive unity feedback, and its input-to-output voltage transfer code configuration is shown in Figure 11.12b. It can be proved that this code configuration yields the transfer code of $D/(1-D)$. However, when switch S_1 is turned on, inductor L_1, capacitor C_1, and inductor L_2 are of resonance, but capacitor C_2 has no resonant path. Therefore, an external inductor L_3 is required to be added to the converter configuration shown in Figure 11.12a, and the converter becomes the one shown in Figure 11.12c. After reconfiguring the circuit connection, we can obtain the new buck–boost converter shown in Figure 11.12d. Note that when switch S_1 is turned off, inductor L_1 is resonant with capacitor C_1, and the rest of reactive components, L_2, C_2, and L_3, are of resonance. This satisfies the resonant principle anytime.

With a pulsating feedback from node N_2 and through a set of L_3C_3 filter, we can have another converter with smooth feedback to the input, as shown in

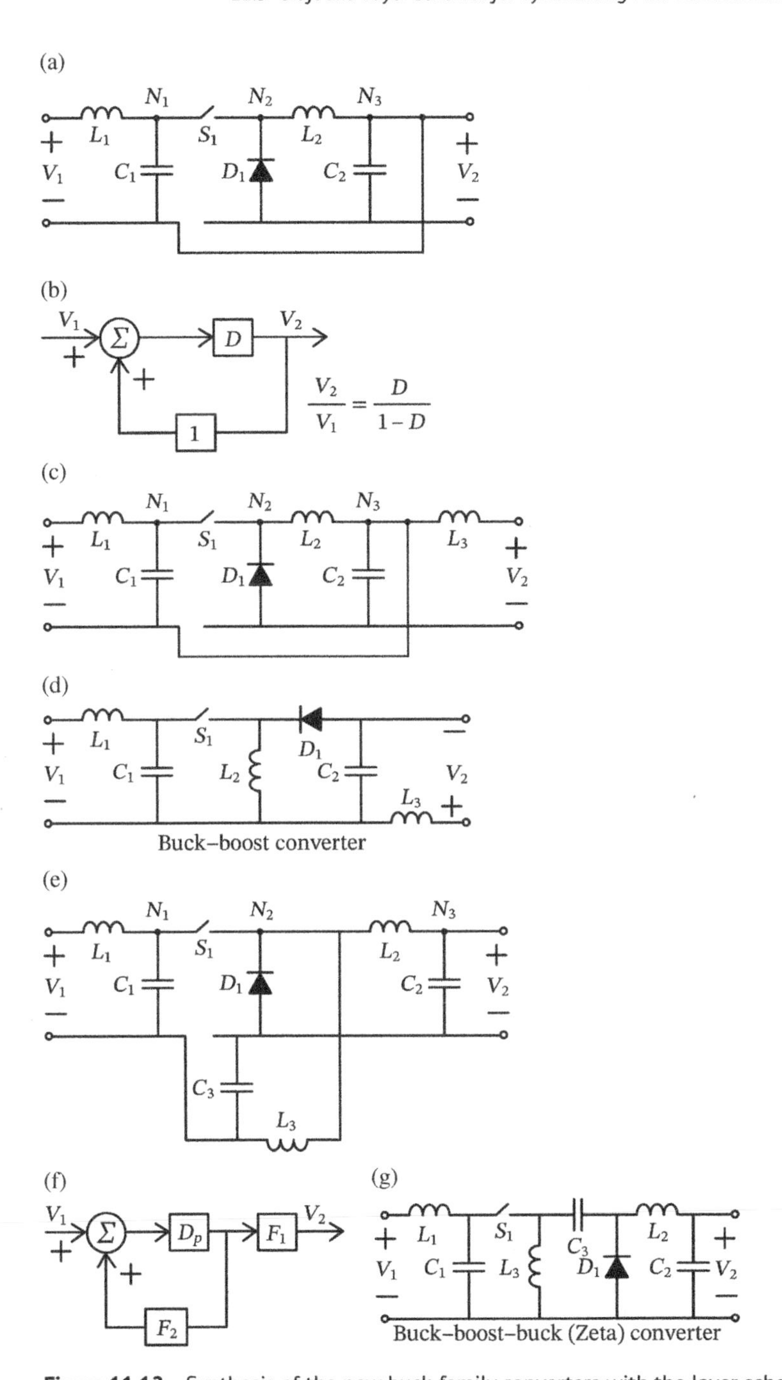

Figure 11.12 Synthesis of the new buck-family converters with the layer scheme.

Figure 11.12e. This converter can be represented by the code configuration shown in Figure 11.12f, in which code D_p denotes the pulsating transfer code of voltage v_1 to the voltage across diode D_1 and F_1 and F_2 are filter gains. If assuming $D_pF_1 = D_pF_2 = D$, we have the input-to-output voltage transfer code $D/(1-D)$. It is the transfer code of the buck–boost–buck converter. And then from the converter configuration shown in Figure 11.12e and with redrawing the component connection, we have the buck–boost–buck converter shown in Figure 11.12g. It can be proved that under either switch S_1 on or off, the reactive components are all of resonance.

As described in Section 6.2.1, the converters in the buck family are supposed to have the same buck-DNA. In the newly derived converters, the buck shown in Figure 11.12a and the buck–boost shown in Figure 11.12c have the same tricodan, switch S_1, diode D_1, and inductor L_2, which is noted as buck-DNA. However, in the buck–boost–buck converter shown in Figure 11.12e, there is no buck-DNA. How can we say that the buck–boost–buck converter is belonged to the buck-family? If adding an external inductor L_x to the branch between node N_2 and the feedback point, we can have the buck-DNA, including S_1, D_1, and L_x, as shown in Figure 11.13. With this extra inductor L_x, its operational principle does not change because three inductors, L_x, L_2, and L_3, share a common node, and the dynamic degree of freedom still keeps in two. Basically, the converter configuration is a universal form of the buck family.

11.3.6 Synthesis of Boost-Family Converters with Layer Scheme

Boost-family converters consist of boost, boost–buck (Ćuk), and boost–buck–boost (sepic) converters. They can be synthesized with the layer scheme. Figure 11.14a shows a new boost converter with a positive unity feedback, and its input-to-output current transfer code configuration is shown in Figure 11.14b. It can be proved that this code configuration yields the transfer code of $D'/(1-D')$,

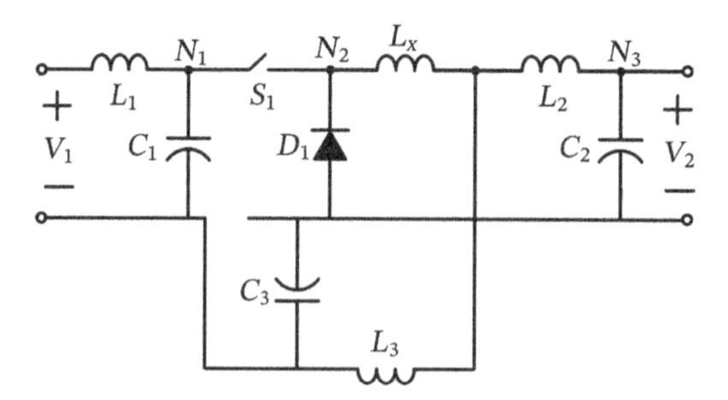

Figure 11.13 The buck–boost–buck converter with an external inductor L_x, showing a universal form of the buck family.

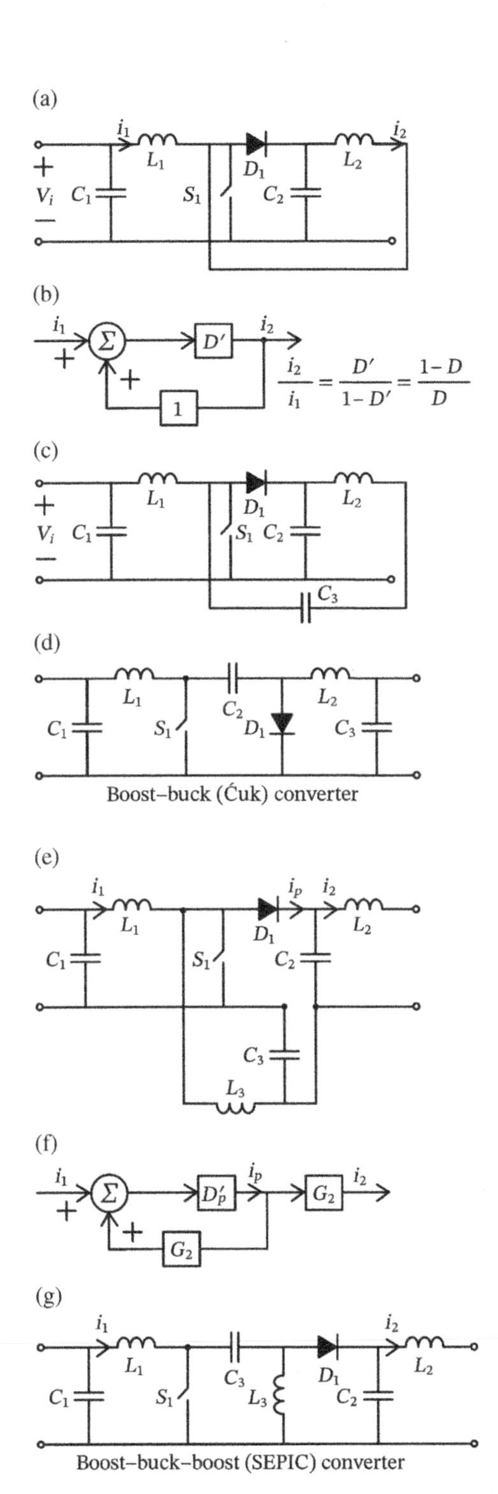

Figure 11.14 Synthesis of the new boost-family converters with the layer scheme.

where D' is equal to $(1-D)$. However, when switch S_1 is turned off, capacitor C_1, inductor L_1, and capacitor C_2 are of resonance, but inductor L_2 has no resonant path. Therefore, an external capacitor C_3 is required to be added to the converter configuration shown in Figure 11.14a, and the converter becomes the one shown in Figure 11.14c. After redrawing the circuit connection, we can obtain the new boost–buck converter shown in Figure 11.14d. Note that when switch S_1 is turned on, capacitor C_1 is resonant with inductor L_1 and the rest of reactive components, C_2, L_2, and C_3, are of resonance. This satisfies the resonant principle anytime.

With a pulsating current feedback from diode D_1 and capacitor C_2 loop and through a set of C_3L_3 filter, we can have another converter with smooth feedback to the input, as shown in Figure 11.14e. This converter can be represented by the code configuration shown in Figure 11.14f, in which code D'_p denotes the pulsating transfer code of current i_1 to the pulsating current i_p through diode D_1, and G_1 and G_2 are filter gains. If assuming $D'_pG_1 = D'_pG_2 = D'$, we have the input-to-output current transfer code $D'/(1-D')$. It is the transfer code of the boost–buck–boost converter. And then from the converter configuration shown in Figure 11.14e and with redrawing the component connection, we have the boost–buck–boost converter shown in Figure 11.14g. It can be proved that under either switch S_1 on or off, the reactive components are all in resonance.

As described in Section 6.2.2, the converters in the boost family are supposed to have the same boost DNA. In the newly derived converters, the boost shown in Figure 11.14a and the boost–buck shown in Figure 11.14c have the same tricodan, switch S_1, diode D_1, and capacitor C_2, which is noted as boost DNA. However, in the boost–buck–boost converter shown in Figure 11.14e, there is no boost DNA. How can we say that the boost–buck–boost converter is belonged to the boost family? If adding an external capacitor C_x to the loop C_2 and C_3, we can have the boost DNA, including S_1, D_1, and C_x, as shown in Figure 11.15. With this extra capacitor C_x, its operational principle does not

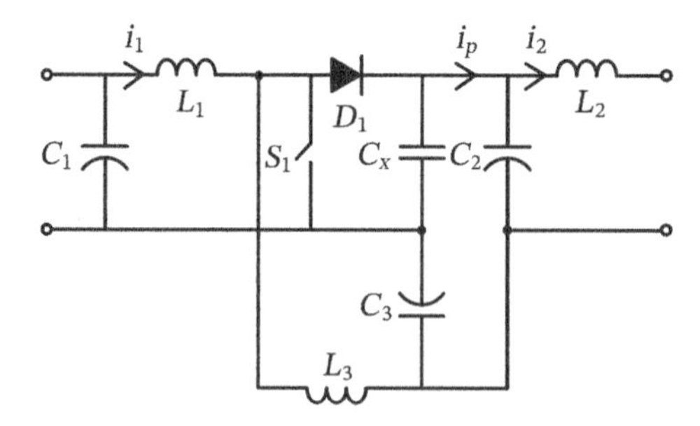

Figure 11.15 The boost–buck–boost converter with an external capacitor C_x, showing a universal form of the boost family.

change because three capacitors, C_X, C_2, and C_3, share a loop, and the dynamic degree of freedom still keeps in two. Basically, the converter configuration is a universal form of the boost family.

11.4 Analogy of Power Converters to DNA

The purpose of conducting an analogy is to learn something from cross fields. For instance, we understand the harmonics in mechanic systems, and we know about resonance in electronic systems. Moreover, we can dig out something that does not exist in our original field. For instance, in a mechanic system, we talk about rigidity. How is it related to an electronic system? It may be analogous to constant current, constant voltage, or constant power in an electronic system. Or it may be analogous to a low quality factor system in an electronic system.

A PWM power converter basically has at least a PWM switch pair, including an active switch and a passive switch, and an *LC* filter pair. When the active switch is turned on, the passive one is reversely biased. On the other hand, when the active switch is turned off, the passive one can conduct. For other types of power converters, the switch pairs also have the same operational principle even though they are realized with all active switches. In general, a PWM switch pair includes an active switch (S) and a passive one (D), and there always exists at least an *LC* filter pair to fulfill the resonant manner all the time. In short, S–D forms a pair and L–C forms the other.

Organisms were not created but evolved, stated in the book entitled *On the Origin of Species*. Descendants are similar to their parents in behavior and characteristics, and there exists something in common, namely, DNA. In DNA, there are four types of bases, namely adenine (A), thymine (T), guanine (G), and cytosine (C), corresponding to the unique four types of nucleotides A, T, G, and C, respectively, which are connected in sequence to form DNA. Moreover, A and T and G and C are always appearing in pair to propagate genetic information. Are converters with various versions derived from certain basic elements through evolution, are there existing similar DNAs in these converters, and are there similar components to A, T, G, and C in the converters? In this discussion, we are looking for the answers to these questions. Before embarking on answering these questions, we will investigate the philosophy of power transfer first.

A Chinese saying "harmony in the family is the basis for success in any undertaking" points out that those who can support each other will benefit mutually, and they will succeed in any affairs undertaken. In "*Qi*" theory, resonance is the principle of circulating blood, consuming much less power (1.7 W) than that of an artificial heart pump (30 W) that is designed based on hemodynamics. What is the philosophy or principle of power transfer in converters that will yield the highest power conversion efficiency? Harmony and resonance describing a state transition without abrupt change can result in the highest effectiveness and efficiency. Power transfer with resonant manner will also come out the highest conversion efficiency that can be readily proved with a second-order *LC* network and a switch. The total electric energy in the *LC* network is always conserved.

James D. Watson's group identified the double-helix structure of DNA in 1953, as conceptually shown in Figure 11.16a, in which A&T and G&C are always present in pairs, respectively. If we stretched the double-helix structure into a straight-line one, it becomes a two-port network like one, as shown in Figure 11.16b. Analogously, the newly derived buck converter represented in two-port network is shown in Figure 11.16c. It can be seen that L&C and S&D always form pairs.

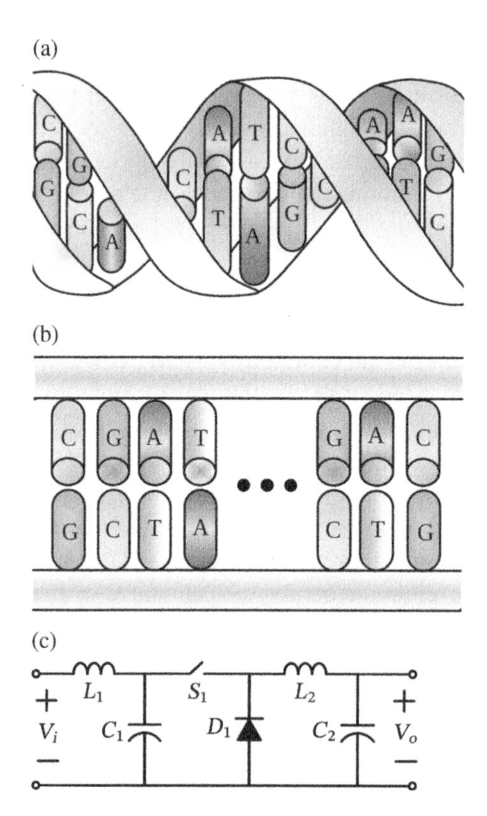

Figure 11.16 Analogy of buck converter to DNA: (a) DNA in double-helix structure, (b) in straight-line structure, and (c) buck converter in two-port network.

11.4.1 Replication

DNA has a mechanism of replication, as illustrated in Figure 11.17a, in which each split nucleotide is made up with its counterpart. For instance, nucleotide A&T is split into A and T, and then they are made up with T and A, respectively, after going through a replication mechanism, because they always form pairs. This mechanism allows each parental carrying only half of DNA information but still can replicate the complete information for heredity purpose.

A new buck converter can have the similar replication mechanism, as shown in Figure 11.17b, in which the split L&C filter is made up with its C&L counterpart and each split S&D switch is made up with its D&S counterpart, forming a new converter. It can be seen that each split branch will form another buck converter, achieving replication mechanism.

Essentially, the A&T and G&C act as codes to form DNA, carrying genetic information, while S&D and L&C are conversion components to transfer power. We can imagine that the code arrangement in a certain pattern for DNA is not just the information itself, but might also propagate other information. It can be said that the codes for DNA is not static information but dynamic one. This could be similar to the components transferring power that is dynamic and of resonance. Moreover, the codes for forming DNA might be in resonant manner, too.

It is worth noting that in protein producing process, a tricodan from an RNA (a transcript of DNA) sequence picks up one type of amino acid to produce protein for actuation. As shown in the universal form of Figure 11.13, *S*, *D*, and *L* form a tricodan to denote a buck-family type of DNA. Similarly, the tricodan, *S*, *D*, and *C* shown in Figure 11.15, is used to denote a boost-family type of DNA. Protein is a kind of actuator, which selected by a tricodan. Analogously, buck-family and boost-family converters are also actuators for power transferring, which are selected by a tricodan formed with *S*, *D*, and *L*, *and S*, *D*, and *C*, respectively.

11.4.2 Mutation

Mutation is one mechanism for evolving new organisms. In a regular DNA, A–T and G–C nucleotide pairs appear in special sequences. However, when there exists mutation, extra nucleotide pairs are inserted into a regular sequence unintentionally or intentionally, resulting in different DNAs. In a regular buck or boost converter, *S*–*D* and *L*–*C* are connected in sequence, as shown in Figure 11.18a and b. However, in Zeta and sepic converters, as shown in Figure 11.18c and d, respectively, the *S*–*D* pair is split into *S* and *D* by inserting an *L*–*C* pair. Although in a DNA sequence the A–T or G–C pair is never split into two separate nucleotides because of covalent bond issues, the *S*–*D* can be split by inserting an *L*–*C* pair intentionally. Different from the mutation of DNA, the mutation in converter

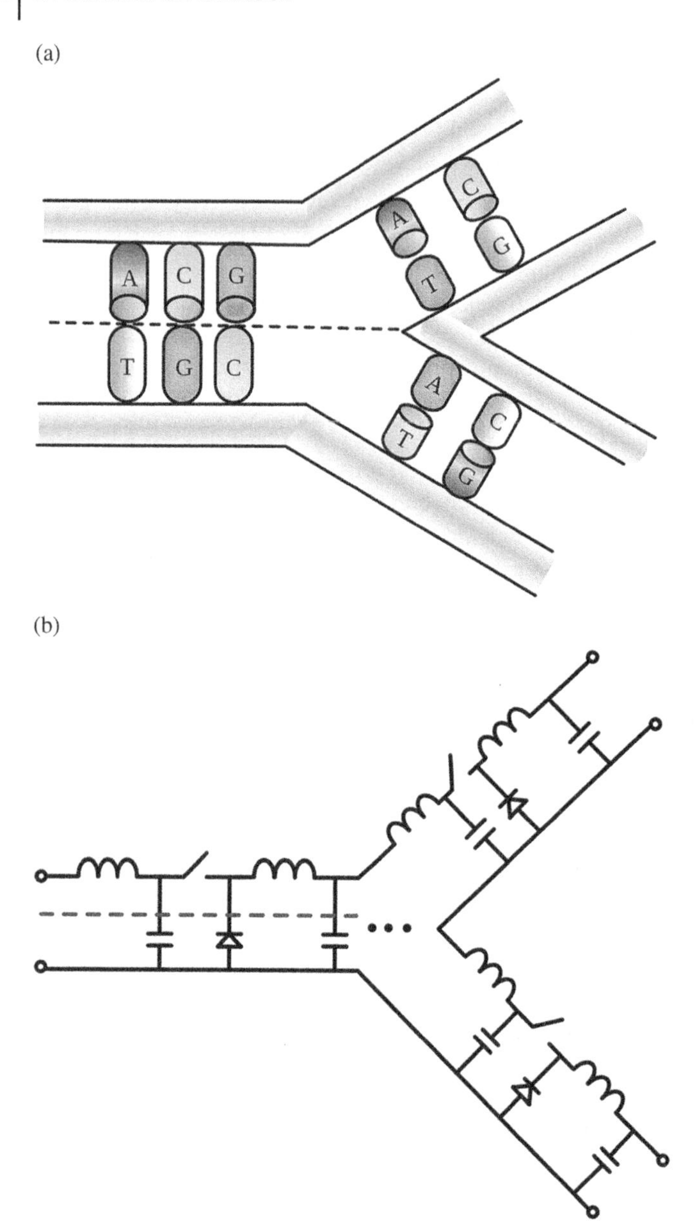

Figure 11.17 Replication of (a) DNA and (b) buck converter.

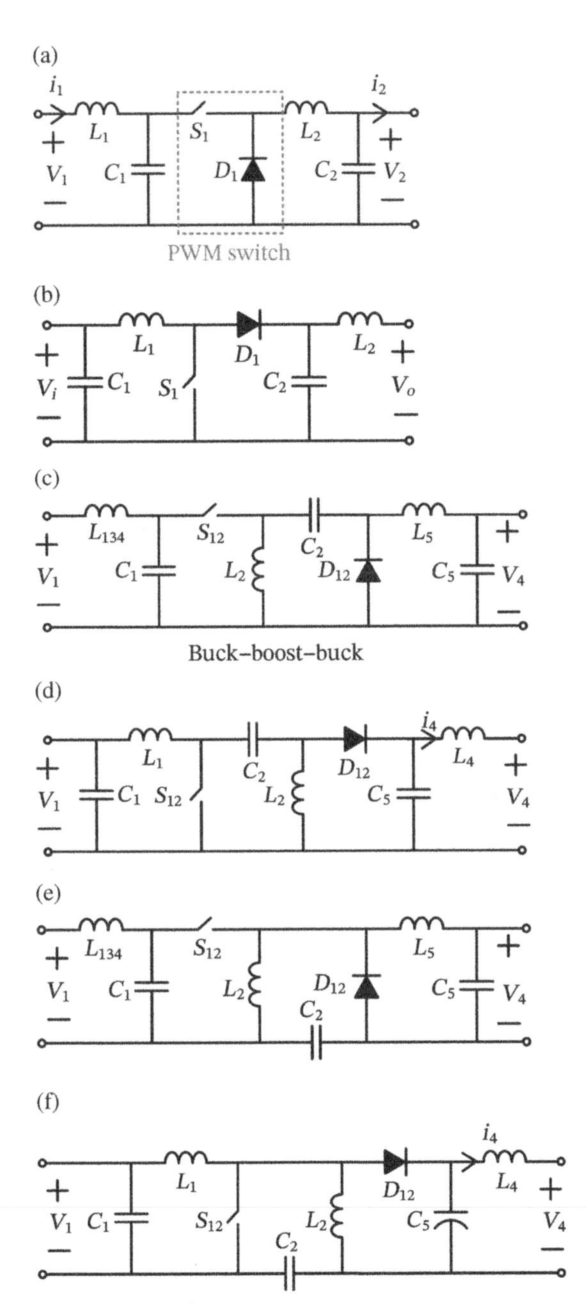

Figure 11.18 Mutation of buck and boost converters to obtain buck–boost–buck and boost–buck–boost converters.

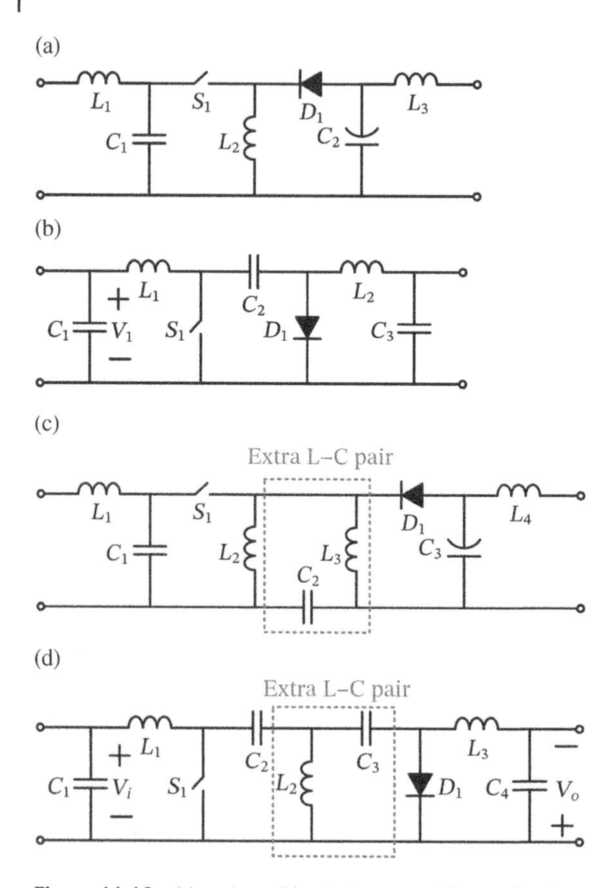

Figure 11.19 Mutation of buck–boost and boost–buck converters to obtain new converters.

evolution is intentional and may be by splitting S&D or L&C. Thus, Zeta and sepic converters can be treated as mutations of buck and boost converters, respectively. It can be seen obviously that the Zeta converter is formed by inserting an L–C pair into S–D pair in the buck converter, so is the sepic converter.

In fact, if we move capacitor C_2 in Figure 11.18c and d from the forward path to the return path, as shown Figure 11.18e and f, there still exists the S–D pair in the two converters. However, different from the buck and boost converters, the L–C pairs in the Zeta and sepic converters look like the internal pairs of the S–D pairs.

For the buck–boost and boost–buck converters shown in Figure 11.19a and b, we can apply the same mutation mechanism to them, and we have their counterparts shown in Figure 11.19c and d, respectively. In Figure 11.19c, the L_3–C_2

pair is inserted between S_1–D_1 pair, while in Figure 11.19d, the L_2–C_3 pair is inserted between the S_1–D_1 pair. In the buck–boost converter shown in Figure 11.19c, there exists an extra L–C pair, and the DC offset voltage across capacitor C_2 is 0V. Thus, the L_3C_2 becomes an extra filter. However, in the boost–buck converter with an extra L–C pair shown in Figure 11.19d, the DC offset voltages of capacitors C_2 and C_3 are V_i and V_o, respectively. The voltage offset of capacitor C_2 in Figure 11.19b is $(V_i + V_o)$, which is different from that shown in Figure 11.19d. Thus, we can consider that the converter shown in Figure 11.19d may be a new one.

11.5 Conclusions

In principle, all of power converters are employing resonant manner to transfer power, which has the highest conversion efficiency. Based on the resonance principle, we discover the original converter, buck converter, from which power converters can be evolved and deduced. In this book, we have presented the fundamentals of circuit theories, the mechanisms of developing power converters, decoding and synthesizing processes, and determination of component stresses of their switches. Hopefully, there will be no more trial and error in developing power converters.

Since Charles Darwin in 1859 initiated an evolution principle, through around one hundred years and many researchers' study, Gregor J. Mendel developed the laws of inheritance in 1866, Boveri-Sutton developed chromosome theory in 1902, and finally James D. Watson discovered the double-helix structure of DNA in 1953 that significantly affected the following genetic engineering innovations. Analogy of power converters to DNA is just an example because we are talking about converter evolution from the original converter, buck converter. Like Charles Darwin, we initiate an evolution of power converters, and we do expect other researchers can follow this step to go further. This does not conclude the work, but just gets started.

For future study, there are several topics needed to investigate further, such as:

1) Code configurations are not unique to a converter topology. How to select a code configuration that can be used for synthesizing a converter directly without trial and error?
2) Switch-voltage stress can be a criterion in selecting a converter topology for power transferring applications. How to determine the stress more systematically?
3) Converters act as protein. There are many types of mechanisms in constructing protein. How to analogize the converters to protein construction?

Over the past 25-year academic careers, we have followed the special mindsets of doing researches, which can be summarized in the following free style of poem.

跳脫本業窠臼 妙趣橫生[1]
跨越領域鴻溝 海闊天空[2]
馳騁學術疆場 創意無窮[3]
究竟天下道理 萬源歸宗[4]

Notes:

1) Jumping out the trapped areas, we will find a lot fun.
2) Crossing the gap between fields, our mind can soar in the sky freely.
3) Based on this kind of mindsets, we can gallop free in academic fields and have unlimited innovations.
4) After realizing the natural laws, we recognize that all of them just deduce from a simple principle.

Further Reading

Ćuk, S. (1979). General topological properties of switching structures. *Proceedings of the IEEE Power Electronics Specialists Conference*, IEEE, pp. 109–130.

C. Darwin, *The Origin of Species*, The Easton Press (MBI, Inc.), Norwalk, CT, 1993.

Erickson, R.W. (1983). Synthesis of switched-mode converters. *Proceedings of the IEEE Power Electronics Specialists Conference*, IEEE, pp. 9–22.

Lin Wang, Y.-Y., Chang, S.-L., Wu, Y.-E. et al. (1991). Resonance-the missing phenomena in hemodynamics. *Circ. Res.* 69: 246–249.

Schrödinger, E. (1944). *What Is Life?* Cambridge University Press.

Severns, R.P. and Bloom, G.E. (1985). *Modern DC-to-DC Switch Mode Power Converter Circuits*. New York: Van Nonstrand Reinhold Co.

Watson, J.D. and Crick, F.H.C. (1953). Molecular structure of nucleic acids: a structure for deoxyribose nucleic acid. *Nature* 171: 737–738.

Wu, T.-F. (2013). The origin of converters. *Proceedings of the IEEE international future energy electronics Conference*, IEEE, pp. 611–617.

Wu, T.-F. (2016). Decoding and synthesizing transformerless PWM converters. *IEEE Trans. Power Electron.* 30 (9): 6293–6304.

Wu, T.-F. and Yu, T.-H. (1998). Unified approach to developing single-stage power converters. *IEEE Trans. Aerosp. Electron. Syst.* 34 (1): 221–223.

Part II

Modeling and Application

12

Modeling of PWM DC/DC Converters

Investigating the dynamic behavior of a PWM DC/DC converter will help devise a system with the desired performance. This usually comprises deriving a small-signal model for the power stage and designing a suitable controller for it. The state-space averaging method is commonly used for establishing the small-signal models as the interested frequency range is about one decade below its switching one. This method averages the power flow to a converter over a switching period and expresses its dynamics in terms of a set of state and output equations. Switch-mode converters, such as buck, boost, buck-boost, etc., have been widely applied to power processing for several decades. To design suitable controllers to govern the dynamic behavior of power converters, accurate models of the converters must be derived.

There are three popular methods which are described as follows: (i) The state-space averaging method is commonly used for establishing small-signal models of converters, which are represented in mathematical expressions or equivalent circuits. (ii) The averaged circuit models of PWM converters are presented by replacing only the switches with their equivalent averaged circuits or pseudocircuit components. They recognized the existence of a PWM switch in each converter, such as buck, boost, etc., which leads to a simplification of the converter modeling.

This method relies on the identification of a three-terminal nonlinear device, called the PWM switch, which consists of only the active and passive switches in a PWM converter. Once the invariant properties of the PWM switch are determined, an averaged equivalent circuit model for the converter can be derived. Therefore, the DC and small-signal characteristics of a large class of PWM converters can then be obtained by a simple substitution of the PWM switch with its equivalent circuit model. The methodology presented is very similar to the analysis of a linear amplifier circuit where the transistor is

Origin of Power Converters: Decoding, Synthesizing, and Modeling, First Edition.
Tsai-Fu Wu and Yu-Kai Chen.
© 2020 John Wiley & Sons, Inc. Published 2020 by John Wiley & Sons, Inc.

replaced by its equivalent circuit model. (iii) The switching flow-graph method is employing the switching flow graph to model converters systematically. A graphical modeling technique, "Switching Flow Graph", is developed to study the nonlinear dynamic behavior of PWM switching converters. Switching converters are variable structure systems with linear subsystems. Each subsystem can be represented by a flow graph. The switching flow graph is obtained by combining the flow graphs of the subsystems through the use of switching branches. The switching flow-graph model is easy to derive, and it provides a visual representation of a switching converter system. In Part II (Chapters 12–15), modeling and application of power converters based on the basic converter units (BCUs) are explored. The systematic and unified approaches to modeling power converters out of the BCUs are proposed based on the two-port network and feedback theory. All of the deriving processes of the PWM converters are achieved completely in one breath. These converters are, consequently, categorized into buck and boost families. Using the proposed approach, not only can one find a general configuration for converters in a family, but one can yield the same small-signal models as those derived directly from the state-space averaging method.

12.1 Generic Modeling of the Original Converter

The original converter, buck converter, is the stem of many non-isolated and isolated PWM converters, such as half-bridge and full-bridge grid-tied converters, neutral-clamped converters, modular multilevel converters, push-pull converters, forward converter, etc. Their operational principles are basically the same as that of the buck converter, of which when the active switch is turned on, power is delivered to the output directly and when the switch is turned off, its inductor current is freewheeling through the output. Therefore, it is worth starting from the modeling of the original converter.

This chapter presents a novel approach to modeling transformerless PWM DC/DC converters using the layer scheme. The typical PWM converters consist of buck, boost, buck-boost, Ćuk, Zeta, and sepic, as shown in Figure 12.1. These converters can be configured into either buck or boost converter, plus linear devices, inductors and capacitors. Thus, the buck and boost converters are recognized as the two basic converter units (BCUs), and the six PWM converters are classified into buck family and boost family. With this classification, the small-signal models of these converters can be readily derived in terms of h-parameter (for buck family) and g-parameter (for boost family). The obtained models are the same as those derived from the direct state-space averaging method.

Although the proposed method is restricted to modeling transformerless converters, it is considered without loss of generality, because most of converters with isolation have their own corresponding transformerless counterparts. For instance, the forward converter is, in principle, equivalent to the buck, so is the flyback to the buck-boost converter.

12.2 Series-Shunt and Shunt-Series Pairs

A method for growing a new tree (converter) is called layer scheme, which is by bending a twig down to ground and covering with soil to induce roots. This is similar to feeding back output voltage or current through buffer to the source. In other words, layering a converter tree does mean as feeding back an output signal. Figure 12.2 shows the basic structure of a feedback network. Rather than showing voltages and currents, Figure 12.2 is a signal-flow diagram, where each X represents either a voltage or a current signal. The BCU network has a forward gain A, and the feedback factor is B. The gain of the feedback network can be obtained and expressed as

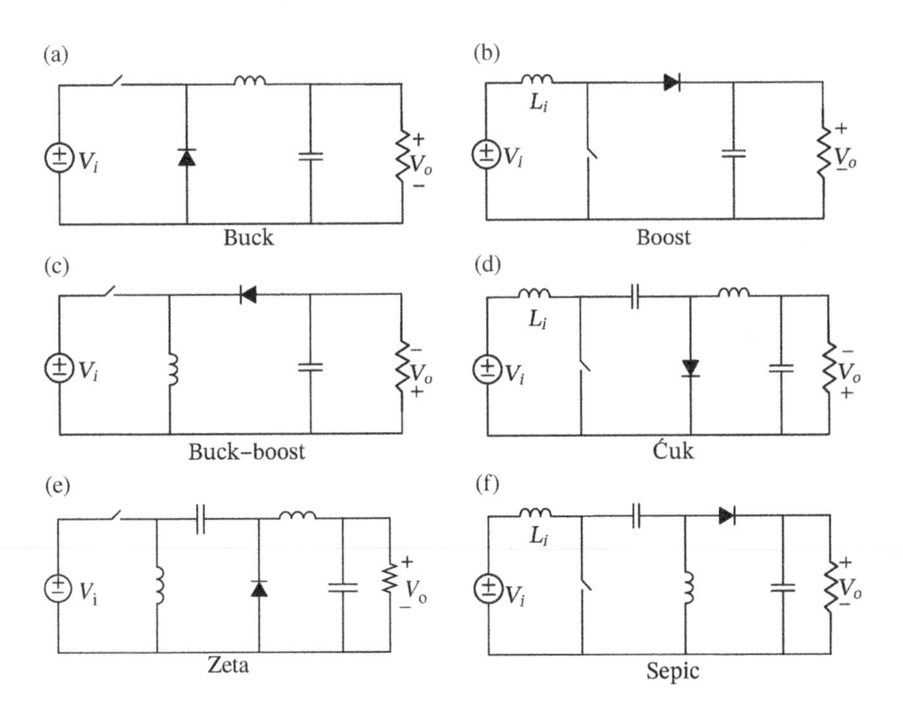

Figure 12.1 The six typical PWM DC/DC converters.

$$A_f = \frac{X_o}{X_s} = \frac{A}{1+AB} \tag{12.1}$$

Based on the quantity to be transferred (voltage or current) and on the desired form of output (voltage or current), the networks can be classified into four categories. The four basic feedback topologies are as follows: (i) voltage-sampling series-mixing (series-shunt) topology, (ii) current-sampling shunt-mixing (shunt-series) topology, (iii) current-sampling series-mixing (series-series) topology, and (iv) voltage-sampling shunt-mixing (shunt-shunt) topology, as shown in Figure 12.3.

PWM DC/DC converters consist of source, switch/buffer, and sink portions, which can be analogous to feeding back output voltage or current through buffer to the source. Figure 12.4a shows a conceptual diagram of the buck BCU from which two possible buck-boost and Zeta with voltage-sampling series-mixing (series-shunt) can be generated, as illustrated in Figure 12.4b and c. Similarly, the same algorithm can be applied to the boost BCU, which leads to the Ćuk and sepic converters, as depicted in Figure 12.4d–f.

Applying the volt-second balance principle to the buck converter, as shown in Figure 12.1a, and operating the converter in the continuous conduction mode (CCM) can yield a steady-state input-to-output voltage transfer ratio of $V_o/V_i = D$ where D is the duty ratio of the controllable switch. For convenience, the buck converter and its input–output relationship are depicted in Figure 12.5a.

Similarly, the input-to-output voltage transfer ratio of the buck–boost converter operating in the CCM can be shown as $V_o/V_i = D/(1-D)$, which also can be derived from that of the buck converter with a unity feedback, as illustrated in Figure 12.5b. The corresponding circuit configuration of this illustration is shown in Figure 12.5c. With a proper rearrangement of the components, the converter shown in Figure 12.5c can be recognized as the buck-boost, as depicted in Figure 12.5d.

Based on the feedback concept, it can be shown that there is another possibility to derive a PWM converter, as drawn in Figure 12.5e. Since at node c the voltage waveform of V_p is pulsating, an inductor filter is required to be inserted at the feedback path for obtaining a smooth voltage V_f with small ripples. The input-to-output voltage transfer ratio of this circuit can be illustrated by the block diagram

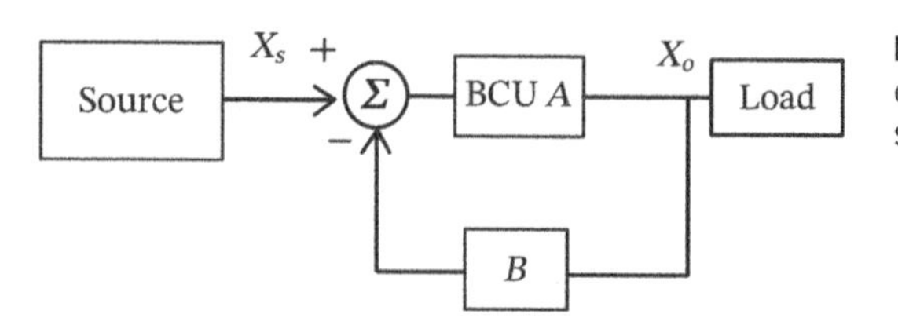

Figure 12.2 General structure of the feedback network with source and load.

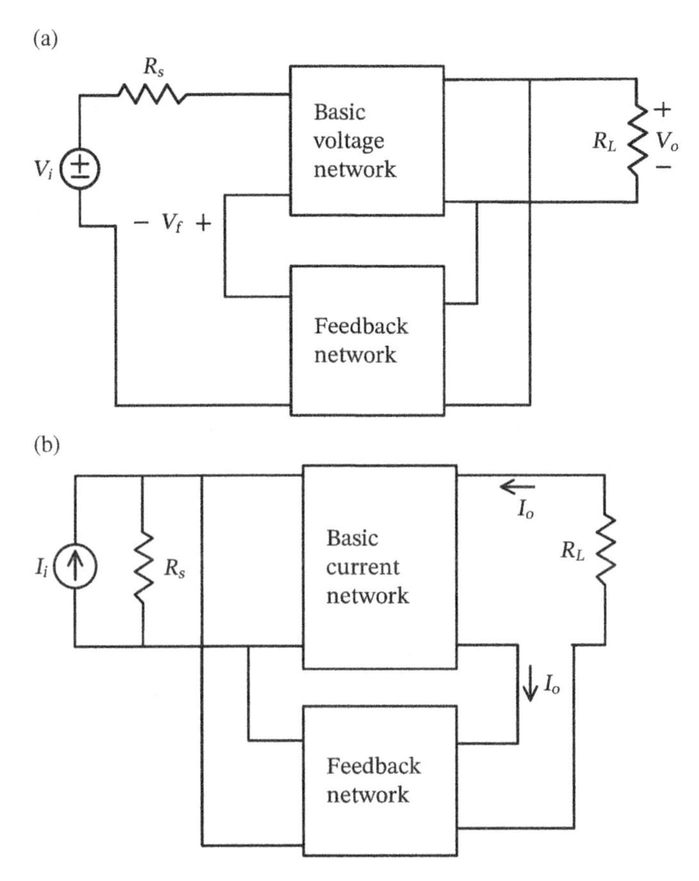

Figure 12.3 The four basic feedback topologies: (a) series-shunt, (b) shunt-series, (c) series-series, and (d) shunt-shunt.

shown in Figure 12.5f, in which D_p denotes the transfer ratio of V_p/V_i conceptually and F_1 and F_2 are the transfer functions of certain low-pass filters. Thus, the V_o/V_i shown in Figure 12.5f can be expressed by

$$\frac{V_o}{V_i} = \frac{D_p F_1}{1 - D_p F_2} \tag{12.2}$$

Observing the buck converter reveals that

$$D_p F_1 = D \tag{12.3}$$

Provided $F_2 = F_1$, the expression in (12.2) becomes

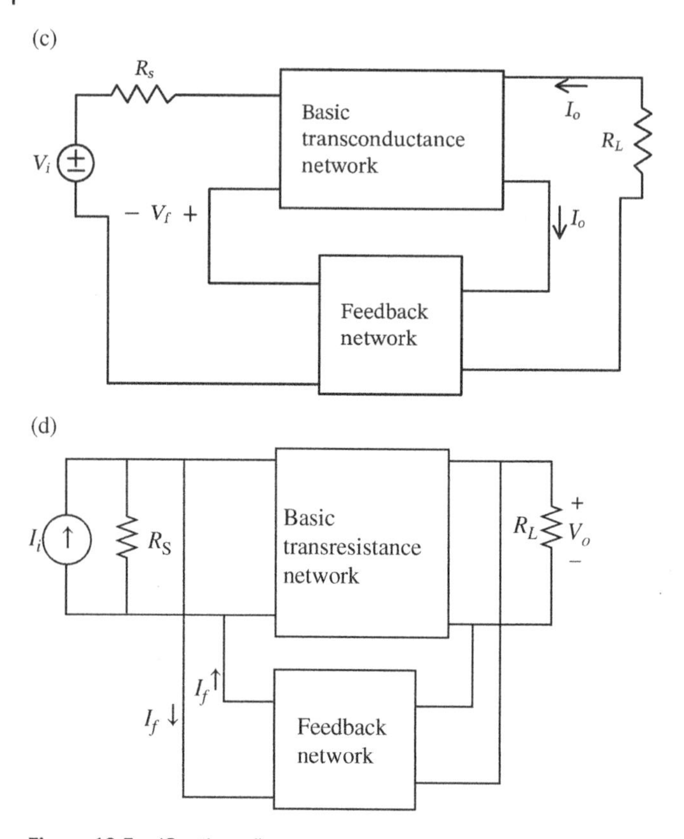

Figure 12.3 (Continued)

$$\frac{V_o}{V_i} = \frac{D}{1-D} \tag{12.4}$$

Above discussion demonstrates that the Zeta converter and its input–output relationship, as depicted in Figure 12.5g, can be derived by rearranging the circuit and expression shown in Figure 12.5e and f, respectively. Buck-boost and Zeta converters can be synthesized from a buck BCU, plus linear devices; thus, they are categorized into the buck family.

Boost, Ćuk, and sepic converters that were used to be derived from the buck, buck-boost, and Zeta converters with the duality theory can be alternatively derived with the layer scheme. The derivation of these converters and their input–output relationships are illustrated in Figure 12.6. Comparing Figures 12.6–12.5 reveals that Ćuk and sepic converters can be derived by replacing the buck BCU with the boost BCU, V_i and V_o with I_i and I_o, respectively, and D with D' $(= 1 - D)$.

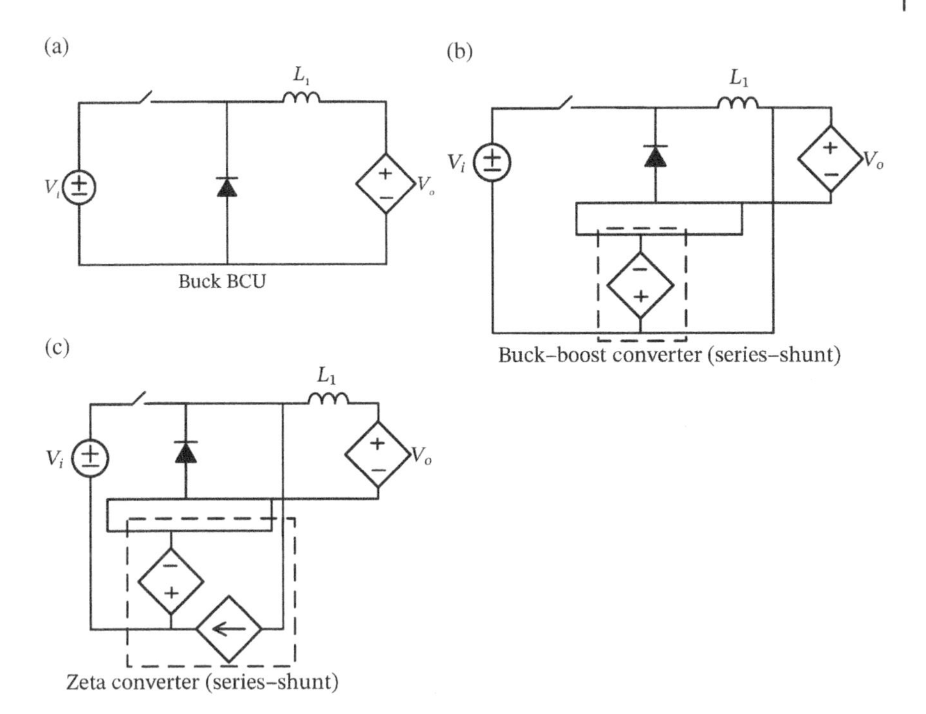

Figure 12.4 Illustration of buck-boost, Zeta, Ćuk, and sepic converters generated from the buck and boost BCUs.

Then, applying the energy conservation principle (assuming the conversion efficiency is 100%), the input-to-output voltage transfer ratio of Ćuk and sepic converters can be determined as

$$\frac{V_o}{V_i} = \frac{D}{1-D} \tag{12.5}$$

In order to conveniently apply the two-port network theory to model the PWM converters, the circuits shown in Figures 12.5c, e and 12.6c, d are modified to the ones depicted in Figure 12.7. Figure 12.7a shows the conventional Zeta converter supplemented with an auxiliary inductor L_X. It can be shown that this circuit does not change the input-to-output transfer ratio of Zeta converter at the steady state because inductor L_X is a reactive device and acts as a unity-gain low-pass filter. Additionally, inductors L_X, L_1, and L_f share a common node N_L exclusively; thus, the order of the circuit is the same as that without the L_X (i.e., $L_X = 0$). This discussion can be analogously applied to the circuit shown in Figure 12.7b, in which capacitors C_x, C_1, and C_f form a loop.

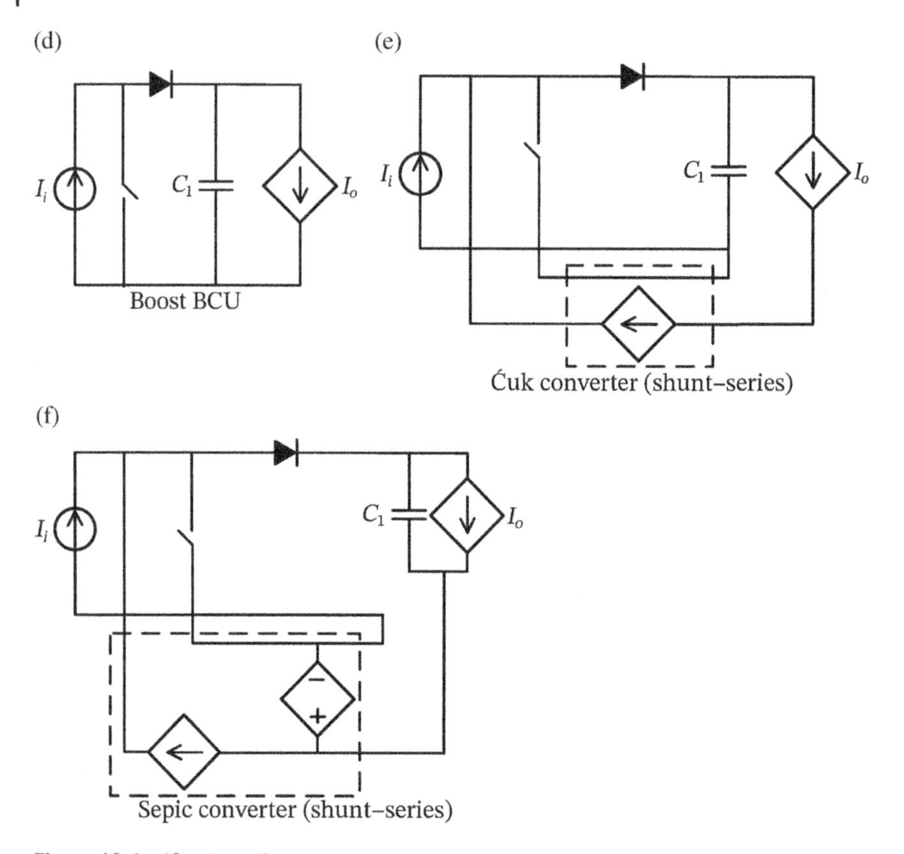

Figure 12.4 (Continued)

Examining the circuit depicted in Figure 12.7a again can reveal that as $L_x = 0$, the circuit is degenerated to Zeta converter; as $L_1 = 0$ and $L_f = 0$, the circuit becomes the buck–boost converter; and when $L_1 = 0$ and taking away the feedback network (C_f and L_f), only is the buck BCU left. Thus, the circuit in Figure 12.7a can be recognized as a general configuration of the converters in the buck family. Likewise, the circuit shown in Figure 12.7b is a general configuration of the converters in the boost family.

12.3 Two-Port Network

The two-port network theory is concerned with the relationships among the voltages and currents at the two ports of a network. Figure 12.8 shows the reference polarities of the terminal voltages and the reference directions of the terminal currents.

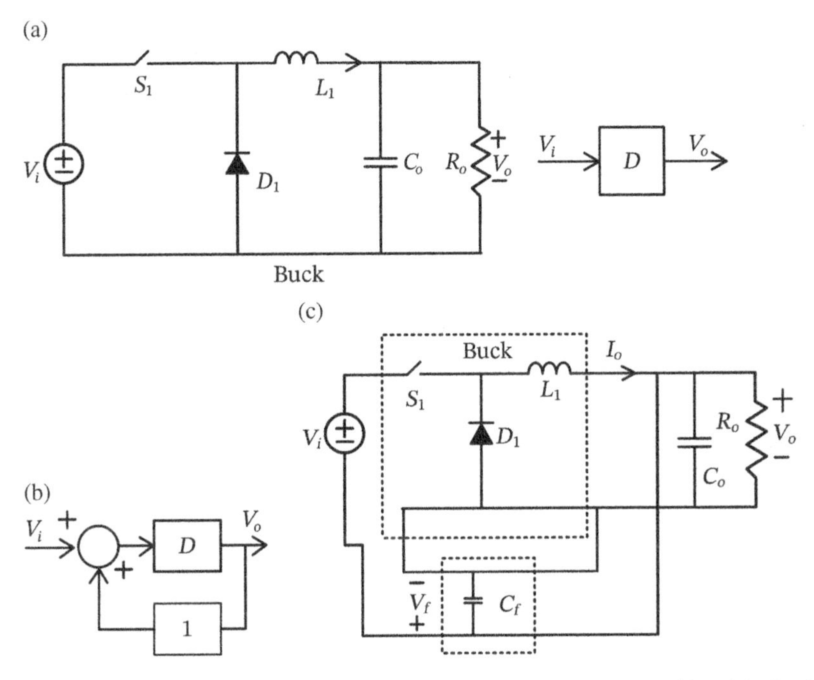

Figure 12.5 Illustration of the buck-boost and Zeta converters derived from the buck BCU.

There are six different ways to combine the four variables, I_1, V_1, I_2, and V_2, and they are represented in matrix forms, as follows:

$$\begin{cases} V_1 = z_{11}I_1 + z_{12}I_2 \\ V_2 = z_{21}I_1 + z_{22}I_2 \end{cases} \Rightarrow \begin{bmatrix} V_1 \\ V_2 \end{bmatrix} = \begin{bmatrix} z_{11} & z_{12} \\ z_{21} & z_{22} \end{bmatrix} \begin{bmatrix} I_1 \\ I_2 \end{bmatrix} \tag{12.6}$$

$$\begin{cases} I_1 = y_{11}V_1 + y_{12}V_2 \\ I_2 = y_{21}V_1 + y_{22}V_2 \end{cases} \Rightarrow \begin{bmatrix} I_1 \\ I_2 \end{bmatrix} = \begin{bmatrix} y_{11} & y_{12} \\ y_{21} & y_{22} \end{bmatrix} \begin{bmatrix} V_1 \\ V_2 \end{bmatrix} \tag{12.7}$$

$$\begin{cases} V_1 = t_{11}V_2 + t_{12}I_2 \\ I_1 = t_{21}V_2 + t_{22}I_2 \end{cases} \Rightarrow \begin{bmatrix} V_1 \\ I_1 \end{bmatrix} = \begin{bmatrix} t_{11} & t_{12} \\ t_{21} & t_{22} \end{bmatrix} \begin{bmatrix} V_2 \\ I_2 \end{bmatrix} \tag{12.8}$$

$$\begin{cases} V_2 = b_{11}V_1 + b_{12}I_1 \\ I_2 = b_{21}V_1 + b_{22}I_1 \end{cases} \Rightarrow \begin{bmatrix} V_2 \\ I_2 \end{bmatrix} = \begin{bmatrix} b_{11} & b_{12} \\ b_{21} & b_{22} \end{bmatrix} \begin{bmatrix} V_1 \\ I_1 \end{bmatrix} \tag{12.9}$$

$$\begin{cases} V_1 = h_{11}I_1 + h_{12}V_2 \\ I_2 = h_{21}I_1 + h_{22}V_2 \end{cases} \Rightarrow \begin{bmatrix} V_1 \\ I_2 \end{bmatrix} = \begin{bmatrix} h_{11} & h_{12} \\ h_{21} & h_{22} \end{bmatrix} \begin{bmatrix} I_1 \\ V_2 \end{bmatrix} \tag{12.10}$$

$$\begin{cases} I_1 = g_{11}V_1 + g_{12}I_2 \\ V_2 = g_{21}V_1 + g_{22}I_2 \end{cases} \Rightarrow \begin{bmatrix} I_1 \\ V_2 \end{bmatrix} = \begin{bmatrix} g_{11} & g_{12} \\ g_{21} & g_{22} \end{bmatrix} \begin{bmatrix} V_1 \\ I_2 \end{bmatrix} \tag{12.11}$$

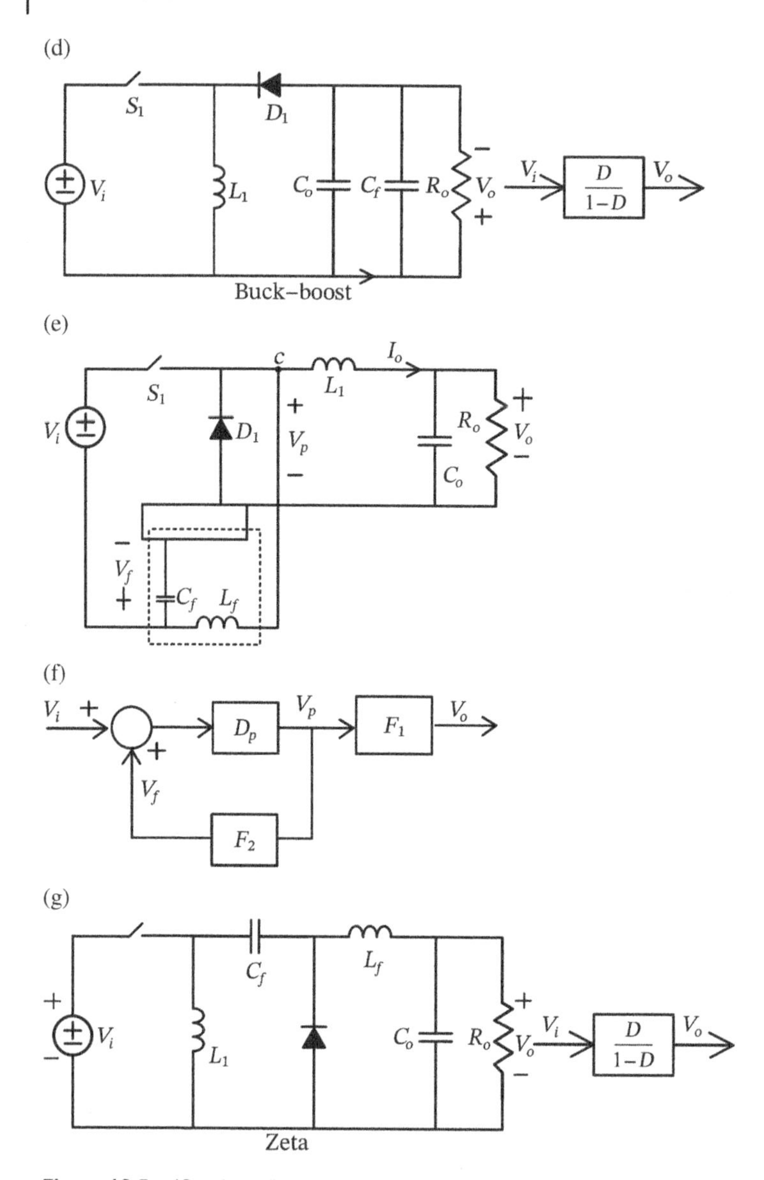

Figure 12.5 (Continued)

The two-port circuits used in the layer scheme are interconnected in two ways: (i) series-shunt and (ii) shunt-series. Figure 12.9 shows the two types of two-port network interconnections with h-parameter and g-parameter for series-shunt and shunt-series connections, respectively.

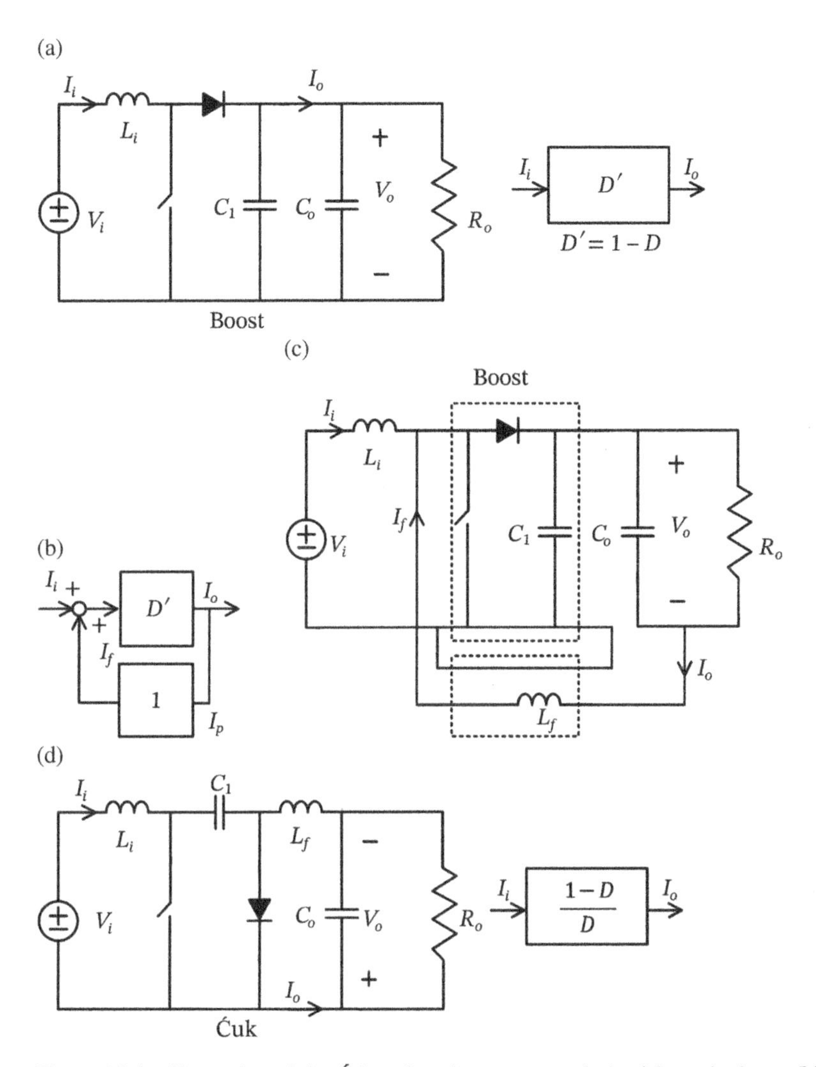

Figure 12.6 Illustration of the Ćuk and sepic converters derived from the boost BCU.

The relationship between port voltages and port currents of the subnetworks shown in Figure 12.9a can be expressed in *h*-parameters.

$$\begin{bmatrix} V_1 \\ I_2 \end{bmatrix} = \begin{bmatrix} H_A + H_B \end{bmatrix} \begin{bmatrix} I_1 \\ V_2 \end{bmatrix} \tag{12.12}$$

For Figure 12.9b, the two-port network with shunt-series feedback can be expressed in *g*-parameters as follows:

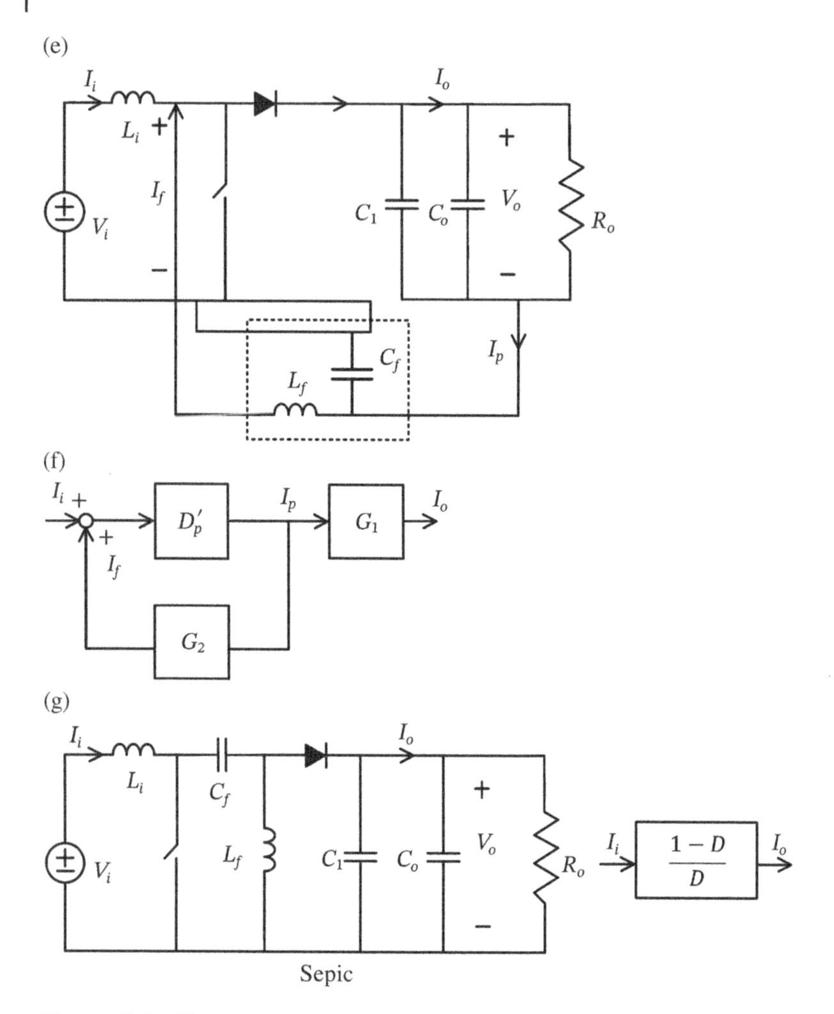

Figure 12.6 (Continued)

$$\begin{bmatrix} I_1 \\ V_2 \end{bmatrix} = \begin{bmatrix} G_A + G_B \end{bmatrix} \begin{bmatrix} V_1 \\ I_2 \end{bmatrix} \tag{12.13}$$

Based on the above discussions, the general converter configurations shown in Figure 12.7 can be conceptually represented by the functional block diagrams, as shown in Figure 12.10, where the source is either a voltage or a current source, the BCU is either a buck or a boost converter, the load is either inductive or capacitive, and the feedback network constitutes *L*–*C* devices. As long as these two configurations are obtained, the work left in modeling the PWM converters is to derive the

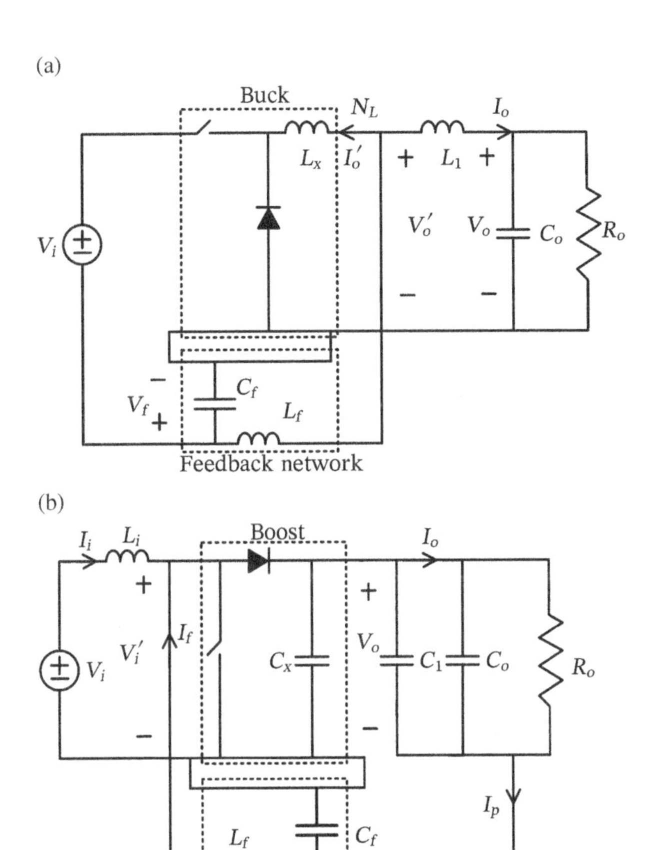

Figure 12.7 General converter forms of (a) the buck family and (b) the boost family.

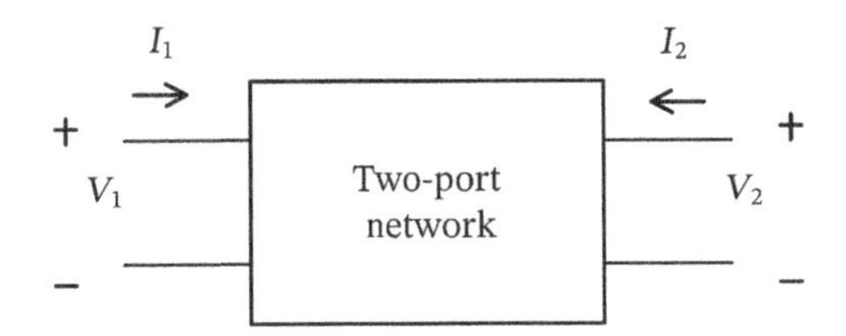

Figure 12.8 The two-port network.

state-space averaged models for the buck and boost converters, plus certain network manipulation. It has been shown that for the buck family with the configuration shown in Figure 12.10a, h-parameter is most suitable, while for the boost family with the configuration depicted in Figure 12.10b, g-parameter is most suitable.

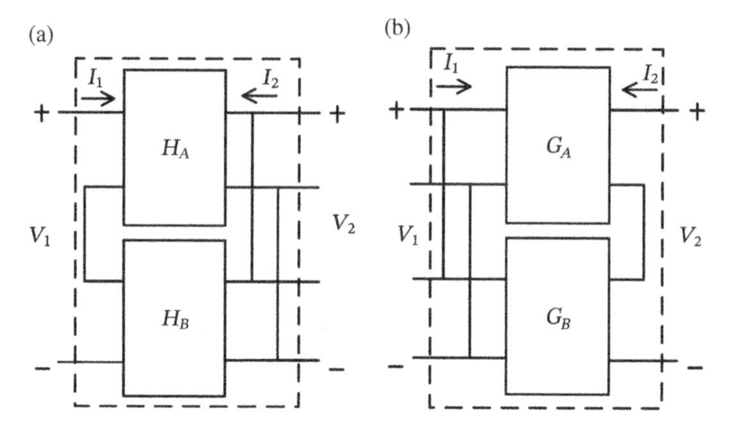

Figure 12.9 The two types of interconnections of feedback two-port networks: (a) series-shunt and (b) shunt-series.

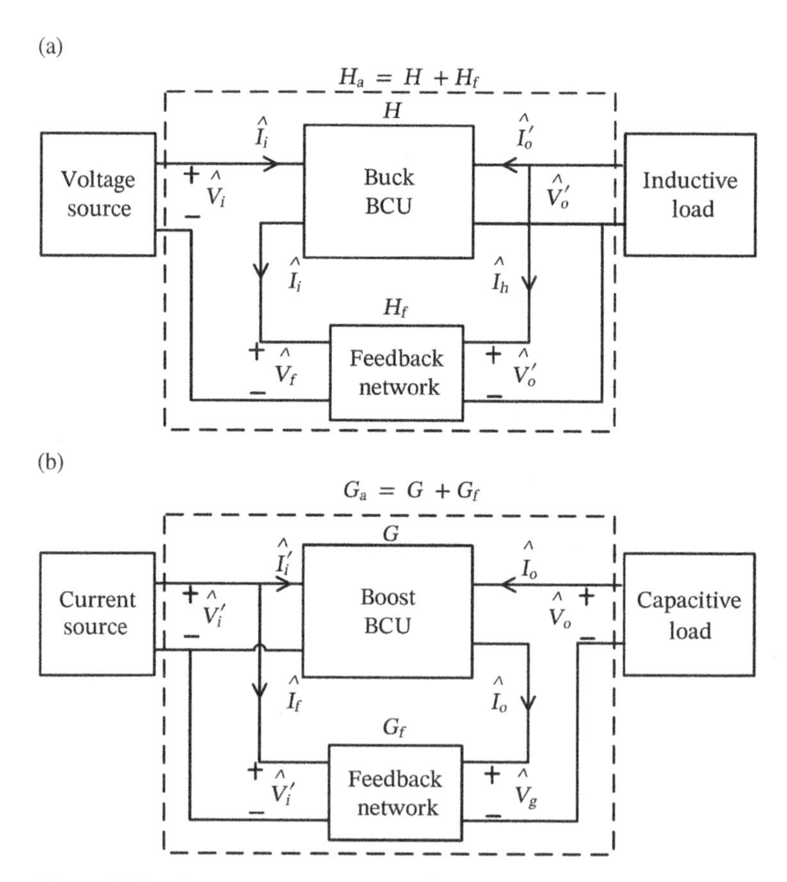

Figure 12.10 Two-port networks with feedback connection for (a) the buck family and (b) the boost family.

12.4 Small-Signal Modeling of the Converters Based on Layer Scheme

Modeling of the buck family and boost family includes two steps. One is to determine the small-signal models of the buck BCU and boost BCU in h- and g-parameters, the other is to derive the small-signal transfer characteristics (SSTCs) of the converters in terms of h- and g-parameters, respectively. The procedure of deriving the small-signal models for the converters is outlined as follows:

1) Determine the h-parameters or g-parameters for the BCUs and feedback networks and sum them up to yield the results for the converters. The h- and g-parameters of the BCUs can be derived from the y-parameter of the BCU, which are illustrated in Figure 12.11. For buck BCU, the following equations can be obtained from the network shown in Figure 12.11a:

$$\hat{V}_i = h_{11}\hat{I}_i + h_{12}\hat{V}_o' + h_{13}\hat{D} \tag{12.14}$$

and

$$\hat{I}_o' = h_{21}\hat{I}_i + h_{22}\hat{V}_o' + h_{23}\hat{D} \tag{12.15}$$

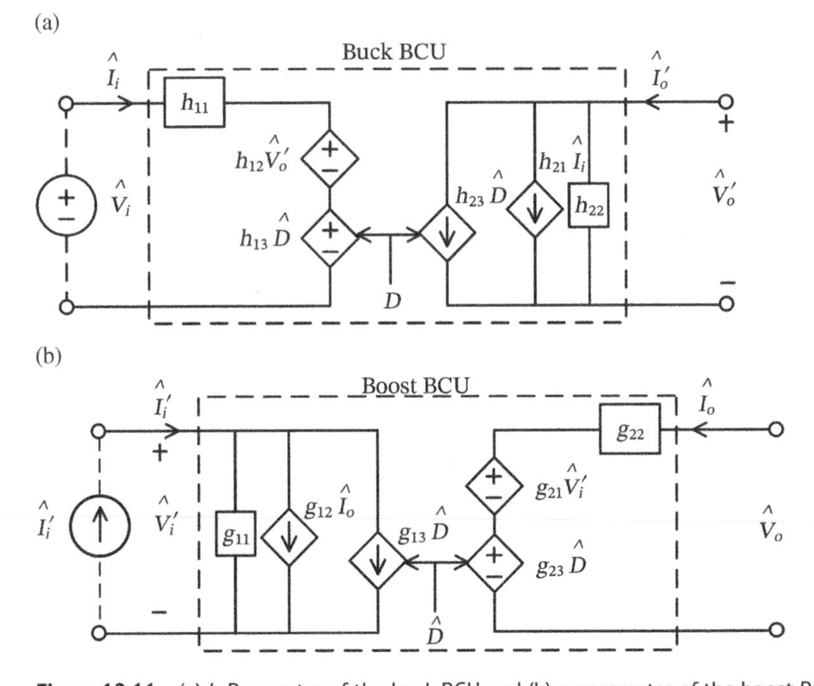

Figure 12.11 (a) h-Parameter of the buck BCU and (b) g-parameter of the boost BCU.

or

$$\begin{bmatrix} \hat{V}_i \\ \hat{I}'_o \end{bmatrix} = H \begin{bmatrix} \hat{I}_i \\ \hat{V}'_o \\ \hat{D} \end{bmatrix}, \tag{12.16}$$

where

$$H = \begin{bmatrix} h_{11} & h_{12} & h_{13} \\ h_{21} & h_{22} & h_{23} \end{bmatrix}$$

Referring to Figure 12.7a, H can be represented by

$$H = \begin{bmatrix} \dfrac{sL_x}{D^2} & \dfrac{1}{D} & \dfrac{sL_x I_o - V_o}{D^2} \\ -\dfrac{1}{D} & 0 & -\dfrac{I_o}{D} \end{bmatrix}, \tag{12.17}$$

where I_o and V_o denote the steady-state output current and voltage, respectively.

For boost BCU, the following equations can be written from the network shown in Figure 12.11b.

$$\hat{I}'_i = g_{11}\hat{V}'_i + g_{12}\hat{I}_o + g_{13}\hat{D} \tag{12.18}$$

and

$$\hat{V}_o = g_{21}\hat{V}'_i + g_{22}\hat{I}_o + g_{23}\hat{D} \tag{12.19}$$

or

$$\begin{bmatrix} \hat{I}'_i \\ \hat{V}_o \end{bmatrix} = G \begin{bmatrix} \hat{V}'_i \\ \hat{I}_o \\ \hat{D} \end{bmatrix}, \tag{12.20}$$

where

$$G = \begin{bmatrix} g_{11} & g_{12} & g_{13} \\ g_{21} & g_{22} & g_{23} \end{bmatrix}.$$

Referring to Figure 12.7b, G can be expressed by

$$G = \begin{bmatrix} \dfrac{sC_x}{D'^2} & -\dfrac{1}{D'} & \dfrac{sC_x V_o - D' I_o}{D'^2} \\ \dfrac{1}{D'} & 0 & \dfrac{V_o}{D'} \end{bmatrix}, \tag{12.21}$$

where $D' = 1 - D$, and $I_o(= -V_o/R_o)$ and V_o denote the steady-state output current and voltage, respectively. In the above equations, the variable with "^" denotes its ac component.

Similarly, the h- and g-parameters of the feedback networks shown in Figure 12.10a and b, respectively, can be derived as

$$\begin{bmatrix} \hat{V}_f \\ \hat{I}_h \end{bmatrix} = H_f \begin{bmatrix} \hat{I}_i \\ \hat{V}_o' \\ \hat{D} \end{bmatrix} \quad (12.22)$$

and

$$\begin{bmatrix} \hat{I}_f \\ \hat{V}_g \end{bmatrix} = G_f \begin{bmatrix} \hat{V}_i' \\ \hat{I}_o \\ \hat{D} \end{bmatrix}, \quad (12.23)$$

where

$$H_f = \begin{bmatrix} h_{f11} & h_{f12} & h_{f13} \\ h_{f21} & h_{f22} & h_{f23} \end{bmatrix}$$

and

$$G_f = \begin{bmatrix} g_{f11} & g_{f12} & g_{f13} \\ g_{f21} & g_{f22} & g_{f23} \end{bmatrix}.$$

Referring to Figure 12.7a and b, H_f and G_f can be derived as

$$H_f = \frac{1}{\Delta H_f(s)} \begin{bmatrix} sL_f & -1 & 0 \\ 1 & sC_f & 0 \end{bmatrix} \quad (12.24)$$

and

$$G_f = \frac{1}{\Delta G_f(s)} \begin{bmatrix} sC_f & 1 & 0 \\ -1 & sL_f & 0 \end{bmatrix}, \quad (12.25)$$

where

$$\Delta H_f(s) = s^2 L_f C_f + 1$$

and

$$\Delta G_f(s) = s^2 L_f C_f + 1.$$

The combined h- or g-parameter of each configuration shown in Figure 12.10 can, therefore, be obtained as follows:

$$H_a = H + H_f = \begin{bmatrix} h_{a11} & h_{a12} & h_{a13} \\ h_{a21} & h_{a22} & h_{a23} \end{bmatrix} \quad (12.26)$$

and

$$G_a = G + G_f = \begin{bmatrix} g_{a11} & g_{a12} & g_{a13} \\ g_{a21} & g_{a22} & g_{a23} \end{bmatrix}, \tag{12.27}$$

where H_a and G_a are the parameters for buck family and boost family, respectively, and

$$\begin{bmatrix} h_{a11} & h_{a12} & h_{a13} \\ h_{a21} & h_{a22} & h_{a23} \end{bmatrix} = \begin{bmatrix} h_{11}+h_{f11} & h_{12}+h_{f12} & h_{13}+h_{f13} \\ h_{21}+h_{f21} & h_{22}+h_{f22} & h_{23}+h_{f23} \end{bmatrix}$$

and

$$\begin{bmatrix} g_{a11} & g_{a12} & g_{a13} \\ g_{a21} & g_{a22} & g_{a23} \end{bmatrix} = \begin{bmatrix} g_{11}+g_{f11} & g_{12}+g_{f12} & g_{13}+g_{f13} \\ g_{21}+g_{f21} & g_{22}+g_{f22} & g_{23}+g_{f23} \end{bmatrix}.$$

The H_a and G_a are dash-blocked in Figure 12.10.

2) Derive the SSTCs of the converters in terms of h- or g-parameter, which include audio susceptibility (A_u), control to output transfer function (F_d), input impedance (Z_i), and output impedance (Z_o). The SSTCs A_u, F_d, Z_i, and Z_o of the converters can be derived using two-port network theory. Referring to Figure 12.7a, the SSTCs of the buck family are determined and expressed by

$$\begin{aligned} A_u &= \left.\frac{\hat{V}_o}{\hat{V}_i}\right|_{\hat{D}=0} = \left.\frac{\hat{V}'_o}{\hat{V}_i} \times \frac{\hat{V}_o}{\hat{V}'_o}\right|_{\hat{D}=0} \\ &= \frac{-Z_{Lh}\left(h_{a11}h_{a23} - h_{a13}h_{a21}\right)}{\Delta H_a Z_{Lh} + h_{a11}} \times \frac{R_o}{s^2 L_1 C_o R_o + sL_1 + R_o}, \end{aligned} \tag{12.28}$$

$$\begin{aligned} F_d &= \left.\frac{\hat{V}_o}{\hat{D}}\right|_{\hat{V}_i=0} = \left.\frac{\hat{V}'_o}{\hat{D}} \times \frac{\hat{V}_o}{\hat{V}'_o}\right|_{\hat{V}_i=0} \\ &= \frac{-Z_{Lh} h_{a21}}{\Delta H_a Z_{Lh} + h_{a11}} \times \frac{R_o}{s^2 L_1 C_o R_o + sL_1 + R_o}, \end{aligned} \tag{12.29}$$

$$Z_i = \left.\frac{\hat{V}_i}{\hat{I}_i}\right|_{\hat{D}} = 0 = h_{a11} - \frac{h_{a21}h_{a11}Z_{Lh}}{h_{a22}Z_{Lh} + 1}, \tag{12.30}$$

and

$$\begin{aligned} Z_o &= \left.\frac{\hat{V}'_o}{\hat{I}_o}\right|_{\hat{V}_i=0,\ \hat{D}=0} \\ &= \left.\frac{\hat{V}'_o}{\hat{I}_o} \times \frac{\hat{V}_o}{\hat{V}'_o}\right|_{\hat{V}_i=0,\ \hat{D}=0} \\ &= \frac{-Z_{Lh} h_{a21}}{\Delta H_a Z_{Lh} + h_{a11}}, \end{aligned} \tag{12.31}$$

where

$$\Delta H_a = h_{a11}h_{a22} - h_{a12}h_{a21}$$

and

$$Z_{Lh} = \frac{s^2 L_1 C_o R_o + sL_1 + R_o}{1 + sC_o R_o}.$$

Referring to Figure 12.7b, the SSTCs of the boost family can be analogously derived as

$$\begin{aligned} A_u &= \left.\frac{\hat{V}_o}{\hat{V}_i}\right|_{\hat{D}=0} = \left.\frac{\hat{V}_o}{\hat{V}_i'} \times \frac{\hat{V}_o'}{\hat{V}_i}\right|_{\hat{D}=0} \\ &= \frac{g_{a21} Z_{Lg}}{\left(1 + g_{a11} Z_g\right)\left(g_{a22} + Z_{Lg}\right) - g_{a12} g_{a21} Z_g}, \end{aligned} \tag{12.32}$$

$$F_d = \left.\frac{\hat{V}_o}{\hat{D}}\right|_{\hat{V}_i=0} = \frac{\left[g_{a23} + \left(g_{a11}g_{a23} - g_{a21}g_{a13}\right)Z_g\right]Z_{Lg}}{\left(1 + g_{a11} Z_g\right)\left(g_{a22} + Z_{Lg}\right) - g_{a12} g_{a21} Z_g}, \tag{12.33}$$

$$\begin{aligned} Z_i &= \left.\frac{\hat{V}_i}{\hat{I}_i}\right|_{\hat{D}=0} = \left.\frac{\hat{V}_i'}{\hat{I}_i} \times \frac{\hat{V}_i}{\hat{V}_i'}\right|_{\hat{D}=0} \\ &= Z_g + \frac{g_{a22} + Z_{Lg}}{\Delta G_a + g_{a11} Z_{Lg}}, \end{aligned} \tag{12.34}$$

and

$$\begin{aligned} Z_o &= \left.\frac{\hat{V}_o}{\hat{I}_o}\right|_{\hat{V}_i=0,\ \hat{D}=0} \\ &= \frac{\left(g_{a22} + \Delta G_a Z_g\right) Z_{Lg}}{g_{a22} + g_{a11} Z_g Z_{Lg} + \Delta G_a Z_g}, \end{aligned} \tag{12.35}$$

where

$$Z_g = sL_i,$$

$$\Delta G_a = g_{a11} g_{a22} - g_{a12} g_{a21},$$

and

$$Z_{Lg} = \frac{R_o}{1 + sC_1 R_o + sC_o R_o}.$$

The SSTCs of the buck family and the boost family which are expressed in s-domain are collected in Tables 12.1 and 12.2, respectively. In the buck family,

Table 12.1 SSTCs of the general converter for the buck family.

SSTCs	Buck family
A_u	$\dfrac{-DR_oD'-DR_oC_fL_fs^2}{R_oD'^2+\left(D^2L_f+L_x+L_1D'^2\right)s+R_o\left(L_1L_oD'^2+C_fL_f+C_oD^2L_f+C_fL_x+C_oL_x\right)s^2+\left(C_fL_1L_f+C_fL_1L_x+C_fL_fL_x\right)s^3+C_fC_oR_o\left(L_1L_f+L_1L_x+L_fL_x\right)s^4}$
F_d	$\dfrac{V_oR_oD'^2-V_o\left(D^2L_f+DL_x\right)s+V_oR_oC_fL_fs^2}{DD'\left[R_oD'^2+\left(D^2L_f+L_x+L_1D'^2\right)s+R_o\left(L_1C_oD'^2+C_fL_f+C_oD^2L_f+C_fL_x+C_oL_x\right)s^2+\left(C_fL_1L_f+C_fL_1L_x+C_fL_fL_x\right)s^3+C_fC_oR_o\left(L_1L_f+L_1L_x+L_fL_x\right)s^4\right]}$
Z_i	$\dfrac{R_oD'^2+\left(D^2L_f+L_x+L_1D'^2\right)s+R_o\left(L_1C_oD'^2+C_fL_f+C_oD^2L_f+C_fL_x+C_oL_x\right)s^2+\left(C_fL_1L_f+C_fL_1L_x+C_fL_fL_x\right)s^3+C_fC_oR_o\left(L_1L_f+L_1L_x+L_fL_x\right)s^4}{D^2+\left(D^2C_fR_o+D^2C_oR_o\right)s+\left(D^2C_fL_1+D^2C_fL_f\right)s^2+\left(D^2C_fC_oL_1R_o+D^2C_fC_oL_fR_o\right)s^3}$
Z_o	$\dfrac{s\left(DL_f+L_x+C_fL_fL_xs^2\right)\left(R_o+L_1s+C_oL_1R_os^2\right)}{R_oD'^2+\left(D^2L_f+L_x+L_1D'^2\right)s+R_o\left(L_1C_oD'^2+C_fL_f+C_oD^2L_f+C_fL_x+C_oL_x\right)s^2+\left(C_fL_1L_f+C_fL_1L_x+C_fL_fL_x\right)s^3+C_fC_oR_o\left(L_1L_f+L_1L_x+L_fL_x\right)s^4}$

Table 12.2 SSTCs of the general converter for the boost family.

SSTCs	Boost family
A_u	$\dfrac{R_oD'\left(D+L_fC_fs^2\right)}{R_oD'^2+\left(L_fD'^2+LD^2\right)s+\left(LC_xR_o+LC_fR_o+LC_oR_oD^2+LC_1R_oD^2+L_fC_1R_oD'^2+L_fC_oR_oD'^2+L_fC_fR_oD'^2\right)s^2+\left(LL_fC_f+LL_fC_x\right)s^3 +\left(LL_fC_fC_oR_o+LL_fC_fR_o+LL_fC_oC_xR_o+LL_fC_xC_1R_o+LL_fC_fC_1R_o\right)s^4}$
F_d	$\dfrac{\left(-R_oD'^2+DLs-C_fLR_os^2D'^2-C_1LR_os^2D'-C_fL_fR_os^2D'^2+C_fLL_fs^3\right)V_o}{-DD'\left(R_oD'^2+\left(L_fD'^2+LD^2\right)s+\left(LC_xR_o+LC_fR_o+LC_oR_oD^2+LC_1R_oD^2+L_fC_1R_oD'^2+L_fC_0R_oD'^2+L_fC_fR_oD'^2\right)s^2+\left(LL_fC_f+LL_fC_x\right)s^3 +\left(LL_fC_fC_oR_o+LL_fC_fC_xR_o+LL_fC_oC_xR_o+LL_fC_xC_1R_o+LL_fC_fC_1R_o\right)s^4\right)}$
Z_i	$\dfrac{R_oD'^2+\left(L_fD'^2+LD^2\right)s+\left(LC_xR_o+LC_fR_o+LC_oR_oD^2+LC_1R_oD^2+L_fC_1R_oD'^2+L_fC_oR_oD'^2+L_fC_fR_oD'^2\right)s^2+\left(LL_fC_f+LL_fC_x\right)s^3 +\left(LL_fC_fC_oR_o+LL_fC_fC_xR_o+LL_fC_oC_xR_o+LL_fC_xC_1R_o+LL_fC_fC_1R_o\right)s^4}{D^2+\left(C_xR_o+C_fR_oD'^2+C_oR_oD^2+C_1R_oD^2\right)s+\left(L_fC_f+L_fC_x\right)s^2 +\left(L_fC_fC_oR_o+L_fC_fC_xR_o+L_fC_xC_oR_o+L_fC_fC_1R_o+L_fC_xC_1R_o\right)s^3}$
Z_o	$\dfrac{sR_o\left(D^2L+D^2L_f+\left(L_fLC_x+L_fLC_f\right)s^2\right)}{R_oD'^2+\left(L_fD'^2+LD^2\right)s+\left(LC_xR_o+LC_fR_o+LC_oR_oD^2+LC_1R_oD^2+L_fC_1R_oD'^2+L_fC_oR_oD'^2+L_fC_fR_oD'^2\right)s^2+\left(LL_fC_f+LL_fC_x\right)s^3 +\left(LL_fC_fC_oR_o+LL_fC_fC_xR_o+LL_fC_oC_xR_o+LL_fC_xC_1R_o+LL_fC_fC_1R_o\right)s^4}$

Table 12.3 SSTCs of the buck-boost converter.

SSTCs	Buck–boost
A_u	$\dfrac{-DR_oD'}{R_oD'^2+L_xs+R_o\left(C_fL_x+C_oL_x\right)s^2}$
F_d	$\dfrac{\left(R_oD'^2-DL_Xs\right)V_o}{DD'\left[R_oD'^2+L_xs+R_o\left(C_fL_x+C_oL_x\right)s^2\right]}$
Z_i	$\dfrac{R_oD'^2+L_xs+R_o\left(C_fL_x+C_oL_x\right)s^2}{D^2\left(sC_oR_o+sC_fR_o+1\right)}$
Z_o	$\dfrac{sL_XR_o}{R_oD'^2+L_xs+R_o\left(C_fL_x+C_oL_x\right)s^2}$

Table 12.4 SSTCs of the Zeta converter.

SSTCs	Zeta
A_u	$\dfrac{-DR_oD'-DR_oC_fL_fs^2}{R_oD'^2+\left(D^2L_f+L_x\right)s+R_o\left(C_fL_f+C_oD^2L_f+C_fL_x+C_oL_x\right)s^2+C_fL_fL_xs^3+C_fC_oR_oL_fL_xs^4}$
F_d	$\dfrac{V_o\left[R_oD'^2-\left(D^2L_f+DL_X\right)s+R_oC_fL_fs^2\right]}{DD'\left[R_oD'^2+\left(D^2L_f+L_x\right)s+R_o\left(C_fL_f+C_oD^2L_f+C_fL_x+C_oL_x\right)s^2+C_fL_fL_xs^3+C_fC_oR_oL_fL_xs^4\right]}$
Z_i	$\dfrac{R_oD'^2+\left(D^2L_f+L_x\right)s+R_o\left(C_fL_f+C_oD^2L_f+C_fL_x+C_oL_x\right)s^2+C_fL_fL_xs^3+C_fC_oR_oL_fL_xs^4}{D^2+\left(D^2C_fR_o+D^2C_oR_o\right)s+D^2C_fL_fs^2+D^2C_fC_oL_fR_os^3}$
Z_o	$\dfrac{s\left(DL_fR_o+L_XR_o+C_fL_fL_xR_os^2\right)}{R_oD'^2+\left(D^2L_f+L_x\right)s+R_o\left(C_fL_f+C_oD^2L_f+C_fL_x+C_oL_x\right)s^2+C_fL_fL_xs^3+C_fC_oR_oL_fL_xs^4}$

let $L_1 = 0$ and $L_f = 0$, and then, a set of SSTCs of the buck-boost converter can be obtained, while let $L_X = 0$, and then, one can derive the SSTCs of the Zeta converter. These SSTCs are collected in Tables 12.3 and 12.4. It can be verified that the SSTCs derived with the layer scheme are the same as those obtained with direct state-space averaging method. The SSTCs of the rest of the converters can be obtained by following the same discussion. Additionally, the SSTCs of the boost-buck (Ćuk) converter and sepic converter are shown in Tables 12.5 and 12.6, respectively.

Table 12.5 SSTCs of the boost–buck (Ćuk) converter.

SSTCs	Boost–buck
A_u	$\dfrac{R_o D'D}{R_o D'^2 + \left(L_f D'^2 + LD^2\right)s + \left(LC_x R_o + LC_o R_o D^2 + L_f C_o R_o D'^2\right)s^2 + LL_f C_x s^3 + LL_f C_o C_x R_o s^4}$
F_d	$\dfrac{\left(-R_o D'^2 + DLs\right)V_o}{-DD'\left(\begin{array}{l} R_o D'^2 + \left(L_f D'^2 + LD^2\right)s + \left(LC_x R_o + LC_o R_o D^2 + L_f C_1 R_o D'^2 + L_f C_0 R_o D'^2\right)s^2 \\ \quad + LL_f C_x s^3 + LL_f C_o C_x R_o s^4 \end{array}\right)}$
Z_i	$\dfrac{R_o D'^2 + \left(L_f D'^2 + LD^2\right)s + \left(LC_x R_o + LC_o R_o D^2 + L_f C_o R_o D'^2\right)s^2 + LL_f C_x s^3 + LL_f C_o C_x R_o s^4}{D^2 + \left(C_x R_o + C_o R_o D^2\right)s + L_f C_x s^2 + L_f C_x C_o R_o s^3}$
Z_o	$\dfrac{sR_o\left(D^2 L + D'^2 L_f + L_f L C_x s^2\right)}{R_o D'^2 + \left(L_f D'^2 + LD^2\right)s + \left(LC_x R_o + LC_o R_o D^2 + L_f C_o R_o D'^2\right)s^2 + LL_f C_x s^3 + LL_f C_o C_x R_o s^4}$

12.5 Quasi-Resonant Converters

In the previous derivation of small-signal models, the PWM converters are operated in CCM. A small-signal model of the converter operated in the discontinuous conduction mode (DCM) can also be derived from the general converter configurations shown in Figure 12.7, but special attention needs to be paid. When the converters operate in the DCM, not only do the BCUs drop into the DCM but also the feedback networks. Thus, a small-signal model of the converter operated in the DCM can be derived only when one can find equivalent circuits for the feedback networks in addition to finding DCM models for the BCUs.

The feedback networks discussed in Section 12.2 can be replaced with some others than the forms shown in Figure 12.7, which might result in different converter topologies. With the proposed analytical method, the feedback network is represented in a general form, either h-parameter or g-parameter. Therefore, the small-signal models of the BCUs with other new feedback networks can also be derived with the proposed method.

The PWM switches of buck and boost BCUs have been replaced with their corresponding quasi-resonant switches and multi-resonant switches, which leads to a set of zero-current switching (ZCS) and zero-voltage switching (ZVS) quasi-resonant converters (QRC), and ZCS and ZVS multi-resonant converters (MRC). Figure 12.12 shows the QRC-BCUs and the MRC-BCUs. It can be observed that substituting these BCUs for the PWM buck and boost BCUs,

Table 12.6 SSTCs of the sepic converter.

SSTCs	Sepic
A_u	$\frac{R_oD'\left(D+L_fC_fs^2\right)}{R_oD'^2+\left(L_fD'^2+LD^2\right)s+\left(LC_fR_o+LC_oR_oD^2+LC_1R_oD^2+L_fC_1R_oD'^2+L_fC_oR_oD'^2+L_fC_fR_oD'^2\right)s^2+LL_fC_fs^3+\left(LL_fC_fC_oR_o+LL_fC_fC_1R_o\right)s^4}$
F_d	$\frac{\left(-R_oD'^2+DLs-C_fLR_os^2D'^2-C_1LR_os^2D'-C_fL_fR_os^2D'^2+C_fLL_fs^3\right)V_o}{-DD'\left(R_oD'^2+\left(L_fD'^2+LD^2\right)s+\left(LC_fR_o+LC_oR_oD^2+LC_1R_oD^2+L_fC_1R_oD'^2+L_fC_0R_oD'^2+L_fC_fR_oD'^2\right)s^2+LL_fC_fs^3+\left(LL_fC_fC_oR_o+LL_fC_fC_1R_o\right)s^4\right)}$
Z_i	$\frac{R_oD'^2+\left(L_fD'^2+LD^2\right)s+\left(LC_fR_o+LC_oR_oD^2+LC_1R_oD^2+L_fC_1R_oD'^2+L_fC_oR_oD'^2+L_fC_fR_oD'^2\right)s^2+LL_fC_fs^3+\left(LL_fC_fC_oR_o+LL_fC_fC_1R_o\right)s^4}{D^2+\left(C_fR_oD'^2+C_oR_oD^2+C_1R_oD^2\right)s+L_fC_fs^2+\left(L_fC_fC_oR_o+L_fC_fC_1R_o\right)s^3}$
Z_o	$\frac{sR_o\left(D^2L+D'^2L_f+L_fLC_fs^2\right)}{R_oD'^2+\left(L_fD'^2+LD^2\right)s+\left(LC_fR_o+LC_oR_oD^2+LC_1R_oD^2+L_fC_1R_oD'^2+L_fC_oR_oD'^2+L_fC_fR_oD'^2\right)s^2+LL_fC_fs^3+\left(LL_fC_fC_oR_o+LL_fC_fC_1R_o\right)s^4}$

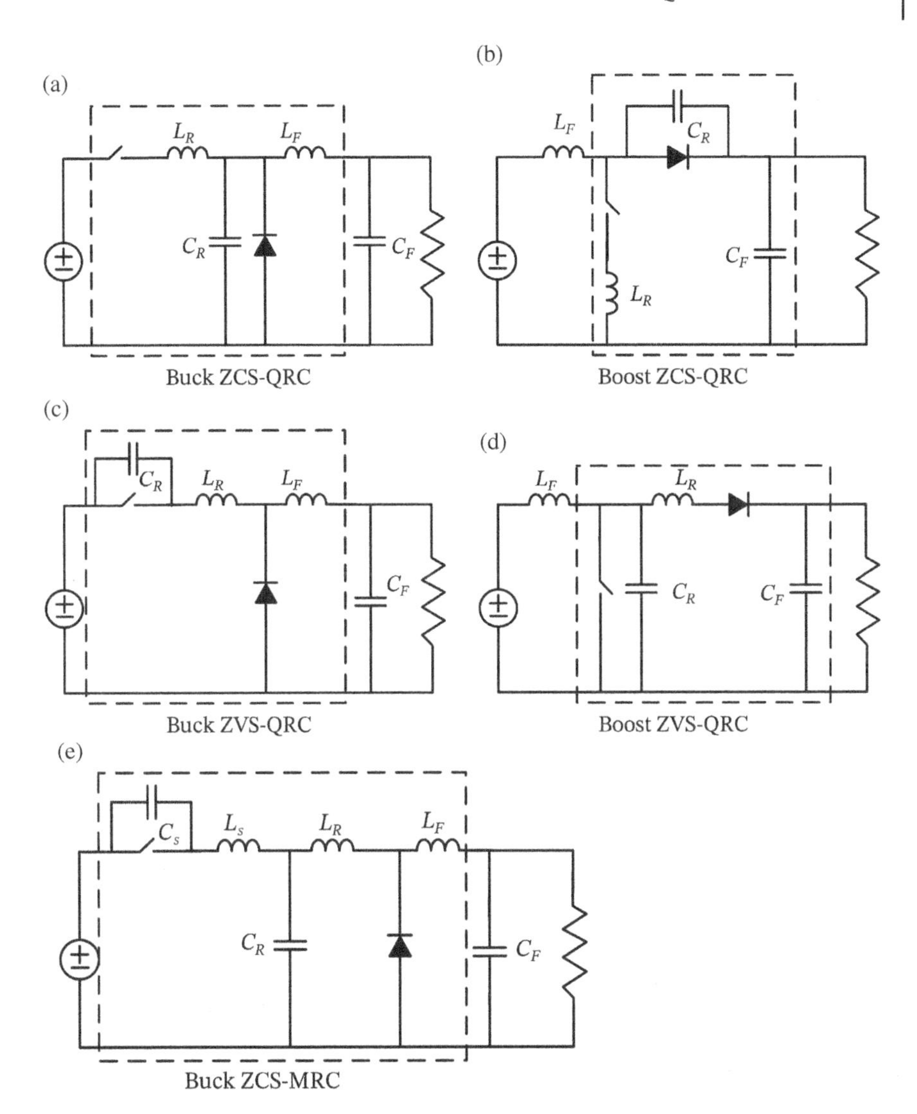

Figure 12.12 Basic converter units of resonant converters: (a) buck ZCS-QRC, (b) boost ZCS-QRC, (c) buck ZVS-QRC, (d) boost ZVS-QRC, (e) buck ZCS-MRC, (f) boost ZCS-MRC, (g) buck ZVS-MRC, and (h) boost ZVS-MRC.

as shown in Figure 12.7, can derive their corresponding QRCs and MRCs. As long as the small-signal models of the QRC-BCUs and MRC-BCUs are obtained, those of the rest resonant converters can, therefore, be derived. Modeling of the QRCs and MRCs is usually much more difficult than that of PWM converters. This work can be alleviated with the proposed layer

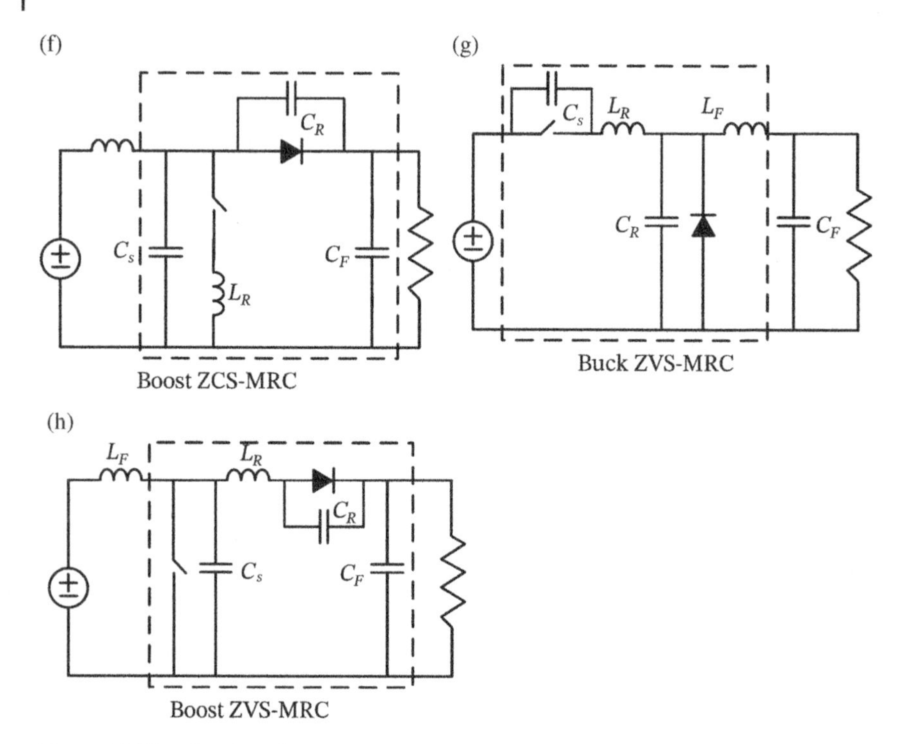

Figure 12.12 (Continued)

approach, and physical insight into the modeling can be identified readily. Modeling of the buck-QRC and boost-QRC can be found from literature. Once they are obtained, the rest of the family members can be derived with the same procedure presented previously.

In practice, there exist parasitic components in converters. Those associated with the switching devices can be included in the converter modeling when using the proposed layer approach. Input filter is indispensable for a converter to suppressing interference. If both the small-signal models of a converter and its cascaded input filter are transformed into t-(transmission) parameters, a model of the overall converter system still can be derived with the help of the layer scheme.

Further Reading

Czarkowski, D. and Kazimierczuk, M.K. (1992). Circuit models of PWM DC–DC converters. *Proceedings of the IEEE National Aerospace and Electronics Conference (NAECON'92)*, Dayton, OH, IEEE (18–22 May 1992), pp. 407–413.

Jovanovic, M.M., Tsang, D.M.C., and Lee, F.C. (1994). Reduction of voltage stress in integrated high-quality rectifier-regulators by variable-frequency control. *Proceedings of the Applied Power Electronics Conference*, IEEE, pp. 569–575.

Lee, Y.S. (1985). A systematic and unified approach to modeling switches in switch-mode power supplies. *IEEE Trans. Ind. Electron.* IE-32: 445–448.

Lehman, B. and Bass, R.M. (1996). Switching frequency dependent averaged models for PWM DC–DC converters. *IEEE Trans. Power Electron.* 11 (1): 89–98.

Lin, B.-T. and Lee, Y.-S. (1997). A unified approach to modeling, synthesizing, and analyzing quasi-resonant converters. *IEEE Trans. Power Electron.* 12 (6): 983–992.

Lin, K.H., Oruganti, R., and Lee, F.C. (1985). Resonant switches – topologies and characteristics. *Proceedings of the Power Electronics Specialists Conference*, IEEE, pp. 62–67.

Liu, K.H. and Lee, F.C. (1984). Resonant switches – a unified approach to improved performances of switching converters. *Proceedings of the Telecommunications Energy Conference*, IEEE, pp. 344–351.

Liu, K.-H. and Lee, F.C. (1990). Zero-voltage switching technique in DC/DC converter. *IEEE Trans. Power Electron.* 5 (1): 293–304.

Middlebrook, R.D. (1985). Topics in multiple-loop regulators and current-mode programming. *Proceedings of the Power Electronics Specialists Conference*, IEEE, pp. 716–732.

Middlebrook, R.D. and Ćuk, S. (1976). A general unified approach to modeling switching converter power stages. *Proceedings of the IEEE Power Electronics Specialists Conference*, IEEE, pp. 18–34.

Middlebrook, R.D. and Ćuk, S. (1997). A general unified approach to modeling DC-to-DC converters in discontinuous conduction. *Proceedings of the IEEE Power Electronics Specialists Conference*, IEEE, pp. 36–57.

Pietkiewicz, A. and Tollik, D. (1987). Unified topological modeling method of switching DC-DC converters in duty-ratio programmed mode. *IEEE Trans. Power Electron.* PE-2 (3): 218–226.

Redl, R. and Sokal, N.O. (1985). Control, five different types, used with the three basic classes of power converters: small-signal AC and large-signal DC characterization, stability requirements, and implementation of practical circuits. *Proceedings of the Power Electronics Specialists Conference*, IEEE, pp. 771–785.

Smedley, K. and Ćuk, S. (1994). Switching flow-graph nonlinear modeling technique. *IEEE Trans. Power Electron* 9 (4): 405–413.

Vorperian, V. (1990a). Simplified analysis of PWM converters using the PWM switch, part I: continuous conduction mode. *IEEE Trans. Aerosp. Electron. Syst.* AES-26: 497–505.

Vorperian, V. (1990b). Simplified analysis of PWM converters using the PWM switch, part II: discontinuous conduction mode. *IEEE Trans. Aerosp. Electron. Syst.* AES-26: 497–505.

Vorperian, V., Tymerski, R., and Lee, F.C. (1989a). Equivalent circuit models for resonant and PWM switches. *IEEE Trans. Power Electron.* 4 (2): 205–214.

Vorperian, V., Tymerski, R., Liu, K.H., and Lee, F.C. (1989b). Generalized resonant switches part I: topologies. In: *High-Frequency Resonant, Quasi-Resonant and Multi-Resonant Converters* (ed. F.C. Lee), 139–145. Virginia Power Electronics Center.

Witulski, A.F. and Erickson, R.W. (1990). Extension of state-space averaging to resonant switches and beyond. *IEEE Trans. Power Electron.* 5 (1): 98–109.

Xu, J. and Lee, C.Q. (1997). Generalized state-space averaging approach for a class of periodically switched network. *IEEE Trans. Circuits Syst.* 44 (11): 1078–1081.

13

Modeling of PWM DC/DC Converters Using the Graft Scheme

In the previous chapter, the PWM converters and their derived QRCs and MRCs are reconfigured into BCUs and linear devices, resulting in simplified modeling and analysis processes. This chapter presents an alternative approach to modeling and analyzing PWM converters, in which the BCUs still can be identified readily.

Over the past two decades, one of the most important advances in characterizing PWM DC/DC converters has been the development of accurate small-signal models for them. These models are useful in designing suitable controllers to govern the dynamic behavior of the converters. The state-space averaging method is commonly used for establishing the small-signal models that are represented in mathematical expressions or equivalent circuits. When using this approach, converters are processed individually and independently, thus lacking connection among them. One paper has presented a scheme to obtain the averaged circuit models of PWM converters by replacing only the switches with their averaged circuit or pseudo-circuit components. In the paper, the authors recognized the existence of a PWM switch in each converter, such as buck, boost, etc., which leads to simplifying the converter modeling. Nevertheless, the derived models do not show close relationship among the converters.

As described in Part I, the graft scheme can be used to integrate multiple converters to form a single-stage converter (SSC). For instance, the buck-boost, Ćuk, sepic, and Zeta converters can be derived from the two BCUs, buck and boost. This scheme has been shown to be useful in developing SSCs and be capable of establishing relationships among the converters in a family. A systematic method for deriving the small-signal models of the PWM converters using the layer scheme was presented in Chapter 12, while the one based on the graft scheme has not been addressed yet.

This chapter presents a systematic and unified approach to modeling transformerless PWM converters based on the graft scheme. With the graft scheme, BCUs can be integrated to become an SSC. Its small-signal model is derived by properly

Origin of Power Converters: Decoding, Synthesizing, and Modeling, First Edition.
Tsai-Fu Wu and Yu-Kai Chen.
© 2020 John Wiley & Sons, Inc. Published 2020 by John Wiley & Sons, Inc.

combining those of the two BCUs and applying the two-port network theory. Thus, relationships among the dynamic models of the single-stage PWM converters can be identified. Derivation of the small-signal models will be discussed in detail in the following sections.

13.1 Cascade Family

Converters are usually connected in cascade to achieve multiple functions. These can be conceptually illustrated by a two-converter system shown in Figure 13.1, where the block diagram comprises two-converter units, namely, root power stage (RPS) and leaf power stage (LPS). For conveniently describing the graft scheme, both RPS and LPS are restricted to be PWM switch-mode DC/DC converters in this chapter. To obtain the merits of compact size, possible high reliability and simple driver design, converters are grafted to form an SSC. In the following, the principle of the graft scheme is briefly described.

By substituting the grafted switches for S_R and S_L, SSCs can be derived from multiple converters connected in cascade/cascode. This has been demonstrated in Part I with the derivation of the well-known PWM DC/DC converters, buck-boost, boost-buck (Ćuk), boost-buck-boost (sepic) and buck-boost-buck (Zeta), from the two BCUs whose schematic diagrams are shown in Figure 13.2.

For instance, grafting boost on buck yields a buck-boost SSC. The derivation of this SSC is illustrated in Figure 13.3. Figure 13.3a shows its input-to-output voltage transfer block diagram when converters are operated in the CCM, and Figure 13.3b shows the buck and boost in cascade connection, corresponding to each block in Figure 13.3a. By grafting S_R and S_L switches and replaced with the I-IIGS, the buck-boost SSC with input-to-output voltage transfer ratio of $V_o/V_s = D/(1-D)$ is obtained and depicted in Figure 13.3c. The detailed derivation can be found in Chapter 5.

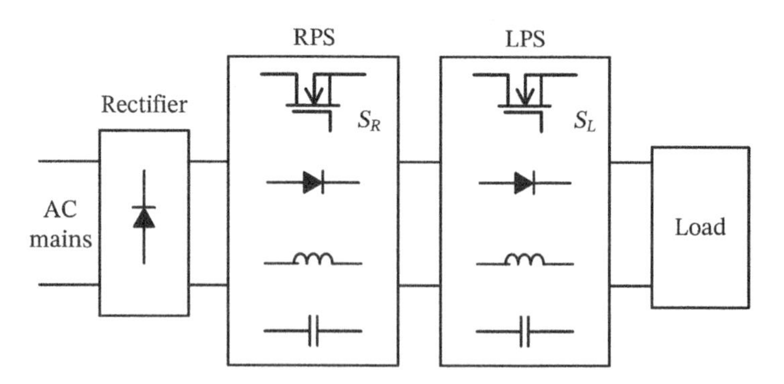

Figure 13.1 Conceptual block diagram of converters in cascade connection.

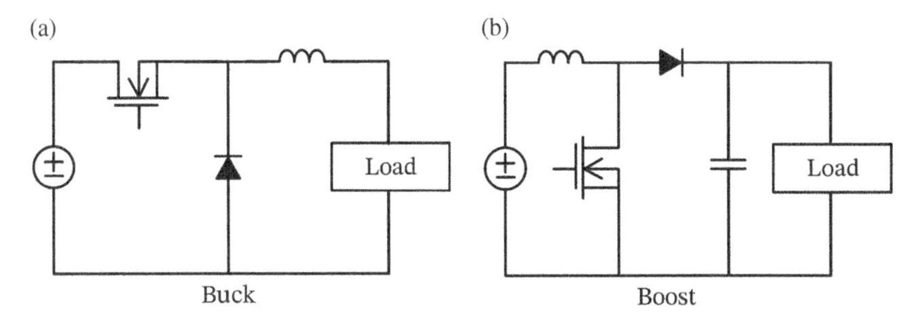

Figure 13.2 Schematic diagrams of the buck and boost converters.

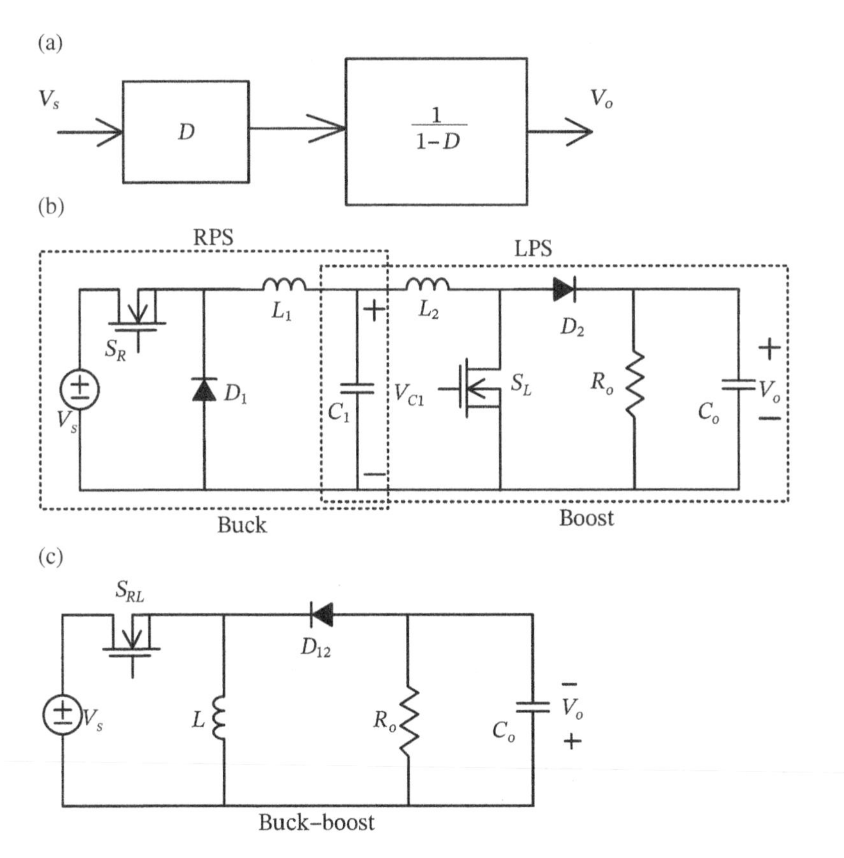

Figure 13.3 Illustration of the buck-boost SSC derived from the buck and boost converters in cascade connection.

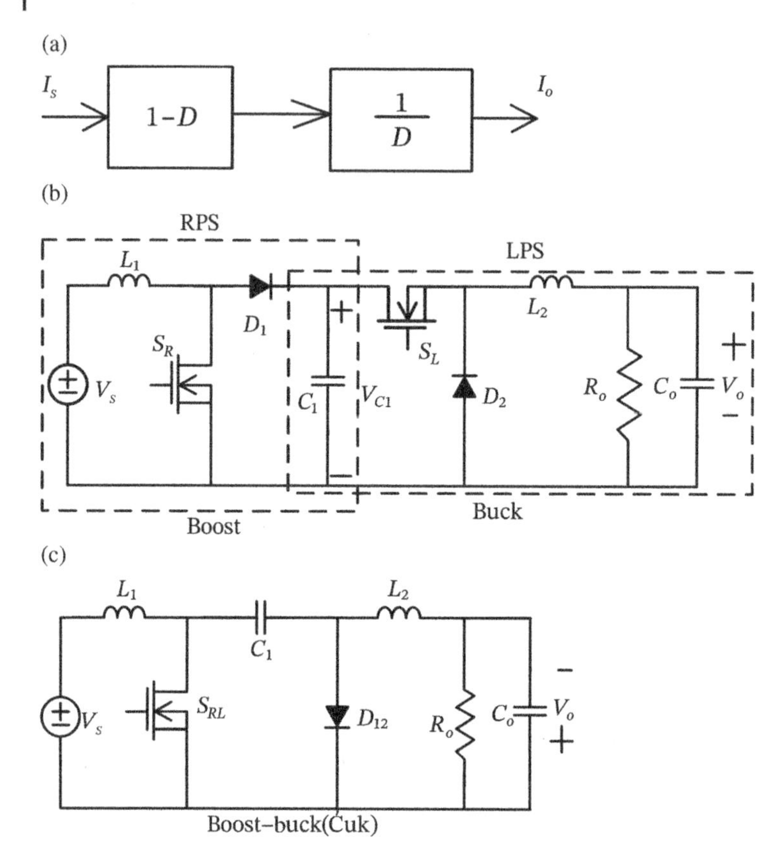

Figure 13.4 Illustration of the boost-buck (Ćuk) SSC derived from the boost and buck converters in cascade connection.

On the other hand, grafting buck on boost will yield a boost-buck SSC (Ćuk converter) with the same transfer ratio of $V_o/V_S = D/(1-\mathrm{D})$ (i.e. $I_o/I_s = (1-\mathrm{D})/D$). The derivation of the SSC is illustrated in Figure 13.4, in which it makes use of the TGS. The configuration of the boost-buck is distinct from that of buck-boost since the roles of RPS and LPS have been exchanged. This procedure can be analogous to that of grafting trees so as the converter-integration technique is named graft scheme.

13.2 Small-Signal Models of Buck-Boost and Ćuk Converters Operated in CCM

As mentioned in Section 13.1, the PWM converters can be generated from the BCUs connected in cascade. Thus, to derive their small-signal models, the converters are most conveniently described by t-(transmission) parameters, which can be readily derived from the y-parameters developed in Chapter 12.

The converters shown in Figures 13.3 and 13.4 are replaced by their corresponding two-port networks, as depicted in Figure 13.5a and b. Note that though the circuit diagrams shown in Figure 13.4 present current source and voltage output, they appear in Figure 13.5b with voltage forms because their output voltage transfer characteristics (i.e. audio susceptibility) are concerned. The procedure of deriving the small-signal models for these SSCs is outlined as follows:

1) Represent the state-space averaging models of the BCUs (buck and boost) in t-parameters.
2) Graft the buck and boost to form a buck-boost or boost-buck SSC. That is, the small-signal models of these converters are derived by multiplying the two t-parameters of buck and boost.
3) Derive the small-signal transfer characteristics of the SSCs in terms of t-parameter. These include audio susceptibility (A_u), control-to-output transfer function (F_d), input impedance (Z_i), and output impedance (Z_o).

From the derivation procedure, the PWM converters can be divided into two categories according to the network topologies, of which buck-boost and boost-buck belong to one category denoted as category 1, while sepic and Zeta are the other one denoted as category 2.

Figure 13.5a and b illustrates the buck-boost and boost-buck converters represented in cascaded two-port networks, respectively. Note that the capacitor buffer is separated from the converters for effectively using the t-parameters. It is

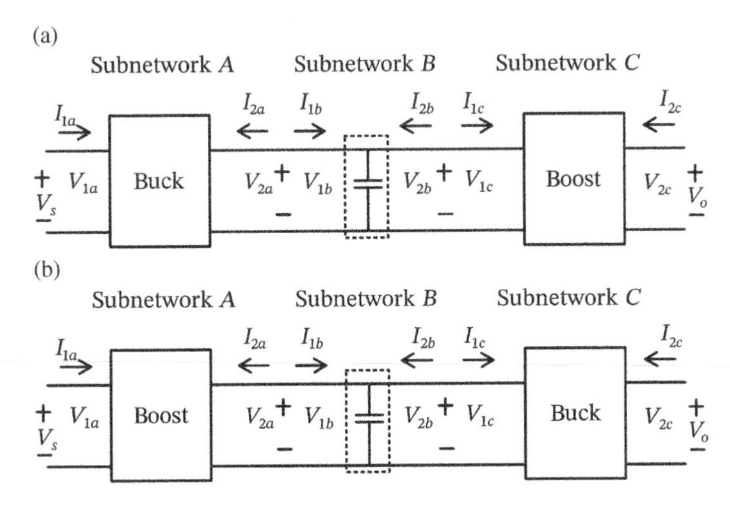

Figure 13.5 Small-signal models of the SSCs represented in cascaded two-port networks: (a) buck-boost and (b) boost-buck (Ćuk).

considered the load of the front stage. The relationships among port voltage, current, and control signal $\hat{D}$ of the subnetworks shown in Figure 13.5a and b can be expressed in t-parameters as follows:

$$\begin{bmatrix} V_{1a} \\ I_{1a} \end{bmatrix} = T_A \times \begin{bmatrix} V_{2a} \\ -I_{2a} \end{bmatrix} + \begin{bmatrix} V_{13} \\ I_{13} \end{bmatrix} \hat{D}, \tag{13.1}$$

$$\begin{bmatrix} V_{1b} \\ I_{1b} \end{bmatrix} = T_B \times \begin{bmatrix} V_{2c} \\ -I_{2c} \end{bmatrix}, \tag{13.2}$$

and

$$\begin{bmatrix} V_{1c} \\ I_{1c} \end{bmatrix} = T_C \times \begin{bmatrix} V_{2c} \\ -I_{2c} \end{bmatrix} + \begin{bmatrix} V'_{13} \\ I'_{13} \end{bmatrix} \hat{D}. \tag{13.3}$$

The y-parameter of a PWM converter is redrawn and shown in Figure 13.6a, and its corresponding t-parameter is illustrated in Figure 13.6b. Since the three subnetworks are connected in cascade, the overall t-parameter of the buck-boost SSC can be represented by

$$\begin{bmatrix} V_{1a} \\ I_{1a} \end{bmatrix} = T_D \times \begin{bmatrix} V_{2c} \\ -I_{2c} \end{bmatrix} + \begin{bmatrix} V_d \\ I_d \end{bmatrix} \hat{D}, \tag{13.4}$$

where

$$T_D = T_A \times T_B \times T_C, \quad V_d = T'_{D11} \times V'_{13} + T'_{D12} \times I'_{13} + V_{13},$$
$$I_d = T'_{D21} \times V'_{13} + T'_{D22} \times I'_{13} + I_{13}, T'_D = T_A \times T_B,$$

and T_{Nij} denotes the element in row i and column j of matrix T_N. The t-parameter of a two-port network is defined as follows:

$$\begin{bmatrix} V_1 \\ I_1 \end{bmatrix} = \begin{bmatrix} T_{11} & T_{12} \\ T_{21} & T_{22} \end{bmatrix} \times \begin{bmatrix} V_2 \\ -I_2 \end{bmatrix}. \tag{13.5}$$

Based on (13.1)–(13.4), one can determine the small-signal transfer characteristics (SSTCs), which are expressed as follows:

$$A_u = \left.\frac{\hat{V}_o}{\hat{V}_s}\right|_{\hat{D}=0} = \frac{Z_L}{T_{D11} \times Z_L + T_{D12}}, \tag{13.6}$$

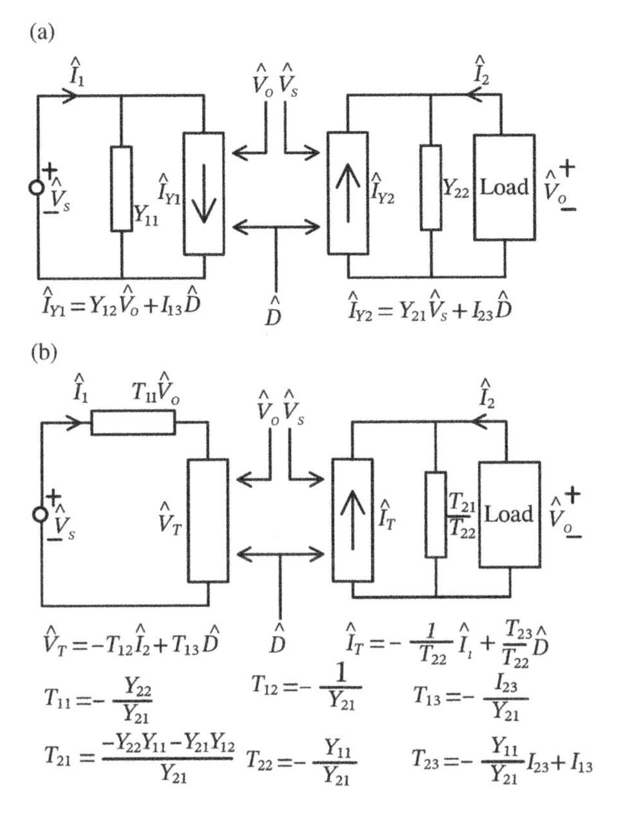

Figure 13.6 (a) *y*-Parameter of a PWM DC/DC converter structure and (b) *t*-parameter of a PWM DC/DC converter derived from *y*-parameter.

$$F_d = \left.\frac{\hat{V}_o}{\hat{D}}\right|_{\hat{V}_s=0} = \frac{V_d \times Z_L}{T_{D11} \times Z_L + T_{D12}}, \tag{13.7}$$

$$Z_i = \left.\frac{\hat{V}_s}{\hat{I}_s}\right|_{\hat{D}=0} = \frac{\left(T_{D12} + T_{D11} \times Z_L\right) \times T_{D12}}{T_{D22} \times \left(T_{D12} + T_{D11} \times Z_L\right) - \Delta T \times Z_L}, \tag{13.8}$$

and

$$Z_o = \left.\frac{\hat{V}_o}{\hat{I}_o}\right|_{\hat{D}=0, \hat{V}_s=0} = \frac{T_{D12} \times Z_L}{T_{D11} \times Z_L + T_{D12}}, \tag{13.9}$$

where

$\Delta T = T_{11}\, T_{22} - T_{12}\, T_{21}$, Z_L is a load connected to the output and the symbols with "^" denote their corresponding small signals. Since the buck-boost and boost-buck SSCs are derived from the two BCUs connected in cascade and operated synchronously, their small-signal models are highly related to those of the BCUs. For instance, the characteristic equation of the control-to-output transfer function (F_d) of the SSCs is equivalent to the product of those of the buck and boost plus the interaction effect. This equation is expressed as follows:

$$T_{A12}C_1T_{C12}C_2R_os^2 + \left(T_{A12}C_1T_{C11}R_o + T_{A12}C_1T_{C12} + T_{A11}T_{C12}C_2R_o\right)s \\ + T_{A11}T_{C11}R_o + T_{A11}T_{C12} + T_{A12}T_{C22}C_oR_os + T_{A12}T_{C22} = 0 \tag{13.10}$$

In (13.10), the last two items represent the interaction effect, which exists because the two BCUs no longer operate independently. These two items are contributed by the T_{D12} shown in the denominator of (13.7).

13.2.1 Buck-Boost Converter

A buck-boost example of deriving the small-signal models for SSCs operated in the CCM is used to illustrate the previously described modeling procedure. When the network shown in Figure 13.5a is loaded with capacitor C_o and resistor R_o ($= Z_L$), the parameter matrices in (13.1)–(13.3) can be determined as

$$T_A = \begin{bmatrix} \frac{1}{D} & \frac{sL_1}{D} \\ 0 & D \end{bmatrix}, \tag{13.11}$$

$$\begin{bmatrix} V_{13} \\ I_{13} \end{bmatrix} = \begin{bmatrix} \frac{-V_{C1}}{D^2} \\ 0 \end{bmatrix}, \tag{13.12}$$

$$T_B = \begin{bmatrix} 1 & 0 \\ sC_1 & 1 \end{bmatrix}, \tag{13.13}$$

$$T_C = \begin{bmatrix} -D' & -\frac{sL_2}{D'} \\ 0 & -\frac{1}{D'} \end{bmatrix}, \tag{13.14}$$

and

$$\begin{bmatrix} V'_{13} \\ I'_{13} \end{bmatrix} = \begin{bmatrix} -\dfrac{\left(sL_2 + R_o D'^2\right)V_o}{R_o D'^2} \\ \dfrac{V_o}{R_o D'^2} \end{bmatrix}, \tag{13.15}$$

where D is the duty ratio of the active switch, D' is equal to $1 - D$, and V_{C1} represents an output voltage to subnetwork A. After substituting (13.11)–(13.15) into (13.4), we obtain

$$\begin{bmatrix} V_{1a} \\ I_{1a} \end{bmatrix} = \begin{bmatrix} \dfrac{D' + s^2 L_1 C_1 D'}{D} & \dfrac{s^3 L_1 L_2 C_1 + sL_1 + sL_2}{DD'} \\ \dfrac{2L_2 DD' + s^2 L_1 L_2 C_1 DD'}{sL_1 L_2} & \dfrac{s^2 L_2 C_1 D + D}{D'} \end{bmatrix} \times \begin{bmatrix} V_{2c} \\ -I_{2c} \end{bmatrix}$$
$$+ \begin{bmatrix} \dfrac{s^3 L_1 L_2 C_1 D V_o - s^2 L_1 C_1 R_o D D'^2 V_o + sL_2 D V_o + sL_1 D V_o - R_o D'^2 D V_o - R_o D'^3 V_o}{R D'^2 D^2} \\ \dfrac{s^3 L_1 L_2 C_1 D - s^2 L_1 C_1 R_o D D'^2 - sL_1 D + 2sL_2 D - 2R_o D D'^2}{sL_1 R_o D'^2} \end{bmatrix} \hat{D}. \tag{13.16}$$

The four transfer characteristics can be consequently derived as

$$A_u = \left.\frac{\hat{V}_o}{\hat{V}_s}\right|_{\hat{D}=0}$$
$$= \frac{-R_o DD'}{s^4 L_1 L_2 C_1 C_o R_o + s^3 L_1 L_2 C_1 + s^2\left(L_1 C_1 R_o D'^2 + L_1 C_o R_o + L_2 C_o R_o\right) + s\left(L_1 + L_2\right) + R_o D'^2} \tag{13.17}$$

$$F_d = \left.\frac{\hat{V}_o}{\hat{D}}\right|_{\hat{V}_s=0}$$
$$= \frac{s^3 L_1 L_2 C_1 D V_o - s^2 L_1 C_1 R_o D D'^2 V_o + sL_2 D V_o + sL_1 D V_o - R_o D'^2 D V_o - R_o D'^3 V}{DD'\left[s^4 L_1 L_2 C_1 C_o R_o + s^3 L_1 L_2 C_1 + s^2\left(L_1 C_1 R_o D'^2 + L_1 C_o R_o + L_2 C_o R_o\right) + s\left(L_1 + L_2\right) + R_o D^2\right]} \tag{13.18}$$

$$Z_i = \left.\frac{\hat{V}_o}{\hat{I}_s}\right|_{\hat{D}=0}$$
$$= \frac{s^4 L_1 L_2 C_1 C_o R_o + s^3 L_1 L_2 C_1 + s^2\left(L_1 C_1 R_o D'^2 + L_1 C_o R_o + L_2 C_o R_o\right) + s\left(L_1 + L_2\right) + R_o D'^2}{D^2\left[s^3 L_2 C_1 C_o R_o + s^2 L_2 C_1 + s\left(C_1 R_o D'^2 + C_o R_o\right) + 1\right]} \tag{13.19}$$

$$Z_o = \left.\frac{\hat{V}_o}{\hat{I}_o}\right|_{\hat{D}=0,\hat{V}_s=0}$$
$$= \frac{s^3 L_1 L_2 C_1 R_o + sL_2 R_o + sL_1 R_o}{s^4 L_1 L_2 C_1 C_o R_o + s^3 L_1 L_2 C_1 + s^2\left(L_1 C_1 R_o D'^2 + L_1 C_o R_o + L_2 C_o R_o\right) + s\left(L_1 + L_2\right) + R_o D'^2} \tag{13.20}$$

Under the LCL filter located between buck and boost converter, the C_1 is set to zero and $L = L_1 + L_2$. Thus, the expressions shown in (13.17)–(13.20) can be further degenerated and represented in (13.21)–(13.24):

$$A_u = \left.\frac{\hat{V}_o}{\hat{V}_s}\right|_{\hat{D}=0} = \frac{-R_o DD'}{s^2 LC_o R_o + sL + R_o D'^2} \tag{13.21}$$

$$F_d = \left.\frac{\hat{V}_o}{\hat{D}}\right|_{\hat{V}_s=0} = \frac{V_o\left(RD'^3 + RDD'^2 - sLD\right)}{DD'\left(s^2 LC_o R_o + sL + R_o D'^2\right)} \tag{13.22}$$

$$Z_i = \left.\frac{\hat{V}_o}{\hat{I}_s}\right|_{\hat{D}=0} = \frac{s^2 LC_o R_o + sL + R_o D'^2}{D^2\left(sC_o R_o + 1\right)} \tag{13.23}$$

$$Z_o = \left.\frac{\hat{V}_o}{\hat{I}_o}\right|_{\hat{D}=0,\hat{V}_s=0} = \frac{sLR_o}{s^2 LC_o R_o + sL + R_o D'^2} \tag{13.24}$$

13.2.2 Boost-Buck Converter

The other boost-buck (Ćuk) example of deriving the small-signal models for SSCs operated in the CCM is used to illustrate the previously described modeling procedure. When the network shown in Figure 13.5b is loaded with capacitor C_o and resistor R_o, the parameter matrices in (13.1)–(13.3) can be determined as

$$T_A = \begin{bmatrix} -D' & -\dfrac{sL_1}{D'} \\ D & -\dfrac{1}{D'} \end{bmatrix}, \tag{13.25}$$

$$\begin{bmatrix} V_{13} \\ I_{13} \end{bmatrix} = \begin{bmatrix} -V_{C1} \\ 0 \end{bmatrix}, \tag{13.26}$$

$$T_B = \begin{bmatrix} 1 & 0 \\ SC_1 & 1 \end{bmatrix}, \tag{13.27}$$

$$T_C = \begin{bmatrix} \frac{1}{D} & \frac{sL_2}{D} \\ 0 & D \end{bmatrix}, \tag{13.28}$$

and

$$\begin{bmatrix} V'_{13} \\ I'_{13} \end{bmatrix} = \begin{bmatrix} -\frac{V_o}{D^2} \\ \frac{V_o}{R_o} \end{bmatrix}, \tag{13.29}$$

where D is the duty ratio of the active switch, D' is equal to $1 - D$, and V_{C1} represents an output voltage to subnetwork A. After substituting (13.25)–(13.29) into (13.4), we obtain

$$\begin{bmatrix} V_{1a} \\ I_{1a} \end{bmatrix} = \begin{bmatrix} \frac{-s^2L_1C_1 - D'^2}{DD'} & \frac{-s^3L_1L_2C_1 - D'^2}{DD'} \\ -\frac{sC_1}{DD'} & -\frac{s^2L_2C_1}{DD'} \end{bmatrix} \times \begin{bmatrix} V_{2c} \\ -I_{2c} \end{bmatrix} + \begin{bmatrix} \frac{s^2L_1C_1V_o}{D'^3} - sL_1V_o - \frac{sL_1V_o}{D'R_o} + D'^2V_oR_1 + \frac{V_o}{D'} \\ \frac{-s^2C_1L_2V_o}{DD'R_o} + \frac{sC_1V_o}{D'^3} + \frac{V_o}{D'^2R_1} \end{bmatrix} \hat{D}. \tag{13.30}$$

The four transfer characteristics can be consequently derived as

$$A_u = \left.\frac{\hat{V}_o}{\hat{V}_s}\right|_{\hat{D}=0} = \frac{-DD'R_o}{s^3L_1L_2C_1 + s^2L_1C_1R_o + sL_2D'^2 + R_oD'^2} \tag{13.31}$$

$$F_d = \left.\frac{\hat{V}_o}{\hat{D}}\right|_{\hat{V}_s=0} = \frac{\left(s^2L_1C_1R_o - sL_1R_oD'^3 - sL_1D'^2 + R_oD'^2 + R_1R_oD'^5\right)DV_o}{\left(s^3L_1L_2C_1 + s^2L_1C_1R_o + sL_2D'^2 + R_oD'^2\right)D'^2} \tag{13.32}$$

$$Z_i = \left.\frac{\hat{V}_o}{\hat{I}_s}\right|_{\hat{D}=0} = \frac{s^2 L_1 C_1 + D'^2}{sC_1} \tag{13.33}$$

$$Z_o = \left.\frac{\hat{V}_o}{\hat{I}_o}\right|_{\hat{D}=0,\hat{V}_s=0} = \frac{sL_2 R_o}{sL_2 + R_o} \tag{13.34}$$

13.3 Small-Signal Models of Zeta and Sepic Operated in CCM

In the buck family, there is one more converter, Zeta, while in the boost family, there is another sepic converter. How to model these two converters is presented in the following. With a proper organization as illustrated in Figure 13.7a and b, the same transfer ratio of $V_o/V_s = D/(1-D)$ but with different configurations can be obtained. In Figure 13.7a, the transfer ratio in Block 1 can be achieved with a buck-boost converter because its input is a voltage source, while that of Block 2 is realized with a buck converter. This results in the Zeta converter shown in Figure 13.7c. Sepic converter can be also developed from a proper configuration of boost-buck and boost converters, which is illustrated in Figure 13.8.

The procedure of deriving the small-signal models for these SSCs is outlined as follows:

1) Represent the state-space averaging models of the BCUs (buck and boost) in *t*-parameters.
2) The small-signal model of the sepic converter is derived from the boost-buck, unity gain buffer, and boost, which are represented in two-port network parameters. Similarly, that of the Zeta converter is derived from buck-boost, unity gain buffer, and buck. These are illustrated in Figure 13.9a and b. In Figure 13.9b, for instance, the two blocks in cascode are first represented in *h*-parameters for obtaining an equivalent model conveniently, and it is expressed in terms of *t*-parameter, which is multiplied with that of boost to derive the model of the sepic.
3) Derive the SSTCs of the SSCs in terms of *t*-parameter. These include audio susceptibility (A_u), control-to-output transfer function (F_d), input impedance (Z_i), and output impedance (Z_o).

As illustrated in Figure 13.9a and b, the small-signal models of Zeta and sepic SSCs can be derived from those of the buck, boost, and unity gain buffer in cascade/cascode connection. First of all, to obtain the equivalent expressions for

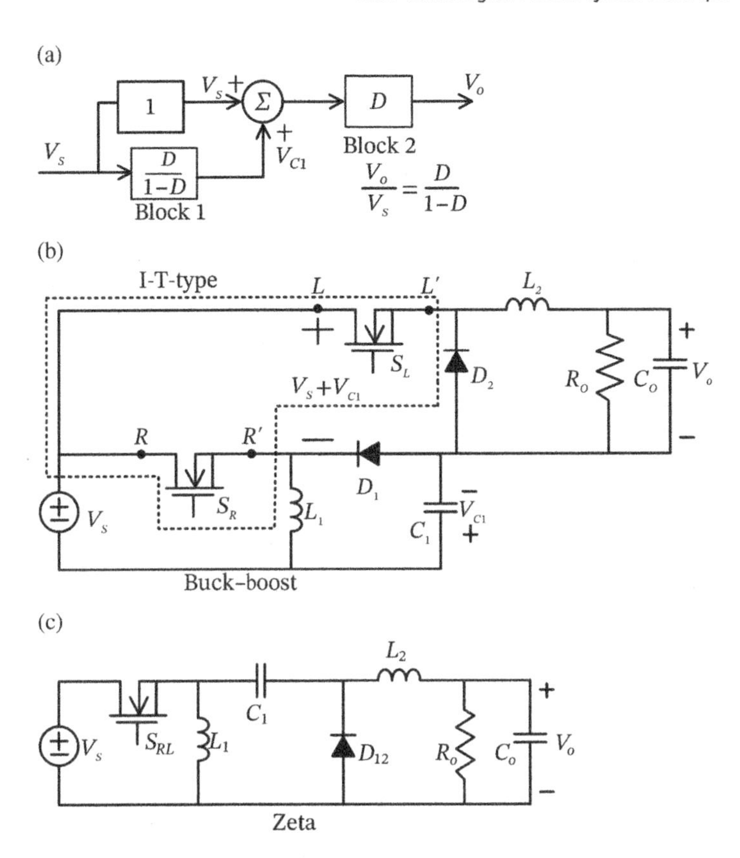

Figure 13.7 Illustration of the Zeta converter derived from the buck-boost, and buck converters in cascade/cascode connection.

the subnetwork A shown in the figures, the previously derived t-parameters of the buck-boost and boost-buck are transformed to g- and h-parameters, respectively, which are shown as follows:

$$\begin{bmatrix} I_{1a} \\ V_{2c} \end{bmatrix} = G_a \times \begin{bmatrix} V_{1a} \\ I_{2c} \end{bmatrix} + \begin{bmatrix} V_g \\ I_g \end{bmatrix} \hat{D} \tag{13.35}$$

and

$$\begin{bmatrix} V_{1a} \\ I_{2c} \end{bmatrix} = H_a \times \begin{bmatrix} I_{1a} \\ V_{2c} \end{bmatrix} + \begin{bmatrix} V_h \\ I_h \end{bmatrix} \hat{D}. \tag{13.36}$$

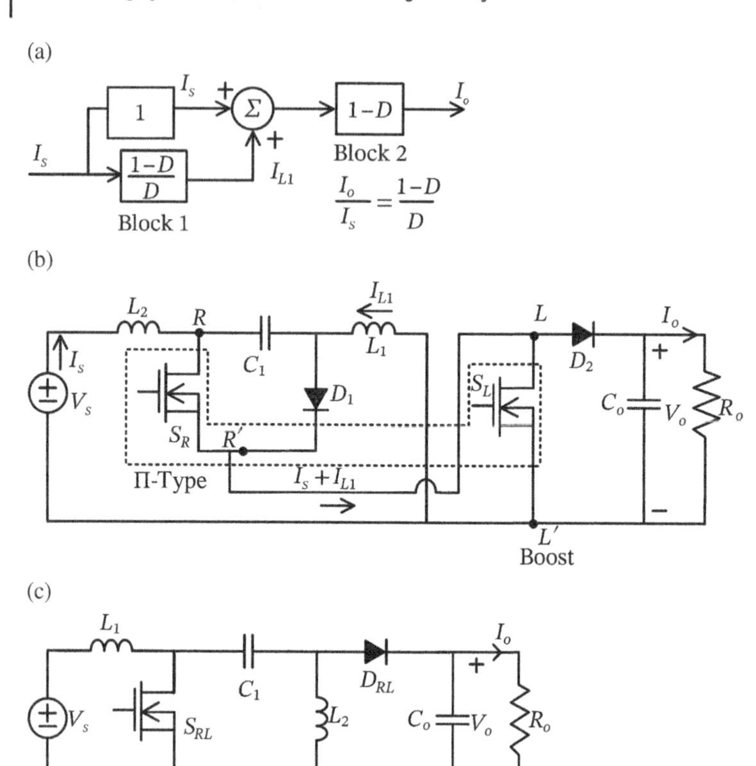

Figure 13.8 Illustration of the sepic converter derived from the boost-buck, and boost in cascade/cascode connection.

Equations (13.35) and (13.36) are combined with unity gain buffers represented in g- and h-parameters to yield expressions for subnetwork A shown in Figure 13.9a and b, respectively. They are given as follows:

$$\begin{bmatrix} I'_{1a} \\ V'_{2c} \end{bmatrix} = G'_a \times \begin{bmatrix} V'_{1a} \\ I'_{2c} \end{bmatrix} + \begin{bmatrix} V_g \\ I_g \end{bmatrix} \hat{D} \tag{13.37}$$

and

$$\begin{bmatrix} V'_{1a} \\ I'_{2c} \end{bmatrix} = H'_a \times \begin{bmatrix} I'_{1a} \\ V'_{2c} \end{bmatrix} + \begin{bmatrix} V_h \\ I_h \end{bmatrix} \hat{D}, \tag{13.38}$$

(a)

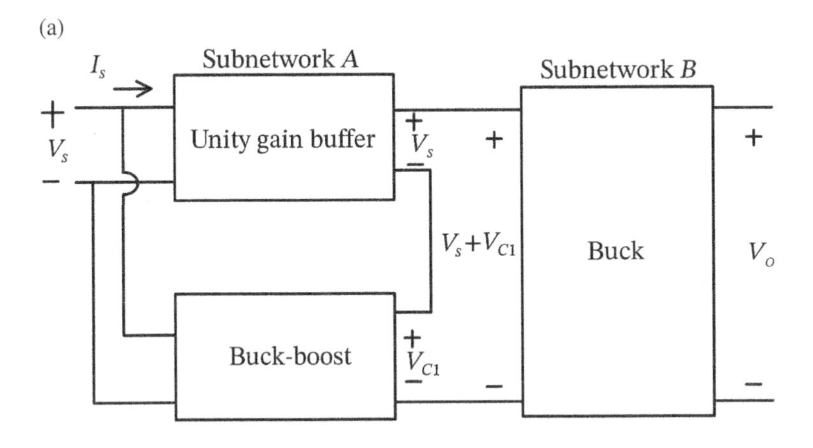

(b)

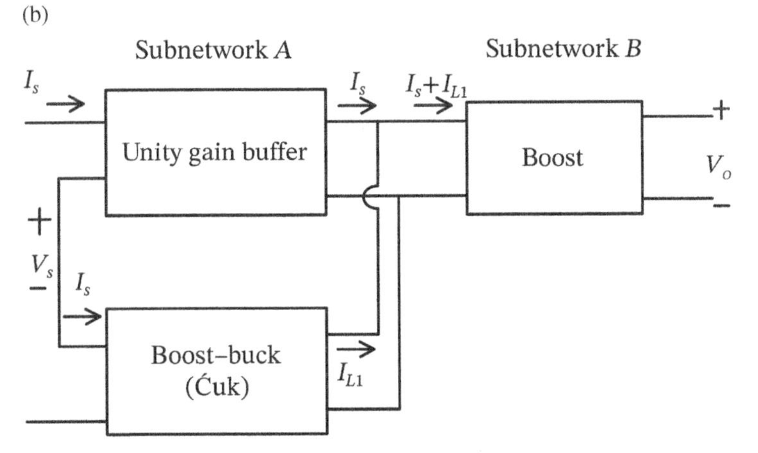

Figure 13.9 Small-signal models of the SSCs represented in cascaded/cascoded two-port networks: (a) Zeta and (b) sepic converters.

where

$$G'_a = \begin{bmatrix} G_{a11} & G_{a12} - 1 \\ G_{a21} + 1 & G_{a22} \end{bmatrix}$$

and

$$H'_a = \begin{bmatrix} H_{a11} & H_{a12} + 1 \\ H_{a21} - 1 & H_{a22} \end{bmatrix}.$$

The above results are transformed into t-parameters, designated as T_S and T_Z, which are multiplied with those of boost and buck to yield the small-signal models for sepic and Zeta, respectively. The general form is represented by

$$\begin{bmatrix} V_{1a} \\ I_{1a} \end{bmatrix} = T_D \times \begin{bmatrix} V_{2c} \\ -I_{2c} \end{bmatrix} + \begin{bmatrix} V_d \\ I_d \end{bmatrix} \hat{D} \tag{13.39}$$

where for Zeta SSC

$$T_D = T_Z \times T_B, V_d = T_{B11} \times V_g + T_{B12} \times I_g + V_g,$$

and

$$I_d = T_{B21} \times V_g + T_{B22} \times I_g + I_g,$$

while for sepic SSC

$$\begin{aligned} &T_D = T_S \times T_B, \\ &V_d = T_{B11} \times V_h + T_{B12} \times I_h + V_h \text{ and} \\ &I_d = T_{B21} \times V_h + T_{B22} \times I_h + I_h \end{aligned}$$

In the above expressions, T_B is the t-parameter of subnetwork B, and T_D denotes the t-parameter of the overall network shown in Figure 13.9a and b. The expressions of the four transfer characteristics of Zeta and sepic SSCs are the same as those shown in (13.21)–(13.24).

13.3.1 Zeta Converter

Modeling of the Zeta SSC is to determine the parameters in (13.37) and (13.39), which are shown as follows:

$$\begin{bmatrix} V_{1a} \\ I'_{2a} \end{bmatrix} = G'_a \times \begin{bmatrix} I_{1a} \\ V_{2a} \end{bmatrix} + \begin{bmatrix} V_g \\ I_g \end{bmatrix} \hat{D} \tag{13.40}$$

and

$$\begin{bmatrix} V_{1a} \\ I_{1a} \end{bmatrix} = T_D \times \begin{bmatrix} V_{2a} \\ -I_{2a} \end{bmatrix} + \begin{bmatrix} V_d \\ I_d \end{bmatrix} \hat{D} \tag{13.41}$$

where

$$G'_a = \begin{bmatrix} \dfrac{C_1 D^2 s^2}{D'^2 + C_1 L s^2} & \dfrac{-D' - C_1 L s}{D'^2 + C_1 L s^2} \\ \dfrac{D' + C_1 L s}{D'^2 + C_1 L s^2} & \dfrac{L s}{D'^2 + C_1 L s^2} \end{bmatrix},$$

$$\begin{bmatrix} V_g \\ I_g \end{bmatrix} = \begin{bmatrix} \dfrac{sLV_o}{D^2\left(D'^2 + C_1 L s^2\right)} \\ \dfrac{D'V_o}{D\left(D'^2 + C_1 L s^2\right)} \end{bmatrix},$$

$$T_D = \begin{bmatrix} \dfrac{D'^2 + C_1 L s^2}{D\left(D' + C_1 L s^2\right)} & \dfrac{s\left(D^2 L + L_1 D'^2 + C_1 L L_1 s^2\right)}{D\left(D' + C_1 L s^2\right)} \\ \dfrac{C_1 D s}{D' + C_1 L s^2} & \dfrac{D\left(1 + C_1 L s^2 + C_1 L_1 s^2\right)}{D' + C_1 L s} \end{bmatrix},$$

and

$$\begin{bmatrix} V_d \\ I_d \end{bmatrix} = \begin{bmatrix} \dfrac{\left(-D'R_o + D^2 L s - C_1 L R_o s^2\right) V_o}{D^2 R_o\left(D' + C_1 L s^2\right)} \\ \dfrac{D^2 D'^2 + C_1 R_o D^2 D' + L R_o D' s - 2 C_1 D^2 D' L s^2 + C_1 D'^2 L R_o s^2 + C_1 L^2 R_o s^3 + C_1^2 D^2 L^2 s^4}{D^2 R_o\left(D' + C_1 L s^2\right)\left(D'^2 + C_1 L s^2\right)} \end{bmatrix}.$$

After substituting (13.4) into (13.6)–(13.9), we can obtain the four transfer characteristics of the Zeta converter as shown in the following:

$$A_u = \left.\frac{\hat{V}_o}{\hat{V}_s}\right|_{\hat{D}=0}$$
$$= \frac{DR_o\left(1 - D + C_1 L_1 S^2 + C_1 L_2 s^2\right)}{D\left(R_o D'^2 + \left(D^2 L + D'^2 L_1\right)s + \left(C_1 L R_o + D^2 L C_o R_o + D'^2 L_1 C_o\right)s^2 + L L_1 C_1 s^3 + C_o C_1 L L_1 R_o s^4\right)} \tag{13.42}$$

$$F_d = \left.\frac{\hat{V}_o}{\hat{D}}\right|_{\hat{V}_s=0}$$
$$= \frac{V_o\left(-R_o + DR_o + D^2Ls - C_1LR_os^2\right)}{D\left(R_oD'^2 + \left(D^2L + D'^2L_1\right)s + \left(C_1LR_o + D^2LC_oR_o + D'^2L_1C_o\right)s^2 + LL_1C_1s^3 + C_oC_1LL_1R_os^4\right)} \tag{13.43}$$

$$Z_i = \left.\frac{\hat{V}_s}{\hat{I}_s}\right|_{\hat{D}=0}$$
$$= \frac{D\left(R_oD'^2 + \left(D^2L + D'^2L_1\right)s + \left(C_1LR_o + D^2LC_oR_o + D'^2L_1C_o\right)s^2 + LL_1C_1s^3 + C_oC_1LL_1R_os^4\right)}{D\left(1 + C_oR_os + C_1R_os + C_1Ls^2 + C_1L_1s^2 + C_oC_1L_1R_os^3 + C_oL_1L_3R_os^3\right)} \tag{13.44}$$

$$Z_o = \left.\frac{\hat{V}_o}{\hat{I}_o}\right|_{\hat{D}=0,\hat{V}_s=0}$$
$$= \frac{R_os\left(D^2L + D'^2L_1 + C_1LL_1s^2\right)}{D\left(R_oD'^2 + \left(D^2L + D'^2L_1\right)s + \left(C_1LR_o + C_oL_1R_o + D^2LC_oR_o + D'^2L_1C_o\right)s^2 + LL_1C_1s^3 + C_oC_1LL_3R_os^4\right)} \tag{13.45}$$

13.3.2 Sepic Converter

Modeling of the sepic is to determine the parameters in (13.37) and (13.39), which are shown as follows:

$$\begin{bmatrix} V'_{1a} \\ I'_{2c} \end{bmatrix} = H'_a \times \begin{bmatrix} I'_{1a} \\ V'_{2c} \end{bmatrix} + \begin{bmatrix} V_h \\ I_h \end{bmatrix} \hat{D} \tag{13.46}$$

and

$$\begin{bmatrix} V_{1a} \\ I_{1a} \end{bmatrix} = T_D \times \begin{bmatrix} V_{2a} \\ -I_{2a} \end{bmatrix} + \begin{bmatrix} V_d \\ I_d \end{bmatrix} \hat{D} \tag{13.47}$$

where

$$H'_a = \begin{bmatrix} \dfrac{s^2L_1C_1 + D'^2}{sC_1} & 1 \\ \dfrac{-s^2L_2C_1 - DD'}{s^2L_2C_1} & \dfrac{-1}{sL_2} \end{bmatrix}$$

$$\begin{bmatrix} V_h \\ I_h \end{bmatrix} = \begin{bmatrix} \dfrac{s^2L_2C_1 - L_1C_1 + DD' - D'^2}{s^2L_2C_1} \\ \dfrac{\left(sC_1R_1R_O - R_1D'^2 + R_oD'\right)V_oD}{s^2L_2C_1} \end{bmatrix},$$

$$\begin{bmatrix} V_g \\ I_g \end{bmatrix} = \begin{bmatrix} \dfrac{sLV_o}{D^2\left(D'^2 + C_1Ls^2\right)} \\ \dfrac{D'V_o}{D\left(D'^2 + C1Ls^2\right)} \end{bmatrix},$$

$$T_D = \begin{bmatrix} \dfrac{D'\left(sC_1\left(s^3L_1L_2C_1 + sL_1 - sL_2\right) + D'^2\left(s^2L_2C_1 + 1\right) - DD'\right)}{s^2L_2C_1 + DD'} & \dfrac{sL_2\left(sC_1\left(s^3L_1L_2C_1 + 2sL_1 - sL_2\right) + D'^2\left(s^2L_2C_1 + 2\right) - DD'\right)}{D'\left(s^2L_2C_1 + DD'\right)} \\ \dfrac{sC_1D'\left(s^2L_2C_1 + 1\right)}{s^2L_2C_1 + DD'} & \dfrac{s^2L_2C_1\left(s^2L_2C_1 + 2\right)}{D'\left(s^2L_2C_1 + D'\right)} \end{bmatrix},$$

and

$$\begin{bmatrix} V_d \\ I_d \end{bmatrix} = \begin{bmatrix} \dfrac{\begin{gathered} V_o\left(sC_1\left(sL_2\left(s^3C_1L_1L_2 + sL_1 - sL_2\right) - s^2L_1L_2\right) + D\left(-sL_2D' - R_oD'^3\left(sL_1 + 1\right) + R_1R_oD'^5\right)\right. \\ + D\left(-sL_2D' - R_oD'^3\left(sL_1 + 1\right) + R_1R_oD'^5\right) + D'^2\left(sC_1R_o\left(s^3L_1L_2C_1 - s^2L_1L_2 + sL_1 - sL_2\right)\right. \\ \left.\left. + L_2\left(s^3L_2C_1 + s\right) - sL_2\right) + R_oD'^4\left(s^2L_2C_1\left(R_1 + 1\right) + 1\right)\right) \end{gathered}}{R_oD'^2\left(s^2L_1C_1 + DD'\right)} \\ \dfrac{\begin{gathered} V_o\left(R_oDD'^2 + sC_1\left(sL_2R_1\left(s^2L_2C_1 + 2\right) + D'\left(sL_2R_1\left(s^2L_2C_1 + 1\right) + sL_2R_o\right)\right.\right. \\ \left.\left.\left. + R_1R_oD'^3\left(s^2L_2C_1 + 1\right)\right)\right)\right) \end{gathered}}{R_1R_oD'^3\left(s^2L_2C_1 + DD'\right)} \end{bmatrix}.$$

After substituting (13.4) into (13.6)–(13.9), we can obtain the four transfer characteristics of the Sepic converter as shown in the following:

$$A_u = \left.\frac{\hat{V}_o}{\hat{V}_s}\right|_{\hat{D}=0} = \frac{R_o D'\left(s^2 L_2 C_1 + DD'\right)}{\begin{array}{l} s^4\left(L_1 L_2^2 C_1^2 + L_1 L_2 C_1^2 R_o D'^2\right) + s^3\left(L_2^2 C_1 D'^2\right) + s^2\left(2L_1 L_2 C_1 - L_2^2 + L_1 C_1 R_o D'^2 - L_2 C_1 R_o D'^2\right. \\ \left. + L_2 C_1 R_o D'^4\right) + s\left(-L_2 DD' + 2L_2 D'^2\right) - R_o DD'^3 + R_o D'^4 \end{array}} \tag{13.48}$$

$$F_d = \left.\frac{\hat{V}_o}{\hat{D}}\right|_{\hat{V}_s=0} = \frac{\begin{array}{l} V_o\left(s^4 L_1 L_2 C_1^2 R_o - s^3\left(L_1 L_2 C_1 D'^2 + L_1 L_2 C_1 R_o D'^3\right) + s^2\left(L_2 C_1 R_o D'^2 + L_2 C_1 R_1 R_o D'^5\right.\right. \\ \left.\left. + L_1 C_1 R_o DD'\right) - s\left(L_1 DD'^3 + L_1 R_o DD'^4\right) + R_o DD'^3 + R_1 R_o DD'^6\right) \end{array}}{\begin{array}{l} s^4 L_1 L_2^2 C_1^2 + s^3\left(L_1 L_2 C_1^2 D'^2 + L_2^2 C_1 D'^2\right) + s^2\left(2L_1 L_2 C_1 - L_2^2 - L_2 C_1 D'^2 + L_1 C_1 D'^2 + L_2 C_1 D'^4\right) \\ + s\left(-L_2 DD' + 2L_2 D'^2\right) - DD'^3 + D'^4 \end{array}} \tag{13.49}$$

$$Z_i = \left.\frac{\hat{V}_s}{I_s}\right|_{\hat{D}=0} = \frac{\begin{array}{l} s^5 L_1 L_2^2 C_1^2 + s^4 L_1 L_2 C_1^2 R_o D'^2 + s^3\left(2L_1 L_2 C_1 - L_2^2 C_1 + L_2^2 C_1 D'^2\right) \\ + s^2\left(L_1 C_1 R_o D'^2 - L_2 C_1 R_o D'^2 + L_2 C_1 R_o D'^4\right) + s\left(-L_2 DD' + 2L_2 D'^2\right) - R_o DD'^3 + R_o D'^4 \end{array}}{s^4 L_2^2 C_1^2 + s^3 L_2 C_1^2 R_o D' + 2s^2 L_2 C_1 + sC_1 R_o D'} \tag{13.50}$$

$$Z_o = \left.\frac{\hat{V}_o}{\hat{I}_o}\right|_{\hat{D}=0,\hat{V}_s=0} = \frac{s^5 L_1 L_2^2 C_1^2 R_o + s^3\left(2L_1 L_2 C_1 R_o - L_2^2 C_1^2 R_o + L_2^2 C_1^2 R_o D'^2\right) + s\left(2L_2 R_o D'^2 - L_2 R_o DD'\right)}{\begin{array}{l} s^5 L_1 L_2^2 C_1^2 + s^4 L_1 L_2 C_1^2 R_o D'^2 + s^3\left(2L_1 L_2 C_1 - L_2^2 C_1 + L_2^2 C_1 D'^2\right) + \\ s^2\left(L_1 C_1 R_o D'^2 - L_2 C_1 R_o D'^2 + L_2 C_1 R_o D'^4\right) + s\left(-L_2 DD' + 2L_2 D'^2\right) - R_o DD'^3 + R_o D'^4 \end{array}} \tag{13.51}$$

Further Reading

Cuk, S. and Middlebrook, R.D. (1977). A new optimal topology switching DC-to-DC converter. *IEEE Power Electronics Specialists Conference*, IEEE, CA (June 1977), pp. 14–16.

Freeland, S.D. (1993). Techniques for the practical application of duality to power circuits. *IEEE Trans. Power Electron.* 7 (2): 374–384.

Jozwik, J.J. and Kazimierczuk, M.K. (1989). Dual sepic PWM switching-mode DC/DC power converter. *IEEE Trans. Ind. Electron.* 36 (1): 64–70.

Lee, Y.S. (1985). A systematic and unified approach to modeling switches in switch-mode power supplies. *IEEE Trans. Ind. Electron.* IE-32: 445–448.

Liu, K.H. and Lee, F.C. (1988). Topological constraints on basic PWM converters. *Proceedings of the IEEE Power Electronics Specialists Conference*, IEEE, Kyoto, Japan (April 1988), Vol. 1, pp. 164–172.

Maksimovic, D. and Cuk, S. (1991). Switching converters with wide DC conversion range. *IEEE Trans. Power Electron.* 6 (1): 151–157.

Middlebrook, R.D. and Cuk, S. (1976). A general unified approach to modeling switching converter power stages. *Proceedings of the IEEE Power Electronics Specialists Conference*, IEEE, pp. 18–34.

Middlebrook, R.D. and Cuk, S. (1997). A general unified approach to modeling DC-to-DC converters in discontinuous conduction. *Proceedings of the IEEE Power Electronics Specialists Conference*, IEEE, pp. 36–57.

Sebastian, J. and Uceda, J. (1987). The double converter: a fully regulated two-output DC–DC converter. *IEEE Trans. Power Electron.* PE-2 (3): 239–246.

Sedra, A.S. and Smith, K.C. (1991). *Microelectronic Circuit*. Saunders College/Holt, Rinehart, and Winston Inc.

Smedley, K. and Cuk, S. (1994). Switching flow-graph nonlinear modeling technique. *IEEE Trans. Power Electron.* 9 (4): 405–413.

Vorperian, V. (1990a). Simplified analysis of PWM converters using the PWM switch, part I: continuous conduction mode. *IEEE Trans. Aerosp. Electron. Syst.* AES-26: 497–505.

Vorperian, V. (1990b). Simplified analysis of PWM converters using the PWM switch, part II: discontinuous conduction mode. *IEEE Trans. Aerosp. Electron. Syst.* AES-26: 497–505.

14

Modeling of Isolated Single-Stage Converters with High Power Factor and Fast Regulation

Due to stringent regulations such as IEC 1000-3-2, there is a need to develop converters that can perform harmonic reduction and power factor correction (PFC), while can achieve isolation and fast DC–DC conversion. Researches have been dedicated in integrating PFC and DC–DC converter stages into an isolated single-stage converter (ISSC) so as to minimize the cost. Among the ISSCs mentioned in the literature, each can usually be divided into a PFC semi-stage and a regulator semi-stage. The PFC semi-stage operating in DCM and with constant on time within half a line period can easily achieve a high power factor without any complex control and leaves freedom for the regulator semi-stage to achieve a fast regulation.

A model of an ISSC is useful in designing a suitable controller to govern its dynamic behavior. The state-space averaging method is commonly used for establishing the small-signal models, which are represented in mathematical expressions or equivalent circuits for the converter operated in either CCM or DCM. Researchers presented a scheme to obtain the averaged circuit models of PWM converters by replacing only the switches with their averaged circuit in either CCM or DCM. In the papers, they recognized the existence of a PWM switch in each converter, such as buck, boost, etc., which leads to simplifying the converter modeling. However, this scheme cannot be applied to model the ISSCs with non-PWM switches, such as a buck-boost converter integrated with a flyback converter. Researchers introduced and discussed the cascade connection of boost and buck converters, which are operated in different conduction modes. Moreover, the proposed graft and layer schemes can be used to systematically integrate these converters to form an ISSC. These two schemes have been shown to be useful in developing SSCs and be capable of establishing relationships among the converters in a family. Systematical methods for deriving the small-signal models of the PWM converters in CCM based on the layer and graft schemes were presented in

Origin of Power Converters: Decoding, Synthesizing, and Modeling, First Edition.
Tsai-Fu Wu and Yu-Kai Chen.
© 2020 John Wiley & Sons, Inc. Published 2020 by John Wiley & Sons, Inc.

Chapters 12 and 13. Yet, the small-signal modeling of the SSC operated in DCM has not been addressed. This chapter presents a systematic and unified approach to modeling SSCs based on the graft scheme, in which the front PFC semi-stage is operated in DCM while the rear regulator semi-stage is in either CCM or DCM. With the graft scheme, power converters can be integrated to form an SSC. The small-signal model of the SSC is derived by combining those of its originally separate converters and applying the two-port network theory. Thus, relationships between the dynamic models of the converters and their derived SSC can be identified. The derivation of the small-signal models will be discussed in detail in the following sections.

14.1 Generation of Single-Stage Converters with High Power Factor and Fast Regulation

Converters are usually connected in cascade to achieve multiple functions. To increase input power factor, reduce harmonic current, and achieve fast regulation, the configuration shown in Figure 14.1 is usually used, where the PFC semi-stage functions as a current shaper and the DC–DC converter is used for fast regulating the output voltage or current. Each of the power stages still functions as a complete power converter and can operate independently. The multistage converter increases the system cost and the number of controllers, deteriorates the system reliability, and occupies more space. Thus, in moderate power applications, converters are integrated to form an SSC.

Several families of ISSCs to achieve unity power factor and tight output regulation can be derived with the graft scheme. Some of them are depicted in Figures 14.2–14.4. Figure 14.2 illustrates the derivation of a buck-flyback ISSC. The buck and flyback converters connected in cascade are shown in Figure 14.2a. Properly repositioning S_R, but without varying its operating principle, can establish a common node for switches S_R and S_L, which is depicted in Figure 14.2b. Investigating the connection of S_R and S_L reveals that the switches are in the

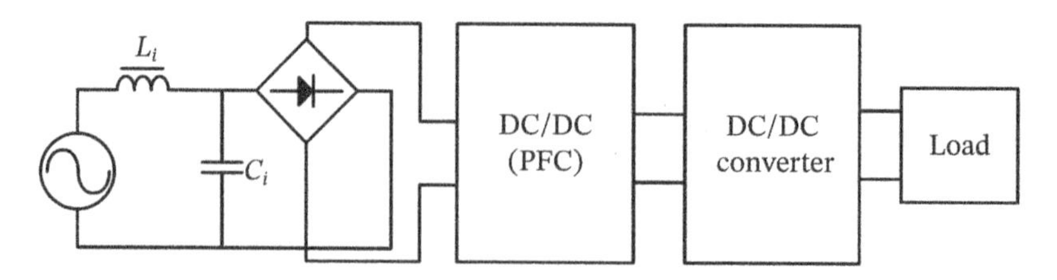

Figure 14.1 Block diagram of a power converter with PFC.

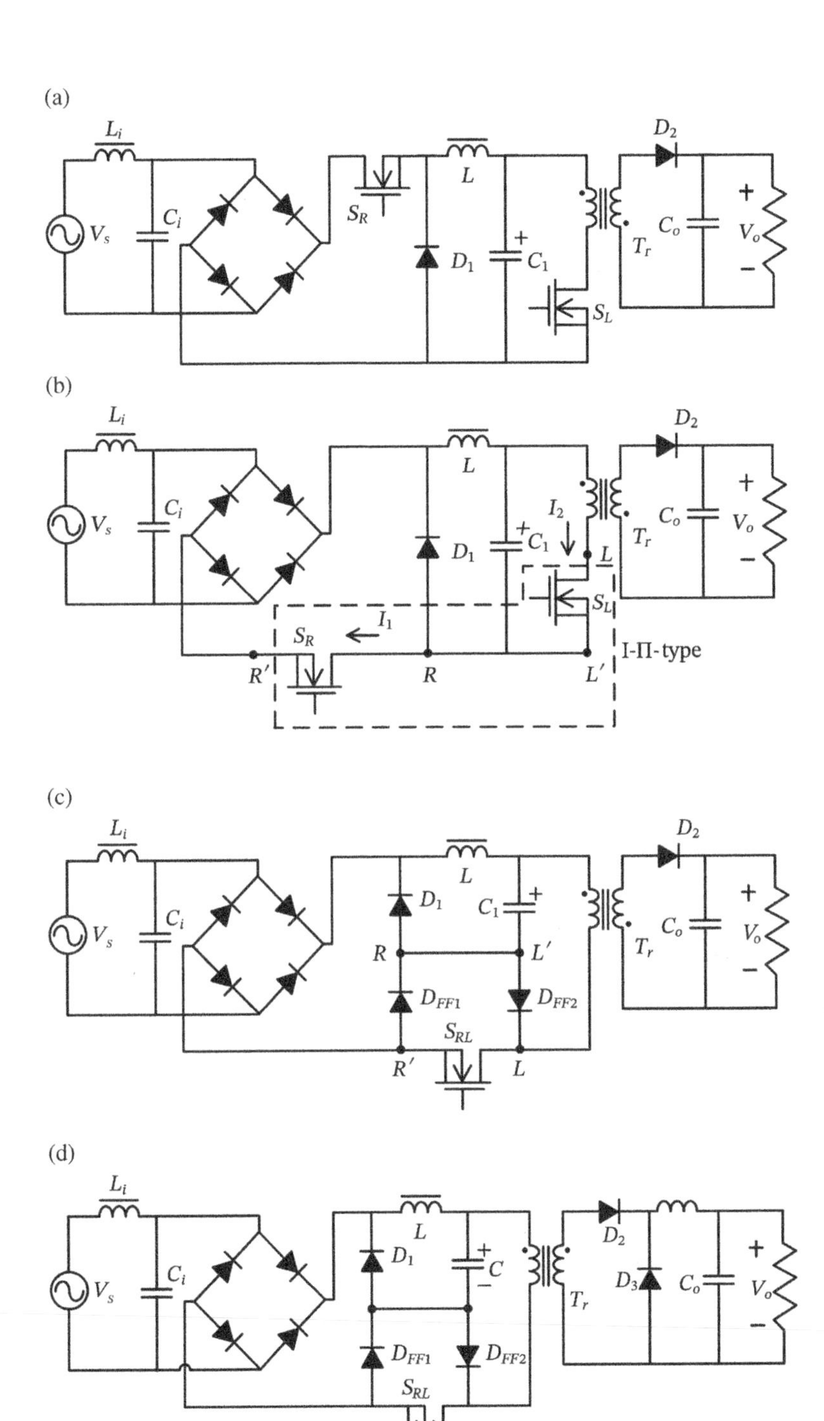

Figure 14.2 The buck ISSC family derived based on the graft scheme. (a) buck+flyback, (b) buck+flyback, (c) buck+flyback, (d) buck+forward.

I-Π-type configuration. Thus, the I-ΠGS is used to replace the S_R and S_L as depicted in Figure 14.2c. Since the relationship between the currents, I_1 and I_2, is inconsistent during the on states for all operation period, the I-ΠGS cannot be further degenerated. Finally, the schematic of the buck-flyback ISSC becomes the one shown in Figure 14.2c. Similarly, a buck integrated with a forward converter can be shown in Figure 14.2d. The rest of the discussed ISSCs in Figure 14.3 can be derived by following the same procedure described above. The ISSC family depicted in Figure 14.4 with a buck-boost as a current shaper presents behaviors in between those of the buck- and boost-ISSC families. The PWM and variable frequency control techniques can be employed simultaneously to control the ISSCs shown in Figures 14.2–14.4 to function with the desired features. The front semi-stage is usually operated in the DCM to achieve a high power factor without the need of a controller, in which the power flow is governed by varying the switching frequency and duty ratio, while the rear semi-stage is to achieve fast regulation and it can be operated in either CCM or DCM. When it is operated in the CCM, the PWM control is employed, while when operated in the DCM, both PWM and variable frequency controls are adopted.

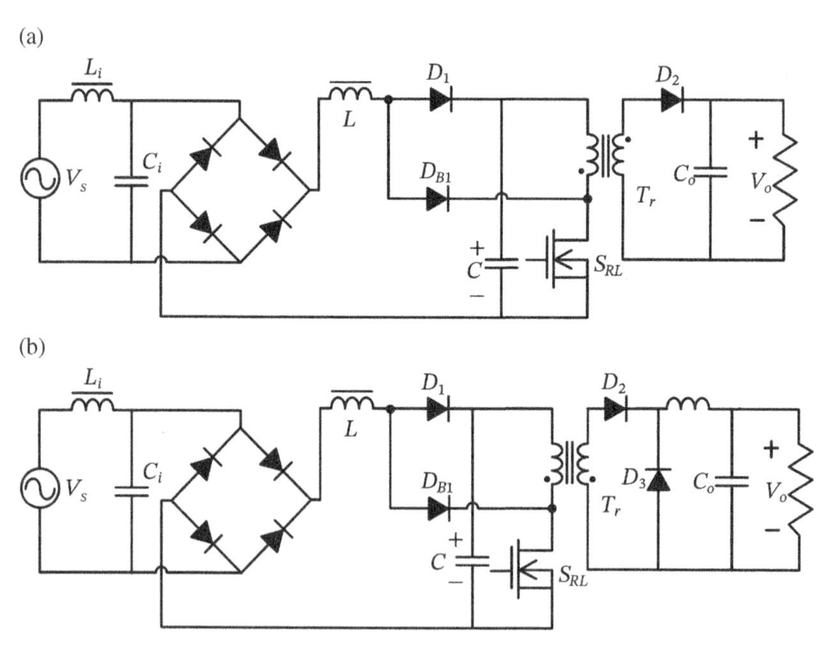

Figure 14.3 The boost ISSC family derived based on the graft scheme. (a) boost+flyback, (b) boost+forward.

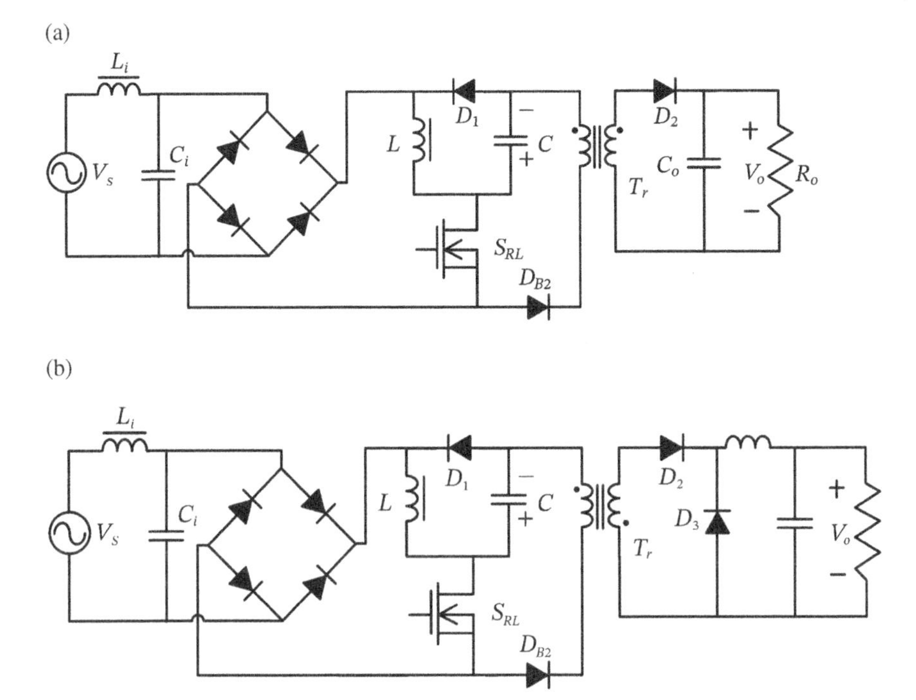

Figure 14.4 The buck-boost ISSC family derived based on the graft scheme. (a) buck-boost+flyback, (b) buck-boost+forward.

14.2 Small-Signal Models of General Converter Forms Operated in CCM/DCM

As mentioned previously, the ISSCs can be generated from the converters connected in cascade with the graft scheme. Thus, to derive their small-signal models conveniently, the converters are represented in t-(transmission) parameters. The ISSCs shown in Figures 14.2–14.4 therefore are replaced by their corresponding two-port networks. For instance, the circuit shown in Figure 14.4a is divided into four subnetworks: line filter subnetwork, PFC subnetwork, buffer subnetwork, and DC/DC regulator subnetwork as depicted in Figure 14.5. The procedure of deriving the small-signal model for these ISSCs is outlined as follows:

1) Derive the small-signal model of the discussed individual converters with the state-space averaging method, in which the PFC subnetwork operates in DCM, while the regulator subnetwork can operate either in CCM or in DCM. To have physical insights and to be convenient for further process, the model is represented in y-parameters.

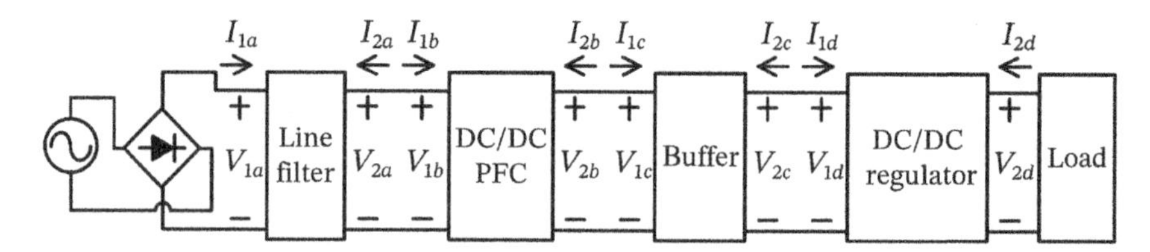

Figure 14.5 Small-signal model of the SSC represented in cascaded two-port networks.

2) Transfer the y-parameters of each subnetwork into t-parameters for convenience in investigating the cascaded subnetworks.
3) Graft the PFC semi-stage and regulator semi-stage to form an SSC. That is, the small-signal models of the ISSC are derived by multiplying the t-parameters of the two semi-stages, of which both PFC and regulator operated in DCM (DCM + DCM) is denoted as category 1, while the PFC in DCM and the regulator in CCM (DCM + CCM) is denoted as category 2. Figure 14.5 illustrates the PFC and regulator represented in cascaded two-port networks. Note that the buffer is considered the load of the front semi-stage and the source of the rear semi-stage and it is separated from the converters for effectively using the t-parameters. The relationships among port voltage, current, and control signal D of the subnetworks shown in Figure 14.5 can be expressed in terms of t-parameters.
4) Derive the small-signal transfer characteristics (SSTCs) of the ISSCs in terms of t-parameters. These include audio susceptibility (A_u), control-to-output transfer function (F_d), input impedance (Z_i), and output impedance (Z_o).

Following the above-described procedure, derivation of the small-signal models is carried out.

Step 1:

The subnetworks are represented in y-parameters and given by

$$\text{Subnetwork } A: \quad \begin{bmatrix} I_{1a} \\ I_{2a} \end{bmatrix} = Y_A \times \begin{bmatrix} V_{1a} \\ V_{2a} \end{bmatrix}, \tag{14.1}$$

$$\text{Subnetwork } B: \quad \begin{bmatrix} I_{1b} \\ I_{2b} \end{bmatrix} = Y_B \times \begin{bmatrix} V_{1b} \\ V_{2b} \end{bmatrix} + \begin{bmatrix} I_{b13} \\ I_{b23} \end{bmatrix} \hat{D}, \tag{14.2}$$

$$\text{Subnetwork } C: \quad \begin{bmatrix} I_{1c} \\ I_{2c} \end{bmatrix} = Y_C \times \begin{bmatrix} V_{1c} \\ V_{2c} \end{bmatrix}, \tag{14.3}$$

and

$$\text{Subnetwork } D: \quad \begin{bmatrix} I_{1d} \\ I_{2d} \end{bmatrix} = Y_D \times \begin{bmatrix} V_{1d} \\ V_{2d} \end{bmatrix} + \begin{bmatrix} I_{d13} \\ I_{d23} \end{bmatrix} \hat{D}. \tag{14.4}$$

The definition and schematics of y-parameters and t-parameters are shown in Figure 13.6a and b, respectively. The y-parameters of the typical PWM converters operated in CCM and DCM are listed in Tables 14.1 and 14.2, respectively. Although only are the y-parameters of the PWM converters tabulated in the tables, it is considered without loss of generality because most of isolated PWM converters are equivalent to their corresponding transformerless ones. For instance, the forward converter is in principle, equivalent to the buck, so is the flyback to the buck-boost.

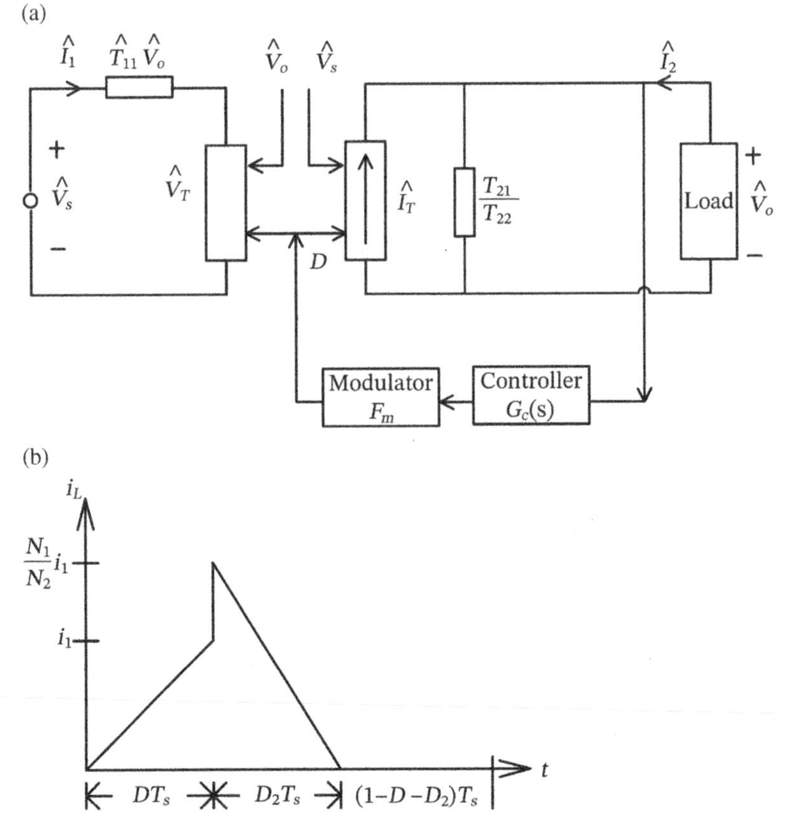

Figure 14.6 (a) An SSC represented in t-parameter and with a single-voltage loop control, (b) the inductor current waveform of the flyback converter.

Table 14.1 y-Parameters of the PWM converters operated in CCM.

	Buck	Boost	Buck-boost
y_{11}	$\frac{D^2}{sL}$	$\frac{1}{sL}$	$\frac{1}{sL}$
y_{12}	$-\frac{D}{sL}$	$-\frac{D'}{sL}$	0
y_{21}	$\frac{D}{sL}$	$\frac{D'}{sL}$	$-\frac{DD'}{sL}$
y_{22}	$\frac{1}{sL}$	$\frac{D'^2}{sL}$	$\frac{D'^2}{sL}$
I_{13}	$\left[\frac{sL+R_O}{sLR_O}\right]V_O$	$\frac{V_O}{sL}$	$\frac{V_o\left(-sLDD'+R_oD'^2\right)}{sLR_oD'^2}$
I_{23}	$\frac{V_O}{sLD}$	$\left[\frac{sL+R_OD'^2}{sLR_OD'}\right]V_O$	$\left[\frac{sLD-R_OD'^2}{sLR_OD'}\right]V_O$

Where $D' = 1-D$.

Table 14.2 y-Parameters of the PWM converters operated in DCM.

	Buck	Boost	Buck-boost
y_{11}	$\frac{M_1^2}{R_o(1-M_1)}$	$\frac{M_2^3}{(M_2-1)R_o}$	$\frac{M^2}{R_o}$
y_{12}	$-\frac{1}{R_o}\left(\frac{M_1^2}{1-M_1}\right)$	$-\frac{1}{R_o}\left[\frac{M_2}{M_2-1}\right]$	0
y_{21}	$-\frac{M_1}{R_o}\left(\frac{2-M_1}{1-M_1}\right)$	$-\frac{1}{R_o}\left[\frac{M_2(2M_2-1)}{M_2-1}\right]$	$-\frac{2M}{R_o}$
y_{22}	$\frac{1}{R_o(1-M_1)}$	$\frac{M_2}{(M_2-1)R_o}$	$\frac{1}{R_o}$
I_{13}	$\frac{2V_o}{R_o}\sqrt{\frac{1-M_1}{2\tau_L}}$	$\frac{2V_o}{R_o}\sqrt{\frac{M_2}{2\tau_L(M_2-1)}}$	$\frac{2\lvert V_o\rvert}{R_o\sqrt{2\tau_L}}$
I_{23}	$\frac{2V_o}{M_1R_o}\sqrt{\frac{1-M_1}{2\tau_L}}$	$\frac{2V_o}{R_o\sqrt{2\tau_LM_2(M_2-1)}}$	$\frac{2\lvert V_o\rvert}{R_o\sqrt{2\tau_L}M}$

Where $\tau_L = \frac{L}{R_oT_S}$ (normalized inductor time constant), $M_1 = \frac{2}{1+\sqrt{1+\frac{8\tau_L}{D^2}}}$ (input-to-output DC transfer ratio), $M_2 \frac{1+\sqrt{1+2D^2/\tau_L}}{2}$, and $M = \frac{D}{\sqrt{2\tau_L}}$.

Step 2:
The t-parameters of the subnetworks are represented as follows:

$$\begin{bmatrix} V_{1a} \\ I_{1a} \end{bmatrix} = T_A \times \begin{bmatrix} V_{2a} \\ -I_{2a} \end{bmatrix} \tag{14.5}$$

$$\begin{bmatrix} V_{1b} \\ I_{1b} \end{bmatrix} = T_B \times \begin{bmatrix} V_{2b} \\ -I_{2b} \end{bmatrix} + \begin{bmatrix} T_{b13} \\ T_{b23} \end{bmatrix} \hat{D}, \tag{14.6}$$

$$\begin{bmatrix} V_{1c} \\ I_{1c} \end{bmatrix} = T_C \times \begin{bmatrix} V_{2c} \\ -I_{2c} \end{bmatrix}, \tag{14.7}$$

and

$$\begin{bmatrix} V_{1d} \\ I_{1d} \end{bmatrix} = T_D \times \begin{bmatrix} V_{2a} \\ -I_{2a} \end{bmatrix} + \begin{bmatrix} T_{d13} \\ T_{d23} \end{bmatrix} \hat{D}, \tag{14.8}$$

where t-parameter matrices T_A, T_B, T_C, and T_D can be obtained from the y-parameters using the relationships between y- and t-parameters listed in Figure 13.6b.
Step 3:
Since the four subnetworks are connected in cascade, the overall t-parameters of the SSC can be represented by

$$\begin{bmatrix} V_{1a} \\ I_{1a} \end{bmatrix} = T_T \times \begin{bmatrix} V_{2d} \\ -I_{2d} \end{bmatrix} + \begin{bmatrix} V_d \\ I_d \end{bmatrix} \hat{D}, \tag{14.9}$$

where

$$\begin{aligned} T_T &= T_A \times T_B \times T_C \times T_D, \\ V_d &= T'_{TR} \times T_{d13} + T'_{TR} \times T_{d23} + T_{A11} \times T_{b13} + T_{A12} \times T_{b2}, \\ I_d &= T'_{TR} \times T_{d13} + T'_{TR} \times T_{d23} + T_{A21} \times T_{b13} + T_{A22} \times T_{b2}, \\ T'_{TR} &= T_A \times T_B \times T_C, \end{aligned}$$

and T_{Nij} denotes the element in row i and column j of matrix T_N.
Step 4:

a) Derivation of the Open-Loop Transfer Characteristics
Based on (14.5)–(14.9), one can determine the open-loop transfer characteristics, which are expressed as follows:

$$A_u = \left.\frac{\hat{V}_o}{\hat{V}_s}\right|_{\hat{D}=0} = \frac{Z_L}{T_{T11} \times Z_L + T_{T12}}, \tag{14.10}$$

$$F_d = \frac{\hat{V}_o}{\hat{D}}\bigg|_{\hat{V}_s=0} = \frac{V_d \times Z_L}{T_{T11} \times Z_L + T_{T12}}, \tag{14.11}$$

$$Z_i = \frac{\hat{V}_s}{\hat{I}_s}\bigg|_{\hat{D}=0} = \frac{T_{T11} \times Z_L + T_{T12}}{T_{T21} \times Z_L + T_{T22}}, \tag{14.12}$$

and

$$Z_o = \frac{\hat{V}_o}{\hat{I}_o}\bigg|_{\hat{D}=0,\hat{V}_s=0} = \frac{T_{T11} \times Z_L + T_{T12}}{T_{T12} \times Z_L}. \tag{14.13}$$

Z_L is a load connected to the output and the symbols with circumflexes "^" denote their corresponding small signals.

b) Derivation of the Closed-Loop Transfer Characteristics

This subsection demonstrates how a single-voltage loop control circuit can be incorporated into the completed ISSC small-signal model. The overall system with a single-voltage loop control is shown in Figure 14.6a. The loop gain T_c can be derived from the block diagram shown in Figure 14.6a. The output impedance Z_{oc} and the line to output transfer characteristic A_{uc} (audio susceptibility) becomes solely defined in terms of the output port elements, while the input port takes part only in determination of the input impedance Z_{ic}. These can be readily confirmed by analyzing the equivalent circuit shown in Figure 14.6a with fundamental linear feedback theory, which leads to

$$T_C = F_D(s)G_C(s)F_m, \tag{14.14}$$

$$Z_{oc} = \frac{Z_o(s)}{1+T_C}, \tag{14.15}$$

$$A_{uc} = \frac{A_u(s)}{1+T_C}, \tag{14.16}$$

and

$$\frac{1}{Z_{ic}} = -\frac{T_C}{1+T_C}\left[\frac{A_u}{F_D} \times \left(I_d - \frac{T_{T22}V_d}{T_{T12}}\right) - \frac{T_{T12}}{T_{T22}}\right] + \frac{1}{1+T_C}\left(\frac{T_{T12}}{T_{T22}} - \frac{T_{T11}T_{T22} - T_{T12}T_{T21}}{T_{T12}}A_u\right), \tag{14.17}$$

where $G_c(s)$ is the transfer function of the controller and F_m is the modulation gain.

14.3 An Illustration Example

An example of a buck-boost + flyback ISSC operated in DCM is used to illustrate the previously described modeling procedure. The specifications and component values of the circuit shown in Figure 14.4a are listed as follows:

$V_s = 110\,\mathrm{V}/60\,\mathrm{Hz}$	$V_o = 48\,\mathrm{V}$	$P_o = 96\,\mathrm{W}$	$L_i = 2.07\,\mathrm{mH}$
$C_i = 120\,\mathrm{nF}$	$L = 150\,\mu\mathrm{H}$	$C = 220\,\mu\mathrm{F}$	$N_1/N_2 = 3.25$
$C_o = 220\,\mu\mathrm{F}$	$R_o = 24\,\Omega$	$f_s = 50\,\mathrm{kHz}$	$L_p = 300\,\mu\mathrm{H}$
$D = 0.4$			

In the derivation of the small-signal models, the flyback converter is treated equivalently as the buck-boost converter except its inductance needs to be properly modified. For a flyback converter, the inductance is equivalent to the primary and secondary magnetizing inductance when the active switch is in the on and off states, respectively. Thus, the average current can be determined from its waveform shown in Figure 14.6b and is given by

$$i_L = \frac{V_S D T_S}{4L_P}\left(1 + \frac{N_1}{N_2}\right) \tag{14.18}$$

where L_p denotes the primary magnetizing inductance of the coupled choke, and N_1/N_2 is the turns ratio of the primary winding to the secondary one. The *y*-parameters of the flyback converter are collected in Table 14.3.

Table 14.3 *y*-Parameters of the flyback converter operated in DCM.

y_{11}	$\frac{M_f^2}{R_o}$
y_{12}	0
y_{21}	$-\frac{2M_f}{R_o}$
y_{22}	$\frac{1}{R_o}$
I_{13}	$\frac{2\lvert V_o\rvert}{R_o}\sqrt{\frac{N_2}{N_1 K_f}}$
I_{23}	$\frac{2\lvert V_o\rvert}{R_o M_f}\sqrt{\frac{N_2}{N_1 K_f}}$

Where $M_f = \frac{2\lvert V_o\rvert}{R_o}\sqrt{\frac{N_2}{N_1 K_f}}$ and $K_f = \frac{N_2}{N_1}\frac{4L_P N_2 f_s}{N_1 R_o (N_1 + N_2)}$.

In the illustration example, since both semi-stages of the ISSC are operated in DCM, loading effect exists in between the stages. That is, the input impedance of the rear semi-stage becomes the load of the front semi-stage. This loading effect complicates the analysis dramatically. The *t*-parameters of the four subnetworks of the buck-boost + flyback ISSC are derived and given by

$$T_A = \begin{bmatrix} 1+s^2L_iC_i & sL_i \\ sC_i & 1 \end{bmatrix}, \tag{14.19}$$

$$T_B = \begin{bmatrix} \dfrac{\sqrt{\tau_{L1}}}{\sqrt{2D}} & \dfrac{\sqrt{\tau_{L1}}R_1}{\sqrt{2D}} \\ \dfrac{D\sqrt{\tau_{L1}}}{2\sqrt{2}f_sL} & \dfrac{D\sqrt{\tau_{L1}}R_1}{2\sqrt{2}f_sL} \end{bmatrix}, \tag{14.20}$$

$$\begin{bmatrix} T_{b13} \\ T_{b23} \end{bmatrix} = \begin{bmatrix} \dfrac{\sqrt{2\tau_{L1}}V_C}{D^2} \\ \dfrac{3V_C}{\sqrt{2\tau_{L1}}R_1} \end{bmatrix}, \tag{14.21}$$

$$T_C = \begin{bmatrix} 1 & 0 \\ sC & 1 \end{bmatrix}, \tag{14.22}$$

$$T_D = \begin{bmatrix} \dfrac{\sqrt{K}}{4D} & \dfrac{R_o\sqrt{K}}{4D} \\ \dfrac{D\sqrt{K}}{KR_o} & \dfrac{D\sqrt{K}}{K} \end{bmatrix}, \tag{14.23}$$

and

$$\begin{bmatrix} T_{d13} \\ T_{d23} \end{bmatrix} = \begin{bmatrix} \dfrac{\sqrt{K}V_o}{2D^2} \\ \dfrac{4V_o}{R_o\sqrt{K}} + \dfrac{2\sqrt{K}V_o}{KR_o} \end{bmatrix}, \tag{14.24}$$

where $\tau_{L1} = \dfrac{f_sL_1}{R_1}$, $R_1 = \dfrac{R_o}{M_f^2}$, $K = \dfrac{(N_1+N_2)R_o}{f_sL_PN_2}$, and M_f is the DC voltage gain of the rear semi-stage (flyback). R_1 represents an equivalent load to subnetwork B, and V_c is the DC-link voltage. Note that the load of the front semi-stage is the input impedance of the rear semi-stage. Thus, the input impedance of the rear semi-stage is substituted for the load term of the front semi-stage. After substituting (14.19)–(14.24) into (14.9), we can obtain the t-parameter matrices of the ISSC, and the four transfer characteristics can be consequently derived. Then, substituting the specifications and component values into the control-to-output transfer function (F_d) can derive the predicted magnitude and phase plots as shown in Figure 14.7a and b, respectively.

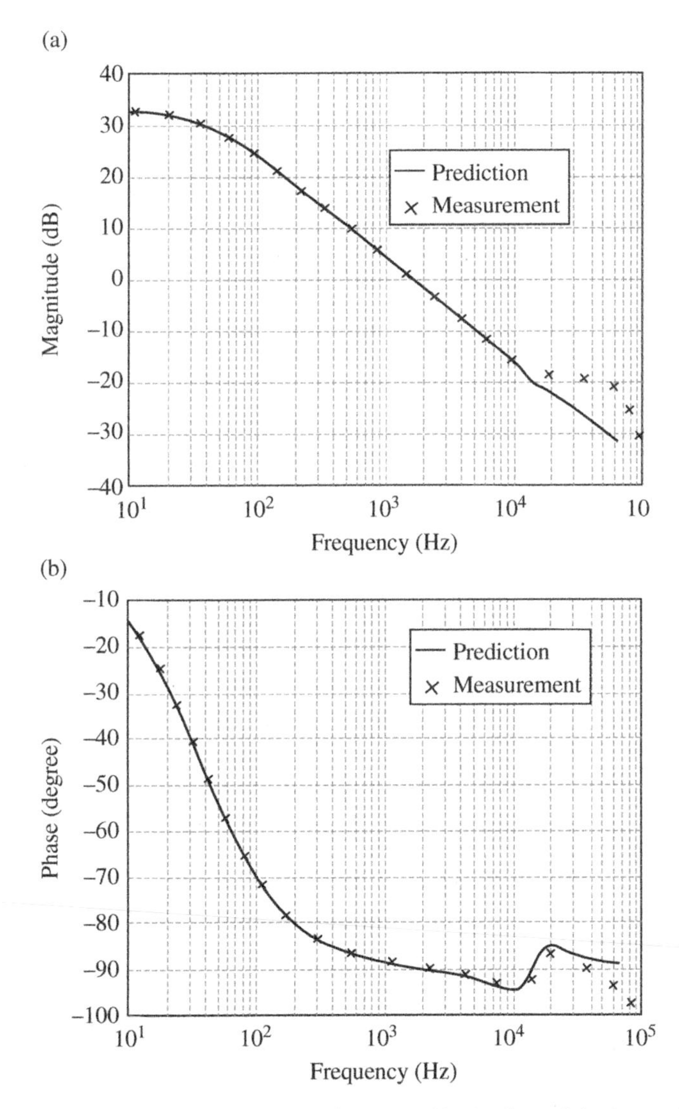

Figure 14.7 The predicted and measured bode plots of the loop gain of the buck-boost + flyback ISSC: (a) magnitude, (b) phase.

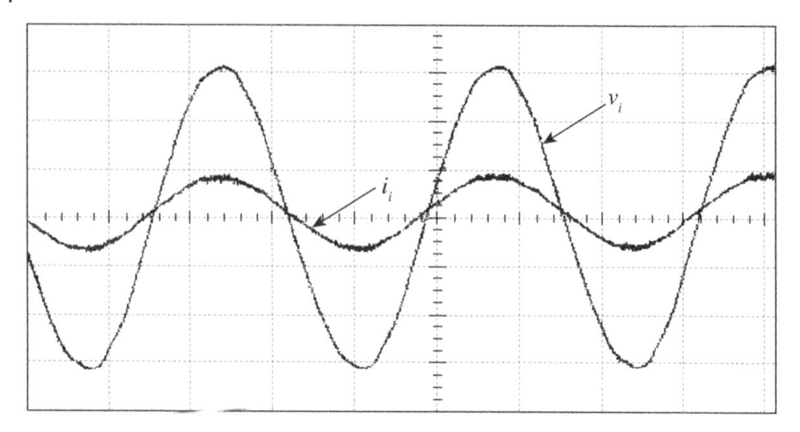

Figure 14.8 The input voltage v_i and current i_i waveforms of the discussed ISSC demonstrate a high power factor (50 V/div, 0.5 A/div, 5 ms/div).

The impedance and gain/phase analyzer were used to measure the frequency response. The measured data are also marked in Figure 14.7. It can be observed that the predicted and the measured are highly agreed. The input voltage and current waveforms of the prototype are also measured and shown in Figure 14.8, which demonstrate a high power factor. When the load is changed and only is the duty-ratio control allowed, it requires that both of the semi-stages (PFC and DC/DC regulator) operate in the DCM to avoid the variation of the DC-link voltage V_C. In most practical applications, the two semi-stages are essentially decoupled by the large energy-storage capacitor C. For this reason, some of the open-loop transfer functions of the PFC semi-stage and the regulator semi-stage can be analyzed without taking into account the influence from the other semi-stage. For instance, it needs only to consider the DC/DC regulator semi-stage (flyback) control-to-output transfer function but not to take into account that of the PFC semi-stage (buck-boost) because PFC stage is decoupled from the rear semi-stage at low frequency. This can be illustrated in Figure 14.9a, in which both the magnitude and phase plots of $\hat{V}_o/\hat{d}$ of the ISSC and the rear semi-stage are highly consistent at low frequency.

This chapter presents an approach based on the graft scheme to derive the small-signal transfer characteristics of ISSCs, which are operated in DCM. With the proposed approach, the small-signal models of the ISSCs can be feasibly derived from those of its originally separate converters plus parameter transformation and manipulation. In particular, relationships among the dynamic models of the converters are readily identified. The derived small-signal models can be readily extended to the ISSCs operated in the CCM with the dead time set to zero. Moreover, the small-signal models of the ISSCs with isolation can be accordingly

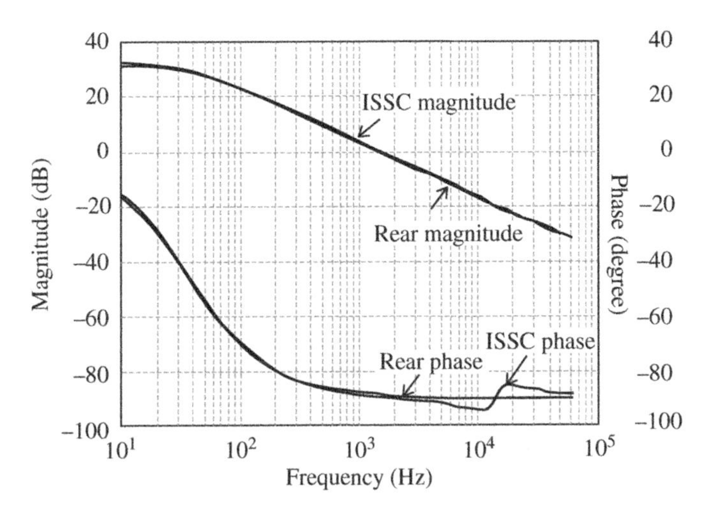

Figure 14.9 The control-to-output magnitude and phase plots of the rear semi-stage and the ISSC.

derived since they can be usually replaced with their transformerless ones. In this chapter, measurements have verified the accuracy and feasibility of the proposed approach to modeling ISSCs.

Further Reading

Cuk, S. (1977). Discontinuous inductor current mode in the optimum topology switching DC-to-DC converter. *Proceedings of the Power Electronics Specialists Conference*, IEEE, pp. 105–123.

Cuk, S. and Middlebrook, R.D. (1977). A general unified approach to modeling switching DC-to-DC converters in discontinuous conduction mode. *Proceedings of the Power Electronics Specialists Conference*, IEEE, pp. 36–57.

Lee, Y.-S. and Siu, K.-W. (1996). Single-switch fast-response switching regulators with unity power factor. *Proceeding of the Applied Power Electronics Conference*, IEEE, pp. 791–796.

Liu, K.-H. and Lin, Y.-L. (1989). Current waveform distortion in power factor correction circuits employing discontinuous-mode boost converter. *Proceeding of the Power Electronics Specialists Conference*, IEEE, pp. 825–829.

Madigan, M., Erickson, R., and Ismail, E. (1992). Integrated high quality rectifier-regulators. *Proceeding of the Power Electronics Specialists Conference*, IEEE, pp. 1043–1051.

Redl, R. and Balogn, L. (1995). Design considerations for single-stage isolated power supplies with fast regulation of the output voltage. *Proceeding of the Power Electronics Specialists Conference*, IEEE, pp. 454–458.

Redl, R., Balogh, L., and Sokal, N.O. (1994) A new family of single-stage isolated power-factor correctors with fast regulation of the output voltage. *Proceeding of the Power Electronics Specialists Conference*, IEEE, pp. 1137–1144.

Vorperian, V. (1990). Simplified analysis of PWM converters using the PWM switch, part II: discontinuous conduction mode. *IEEE Trans. Aerosp. Electron. Syst.* AES-26: 497–505.

Yang, E.-X., Jiang, Y., Hua, G., and Lee, F.C. (1993). Isolated boost circuit for power factor correction. *Proceeding of the Applied Power Electronics Conference*, IEEE, pp. 196–203.

15

Analysis and Design of an Isolated Single-Stage Converter Achieving Power Factor Correction and Fast Regulation

Among the single-stage converters mentioned in the literature, each can usually be divided into a power factor corrector (PFC) semi-stage and a regulator semi-stage. The PFC semi-stage operating in DCM and with constant on-time over half a line cycle can achieve a high power factor and leaves freedom for the regulator semi-stage to achieve a tight- and fast-regulated output. This operation scheme, however, results in higher harmonic current distortion over that operated in CCM. On the other hand, provided the PFC semi-stage operates in the CCM, there exists no freedom for the single-stage converter to achieve both high power factor and fast regulation.

The regulator semi-stage can operate either in DCM or CCM. It has been investigated that when the PFC operates in the DCM while the regulator operates in the CCM, the merits of low conduction loss and low current stress can be attained. Yet, since the DC-link voltage depends on the load current, it is hard to control the output to accommodate a wide load range without dramatically varying the DC-link voltage. With these operation modes, output voltage regulation can be achieved only when both frequency modulation (FM) and PWM are used simultaneously, as conceptually shown in Figure 15.1a. When we operate both the PFC and the regulator in the DCM, as depicted in Figure 15.1b, these two semi-stages can be decoupled by the DC-link capacitor so as high power factor, wide load range, and fast regulation can be achieved but with the penalties of higher conduction loss, higher current stress and larger size of filter components as compared with that operating in the CCM. Therefore, it is mostly suitable for low power applications. For faster dynamic response, current mode control is adopted instead of voltage mode control. In general, both peak current mode and average current mode controls are widely used. The main difference between the two methods is that in the latter, the sensed inductor current signal is averaged and compensated by a current compensation network, while in the former only is

Origin of Power Converters: Decoding, Synthesizing, and Modeling, First Edition.
Tsai-Fu Wu and Yu-Kai Chen.
© 2020 John Wiley & Sons, Inc. Published 2020 by John Wiley & Sons, Inc.

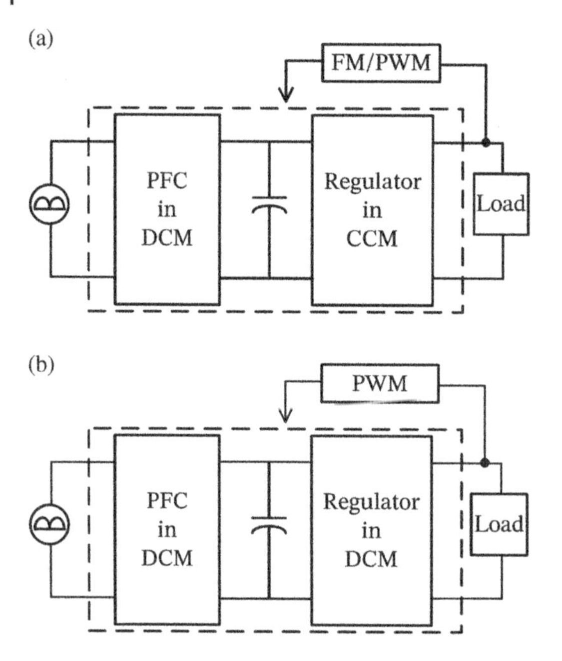

Figure 15.1 Conceptual block diagrams of two different operating modes in PFC and DC regulator.

inductor current sensed and used. Although noise in the average current mode control can be suppressed, the architecture of the system with the average current mode control is complicated. Therefore, peak current mode control is used, and an optimal PI controller designed with Ziegler–Nichols tuning formula is adopted to achieve fast dynamic response, simplicity, and easy implementation. Moreover, an H^{∞} robust control is adopted to design a proper controller for the proposed ISSC to achieve system stability and dynamic performance.

In this chapter, the ISSC derived in Chapter 14 is analyzed and designed into an off-line application. In the designed isolated single-stage converter (ISSC), low input-current harmonic distortion, high power factor, and moderate component stress can be achieved.

15.1 Derivation of the Single-Stage Converter

A typical single-stage converter consists of a PFC and a regulator or an inverter for off-line applications. Derivation of the single-stage converter is an application of the graft scheme. Once a PFC semi-stage and a regulator semi-stage have been chosen, we can apply the graft scheme to integrate the two semi-stages.

15.1.1 Selection of Individual Semi-Stages

Theoretically, all of the DC–DC converters can be used as PFCs. The buck-boost converter is selected in this discussion because it can yield high power factor when operating in the DCM and it can either ascend or descend the input voltage. In addition, the voltage stress imposed on components can be properly controlled by the buck-boost semi-stage. After selecting the PFC semi-stage, a regulator semi-stage needs to be carefully determined. For low power applications, a fly-back converter is more attractive than a boost converter owing to its simplicity and flexibility. In addition, it provides galvanic isolation, short circuit protection, and resolving start-up problems. Moreover, the DC-link voltage (i.e. the input voltage of the flyback converter) is not necessary lower than the output voltage, as in the case with a boost converter. Thus, a buck-boost converter and a flyback converter are selected as the PFC semi-stage and the regulator semi-stage in the proposed ISSC, respectively.

15.1.2 Derivation of the Discussed Isolated Single-Stage Converter

Figure 15.2a shows the schematic diagram of a buck-boost converter and a flyback converter in cascade connection. Note that the dot notations of the two windings are marked in this way owing to the polarity of capacitor C_c. Properly relocating M_2 (as shown in dashed lines) but without varying its operating principle can establish a common node for M_2 and M_3. Since the drain leads of M_2 (after relocation) and M_3 share the same node and they operate synchronously, by adopting the approach presented in previous chapter, an inverted T-type synchronous switch M_{23} can be used to replace them. After the relocation of diode D_{B2} and removement of the other blocking diode, the final version of the ISSC is depicted in Figure 15.2b.

15.2 Analysis of the Isolated Single-Stage Converter Operated in DCM + DCM

The proposed ISSC is the combination of two semi-stages, namely, a buck-boost semi-stage and a flyback semi-stage. Based on the grafted method, the two semi-stages can be analyzed separately. The power stage design that is highly related to operating modes (CCM or DCM) can therefore be achieved. The analysis of each individual semi-stage is presented in this section, while the design equations and component values are determined in next section.

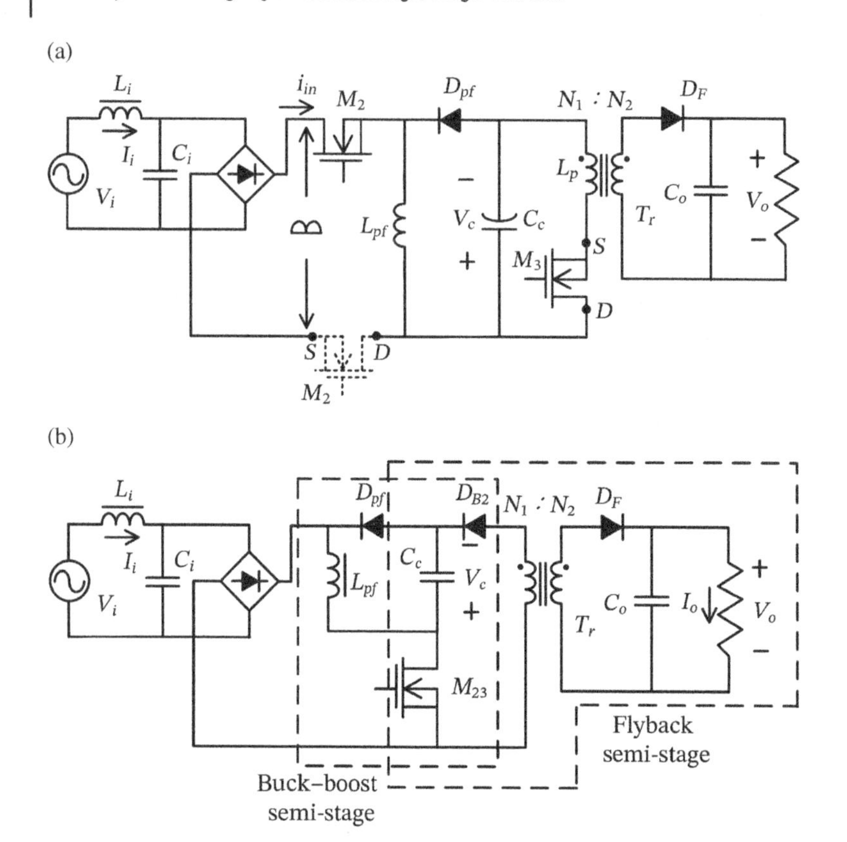

Figure 15.2 The ISSC derived from a buck-boost semi-stage and a flyback semi-stage.

15.2.1 Buck-Boost Power Factor Corrector

The buck-boost semi-stage, as designated in Figure 15.2b, is usually operated in DCM and with constant on-time to achieve a high power factor. Its equivalent circuit is shown in Figure 15.3. The converter peak input current $i_{in(p)}(t)$ can be approximately expressed by the following equation when $T_s \ll 1/f_\ell$:

$$i_{in(p)}(t) = \frac{V_I \left|\sin(2\pi f_\ell t)\right|}{L_{pf}} \cdot D_1 T_s, \quad 0 \le t < \frac{1}{2f_\ell} \tag{15.1}$$

where f_ℓ is the line frequency, V_I is the amplitude of line voltage, T_s denotes the switching period, and D_1 is the duty ratio of switch M_{23}. The waveforms of inductor current i_{Lpf} and input current i_{in} are conceptually illustrated in Figure 15.4a and b, respectively. By employing an L_i–C_i network in series with the AC-voltage

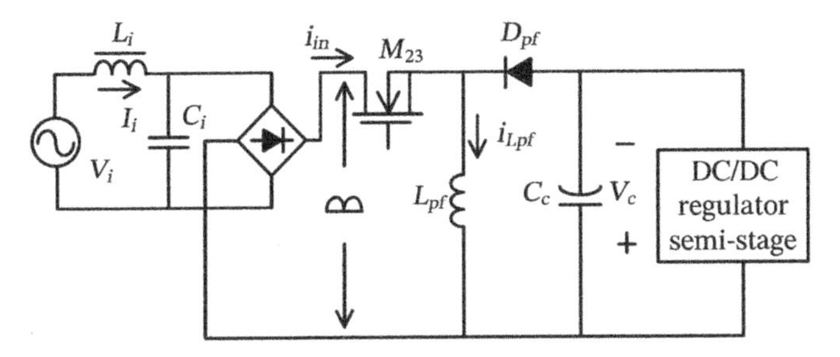

Figure 15.3 The equivalent circuit of the buck-boost semi-stage acts as a PFC.

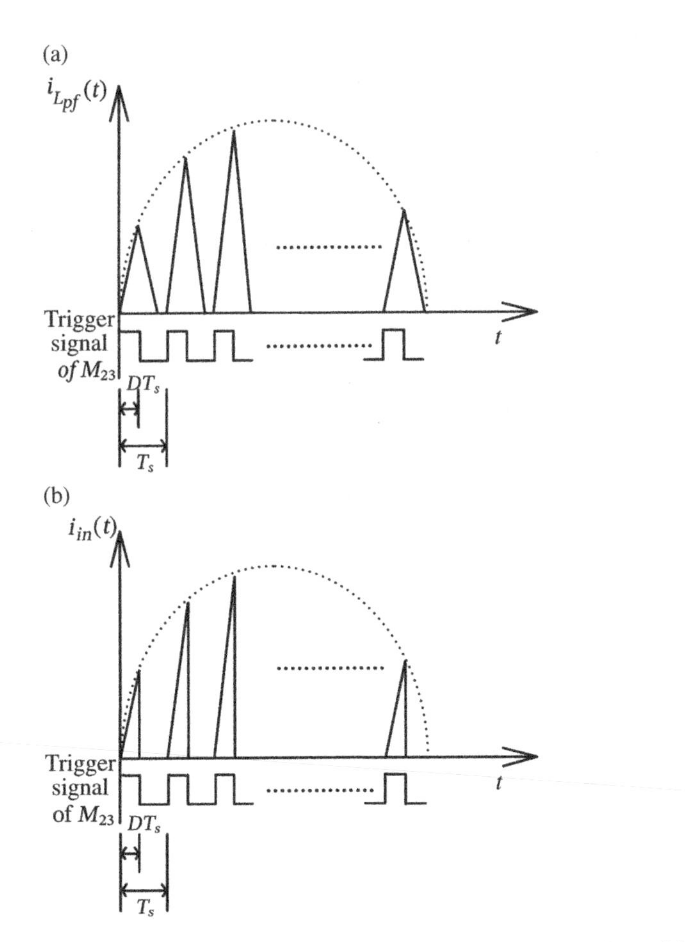

Figure 15.4 Conceptual current waveforms of (a) inductor L_{pf} and (b) active switch M_{23}.

source, as shown in Figure 15.3, the input line current can be smoothed. Since the line current is a low-pass filtered value of $i_{in}(t)$, whose envelope follows the waveform of the line voltage, its waveform will be sinusoidal-like and result in a high power factor. For deriving the component values, the input average power P_i flows into the PFC stage and is evaluated and expressed as

$$P_i = \frac{\frac{D_1}{2L_{pf}f_s}\int_0^{T_\ell} V_I^2 \sin^2\left(2\pi f_\ell t\right)dt}{T_\ell} = \frac{D_1 V_I^2}{4L_{pf}f_s}, \tag{15.2}$$

where $f_s = 1/T_s$ and T_ℓ is the period of the line source.

15.2.2 Flyback Regulator

Figure 15.5 shows the equivalent circuit of the flyback regulator. When both buck-boost and flyback semi-stages operate in the DCM, the DC-link voltage can be controlled to be independent of load current but linearly depends on the line voltage. In this study, both semi-stages are designed to operate in the DCM, and the following equation can be derived as

$$V_c D_1 N_1 = V_o D_2 N_2, \tag{15.3}$$

where D_1 and D_2 are the duty ratios of switch M_{23} and freewheeling diode D_F, respectively, N_1 and N_2 denote the numbers of turns of the primary and secondary windings of transformer T_r, respectively, and V_o is the output voltage. The sum of duty ratios D_1 and D_2 has to be less than unity to ensure a DCM operation in the flyback semi-stage. Design of the flyback semi-stage is in determining the magnetizing inductance L_p seen from the primary side of transformer T_r. The L_p can be determined from (15.2) to (15.3) and the following simultaneous Eqs. (15.4)–(15.6):

$$P_o = V_o I_o, \tag{15.4}$$

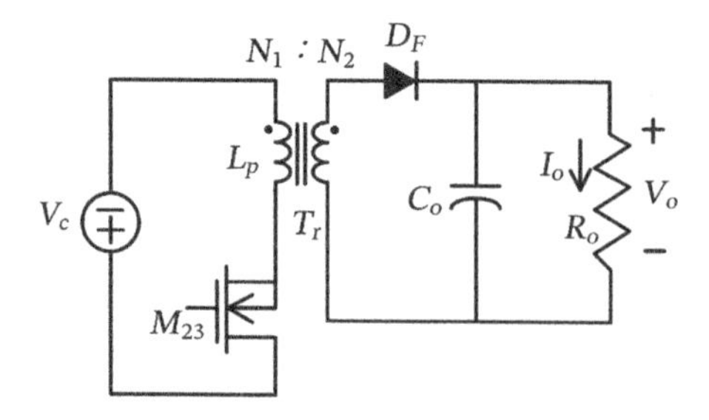

Figure 15.5 An equivalent circuit of the isolated flyback converter.

$$P_f = \frac{V_c^2 D_1^2}{2L_p f_s} \tag{15.5}$$

and let

$$P_o = P_f = P_i, \tag{15.6}$$

where P_f denotes the power flowing into the flyback semi-stage. The DC-link voltage, therefore, can be expressed in terms of the line voltage and the component values, and is uniquely determined by

$$V_c = V_I \sqrt{\frac{L_p}{2L_{pf}}} \tag{15.7}$$

Since DC-link voltage V_c is independent of the load current, the output regulation can be achieved simply by adjusting the duty ratio D_1 so as $P_o = P_f = P_i$.

15.3 Design of a Peak Current Mode Controller for the ISSC

An accurate small-signal model with a peak current mode control is necessary for designing a suitable compensator. Several modeling schemes have been developed, but they resulted in complicated controller design. Recently, a new dynamic modeling approach that employs an equivalent PWM switch model and describes the inductor current with a sampled-data technique was proposed, and it presented convenient results for a controller design. In this discussion, this approach is adopted.

In most practical applications, the PFC and regulator semi-stages are essentially decoupled by the large energy-storage capacitor C_c. For simplifying the modeling of ISSC, only the rear semi-stage is processed. This argument can be judged from the control-to-output bode plots shown in Figure 15.6, in which the magnitude and phase of both plots of the ISSC and the rear semi-stage are relatively close. The electrical specifications and component values of the ISSC are collected in Table 15.1. In the later discussion, the compensator design only takes into account the regulator semi-stage.

Figure 15.7 shows the regulator associated with a peak current mode control where the PI controller is optimally designed to achieve the desired dynamic performance. A control block diagram corresponding to the circuit shown in Figure 15.7 is derived and represented in Figure 15.8 for the convenience of designing a PI controller $G_c(s)$, where the related parameters are listed in

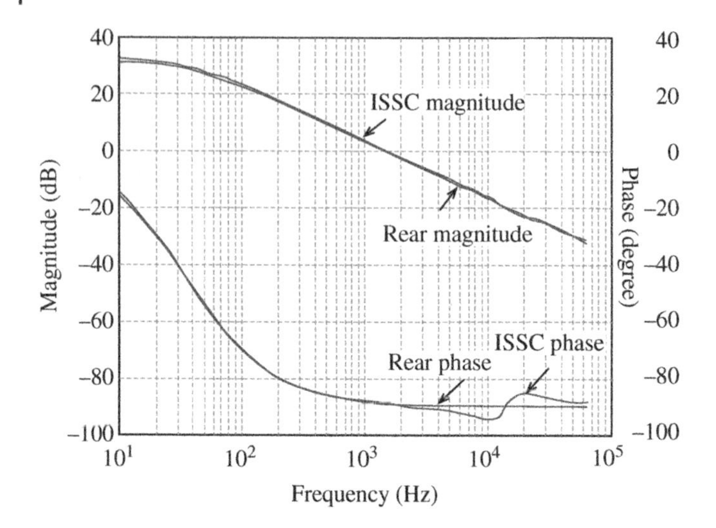

Figure 15.6 The magnitude and phase plots of the control-to-output transfer functions of the rear semi-stage (flyback) and the ISSC.

Table 15.1 Specifications and component values of the proposed ISSC.

Specification	Value	Component	Value
Input voltage	110 V	L_{pf}	150 μH
Output voltage	48 V	L_p	300 μH
Output current	0.4–2 A	C_c	220 μF
Switching frequency	50 kHz	C_o	220 μF
DC-link voltage	156 V	N_1/N_2	3.25
Output voltage ripple	2%	L_i	2.07 mH
		C_i	120 nF

Table 15.2. In the table, K_f is the feedforward gain, K_r is the feedback gain, F_m is the modulation gain, S_n is the slope of inductor current during turn-on, and H_{1r}, H_{2r}, and H_{3r} are the transfer gains as defined in Figure 15.8. To investigate the system dynamics, the control-to-output transfer characteristic is determined and expressed as follows:

$$F_d = \left.\frac{\hat{V}_o}{\hat{d}}\right|_{\hat{V}_c=0} = H_{1r} \cdot H_{3r} \tag{15.8}$$

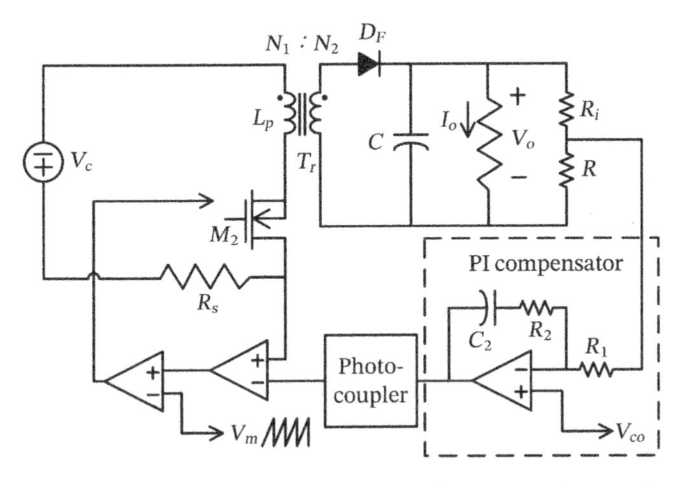

Figure 15.7 Flyback semi-stage with a peak current mode control.

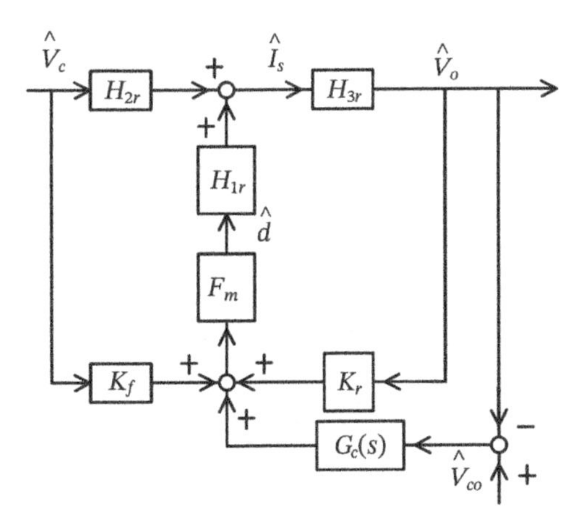

Figure 15.8 A small-signal model of the proposed regulator semi-stage system.

There are many types of compensators that can yield the desired dynamic response and performance. A PID compensator is one of the most popular types because of its simplicity and easy implementation. Consider the transfer function of a PID compensator:

$$G_C(s) = K_p + \frac{K_i}{s} + K_d s, \tag{15.9}$$

where K_p is a proportional gain, K_i is an integral gain, and K_d is a derivative one. In this example, the parameters of the PID compensator are determined

Table 15.2 Summary of gain parameters for the ISSC operated in the DCM.

F_m	$\frac{1}{(S_e + S_n)T_S}$
S_n	$\frac{V_S}{L_P}$
H_{1r}	$\frac{V_C T_S}{4L}\left(1+\frac{N_1}{N_2}\right)$
H_{2r}	$\frac{D}{V_C}$
H_{3r}	$\frac{\frac{4V_O H_4}{C}+\frac{N_1}{N_2}\times\frac{D}{C}\times\frac{V_C}{V_O}}{s+\frac{1}{RC}-4H_4}$, where $H_4 = \frac{LN_2}{RDV_C T_S\left(N_1+N_2\right)}$
K_f	$-\frac{V_C T_S}{L_P}$
K_r	0

Where S_e is the slope of the external ramp.

based on the Ziegler–Nichols tuning method. It tunes the PID parameters according to the critical values at the critical point on the Nyquist plot of the control to output transfer function $\left(\hat{V}_o/\hat{d}\right)$. The critical point is defined as the Nyquist curve intersecting with the negative real axis at (−1,0), and the critical values are the critical gain, k_u, and the critical period, t_u, at which oscillation will sustain.

In the Ziegler–Nichols scheme, the critical gain and the critical period can be generally determined as follows. If the regulator dynamics are not precisely known, a proportional amplifier is connected to the system, and the gain is gradually increased until system oscillation sustains. The gain when oscillation occurs is the critical gain, and the period of the oscillation is the critical period. In the design example, the dynamics of the regulator without compensation can be obtained from the small-signal modeling. Thus, k_u and t_u can be derived directly from the Nyquist plot of the $\hat{V}_o/\hat{d}$.

Equation (15.9) can be rewritten as follows:

$$G_c(s) = K_c\left(1+\frac{1}{T_i s}+T_d s\right), \tag{15.10}$$

where $K_c = K_p$, $T_i = \frac{K_c}{K_i}$, and $T_d = \frac{K_d}{K_c}$.

Table 15.3 Parameters of the P, PI, PID controller tuned with Ziegler–Nichols method.

	PID	PI	P
Proportional gain	$k_c = 0.6k_u$	$k_c = 0.45k_u$	$k_c = 0.125k_u$
Integral time	$T_i = 0.5t_u$	$T_i = 0.85t_u$	
Derivative time	$T_d = 0.125t_u$		

In determining a set of optimal PID parameters (K_c, T_i, and T_d), the performance index (α) of the adopted tuning method is specified as follows:

$$\alpha = \int_0^{\infty} |e|\, dt \tag{15.11}$$

where e is the error between the reference and the feedback. The tuning process is to minimize the above performance index. Thus, the optimal parameters of PID, PI, and P controllers can be determined according to the critical gain, k_u, and the critical period, t_u, as shown in Table 15.3. The derivative part of the PID compensator can improve the transient response of the system, but it may accentuate noise at higher frequencies. Therefore, a PI compensator is chosen and implemented with an operational amplifier and R–C components, which are shown in the dashed line area of Figure 15.7. The parameters, K_p and K_i, of the PI compensator related to the circuit parameters can be determined as $K_p = R_2/R_1 = 2$ and $K_i = R_2/R_1C_2 = 100$ from Table 15.3. Moreover, Figure 15.9 shows the output voltage responses of the system to step duty-ratio changes and with different PI compensators. It can be observed that the system with the determined parameters yields fast response, low steady-state error, and small maximum overshoot, which meet the desired performance.

15.4 Practical Consideration and Design Procedure

15.4.1 Component Stress

The voltage and current stresses imposed on switch components are the primary concern, which are determined as follows. When synchronous switch M_{23} is turned off, it encounters the maximum voltage stress of $V_{M23} = V_c + V_I$. When it is turned on, a current flowing through the synchronous switch M_{23} equals

$$i_{M_{23}} = \frac{N_2}{N_1} I_o + i_{in(p)} \tag{15.12}$$

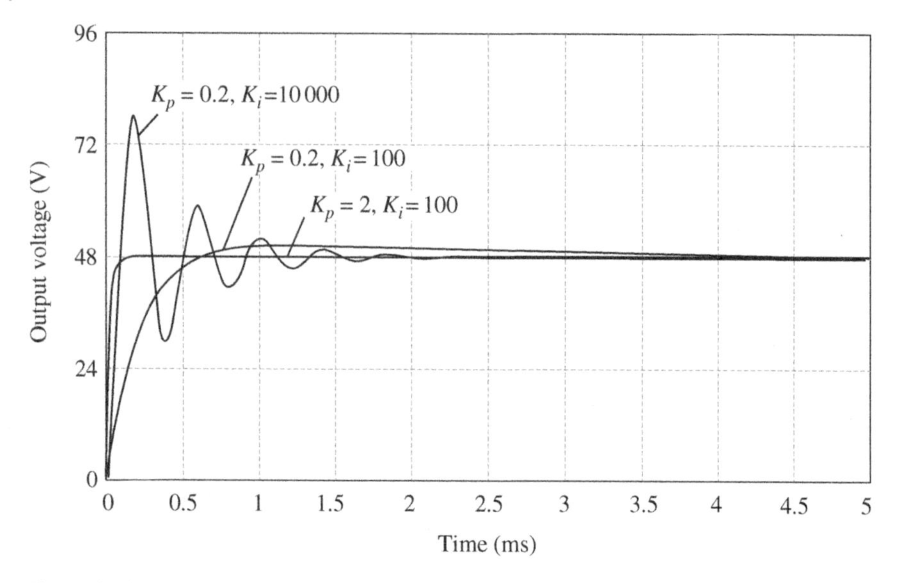

Figure 15.9 Output voltage responses of the ISSC to a step duty ratio change from $D = 0$ to $D = 0.42$ with various PI parameters.

where $i_{in(p)}$ is given in (15.1). Therefore, the maximum current stress is

$$i_{M_{23}} = \frac{N_2}{N_1} I_o + \frac{V_I D T_s}{L_{pf}} \tag{15.13}$$

15.4.2 Snubber Circuit

The voltage stress and current stress listed above are theoretically determined. Indeed, leakage inductance L_l of the transformer is inevitable and it is 4.5 μH in the discussed prototype. The energy stored in the leakage inductance during the switch on state has to release during the switch off state. A snubber, which is indicated in the dashed line box of Figure 15.10, is needed to absorb the released energy. It should be carefully designed to suppress a possibly induced spike so as to reduce voltage stress imposed on switch M_{23}. Additionally, the current flowing through the snubber should be kept as small as possible to sustain high efficiency. For the designed RCD snubber, the R_sC_s time constant is approximately chosen as half the switching on-time interval (i.e., 3.6 μs).

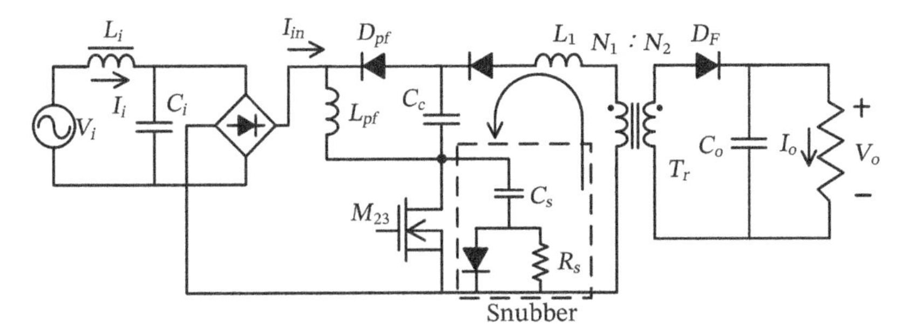

Figure 15.10 The ISSC with an RCD snubber.

15.4.3 Design Procedure

The design procedure of the proposed single-stage converter with high power factor is summarized as follows:

1) Determine the duty ratios of D_1 and D_2 of switch M_{23} and freewheeling diode D_{pf}, respectively. The sum of D_1 and D_2 must be limited within unity to ensure a DCM operation in the flyback semi-stage.
2) Specify DC-link voltage V_c and switching frequency f_s for calculating the inductance L_{pf} of the buck-boost semi-stage using (15.2).
   Determine the magnetizing inductance L_p in the primary side of transformer T_r from (15.7) and determine the turns ratio of T_r from the following equation:

$$\frac{V_c}{V_o} = \frac{N_1}{N_2} \tag{15.14}$$

   Choose an appropriate output filter capacitance C_o to meet the specification of output ripple voltage (ΔV_o). That is, C_o needs to satisfy the following inequality:

$$C_o \geq \frac{P_o}{2\pi \cdot 120 \cdot V_o \cdot \Delta V_o}, \tag{15.15}$$

   and let the DC-link capacitor C_c equals the C_o. Generally, the proposed system can exhibit fast transient when the DC-link capacitor (C_c) is relatively larger in size than the output capacitor (C_o).

3) The design of a compensator can be accomplished with the guidelines mentioned in Section 15.4. A peak current mode control IC can be used to simplify the feedback compensator design.
4) Choose an appropriate current sensing resistor. For suppressing the sensed current spikes, a low-pass filter is located in series with the current sensing resistor. The *RC* time constant of the low-pass filter can be set approximately equal to spike duration. Because the duty cycle of the switch is shorter than 50%, slope compensation is not needed in the discussed system.

15.5 Hardware Measurements

According to the design procedure, a prototype is designed and implemented with the specifications listed in Table 15.1. The maximum duty ratios are selected as $D_{1(max)} = D_{2(max)} = 0.42$ to ensure both semi-stages operating in the DCM. The DC-link voltage V_c is selected as $110\sqrt{2}\left(=156\right)$ V, which results in a reasonable voltage stress imposed on switches.

The controller is implemented with a peak current mode control PWM IC (UC 3845). Figure 15.11a shows the transient response of the line current to periodic step load changes between 50 and 100% of the full power. It can be observed from the waveforms that the sinusoidal line current can follow the change without noticeable delay. In addition, Figure 15.11b shows the response of the output voltage to the same changes. The waveforms show that a fast regulation can be achieved with an insignificant steady-state error. The input voltage and current waveforms under the operating conditions of $V_i = 90\,\text{V}$ and $I_o = 0.4\,\text{A}$ are shown in Figure 15.11c. It can be seen that the current waveform follows that of the voltage sinusoidally. The measurements of power factor, total harmonic distortion (THD), and harmonics of the line current are listed in Table 15.4. Figure 15.12a shows the measured efficiency, which is about 70%, under different input voltages $V_i = 90\text{–}132\,\text{V}$ and full load. In addition, the measured power factor and input current THD under the same condition are shown in Figure 15.12b and c, respectively. It can be seen that the input current THD is almost independent of the input voltages, and the power factor is close to unity. The measured efficiency obtained as a function of the output power can be seen in Figure 15.12d. Figure 15.12e and f show the power factor and input current THD under different output power ratings and $V_i = 110\,\text{V}$, respectively. In the proposed ISSC, a high power factor (0.999) and a low input current THD can be achieved for different input voltages and output power ratings.

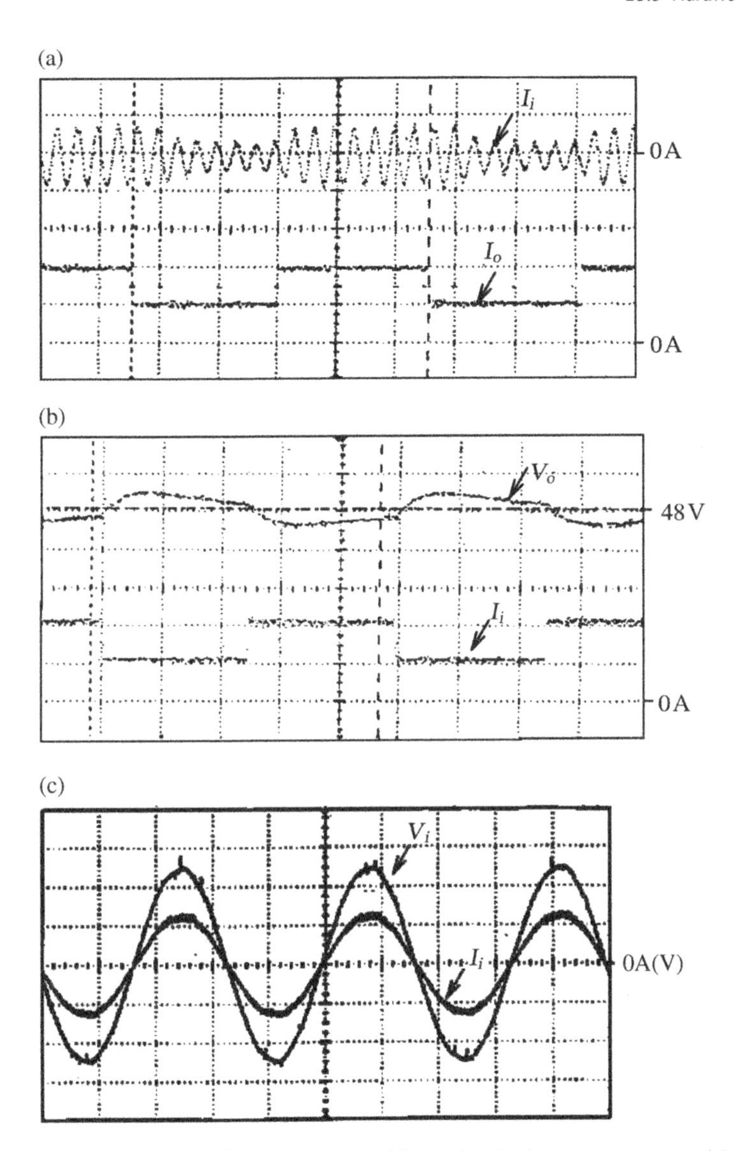

Figure 15.11 Oscillogram measured from the single-stage converter (a) the line current to periodic step changes of the loads between 50 and 100% of the full load (I_i: 2 A/div, I_o: 1 A/div, 5 ms/div), (b) the output voltage to periodic step changes of the loads between 50 and 100% of the full load (V_o: 1 V/div, I_o: 1 A/div, 5 ms/div), and (c) the measured input voltage and current at V_i = 90 V and I_o = 0.4 A (V_i: 50 V/div, I_o: 1 A/div, 5 ms/div).

Table 15.4 List of the power factor and current harmonics of the ISSC.

Power factor	0.999 (%)
THD of input current	4.019
3rd harmonic	3.268
5th harmonic	0.871
7th harmonic	0.927
9th harmonic	1.181
11th harmonic	0.429
13th harmonic	0.265

15.6 Design of an H^{∞} Robust Controller for the ISSC

15.6.1 H^{∞} Control

To improve the system dynamics and robust performance, the ISSC with a robust controller is also presented in this chapter. A block diagram used to illustrate the proposed H^{∞} robust control is depicted in Figure 15.13, in which the uncertainly plant is with two uncertainties, the variation of component value, and load variation. The effects due to the both variations are the same in the proposed ISSC. Therefore, the variation of the component values is combined with the variation of the load. Besides, the line voltage is considered a disturbance to the ISSC. To achieve a better dynamic response, a robust controller is adopted to handle the ISSC under various operating conditions.

A standard H^{∞} optimal control problem is conceptually illustrated by the block diagram shown in Figure 15.14, where w represents an external input (disturbance), y is a measured signal, u is a control signal to the plant, and z is a signal to be regulated. The augmented plant $P(s)$ includes not only the conventional plant but also the weighting functions used for specifying the desired performance. It can be conceptually illustrated by Figure 15.15. The relationship between input and output vectors, thus, can be linked together through $P(s)$, and is given as follows:

$$\begin{bmatrix} z \\ y \end{bmatrix} = \begin{bmatrix} P_{11}(s) & P_{12}(s) \\ P_{21}(s) & P_{22}(s) \end{bmatrix} \begin{bmatrix} w \\ u \end{bmatrix} \tag{15.16}$$

Defining

$$F_l(P,K) = T_{zw} = \left[P_{11} + P_{12}K\left(I - P_{22}K\right)^{-1} P_{21} \right], \tag{15.17}$$

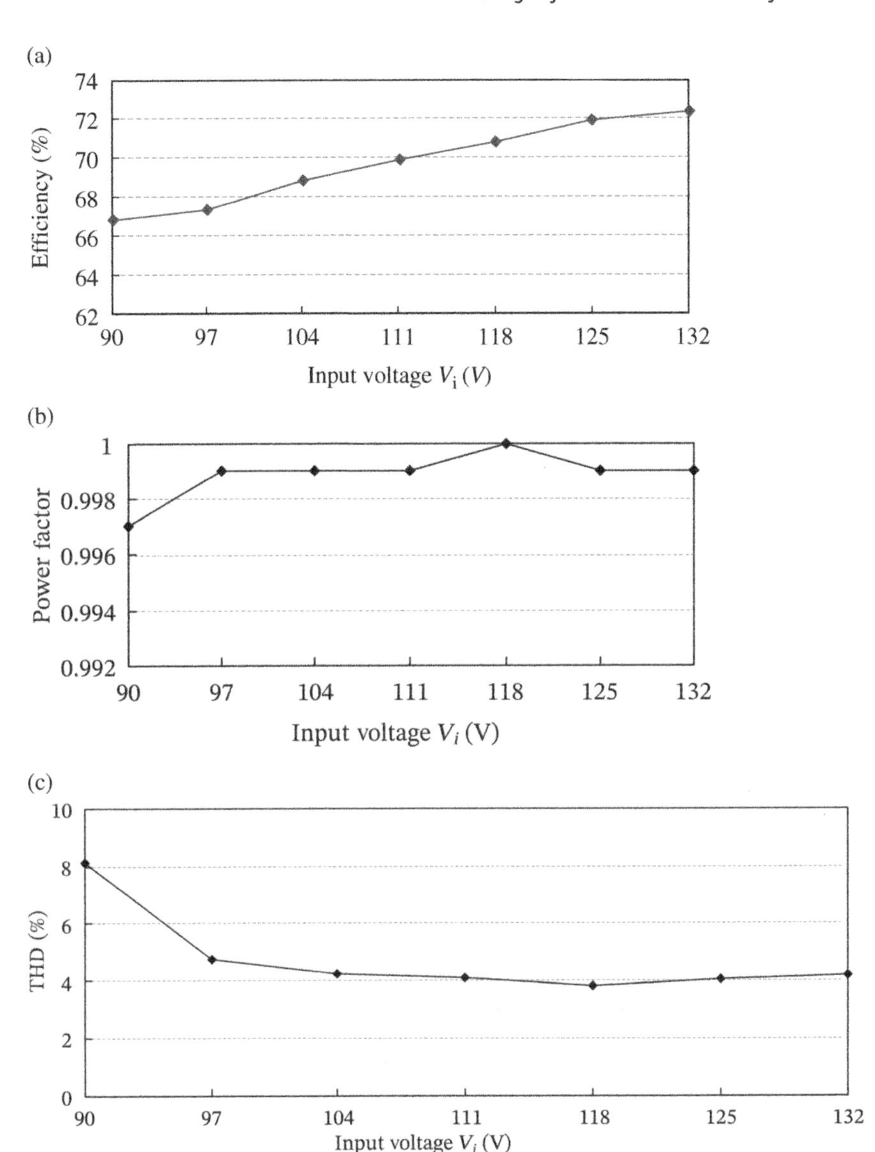

Figure 15.12 (a) Measured efficiency versus input voltages, (b) measured power factor versus input voltages, (c) measured THD versus input voltages, (d) measured efficiency versus output power, (e) measured power factor versus output power, and (f) measured THD versus output power.

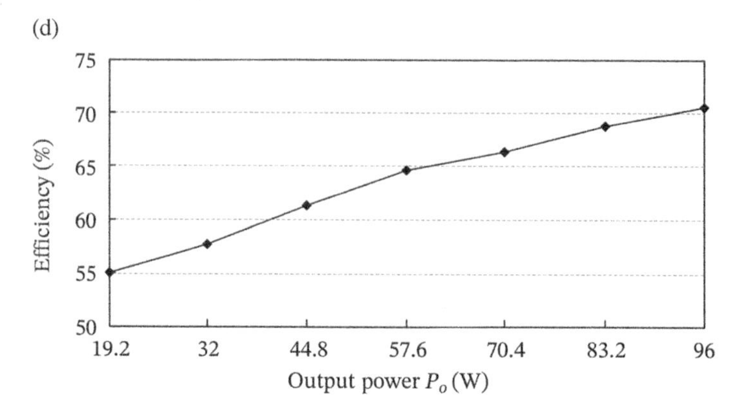

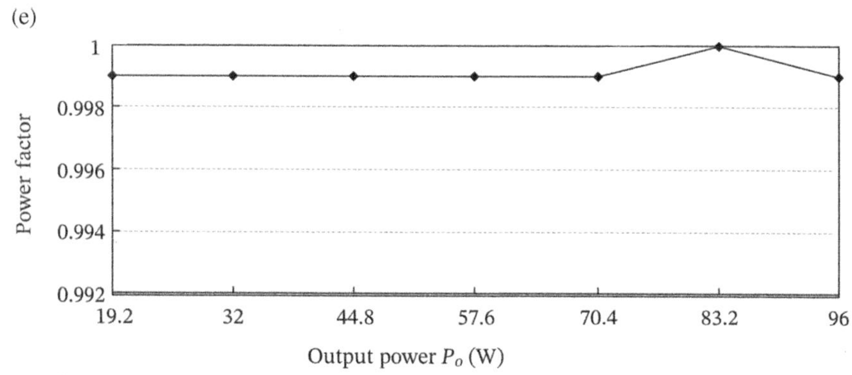

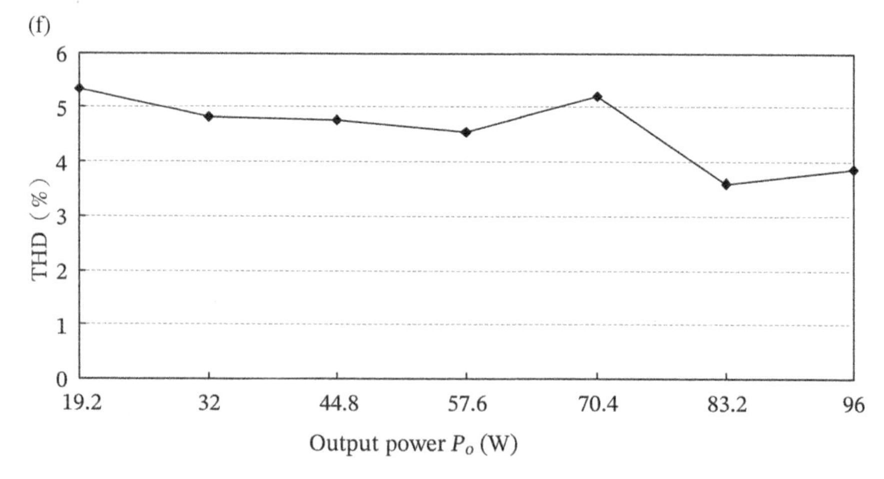

Figure 15.12 (Continued)

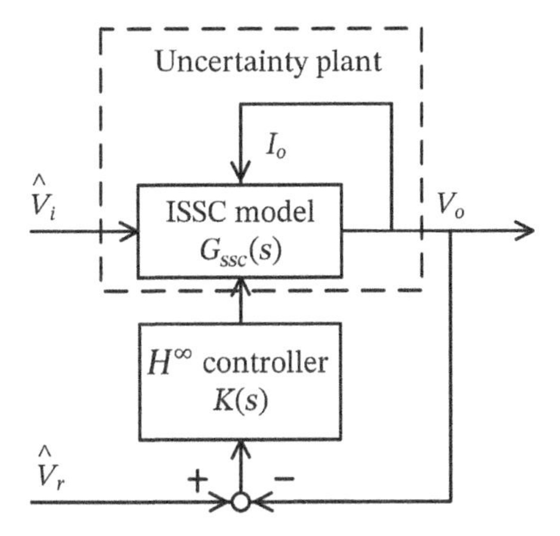

Figure 15.13 Control block diagram of the proposed ISSC with an H^{∞} controller.

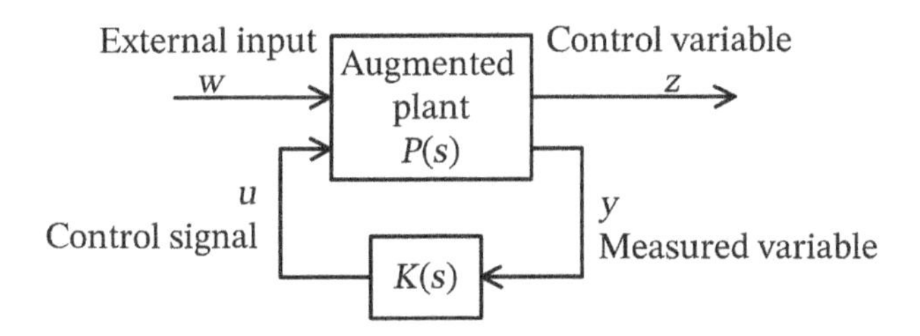

Figure 15.14 Configuration of a standard H^{∞} control problem.

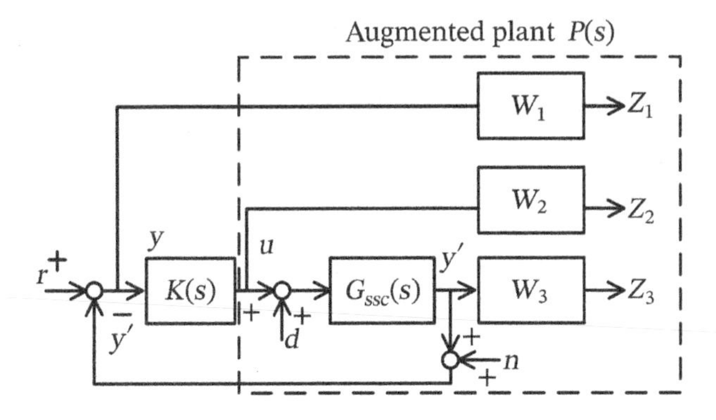

Figure 15.15 Illustration of an augmented plant with a robust control.

and

$$u = Ky \tag{15.18}$$

as an augmented transfer matrix of external input w to output z, an H^{∞} optimal control problem is then to design a stabilizing controller, K, so as to minimize the ∞-norm of T_{zw}, as shown in (15.19). The criterion is mathematically represented as

$$\min_{K \in H^{\infty}} \left\| T_{zw} \right\|_{\infty} = \min_{K \in H^{\infty}} \left\{ \sup_{R_e(s)>0} \bar{\sigma}\left[T_{zw}(s) \right] \right\} \tag{15.19}$$

In general, T_{zw} is formed with sensitivity function (S), complementary sensitivity function ($T = I - S$) of the desired closed-loop control, and weighting function (W); and R measures the control effort. T_{zw} is minimized over a specified frequency range to achieve the optimal performance. Thus, the standard problem can be expressed as

$$\left\| T_{zw} \right\|_{\infty} = \left\| \begin{matrix} W_1 S \\ W_2 R \\ W_3 T \end{matrix} \right\|_{\infty} < r, \tag{15.20}$$

where W_1 and W_3 are frequency-dependent weights that represent the desired design specifications, and $r > 0$ is the cost coefficient. When $\| W_1 S \|_{\infty}$ is minimized, tracking error is also minimized, and when $\| W_3 S \|_{\infty}$ is minimized, the robustness can be optimally achieved.

15.6.2 An Illustration Example of Robust Control and Hardware Measurements

An example of a buck-boost + flyback ISSC operated in the DCM with a proper H^{∞} robust controller is used to illustrate the previous discussion, including the analysis, modeling, and design of the ISSC. The design specifications of the circuit are listed as follows:

A) Design specifications:
  1) Minimize the sensitivity to line voltage and load variations.
  2) Achieve zero steady-state error.
  3) Output voltage overshoot is less than 2% of the regulated value.
  4) Settling time is less than 100 μs.
  5) Bandwidth is higher than 5 kHz.

B) Design procedure:
   1) Derive the small-signal transfer characteristics of the ISSC, including the control-to-output and input-to-output transfer functions represented in the two-port networks. These are used in designing a robust controller and expressed as follows:

$$F_d(s) = G_{SSC}(V_i, I_o) = \left.\frac{V_o}{d}\right|_{V_i=0} = \frac{1.948\times10^5 s^4 + 13.98\times10^9 s^2 + 7.839\times10^{13} s + 2.87197\times10^{16}}{\sqrt{15}\times\left(s^4 + 89308.2s^3 + 4.06\times10^9 s^2 + 1.69\times10^{12} s + 6.18\times10^{13}\right)} \tag{15.21}$$

and

$$A_u(s) = G_{au}(s) = \left.\frac{V_o}{V_i}\right|_{d=0} = \frac{1.21093\times10^{14}}{\sqrt{15}\times\left(s^4 + 89308.2s^3 + 4.06\times10^9 s^2 + 1.69\times10^{12} s + 6.18\times10^{13}\right)} \tag{15.22}$$

The above transfer functions are derived corresponding to a nominal load of $R_o = 36\,\Omega$. Figures 15.16 and 15.17 show the magnitude–frequency plots

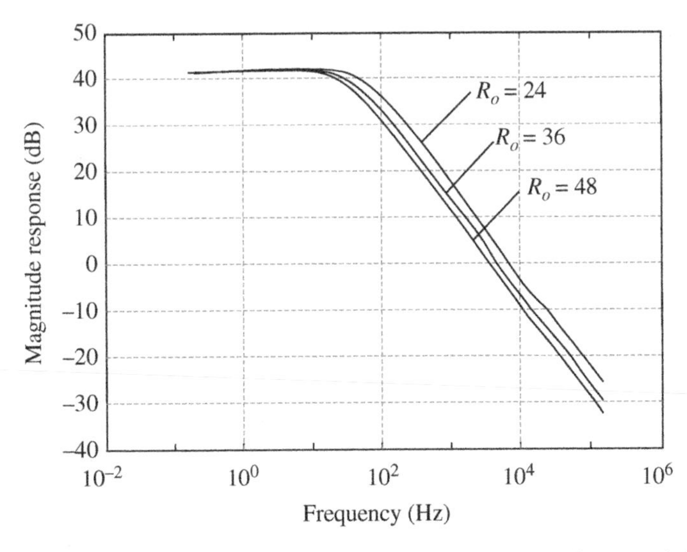

Figure 15.16 The magnitude–frequency plots of $F_d(s)$ under several different load conditions.

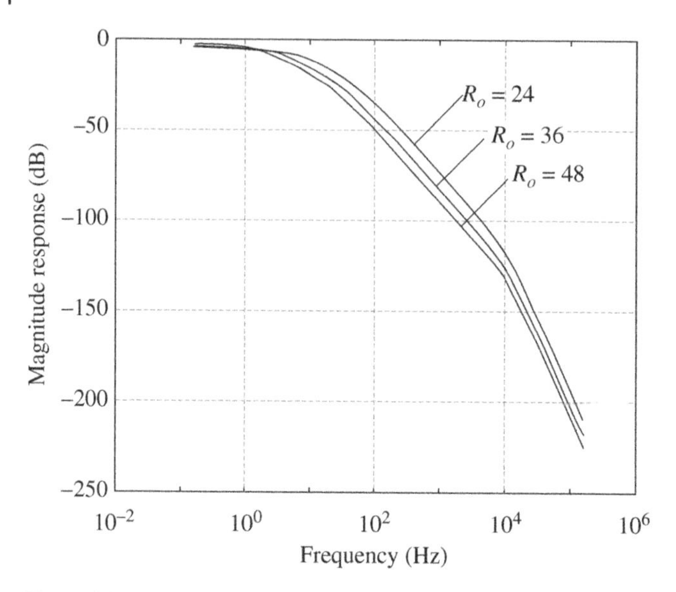

Figure 15.17 The magnitude–frequency plots of $A_u(s)$ under several different load conditions.

of $F_d(s)$ and $A_u(s)$ with respect to several different load conditions ($R_o = 24\,\Omega$, $36\,\Omega$ and $48\,\Omega$), respectively.

2) Augment the plant G_{ssc} with proper weighting functions $W_1(s)$ and $W_3(s)$ based on the desired performance indexes. The augmented plant, $P(s)$, can be conceptually illustrated by Figure 15.15. The minimization of sensitivity function S over low to middle frequencies is to improve tracking capability and disturbance attenuation, while the minimization of complementary sensitivity function T over high frequencies is to provide robust stability in the presence of sensor noise and multiplicative perturbation arising from unmodeled (high-order) dynamics, neglected nonlinearities, and plant-parameter variations. Thus, $W_1(s)$ is a typical low-pass filter, shaping the sensitivity function S at low frequencies to reject disturbances and reduce tracking errors. Weighting function $W_3(s)$ is usually chosen to be a high-pass filter, shaping T at high frequencies to minimize the instability effects. The weighting functions $W_1(s)$, $W_2(s)$, and $W_3(s)$ in (15.20) are selected as

$$W_1(s) = \frac{100 \times \left(\dfrac{s}{15000} + 1\right)}{s + 10^{-6}}, \tag{15.23}$$

$$W_2(s) = 0, \tag{15.24}$$

and

$$W_3(s) = \frac{0.56 \times \left(\dfrac{s}{8500} + 1 \right)}{s + 10^{-6}}. \tag{15.25}$$

3) Find a robust controller $K_{\infty}(s)$, which can be derived with MATLAB Robust Control Toolbox:

$$K_{\infty}(s) = \frac{12.5s^3 + 1.1185 \times 10^6 s^2 + 5.088 \times 10^{10} s + 2.12 \times 10^{13}}{s^3 + 7.19 \times 10^4 s^2 + 4.03 \times 10^9 s + 1.47 \times 10^{11}} \tag{15.26}$$

such that

$$\begin{bmatrix} W_1 S \\ W_3 T \end{bmatrix}_{\infty} \leq 1 \tag{15.27}$$

Figures 15.18 and 15.19 show that $W_1(s)$ and $W_3(s)$ can shape S and T to achieve the desired performance and meet the robustness specifications of the designed system. The above example is designed for the ISSC operated at $f_S = 50\,\text{kHz}$ and 75% of the full load condition ($I_o = 1.5\,\text{A}$). It can be verified that the desired system specifications are all met in the proposed ISSC. To further investigate the system

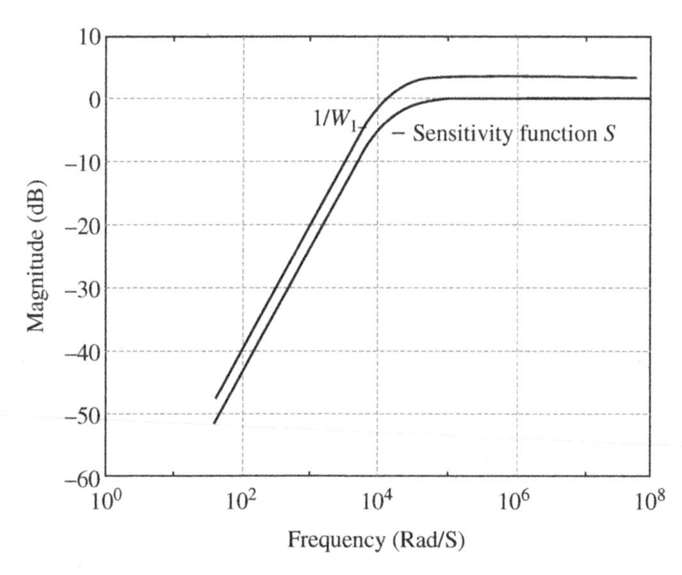

Figure 15.18 The magnitude versus frequency plots of sensitivity function S and weighting function $1/W_1$.

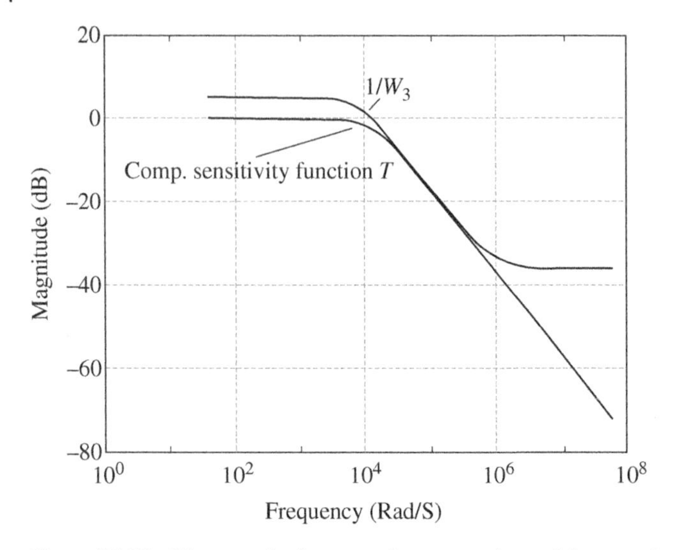

Figure 15.19 The magnitude versus frequency plots of the complementary sensitivity function T and weighting function $1/W_3$.

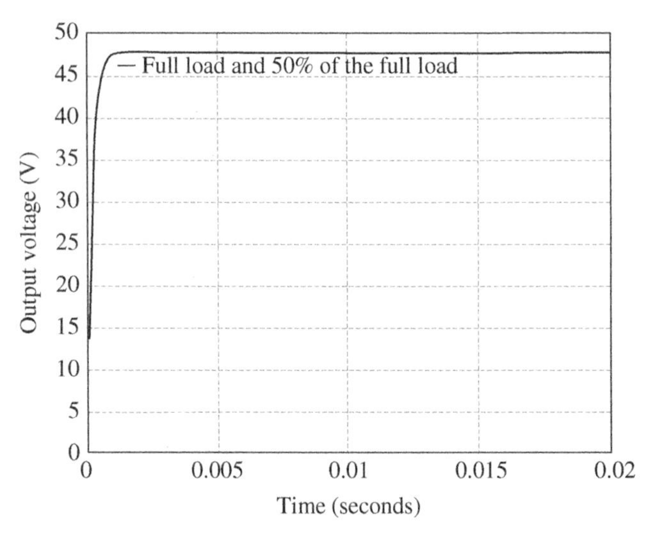

Figure 15.20 Step responses of the ISSC to the load changes from 0 to 50% and from 0 to 100% of the full load.

robustness when load current and switching frequency vary over a certain range, step response to load changes from 0 to 50% and from 0 to 100% of the full load are performed and shown in Figure 15.20 (by MATLAB simulations). It can be observed from the figures that the regulation and robustness can be sustained with the designed controller $K_\infty(s)$.

Figure 15.21a and b show the measured results of output voltage and line current to step load changes between 50 and 100% of the full load, respectively. The waveform of the output voltage shows that a fast regulation can be achieved with an insignificant steady-state error, and the line current waveform is sinusoidal-like, leading to a high power factor.

An ISSC derived in this chapter is adopted to achieve high power factor and fast regulation. The two semi-stages of the ISSC are designed separately so as the analytical work can be reduced. Operating both semi-stages in the DCM can minimize the control effort, while limit the load range and increase component stress

(a)

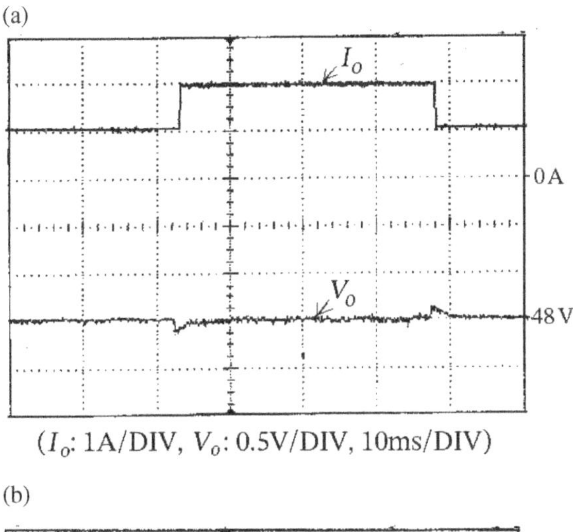

(I_o: 1A/DIV, V_o: 0.5V/DIV, 10ms/DIV)

(b)

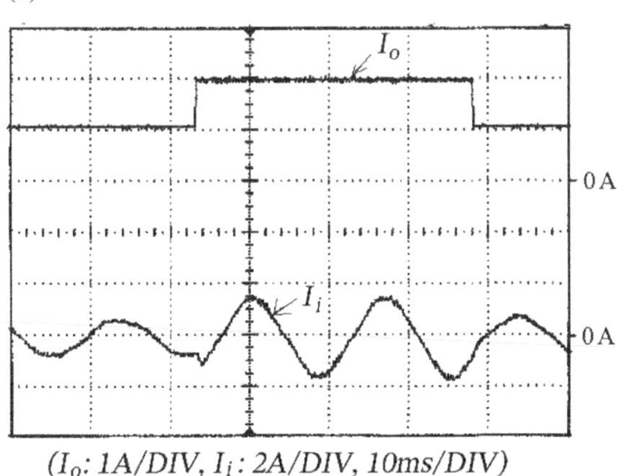

(I_o: 1A/DIV, I_i : 2A/DIV, 10ms/DIV)

Figure 15.21 Transient response of the ISSC with a robust controller: (a) the output voltage and (b) the line current to step changes of the loads between 50 and 100% of the full load.

slightly. In addition, the small-signal model of the proposed converter is developed, from which an optimal PI compensator is designed for the converter system with peak current mode control. Experimental results of a peak current mode control have shown that fast dynamic response, good output voltage regulation, low harmonic distortion, and high power factor can be achieved with the proposed single-stage converter and control scheme. Moreover, the proposed H^{∞} controller can accommodate the load variation, input voltage, and control disturbance. Experimental measurements of the system with a robust control have shown that steady-state errors and overshoots can be improved as compared with that with a peak current mode control.

Further Reading

Cuk, S. (1977). Discontinuous inductor current mode in the optimum topology switching DC-to-DC converter. *Proceedings of the Power Electronics Specialists Conference*, IEEE, pp. 105–123.

Cuk, S. and Middlebrook, R.D. (1977) A general unified approach to modeling switching DC-to-DC converters in discontinuous conduction mode. *Proceedings of the Power Electronics Specialists Conference*, IEEE, pp. 36–57.

Deshpande, P.B. and Ash, R.H. (1981). *Computer Process Control*. USA: ISA Publication.

Huber, L. and Jovanovic, M.M. (1997). Single-stage, single-switch, isolated power supply technique with input-current shaping and fast output-voltage regulation for universal input-voltage-range applications. *Proceedings of the Applied Power Electronics Conference*, IEEE, pp. 272–280.

Kornetzky, P., Wei, H., and Batarseh, I. (1997a). A novel one-stage power factor correction converter. *Proceedings of the Applied Power Electronics Conference*, IEEE, pp. 251–258.

Kornetzky, P., Wei, H., Zhu, G., and Batarseh, I. (1997b). A single-switch AC/DC converter with power factor correction. *Proceedings of the Power Electronics Specialists Conference*, IEEE, pp. 527–535.

Lee, Y.S., Siu, K.W., and Lin, B.T. (1997). Novel single-stage isolated power-factor-corrected power supplies with regenerative clamping. *Proceedings of the Applied Power Electronics Conference*, IEEE, pp. 259–265.

Liu, K.-H. and Lin, Y.-L. (1989). Current waveform distortion in power factor correction circuits employing discontinuous-mode boost converter. *Proceedings of the Power Electronics Specialists Conference*, IEEE, pp. 825–829.

Qian, J., Zhao, Q. and Lee, F.C. (1998). Single-stage single-switch power factor correction (S4-PFC) AC/DC converters with DC bus voltage feedback for universal line applications. *Proceedings of the Applied Power Electronics Conference*, IEEE, pp. 223–229.

Redl, R. and Balogn, L. (1995). Design considerations for single-stage isolated power supplies with fast regulation of the output voltage. *Proceeding of the Power Electronics Specialists Conference*, IEEE, pp. 454–458.

Vorperian, V. (1990). Simplified analysis of PWM converters using the PWM switch, part II: discontinuous conduction mode. *IEEE Trans. Aerosp. Electron. Syst.* AES-26: 497–505.

Wu, T.-F. and Chen, Y.-K. (1996). A systematic and unified approach to modeling PWM DC/DC converters based on the grafted scheme. *Proceedings of the Industrial Electronics, Control, and Instrumentation Conference*, IEEE, pp. 1041–1046.

Yang, E.-X., Jiang, Y., Hua, G., and Lee, F.C. (1993). Isolated boost circuit for power factor correction. *Proceeding of the Applied Power Electronics Conference*, IEEE, pp. 196–203.

Ziegler, J.G. and Nichols, N.B. (1942). Optimal setting for automatic controllers. *Trans. ASME* 65: 433–444.

Index

a
Active clamp 49
Active lossless 220–230
Active rectifier 203
Active soft-switching 234–240, 245–247
Active switch 103–109, 174, 177
Ampere-second balance 49, 170, 198, 272
Analytical work 391
Anti-parallel diode 222, 229, 230
Audio-susceptibility 318, 333, 340, 356, 360
Auxiliary switch 222, 223, 225–227, 229, 230, 245
Averaged circuit model 301, 329, 351
Average voltages 50, 52, 53, 175, 192, 209

b
Bandwidth 386
Bipolar mode 201
BJT 5, 6, 67
Boost buck BCU 247, 249, 250
Boost-type 14, 17, 203
Boost ZVS-QRC 14, 325
Buck-type 17, 203, 207
Buffer 44, 46

c
Capacitive 221, 242, 250, 312
Capacitor-clamped converter (CCC) 209
Capacitor splitting 49–51, 191, 192
Cascade/cascode
  boost-buck 73, 332, 342
  buck-boost 63, 336–338, 358
  sepic 346–348
  transfer code 86
  zeta 344–346
Class-E 125–127
CLC filter 279
Closed-loop 360, 386
Common D-S 22, 105–107, 111, 121, 130, 168, 171, 176, 188
Common node 62, 108–110, 137
Continuous conduction mode (CCM) 31, 34, 78, 98, 304
Control-gate 62
Control-to-output transfer function 318, 333, 336, 340, 356, 363, 364, 374, 376, 387
Converter Circuits 59, 101, 282
Converter layer technique (CLT) 25
Cross 89–91
Ćuk-buck 247, 249
Current doubler 60
Current-fed z-source converter 161–164, 261–262
Current sensing 380
Current sink 40, 41, 274, 275
Current stresses 49, 223, 227, 245, 255, 367, 377, 378
Current transfer 73, 74, 77, 78, 112, 116, 139, 140, 215, 282, 286, 288

d
DC-current offsetting 43–49
DC-link voltage 362, 364, 367, 369, 372–374, 379
DC offset 151, 183, 273, 295
DC transformer
  current doubler 59, 60
  full-bridge 60
  half-bridge 60
  push-pull 58, 61
  voltage doubler 59, 60
DC-voltage offsetting 44–47
Decoding 77–94
Deduction 146–151
Diode grafting 26, 43, 62, 67–71
Discontinuous conduction mode (DCM) 34, 92, 110, 255, 323

Origin of Power Converters: Decoding, Synthesizing, and Modeling, First Edition.
Tsai-Fu Wu and Yu-Kai Chen.
© 2020 John Wiley & Sons, Inc. Published 2020 by John Wiley & Sons, Inc.

DNA 137, 138, 271, 291
Duality 40–41, 66, 274–277
Duty cycle 380
Duty ratio 13, 19, 22, 59, 72, 379, 380

e
Electro-magnetic interference (EMI) 7
Electronic systems 289
Energy recovery 216, 218, 220, 227

f
Feedback
  series-shunt 303–308
  shunt-series 303–308
  transfer code 82–86
Feedback configuration 82–83, 156
Feedback-path gain 37
Feedforward 83–85
Flow-graph model 302
Flyback 15, 56, 374, 379
Flywheeling capacitor converters (FCC) 27, 199
Forward 9, 15, 56, 90, 96, 151, 156, 168, 177
Four-wire 205
Freewheeling 6, 32, 43, 59, 216–220, 226, 227, 268, 302, 372, 379
Frequency modulation 367
Frequency operation 15, 227, 240
Full-bridge 15, 58, 60, 202, 205, 210, 211
Full load 380, 381, 389–391

g
Gain-D plots 79–81
Graft
  bi-directional flyback 124–125
  Boost-buck (Ćuk) 116–119, 338–340
  boost-buck-boost (sepic) 117, 135, 282–284
  buck-boost 114–116, 278–279
  buck-boost-buck (Zeta) 133, 280–282
  Class-E 125–127
Grafted switches
  inverse Π-type 23, 64, 66, 106
  inverse T-type 23, 64, 65, 105
  T-type 23, 52, 64–66, 105, 204
  Π-type 23, 64, 66, 106
Graft switch technique (GST) 22, 24, 25

h
H^∞ 382–392
Half-bridge 15, 60, 124–125, 128–130, 202, 205, 208, 210
Hard-switching 26, 27, 215, 230, 234, 240, 271
Heat sink 5
High efficiency 5, 52, 216, 378
Hybrid converters 12

i
I-Buck 167, 169, 172, 197
I-Buck-boost 176–177
  Ćuk 177–178
Ideal switch 7
IGBT 6, 67, 226
Inductive 242, 312
Inductor splitting 26, 43, 49–51, 144, 154, 177
Input impedance 318, 333, 340, 356, 360, 362, 363
Input-to-output transfer ratio 19, 99
Interleaved 127–130
Irreducible quadratic factors 87
Isolated PWM converter
  flyback 16
  forword 16
  push-pull 16
Isolated regulator 6
Isolation transformer 16, 170
ISSC 351, 373–377, 382–392

j
Junction 216, 226

k
KVL 46, 261, 264–266
KY converter 9

l
Layer
  boost-buck 73
  boost family 138–142, 302, 308
  buck-boost 24, 72, 304
  buck family 135–138, 284–286
  Ćuk 332
  sepic 311, 322
  zeta 149, 150, 344–346
LC filters 18, 58, 98, 136, 138, 146–148, 255, 258, 272–274
LCL filter 101, 111, 200, 203, 234, 278, 338
LC resonance 219, 225, 226
Leaf power stage (LPS) 330
Leakage inductor 57, 378
Linear components 50, 62, 74, 94, 192
Line voltage 370, 372, 373, 382, 386
Load range 222, 227, 367, 391
Load variations 4, 382, 386, 392
Long division 86, 88–90, 114, 155, 162
Low efficiency 5
Low-pass filtered 305, 307, 372, 380, 388

m
Magnetic components 194
Magnetic coupling 55–57, 193–194
MATLAB 389, 390
Maximum duty 380

Measurements 380–382, 386–392
Minimize 351, 377, 386, 388, 391
Minimum-phase control system 270
Modular multi-level (MMC) 9, 27, 199, 207, 302
MOSFET 103, 105, 108, 109, 222, 226
Multi-level converter 199–212
Multiple diodes 71, 263
Multi-stage 199–212
Mutation 25, 26, 291–295

n
Natural frequency 15
N-cell 96–97
Near-zero-current switching (NZCS) 216–218
Near-zero-voltage switching (NZVS) 218–220
Negative sequence outputs 203
Neutral-point clamped (NPC) 9, 27, 199
Neutral wire 206
Nominal load 387
Non-fundamental PWM 263
Nonlinear components 62, 94
Non-minimum-phase systems 34
Non-PWM 7–9

o
Open-Loop 359–360, 364
Optimal performance 386
Origin 25, 31–41
Output current transfer 73, 74, 78, 112, 139, 140, 286, 288
Output impedance 318, 333, 340, 356, 360
Overshoots 377, 386, 392

p
Parallel 85–86, 119
Parasitic capacitance 216, 222
Passive diodes 12, 206, 256, 259, 261, 262, 267, 274
Passive lossless 216–220
Passive soft-switching 230–234, 240–245
Passive switch 245
P-cell 18, 95–97
Peak-current control 28
Peak-current mode 367, 368, 373–377, 380, 392
Pi canonical cell 18, 97–98
PI controller 368, 373
Positive feedback 83
  feedback gain 83, 84, 170, 374
Positive half cycle 199, 203, 206
Power factor corrector (PFC) 6, 367
  boost family 351
  buck-boost 370–372
  buck family 127, 351, 352
Power ratings 380
Power semiconductor 226
Primary winding 361
Prime factors 87
Pulsating 46
Push-pull 9, 15, 59
PWM converter
  boost converter 8, 9, 17, 18
  buck–boost converter 9, 21–25, 96, 154
  buck converter 9, 15, 34, 153–154
  Ćuk converter 9, 279–280
  sepic converter 9, 157–159, 164–165
  Zeta converter 9, 160–161, 165–166, 188–189
PWM power converters 9–10, 31, 43, 51, 95, 289

q
Quasi-resonant
  boost ZCS-QRC 325
  buck ZCS-QRC 325
  buck ZVS-QRC 325
Quasi z-source converter 13, 154, 162–166, 262

r
RCD snubber 378, 389
RC time 380
Recovery currents 216, 217
Recovery time 108
Rectifier 5
Rectifier diodes 13, 223, 227, 245
Reduce current 59, 85
Regulator
  boost family 351
  buck family 351, 352
  flyback 372–373
  semi-stage 351, 352, 356, 364, 367–369, 373, 375
Replication 291
Resonant capacitor 222, 223, 225, 226, 229, 230
Resonant frequency 14, 78, 91, 211
Resonant inductor 222, 223, 225, 226, 229, 230
Resonant network 16, 53, 92, 210, 212, 222, 223, 227
Right-hand side (RHS) 124, 158, 160, 165, 256
Ripple 4, 21, 59, 304
Ripple voltage 85, 210, 379
Robust control 28, 368, 382, 386, 387, 389, 392
Root power stage (RPS) 330

s
S–D node 116, 119, 130, 174, 189
Secondary side 16, 55–59, 124, 194
Secondary winding 15–17, 55–57, 59, 372
Sensitivity function 386, 388–390
Sepic–buck 247, 250
Series-parallel resonant 124
Settling time 386
Single-phase converter
  full-bridge 201
  half-bridge 202
  NPC 203
  T-type 203
Single-pole double-throw (SPDT) 22, 101

Single-stage converter (SSC) 28, 108, 110, 124, 127, 329, 352–355, 367–392
Single-voltage 357, 360
Small-signal model 242, 301, 302, 315–323, 325, 326, 329, 330, 332–348
Small-signal transfer characteristics (SSTCs) 28, 315, 333, 334, 356, 387
Snubber 216, 217, 378
Snubber inductor 216, 217, 219, 222
Soft-switching cell
  active soft-switching PWM
    boost 222, 234
    buck 230, 234
  converter 230
  passive NZCS 231
  passive NZVS 231
  passive soft-switching PWM converter 230–234, 240–245
  ZCS-PWM 230
  ZVS-PWM 223, 226
Source–load 26, 32, 33
Space-vector PWM (SVPWM) 203
State-space averaging model 333, 340
Steady state error 377, 380, 391, 392
Step-down 119–120, 167–178, 263–265, 267–268
Step-up 120–121, 178–185, 268–270
Step-up/step-down 264–267
Stored energy 7, 9, 33, 217
Switch current 216, 218, 220, 226, 227, 229
Switched capacitor 20, 98–100, 166–185
Switched inductor 20, 98–100, 166–185
Switching frequency 5, 15, 78, 91, 194, 210, 211, 215, 220, 354, 379, 390
Switching losses 6, 216, 220, 226, 227, 230, 240, 245
Switchmode power 59, 101
Switch-voltage stress
  boost converter 257–258
  buck–boost converter 258–259
  Buck converter 278
  Ćuk converter 259
  high step-down converter 263–264, 267–268
  high step-down/step-up converter 264–265
  high step-up converter 268–270
  quasi-resonant converter 323–326
  sepic converter 259
  Zeta converter 259
  z-source converter 260–262
Synchronous switch 21, 22, 27, 101–103, 230, 369, 377
Synthesizing 95, 133, 154, 157, 230
System efficiency 222

t
Three-phase converter
  full-bridge 203
  half-bridge 203
  NPC 203
  Vienna rectifier 206
Three-phase outputs 203
Three-terminal 112, 301
Topological 274–277
Total harmonic distortion (THD) 380, 383
Transfer code (TC)
  cascade 82
  feedback 82–83
  feedforward 83–85
  high step-down converter 167–178, 267–268
  multivariables 91–93
  parallel 85–86
Transfer functions 305, 318, 333, 336, 340, 356, 360, 363, 364, 374–376, 387
Transfer ratio 77–93
Turns off 34, 35, 203, 206, 216
Two-port network
  g-parameter 310
  h-parameter 310
  series-shunt 310
  shunt-series 310
  t-parameter 357
  y-parameter 357

u
Unity-gain feedback 37, 83
Unity-gain feedforward 83, 84, 88, 90–92, 157

v
Variable frequency control 6, 354
Voltage doubler 59, 60
Voltage feedback 74, 140, 143, 145, 146, 195, 196
Voltage offset 44, 274, 295
Voltage overshoot 386
Voltage ripple 4, 85, 210
Voltage source 44
Voltage spike 32, 216, 220
Voltage waveforms 15, 53, 55, 223, 256, 304
Volt-second balance 20, 21, 35, 36, 39, 46, 50, 55, 57, 77, 86, 99, 100, 138, 142, 192, 194, 266, 304

w
Wire 55, 203, 205, 206, 220, 277

z
ZCS-PWM 227, 230
Zero-current switching (ZCS) 226–230
Zero-voltage transition (ZVT) 218, 225
Zeta-buck 247, 248
Z-source converter 13, 154–166, 260–262
ZVS-PWM 223, 245

Printed and bound by CPI Group (UK) Ltd, Croydon, CR0 4YY